Polar Lows

Mesoscale Weather Systems in the Polar Regions

Polar Lows provides a comprehensive review of our understanding of the small, high latitude weather systems known as polar lows. These often vigorous depressions are a hazard to maritime operations and high latitude communities, yet have only been investigated in detail since the 1960s. In this volume the authors describe the climatological distribution of these lows, the observational investigations into their structure, the operational forecasting of polar lows and the theoretical research into why they develop. They also discuss the experiments carried out with high resolution numerical weather forecast models that have demonstrated how some polar lows can be predicted a day or more in advance.

The book has been written by a number of experts within the field and has been carefully edited to form an integrated, cohesive volume. It will be of value to meteorologists and climatologists with an interest in the polar regions, as well as professional weather forecasters concerned with these areas. It may also be used as a supplementary text on graduate courses concerned with high latitude geography, meteorology, and climatology.

ERIK RASMUSSEN began his career as a meteorologist at the age of 17, in 1952, as flight forecaster in Kastrup Airport, Copenhagen. Following a Masters Degree in 1964 he worked at the Danish Meteorological Institute from 1964 to 1968. Since then he has been affiliated to the Geophysical Department in the University of Copenhagen, where he is currently Professor Emeritus. Professor Rasmussen received a Ph.D. in 1974 for his work on spectral models, and in 1989, the D.Sc. degree for his work on polar lows. He has also spent time working for the World Meteorological Organization in Barbados and, on sabbatical, at Colorado State University and at the Navy Postgraduate School, California, USA. Professor Rasmussen was a co-editor of *Polar and Arctic Lows* (1989) and has also contributed to *Images in Weather Forecasting* (Cambridge University Press, 1995), *Encyclopedia of Climate and Weather* (1996), and *Storms* (1999).

JOHN TURNER has been a research meteorologist at the British Antarctic Survey since 1986 and currently leads the project Variability of the Antarctic Climate System. His present research interests include how ENSO affects the Antarctic climate, Antarctic weather forecasting, high latitude precipitation, and polar lows. Prior to joining BAS, he worked at the Meteorological Office from 1974 to 1986 where he was involved in numerical weather prediction and satellite meteorology. Dr Turner is currently President of the International Commission on Polar Meteorology and Chairman of the Physical Sciences Standing Scientific Group of the Scientific Committee on Antarctic Research. He has been a Fellow of the Royal Meteorological Society since 1974 and served on the Society's Council. Dr Turner edited the journal *Weather* for three years and is also the co-author of *Antarctic Meteorology and Climatology* (Cambridge University Press, 1997).

Polar Lows
Mesoscale Weather Systems in the Polar Regions

EDITED BY

ERIK A. RASMUSSEN
Emeritus Professor, Department of Geophysics, University of Copenhagen

and

JOHN TURNER
Physical Sciences Division, British Antarctic Survey

PUBLISHED BY THE PRESS SYNDICATE OF THE UNIVERSITY OF CAMBRIDGE
The Pitt Building, Trumpington Street, Cambridge, United Kingdom

CAMBRIDGE UNIVERSITY PRESS
The Edinburgh Building, Cambridge CB2 2RU, UK
40 West 20th Street, New York, NY 10011-4211, USA
477 Williamstown Road, Port Melbourne, VIC 3207, Australia
Ruiz de Alarcón 13, 28014 Madrid, Spain
Dock House, The Waterfront, Cape Town 8001, South Africa

http://www.cambridge.org

© Cambridge University Press 2003

This book is in copyright. Subject to statutory exception
and to the provisions of relevant collective licensing agreements,
no reproduction of any part may take place without
the written permission of Cambridge University Press.

First published 2003

Printed in the United Kingdom at the University Press, Cambridge

Typefaces Lexicon No. 2 10/14 pt and Lexicon No. 1   *System* LaTeX 2_ε [TB]

A catalogue record for this book is available from the British Library

Library of Congress Cataloguing in Publication data

Polar lows / edited by E.A. Rasmussen, J. Turner.
   p. cm.
Includes bibliographical references and index.
ISBN 0 521 62430 4
1. Meteorology.   I. Rasmussen, Erik A.   II. Turner, J. (John), 1953–
QC994.75 .P645   2003   551.55′13′0911–dc21   2002073925

ISBN 0 521 62430 4 hardback

The publisher has used its best endeavours to ensure that the URLs for
external websites referred to in this book are correct and active at the time of
going to press. However, the publisher has no responsibility for the websites
and can make no guarantee that a site will remain live or that the content is or
will remain appropriate.

Contents

List of contributors *page* vii
Preface ix

1 **Introduction** 1
   1.1 Polar lows and other mesoscale lows in the polar regions 1
   1.2 A brief historical review 3
   1.3 Definition 10
   1.4 Nomenclature 12
   1.5 Classification 14
   1.6 Cloud signatures 16

2 **Climatology** 52
   2.1 The Arctic 52
   2.2 The Antarctic 108

3 **Observational studies** 150
   3.1 The Arctic 150
   3.2 The Antarctic 246

4 **Theoretical investigations** 286
   4.1 Introduction 286
   4.2 Baroclinic instability 287
   4.3 Barotropic instability 317
   4.4 Potential vorticity thinking 321
   4.5 The role of thermal instability in polar low formation and maintenance 335
   4.6 Further theoretical considerations 393
   4.7 Summary and concluding remarks 406

5 **Numerical simulation** 405
   5.1 The Arctic 405
   5.2 The Antarctic 460

6 **Forecasting of polar lows** 501
   6.1 Aspects common to both polar regions 501
   6.2 Aspects of forecasting specific to the Arctic 510
   6.3 Aspects of forecasting specific to the Antarctic 565
   6.4 Future work required to improve forecasts 573

7 **Conclusions and future research needs** 575
   7.1 The spatial distribution of polar lows and other high latitude mesoscale vortices 575
   7.2 Possible climatic effects of polar lows 576
   7.3 Observational and modelling studies 577
   7.4 Formation and development mechanisms 577
   7.5 Forecasting of polar lows 579
   7.6 Final thoughts 579

References 580
Index 605

Contributors

A. M. CARLETON
College of Earth and Mineral Sciences
The Pennsylvania State University
302 Walker Building
University Park
PA 16802-5011
USA

S. GRØNÅS
Allegt. 70
Geofysisk Institutt
University of Bergen
5007 Bergen
Norway

G. HEINEMANN
Meteorologisches Institut der Universität Bonn
Auf dem Hugel 20
53121 Bonn 1
Germany

K. NINOMIYA
Frontier Research System for Global Change
3173-25 Showamachi
Kanazawa-ku
Yokohama
236-0001
Japan

E. A. Rasmussen
Hoejbakkestraede 2
DK 2620 Albertslund
Denmark

E. Raustein
Allegt. 70
Geofysisk Institutt
University of Bergen
5007 Bergen
Norway

I. Renfrew
British Antarctic Survey
High Cross, Madingley Road
Cambridge
CB3 0ET
UK

B. Røsting
Norwegian Meteorological Institute (DNMI)
P. O. Box 43
Blindern
N-0313 Oslo
Norway

J. Turner
British Antarctic Survey
High Cross
Madingley Road
Cambridge
CB3 0ET
UK

A. van Delden
Institute for Marine and Atmospheric Sciences
Utrecht University
Princetonplein 5
3584 CC Utrecht
The Netherlands

Preface

Since the first detailed investigations of polar lows and other high latitude, mesoscale weather systems were carried out in the late 1960s there have been major advances in our knowledge regarding the nature of such systems and the mechanisms behind their formation and development. High resolution satellite imagery has shown how frequently such lows occur in both polar regions and has illustrated the very wide range of cloud signatures that these systems possess. Great strides have also been made in representing these weather systems in numerical models. With their small horizontal scale, it proved difficult to represent the lows in the early modelling experiments, but the new high resolution models with good parameterizations of physical processes have been able to replicate a number of important cases, despite the lack of data for use in the analysis process.

Although case studies of mesoscale lows have been undertaken for many years, recent research has been able to draw on many new forms of data, especially from instruments on the polar orbiting satellites. Scatterometers have provided fields of wind vectors over the ice-free ocean, passive microwave radiometers have allowed the investigation of the precipitation associated with the lows, and new processing schemes for satellite sounder data have given information on their three-dimensional thermal structure. In addition, aircraft flights through polar mesoscale lows have provided high resolution, three-dimensional data sets on the thermal and momentum fields. In recent years meteorologists have therefore had an unprecedented amount of data with which to conduct research into these important high latitude weather systems.

Although many research papers have been published on polar lows since the 1960s, a major advance in research into such systems was the publication of the book *Polar and Arctic Lows* in the late 1980s (Twitchell *et al.*, 1989). This volume came at the end of an intense period of research focused around the Norwegian Polar Lows Project, which had observational, climatological and modelling elements, which were carried out in Europe and the USA. The book contained papers by a number of workers active in polar lows research and brought together many of the new research ideas that emerged during the 1980s. However, since that time there have been many advances in the study of polar lows and other mesoscale weather systems. Whereas most of the early research was concerned with the Arctic, during the 1990s there have been a number of studies of Antarctic weather systems, including aircraft investigations. There has also been a noticeable broadening of the investigations to include the whole spectrum of observed systems and not just the intense 'polar lows' that the early studies examined. Finally, recent investigations have often made use of both observational data and the output of numerical models, which has provided very comprehensive pictures of the vortices examined. This current volume therefore grew out of our desire to provide a comprehensive review of the mesoscale weather systems that are found in the two polar regions in light of the great deal of research that has been carried out over the last ten years.

The book has been written by a number experts in the field of high latitude mesoscale weather systems, many of whom are members of the European Geophysical Society's Polar Lows Working Group. This group is a focus for much of the polar lows research in Europe and includes scientists with interests in the modelling, observational and climatological aspects of the subject. In order to have the best possible coverage of the current research, the decision was taken to have a fairly large number of authors for the various chapters, but with considerable editing of the volume so that a coherent picture of the subject could be presented. Primary goals for the book were to try and integrate the theoretical and observational aspects of the subject and to bring together Arctic and Antarctic investigations. There is therefore extensive cross-referencing between the observational, modelling and theoretical sections where we have tried to point to well-observed examples of vortices that help to illustrate the presently understood theoretical background to the development and signatures of these lows. Of course there are still large gaps in our understanding of polar lows and there are frequent references to future research needs.

The book has been organized to reflect the different facets of research that are taking place. In the Introduction (Chapter 1) we provide details of the scope of the book and discuss the historical background to the subject and the bewildering variety of names used over the years to describe the phenomena observed. Chapter 2 deals with the climatological occurrence of the lows, based largely on satellite imagery. An important goal here was to try and relate the spatial and temporal variability of the lows with some of the major climatological cycles, such as the El Niño–Southern Oscillation and the North Atlantic Oscillation. Chapter 3 covers the observational investigations that have been carried out and attempts to summarize this work into a picture of the underlying mechanisms that are responsible for the observed form of the vortices. Chapter 4 examines the various theoretical ideas that have been proposed to explain the observed weather systems. Where possible, pointers are provided to cases that seem to fit the theoretical models proposed. Within this chapter we have assumed that the reader is familiar with the fundamentals of dynamical meteorology and have therefore not dealt extensively with the basics of, for example, baroclinic instability. On the other hand, we have discussed new concepts, such as potential vorticity, in more detail as they relate to polar lows. Chapter 5 deals with recent numerical experiments that have attempted to simulate various cases in the Arctic and Antarctic. Chapter 6 examines the practical aspects of forecasting mesoscale lows using model output and observational data. In Chapter 7 we attempt to draw together work presented in the earlier chapters to assess our current understanding of high latitude mesoscale weather systems. We also compare the systems observed in the two polar regions and suggest possible topics for future research.

Many individuals and organizations have contributed to the preparation and production of this volume. The Satellite Receiving Station at the University of Dundee, Scotland provided many excellent satellite images illustrating polar lows in the Arctic. Professor David Bromwich (Byrd Polar Research Center, Ohio State University) provided a number of the Southern Hemisphere DMSP and AVHRR satellite images used in Chapters 1 and 2. A large number of the figures previously published in journals were scanned and optimized for publication by Jean Sinclair, Jayne Goodman and Mark Andrews while employed by the British Antarctic Survey. Many of the surface and upper-air charts used were redrawn by Mark Andrews and Nick McWilliam at the British Antarctic Survey. Izaak Santoe and Jaco Bergenhenegouwen drew several of the figures used in Chapter 4. Sander Tijm wrote the program which was used to illustrate hydrostatic adjustment to horizontally inhomogeneous heating in Chapter 4. Jens Rytter prepared the CAPE figures used in Chapter 4. Assistance was provided by Duane Carpenter, Chantal Claud, Mark Fitch, Kristina Katsaros, Lynn McMurdie, Nelly Mognard, and Yudong Song in the research presented in Section 2.1.

Finally, we would like to thank the many organizations and funding agencies that have supported the research presented here, including the US National Science Foundation (the work presented in Section 2.1 was funded by grants SES-86-03470, OPP-88-16912 and OPP-92-19446), the British Antarctic Survey, Natural Environment Research Council, UK, Climate System Research, University of Tokyo and the University of Copenhagen.

J. TURNER, E. A. RASMUSSEN AND A. M. CARLETON

1

Introduction

1.1 Polar lows and other mesoscale lows in the polar regions

In this volume we are concerned with the whole range of mesoscale lows with a horizontal length scale of less than *c.* 1000 km that occur in the Arctic and Antarctic poleward of the main polar front or other major frontal zones. However, much of the interest will be focused on the more intense systems, the so-called polar lows. The term mesocyclone covers a very wide range of weather systems from insignificant, minor vortices with only a weak cloud signature and no surface circulation, to the very active maritime disturbances known as polar lows, which in extreme cases may have winds of hurricane force and bring heavy snowfall to some areas. Clearly it is very important to be able to forecast these more active systems since they can pose a serious threat to marine operations and coastal communities when they make landfall.

Although it has been known for many years in high latitude coastal communities that violent small storms could arrive with little warning, it was only with the general availability of imagery from the polar orbiting weather satellites in the 1960s that it was realized that these phenomena were quite common. The imagery indicated that the storms developed over the high latitude ocean areas (generally during the winter months) and tended to decline rapidly once they made landfall. Much of the early interest in polar lows came from meteorologists in the Scandinavian countries and the British Isles, since coastal districts in these areas were particularly prone to being affected by polar lows during the winter months. The early satellite imagery provided a means of forecasting the arrival of the storms at least a few hours ahead and also initiated investigations into the climatological occurrence of the lows.

During the 1970s it was realized that intense polar lows tended to develop over ocean areas where the sea surface temperatures were relatively high during outbreaks of cold, Arctic air. Such conditions promote strong, deep convection and the satellite imagery almost always shows cumulonimbus clouds associated with the more active systems. In fact, meteorologists noted the similarity between some polar lows and tropical cyclones. However, other polar lows appeared more like small frontal depressions, so prompting a long debate in the scientific literature (which, as later chapters will make clear, is still ongoing) regarding the mechanisms that are behind the various types of development observed.

While the early studies of polar lows were concerned with systems over the northeast Atlantic and the Barents and Norwegian Seas, meteorologists soon noted that similar vortices were to be found in other parts of the world, including the North Pacific, the Sea of Japan and the Labrador Sea. Mesoscale lows in these areas varied in intensity, but they were clearly very similar to the polar lows that had been observed over the northeast Atlantic and in the Scandinavian region. In most of these areas there is a large difference between the sea surface temperature and near-surface air temperature, pointing to the importance of air–sea interactions in the development and maintenance of the vortices.

During the 1970s and 1980s there was interest in whether polar low-like systems were to be found in the high latitude areas of the Southern Hemisphere and a number of studies based on satellite imagery were carried out. It was found that there were indeed many mesoscale vortices over the Southern Ocean and also over the sea ice near to the coast of the Antarctic, but the lows at more southerly latitudes appeared to be generally weaker than their Northern Hemisphere counterparts. However, in the area around New Zealand and probably in other regions as well there were more active systems with deep convection that appeared more like the active northern systems. The term 'mesocyclone' then came into use to describe the ubiquitous rather weak mesoscale lows of the Southern Hemisphere. The difference between Northern Hemisphere and Southern Hemisphere mesoscale lows is examined in later chapters, but is felt to be a result of the different oceanic conditions in the two polar regions where the flow is much more zonal in the south and does not promote the large air–sea temperature differences that are found in the Arctic.

Early Northern Hemisphere studies were concerned mainly with the very active polar lows, but recent investigations in the Arctic have also documented the very large number of minor vortices that seem to be a year-round feature of the ocean areas of the Arctic. Minor mesoscale vortices therefore seem to be a feature of both polar regions.

While most of the early research into polar lows consisted of observational studies, attempts were made during the 1980s to represent these lows in atmospheric models. The first results were generally poor because the models did not have a resolution high enough to resolve these systems, which often have a diameter of only a few hundred kilometres. There were also difficulties because of the poor representation of convection in many of the early models, and theoretical and observational considerations suggested that convection was a very important factor in many of the developments. However, throughout the 1990s there have been many advances in modelling and the latest generation of models with resolutions of 50 km and higher, and a good representation of physical processes, are having more success in simulating some polar mesoscale weather systems. Forecasting polar lows and the weaker mesoscale vortices still presents many challenges, but the indications are that improvements in modelling will give further advances in the coming years.

This brief introduction to polar lows and other high latitude mesoscale weather systems has put forward various ideas that are covered in much more detail in the following chapters. The study of polar lows is still relatively new and there remain many gaps in our theoretical understanding of the development of the lows and aspects of the observational data that cannot be explained. In a summary in the *Bulletin of the American Meteorological Society* of a workshop on 'Arctic lows', William W. Kellogg and Paul F. Twitchell (Kellogg and Twitchell, 1986) wrote:

> The history of meteorology is replete with instances of some phenomenon in the atmosphere that defies an adequate description. We know that something exists, sometimes with disastrous consequences to people and their possessions, but its origins and evolution and characteristics are only vaguely understood. Furthermore, it may even be hard for meteorologists to agree what to call it.

Kellogg and Twitchell were describing the situation facing meteorologists around 1980 concerning the meteorological phenomenon known as polar lows (occasionally called 'Arctic lows' by Kellogg and Twitchell). Their statement was very precise and actually it was not until 1994 that some meteorologists, after a considerable debate, agreed upon a definition of polar lows (see editorial comment by A. Carleton in the *Global Atmosphere Ocean System* special issue on cold air mesoscale cyclones in the Arctic and Antarctic, Vol. 4, Nos. 2–4, 1996) and even this definition could be developed further. However, it is hoped that the following chapters will provide a comprehensive description of our current understanding of these fascinating weather systems and shed a little light on their relationship to other depressions.

1.2 A brief historical review

Our knowledge of polar lows and mesocyclones has come almost entirely during the period for which we have data from satellites since, by virtue of their small horizontal scale, it was rarely possible to analyse these lows on conventional weather charts using only the data from the synoptic observing network. However, the effects of intense polar lows have been felt by coastal communities and seafarers since the earliest times and there are many tales in the Scandinavian countries of sailors encountering small violent storms. These weather systems were thought to be responsible for the loss of many small vessels over the centuries, although the nature of the storms was not understood and their arrival could not be predicted. The effects of many of the polar lows were also felt during the winter months in coastal areas, such as along the northern coast of Norway where the weather could deteriorate very rapidly with winds increasing to gale force and heavy snow occurring in relatively limited areas. While Norwegian weather forecasters were aware of the existence of these lows, it was nearly impossible to forecast them.

Without specifically mentioning polar lows, Sumner (1950) in a study of the role of vertical stability for synoptic developments, concluded that '… there is every justification for supposing that tropical cyclones and a number of small hurricane-like centres, which develop in higher latitudes, are the result of the instability in depth i.e. saturation with a lapse greater than the saturated adiabatic through a deep layer.' The type of 'hurricane-like centre' envisaged by Sumner is illustrated in Figure 1.1. One of the earliest references to what became known as polar lows was made by Peter Dannevig, who wrote about 'instability lows' over the sea areas around Norway in a book for pilots (Dannevig, 1954). Dannevig produced a schematic weather chart showing the relationship of these vortices to the typical airflow around Norway during a polar outbreak (Figure 1.2). A satellite image of such an outbreak of polar lows is shown in Figure 1.3. He also considered the possible mechanisms behind their formation and suggested that these lows could develop in the same way as tropical cyclones.

Another early (German) reference to polar lows is given by Scherhag and Klauser (1962). Sherhag and Klauser described '*das polartief*' as a young, active and, mainly in height, well-developed cyclone with a marked pressure and temperature minimum. According to Sherhag and Klauser, the lows were best developed over warm sea surfaces. The surface circulations were thought to form because of vertical exchange of momentum in the strongly unstable air masses within which the lows formed. '*Das polartief*' was described as following the general flow in the region where it formed, initially containing no fronts. Fronts, however, could eventually form as the low passed surfaces with a varying temperature.

'Arctic instability low' was the name used for polar lows in Norway up to the 1980s. Concerning the motivation for the application of this name to the

1.2 A brief historical review 5

Figure 1.1. An infra-red satellite image for 0853 GMT 18 December 1994 showing a large, synoptic-scale low with a 'merry-go-round' structure, including a central 'hurricane-like' mesoscale vortex (indicated by the long arrow) of the type alluded to by Sumner (1950). Minor vortices (indicated by short arrows) circulate around the central vortex. (Image courtesy of the NERC Satellite Receiving Station, University of Dundee.)

small-scale lows in question, Rabbe (1975) explained that 'since the lows occur in cold unstable air masses they lend themselves to be called "Arctic instability lows".' In this early paper Rabbe presented several examples of polar lows around Norway and discussed their formation using the vorticity equation. Concerning the energy source for the lows, Rabbe pointed towards heating of the atmosphere by the ocean, noting that the energy transfer from the sea to the atmosphere reached extremely high values in connection with such

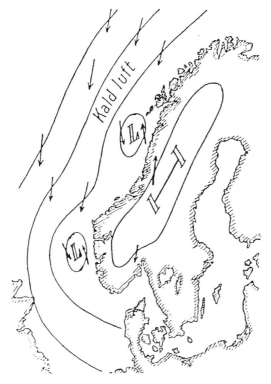

Figure 1.2. Dannevig's 1954 schematic surface chart showing two polar lows (called 'instability lows' by Dannevig) within a northerly outbreak of polar air near the Norwegian coast.

developments. Rabbe also included a number of examples of the near impossibility at that time of forecasting these dangerous, small-scale lows.

Since the 1960s British meteorologists have taken a keen interest in mesoscale weather systems in polar airstreams since such systems could give extensive snowfall across the British Isles, especially in Scotland. In Britain such systems were called 'cold air depressions' (Meteorological Office, 1962) and forecasters were clearly aware of their existence and importance. This interest resulted in a number of preliminary descriptive accounts of these lows in the literature. The earliest known case studies of polar lows by British meteorologists were published in the British magazine *Weather* in the 1960s and 1970s (Harley, 1960; Stevenson, 1968; Lyall, 1972), based mainly on routine surface observations. Lyall showed a Nimbus 3 satellite picture of the clouds associated with a polar low that occurred on 5 January 1970. The picture, which is probably the first published satellite image of a polar low, showed that the very active polar low (Suttie, 1970) was associated with a small, comma-shaped cloud. While these early studies were true observational studies, many

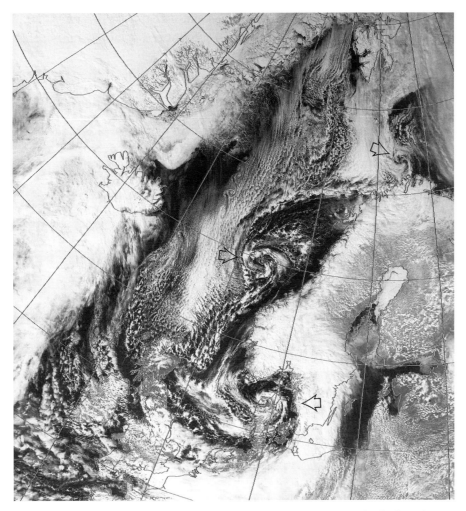

Figure 1.3. A visible wavelength satellite image of an outbreak of polar lows down the coast of Norway and Denmark. Three polar lows (indicated by arrows) have formed within the outbreak at, respectively, North Cape, the Norwegian Sea at 65° N, and over Denmark. The image was taken by NOAA 9 at 1308 GMT 27 April 1985. (Image courtesy of the NERC Satellite Receiving Station, University of Dundee.)

recent studies combine the observational aspect with results from numerical models.

The Stevenson (1968) study described a polar low that crossed southern England, giving 11 inches of snow around Brighton and causing major traffic disruption. However, no attempt was made to account for the development of the low. Nevertheless, this case formed the basis for the much more detailed investigation carried out by Harrold and Browning (1969) who studied the structure of the low using frequent radiosonde ascents and surface synoptic observations. This paper was the first full account of a polar low in the refereed scientific literature.

The arrival of imagery from the polar orbiting satellites during the 1960s provided a major advance in the study of polar lows. In the early years of meteorological satellites the only data available were infra-red and sometimes visible, hard copy images. Since then a variety of satellite data have been available, including satellite sounder measurements, scatterometer data for estimation of surface winds over the sea, and microwave data.

The imagery available from high latitudes indicated the high frequency with which polar lows developed and the wide range of cloud signatures associated with these lows. The imagery for the Arctic showed many minor mesoscale vortices well north of the main polar front, as well as the less common, intense vortices that became known as polar lows. Anderson *et al.* (1969), in their early manual on interpreting satellite imagery, referred to comma clouds, a special class of polar low, which is discussed in detail in Section 3.1.3.

In parallel with the early observational studies of polar lows, consideration was also being given to the mechanisms responsible for the development of the vortices. Early theoretical studies considered two possible mechanisms. Harrold and Browning (1969) suggested that the lows formed as a result of baroclinic instability, with Mansfield (1974) and Duncan (1977) further developing these ideas during the 1970s. In contrast, Økland (1977) and Rasmussen (1977, 1979) proposed that these polar lows developed as a result of Conditional Instability of the Second Kind (CISK) (see Section 4.5). Since the 1970s it has become clear that a very wide range of polar lows and mesocyclones develop in the polar regions and that both baroclinic and convective processes are involved in the development of these lows or can both be involved during the lifetime of a single polar low. A full discussion of the theories that have been proposed for the development of polar lows and other vortices is presented in Chapter 4.

A major handicap to research on polar lows has always been the lack of data in the high latitude areas, both for investigation of the structure of the lows and for the preparation of numerical analyses from which models could be run. A significant advance came with the Norwegian Polar Lows Project (Lystad, 1986; Rasmussen and Lystad, 1987), which took place between 1983 and 1985. Driven by the possible destructive effects of polar lows on the gas and oil drilling activities in the northern North Sea, this international project had observational and modelling elements that sought to improve our understanding and capability to forecast these vigorous storms. Within the project, the first aircraft observations were collected within a polar low (Shapiro *et al.*,

1987), a climatology of polar lows was prepared (Wilhelmsen, 1985), and advanced modelling studies were carried out (Nordeng, 1987).

During the 1980s attention turned also to the polar lows that occurred in areas other than the Atlantic sector of the Arctic. Inspection of satellite imagery showed that many polar lows were also to be found in the North Pacific (Businger, 1987; Douglas *et al.*, 1991) and in the Sea of Japan (Ninomiya, 1989). A number of these lows were quite intense and similar to the polar lows of the Norwegian/Barents Seas that had been studied for more than a decade.

In the 1980s researchers also considered the mesocyclones that were to be found around the Antarctic continent. In probably the first paper on polar lows over the Southern Hemisphere, 'An observational study of polar air depressions in the Australian region', Auer (1986) discussed the occurrence, evolution and maintenance of polar air depressions (polar lows) in the Australian region. Since the availability of routine satellite imagery, Australian meteorologists had, according to Auer, long been aware of subsynoptic-scale storms that develop rapidly over the Tasman Sea. As a useful forecasting tool Auer recommended the use of a temperature index called the Polar Depression Index (PDI), simply defined as the temperature surplus (or deficit) of a saturated parcel of air warmed to the sea surface temperature and lifted moist adiabatically to 500 hPa and compared to the environmental temperature at that level. Auer also stressed the importance of the upper-air flow geometry, noting amongst other things that 86% of the moderate to strong polar air depressions found by him were identified with medium amplitude short-waves or closed circulations.

With the availability of high resolution satellite imagery it had become clear that many mesoscale vortices were present over the Southern Ocean (Turner and Row, 1989; Heinemann, 1990), although, as in the Northern Hemisphere, very few systems were found over the land areas and the high ice cap of the Antarctic continent. However, many vortices were found on the low-lying ice shelves (Carrasco and Bromwich, 1991) as a result of the baroclinic conditions that exist in these areas. Although conventional synoptic data are rather limited around the Antarctic, the observations collected when vortices crossed observing stations suggested that most Southern Hemisphere mesocyclones were rather weak. Some systems had surface wind speeds of more than gale force but few vortices have been discovered with deep convective cloud and winds of up to around 30 m s^{-1}, such as are occasionally found in parts of the Arctic. As is discussed in later chapters, this is probably a result of the generally more stable atmospheric conditions found around the Antarctic compared to the Arctic.

In the 1990s the availability of large amounts of satellite imagery for both polar regions had allowed the production of mesocyclone climatologies describing the frequency and form of vortices found in many areas, although consistent broad-scale climatologies are not yet available for either polar region. Our understanding of the structure of mesocyclones was further advanced by the first instrumented aircraft flights through a mesocyclone in the Antarctic (Heinemann, 1996b) and the use of new forms of satellite data, such as the surface wind vectors available from scatterometers (Claud *et al.*, 1993; Marshall and Turner, 1997a). Further advances in our understanding of mesocyclone formation and development have also come about through the use of sophisticated numerical models simulating individual lows (see Chapter 5), as well as new theoretical investigations based on the use of simple models (Craig and Cho, 1989).

At the time of writing, there is still a great deal of research taking place into mesoscale weather systems in the polar regions. The European Geophysical Society's Polar Lows Working Group is a major focus for research and has been instrumental in organizing international, combined modelling/observational studies into selected cases. As will be apparent in the following chapters, we have made many advances in our understanding of these systems over the last three decades, but there are still many outstanding questions that require continued research.

1.3 Definition

Mesoscale vortices at high latitudes have been known by a variety of different names, among which the term 'polar low' is the most common. Other terms used include Arctic instability low, polar air depression, mesocyclone, mesoscale vortex, mesoscale cyclone, Arctic hurricane and polar airstream cyclone.

Polar lows are generally characterized by severe weather in the form of strong winds, showers and occasionally heavy snow, which have sometimes resulted in the loss of life, especially at sea. The severity of these systems is reflected in the term 'Arctic hurricane', which has been used for especially intense polar lows.

The difficulty of formulating a brief, unambiguous polar low definition is partly due to the fact that a variety of forcing mechanisms can play a role and may all be important for the development of these systems. Depending upon the relative importance of the forcing mechanism, different types of polar lows may form, leading to the idea of a 'spectrum' of mesoscale cyclones including both purely baroclinic as well as purely 'convective' systems, i.e. systems for which the main energy source is latent heat released in deep convection. A practical definition must include all the different types and also reflect the fact that the polar low traditionally is considered as an intense and vigorous phenomenon.

Dannevig (see Section 1.2) defined instability lows (polar lows) as small but intense vortices which form in cold air outbreaks over the sea, occasionally accompanied by strong winds (of gale or storm force) and strong precipitation.

In *The Handbook of Weather Forecasting* (Meteorological Office, 1964) the term polar low was taken to refer to 'fairly small-scale cyclones or troughs embedded in a deep northerly current which has recently left northerly latitudes'. However, the 1972 edition of *The Meteorological Glossary* (Meteorological Office, 1972) referred only to a polar-air depression, which was considered to be 'A secondary depression of a non-frontal character which forms, more especially in winter, within an apparently homogeneous polar air mass'. Following this idea, Carlson (1991) defined a polar low as being 'usually a *non-frontal low* that occurs to the rear of the cold front …'.

Reed (1979) identified the polar low with the so-called comma cloud (see Section 3.1.3), which typically develops in baroclinic regions, poleward, but relatively near the polar front. Reed's paper started a debate on whether comma clouds should be considered as 'real' or 'true' polar lows and highlighted the need for a more precise definition. Businger and Reed (1989a, b) suggested 'a broad definition' in order to group the various cases reported in the literature together, defining the polar low as 'any type of small synoptic- or subsynoptic-scale cyclone that forms in a cold air mass poleward of major jet streams or frontal zones and whose main cloud mass is largely of convective origin'. It should be noted that some significant polar low developments can be found that do not have convective cloud associated with them (see the case of the polar low that occurred on 2 March 1989 discussed in Section 4.2).

A slightly different definition was suggested by Rasmussen *et al.* (1993), defining the polar low as 'a small, but fairly intense maritime cyclone that forms poleward of the main baroclinic zone (the polar front). The horizontal scale of the polar low is approximately between 200 and 800 km and the winds around it of gale force or above'. This latter definition specifically points out that polar lows are maritime systems. Also, it takes into account that polar lows traditionally have been associated with the occurrence of relatively strong surface winds, around or above gale force. Small vortices ('cloud vortices') have a tendency to form in regions where cyclonic vorticity is maximized, such as along troughs or around cyclonic shear lines within the low-level flow. Such vortices are not necessarily accompanied by a significant pressure perturbation in the form of a trough or a closed circulation and/or any significant wind. Omitting the wind force requirement means that in principle *any*, however insignificant, cyclonic disturbance (cloud vortex) within cold air masses, should be called a polar low, including the type of cloud vortices mentioned earlier for which the term polar low is clearly inappropriate. The requirement of a relatively strong wind speed excludes the small and frequently very weak cyclones that often can be observed close to the Arctic and Antarctic coasts/sea ice edges, and also, to a large extent, the comma clouds, which are often associated with relatively weak surface circulations.

A definition endorsed by the European Geophysical Society's Polar Lows Working Group (Paris 1994), states: 'The term "*polar mesoscale cyclone*" (polar mesocyclone) is the generic term for all meso-α and meso-β cyclonic vortices poleward of the main polar front (scale definition according to Orlanski, 1975). The term "*polar low*" should be used for intense maritime polar mesoscale cyclones with scales up to 1000 km with a near-surface wind speed exceeding 15 m s^{-1}'. An advantage of the Paris definition is that it distinguishes clearly between weak vortices such as, for example, those often observed along the Antarctic coast, and the more vigorous polar lows from the Northern Hemisphere. A drawback is the rather cumbersome formulation and the fact that the wind speed requirement is probably *too* severe. On the other hand, Fett *et al.* (1993), referring to the distinction made in tropical meteorology between an easterly wave and a tropical depression, used an even more restrictive definition defining a cold air vortex as a polar low only when the wind speed was 18 m s^{-1} or above.

As a useful compromise between the different definitions mentioned above we will use the following definition of a polar low in this volume:

> *A polar low is a small, but fairly intense maritime cyclone that forms poleward of the main baroclinic zone (the polar front or other major baroclinic zone). The horizontal scale of the polar low is approximately between 200 and 1000 kilometres and surface winds near or above gale force.*

The definition is fairly general, and there are no requirements about the existence (or rather non-existence) of fronts or the prevailing cloud type, as in Businger and Reed's definition from 1989. The definition above can be extended if necessary by specifying the dominant physical mechanism responsible for the development of the low, such as, for example a 'baroclinic polar low' or a 'convective polar low', the latter being driven primarily by organized convection.

1.4 Nomenclature

Closely connected to the definition problem is the difficulty of nomenclature. As mentioned in the preceding sections, one of the pioneers of polar low research, Peter Dannevig, used the term 'instability low' reflecting his point of view that the small-scale lows were the result of organized release of convective instability. This term, however, had to yield to the expression 'polar low' used by British meteorologists for the small-scale systems in question.

In order to stress the rapid development and the similarity between some polar lows and tropical systems (hurricanes), Rasmussen (1983) used the term 'extra-tropical hurricane' and 'Arctic bomb' for an impressive polar low development around 27 January 1982 (see the discussion in Section 3.1.4, The 'polar low spectrum'). Emanuel and Rotunno (1989) argued that 'at least some polar lows are indeed Arctic hurricanes', and also Businger and Baik (1991) used the term 'Arctic hurricane' for intense polar lows. In Businger and Baik the Arctic hurricane was defined as 'a polar low with symmetric signature and threshold winds greater than or equal to 25 m s^{-1}, in which surface fluxes play the dominant role in the structure and sustenance of the mature storm'. Reed (1992), on the other hand, raised objections to the use of the term 'Arctic hurricane' arguing that in no cases had the surface winds exceeded the 33 m s^{-1} speed required of a hurricane, and, referring to the fact that similar systems have been observed over the Mediterranean (e.g. Rasmussen and Zick, 1987), that these systems are not peculiar to the polar regions. Reed's criticism, however, seems a little misplaced since none of the authors cited above have claimed that a complete similarity existed between hurricane-like systems in the Arctic and in the tropics, but rather have pointed out certain basic similarities, such as the existence of a warm core, the role of deep convection in the dynamics of the system, etc. Also, at least on two occasions, surface winds associated with a polar low have exceeded the 33 m s^{-1} threshold for hurricane force winds (the polar low on 25 April 1985 described by Lystad, 1986, p. 38, and the case discussed in Section 6.2.6).

Reed's 1979 paper, 'Cyclogenesis in polar airstreams', marked the starting point of a debate over whether comma clouds were 'true' polar lows, the latter being defined as the systems observed further poleward near the ice edge and often associated with a spiraliform cloud signature. This sometimes heated debate has been resolved through the general acceptance of the definitions discussed earlier through which a polar low is defined by some objective characteristic independent of the basic dynamics of the system and/or of their cloud field structure as seen from a satellite perspective.

Polar low research during the early 1980s was almost exclusively focused on the Northern Hemisphere systems. During the late 1980s and the 1990s, however, increasing interest was focused on Southern Hemisphere phenomena (see Section 1.2). While Southern Hemisphere systems just north of the polar front resemble their Northern Hemisphere counterparts, higher latitude systems near the Antarctic sea ice/coast line, differ in important respects from Northern Hemisphere polar lows. Most significantly, most of the vortices observed near the Antarctic ice edge are much weaker and of small horizontal scale compared to the Northern Hemisphere polar lows. Further, the small, weak cyclonic systems observed near Antarctica are generally characterized by low-level, stratiform cloud and not by the deep convection characteristic of Northern Hemisphere polar lows. In order to avoid confusion between the weak Southern Hemisphere systems and the generally much more vigorous Northern Hemisphere polar lows it soon became customary to denote the Southern Hemisphere systems by terms such as mesocyclone, Antarctic mesocyclone, mesoscale vortex etc. This way of distinguishing between the different systems was formalized through the 'Paris definition' mentioned above.

1.5 Classification

The problem of nomenclature of polar low and other cold air mass, small-scale disturbances is closely connected to the problem of *classification* of these systems.

Troup and Streten (1972) and Streten and Troup (1973) carried out one of the first investigations into the relationship between cloud patterns in satellite imagery and the synoptic environment and proposed a number of vortex types based on their appearance in the imagery. Their type 'A' signature was essentially a cold air development/polar low occurring in southerly flow around the Southern Ocean.

A further early morphological classification based on the appearance of the polar low as seen from a satellite perspective (Rasmussen, 1981, 1983), distinguished between 'real' or 'true' polar lows and comma clouds, the first terms denoting high latitude systems characterized by deep convection and spiraliform cloud systems, in contrast to the comma-shaped systems closer to the main baroclinic zones (see Section 1.6, Cloud signatures).

Forbes and Lottes (1985) presented a detailed morphological scheme for classification of mesoscale vortices, based on a systematic study of organized cloud systems in satellite imagery for the northeast Atlantic, the Norwegian Sea and the North Sea covering the winter 1981–82. Cloud configurations accompanying polar vortices were classified into nine categories, and only vortices with a pressure perturbation estimated to be of about 6 hPa were considered sufficiently intense to 'merit the title of a polar low'. It was concluded that satellite imagery can be used to identify and distinguish between polar lows and insignificant polar vortices and that the cloud configurations in early stages of the vortex lifetime give some clues regarding which vortices will subsequently develop.

A widely used classification system differentiating between three elementary types of polar low development based on associated distinctive synoptic patterns was suggested by Businger and Reed (1989a, b). The classification comprised three basic types: (1) the short-wave/jet-streak type, characterized by a secondary vorticity maximum and PVA (positive vorticity advection) aloft, deep, moderate baroclinicity, and modest surface fluxes; (2) the Arctic-front type, associated with ice boundaries and characterized by shallow baroclinicity and strong surface fluxes; and (3) the cold-low type, characterized by shallow baroclinicity, strong surface fluxes, and deep convection. The first type is identical with the so-called comma clouds (Section 3.1.3) and characterized by PVA aloft. Actually, upper-level PVA (and the cold upper-level temperatures within the region of the short wave trough) seems essential for a large number of significant polar low developments, including the systems placed by Businger and Reed in the second category or group, the Arctic front type.

Concerning the classification proposed by Businger and Reed (1989a, b), Grønås and Kvamstø (1995) argued the following: 'Our findings suggest a modification of the definition of the *Arctic front polar low* class in the classification of polar lows proposed by Businger and Reed (1989a). They found that this type is characterized by shallow baroclinicity and strong surface fluxes. We chose to call this class *Arctic outbreak polar lows* since they do not form at the leading edge of the Arctic front. It seems evident that a mobile upper disturbance is also active.' And, they continue: 'The presence of an upper disturbance, using the potential vorticity concept, should therefore be included in the definition.'

Rasmussen *et al.* (1993) proposed a basic classification of *primary* and *secondary polar lows* resembling the earlier mentioned classification of real (true) polar lows and comma clouds, but based on physical/dynamical considerations rather than on the cloud field structure. According to this classification, a polar low that forms as the result of a southward migration of an upper-level cold core vortex, originating within the large-scale circumpolar vortex, should be called a 'primary polar low' (a corresponding definition can of course most likely be used in the Southern Hemisphere). Primary polar lows associated with upper-level cold core disturbances will generally be characterized by deep convection and 'real' or 'true' polar lows can indeed be defined as being primary polar lows.

An attempt is presented in Section 3.1.4 to extend and improve the Businger–Reed classification scheme based on a study of a number of polar lows examined by Wilhelmsen (1985). According to this scheme most (nearly all) polar lows can be placed within one of the seven categories shown in Table 3.1 (p. 159). The 'classification' should not be considered as a 'dynamical classification' based on the mechanisms involved in the formation of the lows, but rather as *a convenient way to group polar lows within a relatively small number of categories.*

1.6 Cloud signatures

1.6.1 Introduction

Satellite data can provide a great deal of information about the form and composition of the cloud associated with mesoscale cyclones/polar lows from which information can be inferred about the air masses and physical processes involved in the formation of the systems. The visible and infra-red imagery shows the locations of the lows and provides data on the cloud-top temperatures. If a nearby radiosonde profile is available or an estimate of the atmospheric profile is obtained from a numerical model, then the actual height of the cloud tops can be estimated. Passive microwave data from satellites allow the estimation of integrated water vapour (IWV) across the system and integrated cloud liquid water (CLW) in the cloudy regions. These two quantities are particularly valuable since they provide information on the whole atmospheric column and not just for the top of the cloud. For example, the CLW data can provide information on the frontal structure not apparent in the conventional imagery (Lieder and Heinemann, 1999).

All these types of data allow an estimation of the diameter of the lows to be determined, although these cannot easily be related to the surface pressure pattern without other data being available. The imagery shows that mesoscale cyclones at high latitudes have a very wide range of diameters from a few tens of kilometres up to the maximum of the mesoscale taken here as 1000 km.

The imagery is also very useful for determining the lifetime of mesoscale cyclones, which although often existing for less than 24 h, can be observed to remain as distinct features for two or three days if the large-scale synoptic activity in the area is limited (Turner *et al.*, 1993b, 1996a). Over the southwestern Ross Sea, Bromwich (1991) found that the mesoscale cyclones observed over a two-year period existed for an average of 29 h, a much longer period than found for systems around the Antarctic Peninsula (6–12 h) in the study by Turner *et al.* (1996a). The vortices over the Ross Sea probably exist for such a long period because of the relatively quiet synoptic conditions at this location well south of the circumpolar trough. The mesoscale cyclones that were classified as 'significant' with winds of at least 7.5 m s^{-1} had a longer mean lifetime of 35 h. In the Northern Hemisphere, early case studies (Rasmussen, 1985a) showed that occasionally polar lows could exist for around two days provided they remained over the sea.

The work of Harold *et al.* (1999a) examining mesoscale cyclones in the North Atlantic section of the Arctic via AVHRR (Advanced Very High Resolution Radiometer) satellite imagery, has shown that many vortices can be observed year-round, but with a maximum in the winter. Eighty-six per cent had a diameter of less than 400 km and most had a lifetime of less than one day (see Section 2.1).

Mesoscale cyclones and polar lows owe their existence to specific physical forcing mechanisms. Depending on the nature of these mechanisms, the cloud fields of the different types of vortices will show certain basic, characteristic features through which the cyclones can be identified.

Examination of visible and infra-red satellite images has shown the existence of several characteristic cloud signatures of mesoscale vortices at higher latitudes. Two of these, the 'comma cloud' and the 'spiraliform vortex', are the most commonly observed types (Carleton *et al.*, 1995). In the following we consider the most important, characteristic cloud signatures for mesoscale cyclones/polar lows seen on satellite imagery.

1.6.2 The comma cloud

The majority of comma clouds are caused by a region of upper-level PVA ahead of a short-wave trough (see Section 3.1.3), the tail of the comma cloud being aligned along the trough axis. Comma clouds, such as the one illustrated in Figure 1.4 showing a typical signature over Iceland around 1000 km west of the polar front, are usually found at middle latitudes close to the main baroclinic zone. Occasionally, though, they may also be seen over higher latitudes, as illustrated on Figure 1.5 showing a characteristic comma cloud over the Barents Sea. Comma clouds are not limited to the Northern Hemisphere, and Figure 1.6 shows an example from the Southern Hemisphere. The comma cloud tail often marks the leading edge of a cold air outbreak with cumulus cloud, sometimes in the form of cloud streets, behind it (see Figure 1.4), and is often represented on synoptic charts as a short 'secondary' cold front. This mesoscale cloud vortex type is the most frequently observed mesocyclone signature over the Southern Hemisphere extratropics (Zillman and Price, 1972; Carleton and Carpenter, 1990), although in satellite climatologies that use relatively coarse resolution imagery (Streten and Troup, 1973; Carleton, 1979) this dominance is partly a function of the larger mean size of this vortex type compared with other mesocyclone signatures. For a detailed discussion of comma clouds see Section 3.1.3, Comma clouds.

1.6.3 The 'Spiraliform' signature

The second major cloud form seen on satellite imagery is the 'spiraliform' signature first described for the northeast North Atlantic by Rasmussen

18   1 Introduction

Figure 1.4. A visible wavelength satellite image showing a comma cloud over Iceland around 1000 km west of a synoptic-scale cloud band associated with the polar front. The comma tail was leading an outbreak of cold polar air characterized by cellular, rather shallow convection within the region of subsiding air behind an upper-level trough. Taken by NOAA 9 at 1528 GMT 8 March 1988. (Image courtesy of the NERC Satellite Receiving Station, University of Dundee.)

(1981). Figure 1.7 taken from his paper shows a spiraliform mesoscale cyclone over the sea between Iceland and the Faeroe Isles. Interestingly, a comma cloud situated over the northwestern part of the British Isles can be seen on the same image, which illustrates the striking difference between the two types of disturbance. The spiraliform system seen on Figure 1.7 initially formed south of Iceland and from there moved across the northeastern Atlantic into the North Sea, where it dissipated after making landfall on the German coast. The low,

1.6 Cloud signatures 19

Figure 1.5. An infra-red satellite image showing a rare high latitude comma cloud over the Barents Sea close to Novaya Zemlya. The image was taken by NOAA 11 at 1244 GMT 20 October 1992. (Image courtesy of the NERC Satellite Receiving Station, University of Dundee.)

20   1 Introduction

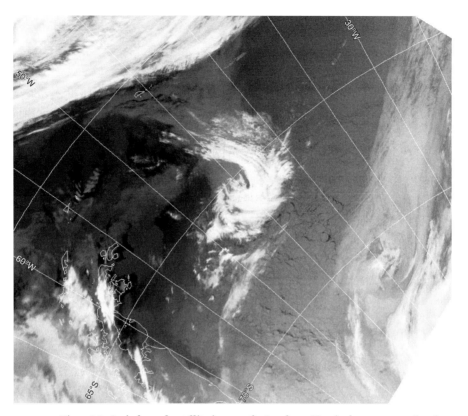

Figure 1.6. An infra-red satellite image of a Southern Hemisphere comma cloud over the northern Weddell Sea at 1726 GMT 6 October 1995.

which developed a warm core, could be followed on the synoptic charts for about two days over a distance around 2000 km.

Spiraliform systems are characterized by one or more spiral bands of convective cloud around the circulation centre, as illustrated in Figures 1.7 and 1.8, the latter figure showing a spiraliform polar low over one of the major genesis regions for such systems over the northern part of the Norwegian Sea. Spiraliform mesoscale cyclones occasionally have an 'eye-like' cloud-free or nearly cloud-free area at the centre of the low, as seen on Figures 1.7 and 1.8 (see also Figure 5.4, 'the most beautiful polar low'). On Figure 1.8 two major spiral arms can be seen, one being the southern extension of the Svalbard boundary layer front (BLF) (indicated by small arrows) entering the central part of the low from the west (BLFs play an important role in the development of many mesoscale cyclones and polar lows in the region and will be further discussed in Section 3.1.4). The other cloud band forms an arc north of the centre marking the boundary between an outbreak of unmodified Arctic air flowing towards the

1.6 Cloud signatures 21

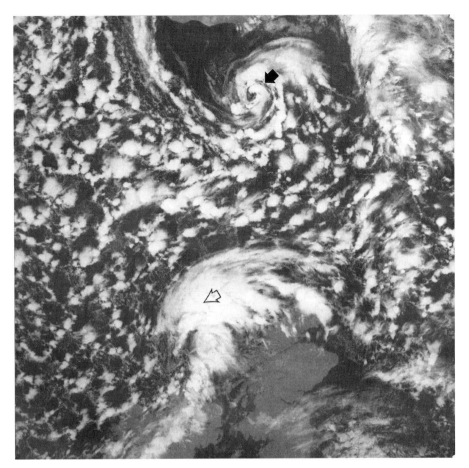

Figure 1.7. An infra-red satellite image taken by NOAA 5 at 1932 GMT 25 November 1978, showing a spiraliform polar low (indicated by a black arrow) southeast of Iceland and a comma cloud (indicated by an open arrow) over Scotland. (Image courtesy of the Satellite Receiving Station, University of Dundee.)

low and a modified warmer air mass further south. The form of the low shown on Figure 1.8 is typical of many spiral structured polar lows at high latitudes and may be explained by reference to Figure 1.9. The figure illustrates how a vortex superimposed upon a baroclinic field will form regions of increased temperature gradients, i.e. frontal zones. These zones have positions corresponding to the warm and cold fronts in a typical mid-latitude cyclone. Since the air in polar regions, where the spiral polar lows generally develop, will be conditionally unstable, the clouds along the frontal zones will predominantly be of a convective nature. Also, the region between the two frontal zones will often be filled partly or wholly by convective cloud.

22   1 Introduction

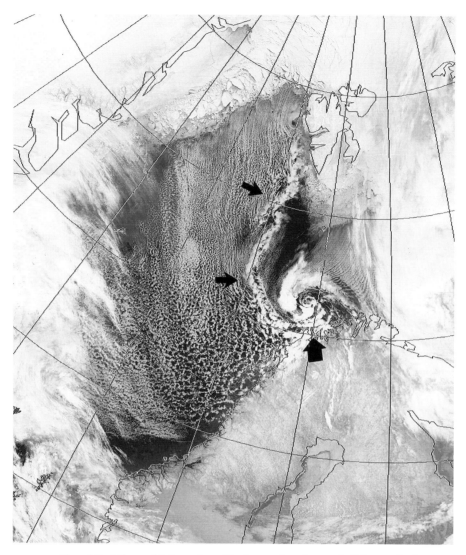

Figure 1.8. A thermal infra-red satellite image showing a spiraliform polar low (indicated by a large arrow) over the sea close to North Cape. The figure also shows a well-defined *boundary layer front* (indicated by small arrows) from west of Spitsbergen and south along 10° E. Taken by NOAA 6 at 1756 GMT 26 March 1981. (Image courtesy of the NERC Satellite Receiving Station, University of Dundee.)

On Figure 1.10a, from 27 January 1982, *two* small scale vortices are seen, one in the far north near North Cape, and another polar low-like system much further south over the Mediterranean between Sicily and Greece. The strongly asymmetric cloud spiral associated with the northern system is typical of many polar lows of the so-called 'cold low type', which often form

1.6 Cloud signatures 23

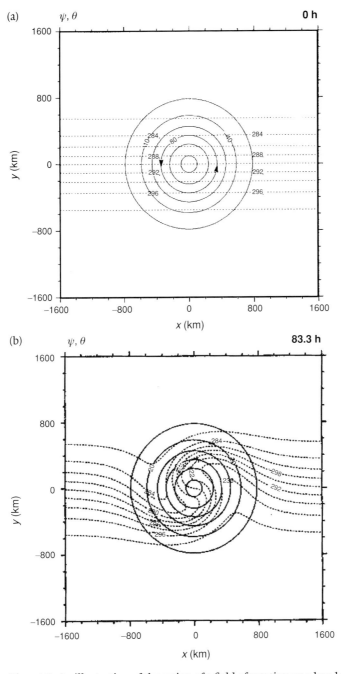

Figure 1.9. An illustration of the action of a field of rotation on a low-level temperature gradient. (a) Initial field, (b) about 3 days later (83.3 h). The isotherms are rotated with zones of intensified temperature gradients being formed south and southwest, and north and northeast of the vortex (from Bluestein, 1993).

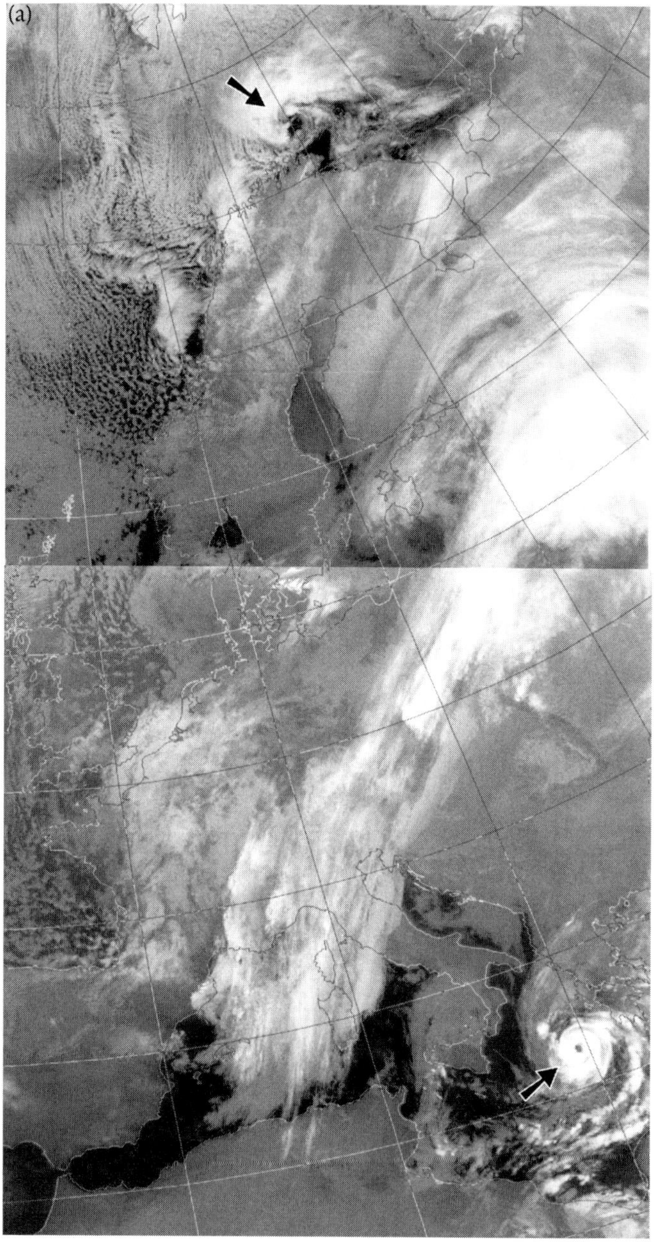

24   1 Introduction

Figure 1.10. (a) An infra-red satellite image showing a polar low near North Cape and a polar low-like mesoscale cyclone over the Mediterranean. The positions of the two vortices are indicated by arrows. The extensive northeast–southwest oriented cloud band across central Europe indicates the location of the polar front. The image was taken by NOAA 7 at 0237 GMT 27 January 1982. (Image courtesy of the NERC Satellite Receiving Station, University of Dundee.)
(b) 500 hPa chart for 0000 GMT 27 January 1982. The positions of the two polar lows shown on (a) are indicated by crosses.

(b)

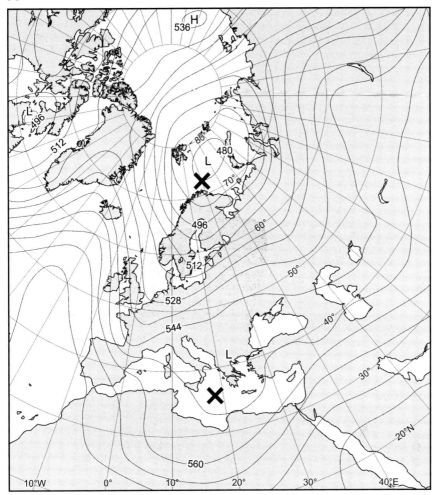

Figure 1.10 (cont.).

close to or within the central region of the circumpolar planetary vortex, as illustrated by Figure 1.10b (see also Sections 2.1.1 and 3.1.4). The system over the Mediterranean, which formed in a nearly barotropic environment, showed a striking similarity to a small tropical cyclone, including a diminutive eye, and strong surface winds near the centre of the system. The system formed within a cold large-scale, cut-off low over the Mediterranean and, in contrast to the situation pertaining to the northern system, there was a very weak upper-level flow (Figure 1.10b). Mediterranean polar low-like disturbances, such as the one seen on Figure 1.10a, will be discussed further in Section 3.1.8.

26 1 Introduction

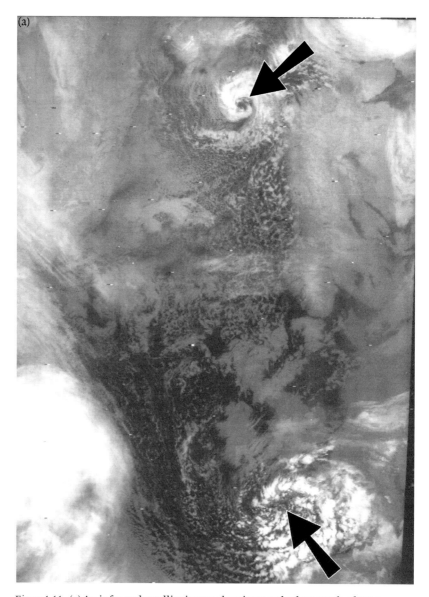

Figure 1.11. (a) An infra-red satellite image showing a polar low south of Bear Island and a cold core low over the Bay of Biscay (both vortices are indicated by arrows). The image was taken by NOAA 7 at 0423 GMT 13 December 1982. (Image courtesy of the Danish Meteorological Institute, Copenhagen.) (b) The 500 hPa chart for 0000 GMT 13 December 1982. The positions of the vortices shown on (a) are indicated by crosses. (c) The surface synoptic analysis for 0000 GMT 13 December 1982.

(b)

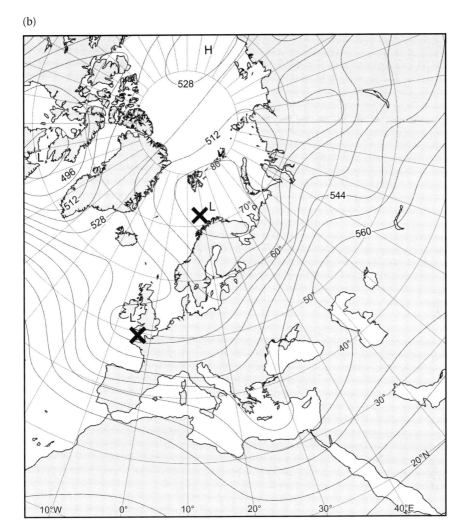

Figure 1.11 (cont.).

Figure 1.11a, from 13 December 1982, illustrates two more types of spiraliform system at different latitudes. The northern cloud spiral shows a polar low during early stages of development near Bear Island. The vortex around which the cloud spiral formed was a cold-core, vertically aligned system originating over the ice-covered regions far to the northeast (for further details see Section 3.1.4). After moving over the open (ice-free) ocean south of Svalbard the vortex penetrated to the surface resulting in the formation of the cloud pattern seen on Figure 1.11a (cf. the mechanism illustrated on Figure 1.9).

The southern spiral, which was also associated with a cold core low, was composed of more loosely organized spiral segments of convective clouds.

(c)

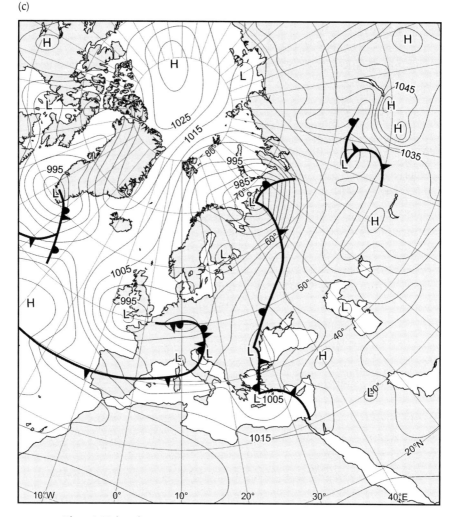

Figure 1.11 (cont.).

The northern spiral formed in one of the centres of the planetary, polar vortex, while the cloud spiral further south was associated with a trough in the westerlies (Figure 1.11b). For a correct interpretation of the mechanisms behind the formation of the two systems and their similarities/differences, it should be noted that the southerly 500 hPa trough had formed through a superposition of a vortex and a basic westerly flow. The vortex was clearly recognizable on the surface chart (Figure 1.11c) as a well-developed low. As discussed in Section 3.1.3, upper-level troughs in the westerlies are normally associated with *comma clouds* owing to differential positive vorticity advection between the upper and lower levels (see the omega equation, Eqn. 4.13b). In this case, however, the baroclinicity in the environment around the cold low was small. Because of this, organized ascent due to the baroclinic effect was limited, and processes within the boundary layer most likely became the dominant agents for determining the spiraliform structure of the cloud field.

In other cases, frequently when the cold core low is still situated over a cold snow/ice-covered surface, organized stable ascent takes place around upper-level cold core lows leading to the formation of a system of stratified, stable cloud. Once the system moves over the open sea, cumuliform clouds, embedded within the system of stratiform cloud, may form as well (Figure 1.12).

In the 1980s, mesoscale cyclones (polar lows) associated with comma and spiraliform cloud signatures were also identified over the higher latitudes of the Southern Hemisphere (Carleton and Carpenter, 1989a, 1990). Spiraliform vortices, in contrast to comma clouds, tend to occur deep within the cold air masses. Whilst Carleton (1985a), amongst others, found relatively large numbers of deeply convective spiraliform polar lows in the northern North Atlantic in winter, the analysis of many months' high resolution imagery by him has not identified many examples of this type in the Southern Hemisphere, in contrast to the larger comma cloud systems. Deeply convective spiraliform systems comprise less than 25% of all spiraliform systems occurring over latitudes poleward of about 50°S, and only 10% of all mesoscale cyclones observed (Carleton, 1995, [his] table 5). Turner *et al.* (1993a) ascribe this difference in convective development within Antarctic mesoscale cyclones to the stronger static stability over the ice sheet or ice-free but cold open water that occurs in summer. When spiraliform mesoscale cyclones are observed on satellite imagery for the southern oceans they are more likely (about 57%) to comprise stratocumulus clouds at low altitudes (Carleton, 1995). Accordingly, an example of this more commonly occurring type of spiraliform system is shown in Figure 1.13. Spiraliform mesoscale cyclones can have an extensive, dominant single cloud band away from the centre that marks the limit of a cold air outbreak, with additional convective bands or cloud areas (see Figure 1.14).

Many high latitude polar lows are triggered by cold, upper-level, short-wave troughs as the tracks of the systems take them over an ice-free ocean. The precise mechanism through which an upper-level trough affects the development of a surface mesoscale cyclone (polar low) varies from case to case. Strong PVA ahead of the trough axis (causing ascent due to dynamic forcing: see Eqn. 4.13b) and decreased vertical stability around the (cold) troughs (promoting deep

30 1 Introduction

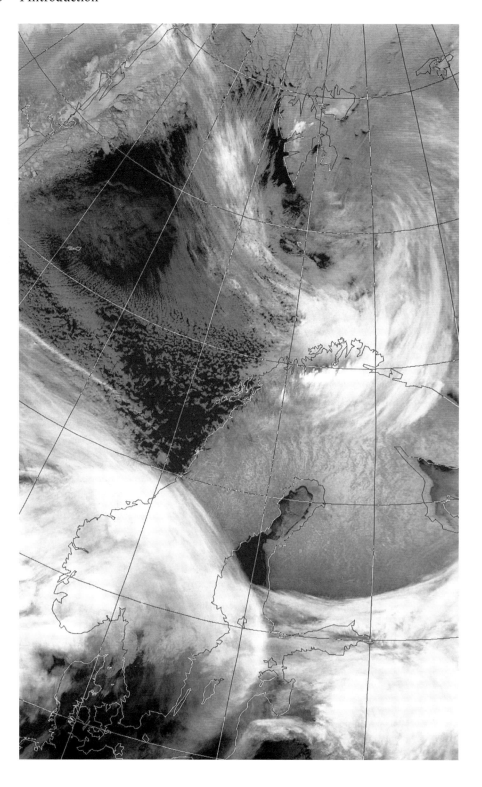

1.6 Cloud signatures   31

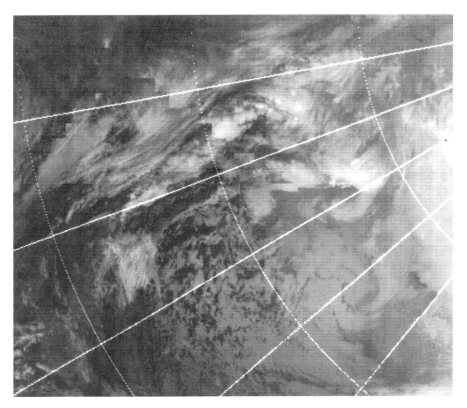

Figure 1.13. A thermal infra-red satellite image mosaic (5.4 km resolution) showing a spiraliform mesoscale cyclone consisting of low-level stratiform cloud that formed near the sea ice edge off Adélie Land. Comma cloud mesoscale cyclones comprising deeper clouds are situated just eastward. Taken at 0150 GMT 21 August 1988. Produced from the USAF DMSP (Defense Meteorological Satellite Program) film transparencies archived for NOAA/NESDIS at the University of Colorado, CIRES/National Snow and Ice Data Center, Campus Box 449, Boulder, CO 80309, USA.

convection) are both conducive to the formation of a low-level circulation. Occasionally polar lows develop *near the centre* of an upper-level, cold large-scale vortex, as seen from examples shown in the preceding, i.e. the northernmost vortices as seen in Figures 1.10 and 1.11. In such cases the upper-level PVA will be small and the formation of a mesoscale cyclone (polar low) will be due mainly

Figure 1.12. A thermal infra-red satellite image showing a cloud field of mainly stratified clouds at different levels associated with an upper-level cold core cyclone over the Barents Sea. The image was taken by NOAA 7 at 1255 GMT 25 January 1982. (Image courtesy of the NERC Satellite Receiving Station, University of Dundee.)

32   1 Introduction

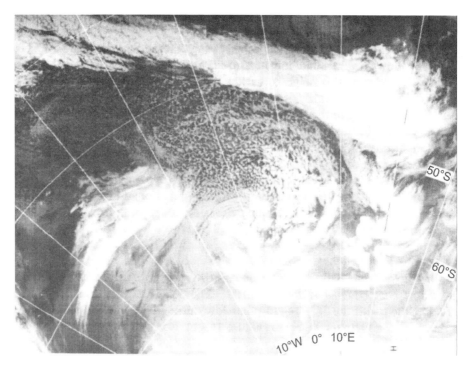

Figure 1.14. An infra-red satellite image showing a spiraliform mesoscale cyclone with multiple convective cloud bands. A DMSP image for 2044 GMT 14 July 1988 (from Claud *et al.*, 1996, *The Global Atmosphere and Ocean System*, Taylor & Francis Ltd (http://www.tandf.co.uk/journals)).

to reduced vertical stability caused by the very low upper-level temperatures near the centre. These cold low-type polar lows, which are quite common, will generally be of the spiraliform type and were included in the 1989 classification of Businger and Reed as the 'cold-low type'. An example of a cold core low situated over the Davis Strait in the central region of an upper-level, cold core vortex with 500 hPa temperatures around $-43\,°C$ is shown on Figure 1.15. A sequence of satellite images illustrating the formation of a mesoscale cyclone under the central part of an upper-level cold low over the Hudson Bay is shown as Figures 1.15b–d. In this case, only a rather weak system developed and the clouds did not show any significant spiral structure (for a detailed discussion of this development see Section 3.1.9, and for a more general discussion of cold lows see Section 3.1.4).

A final example of a spiraliform cloud signature is shown on Figure 1.16. This pronounced spiral was associated with a well-developed surface low, formed as an upper-level, cold low on a southerly track east of Greenland passed over the sea around Iceland.

1.6 Cloud signatures 33

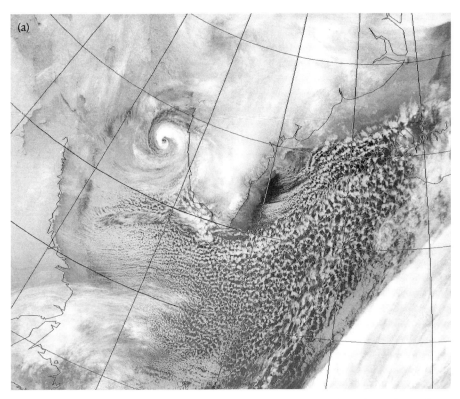

Figure 1.15. (a) An infra-red satellite image showing a polar low situated over the Davis Strait in the central region of an upper-level cold core vortex with 500 hPa temperatures around −43 °C. Parts of the cloud band associated with the polar front are seen at the lower right corner of the image. The image was taken by NOAA 11 at 1540 GMT 18 January 1989. (Image courtesy of the NERC Satellite Receiving Station, University of Dundee.) (b) an infra-red satellite image for 2307 GMT 9 November 1985 showing an incipient mesoscale cyclone under the central part of an upper-level cold low entering the western part of the Hudson Bay (from Parker *et al.*, 1991). (c) an infra-red satellite image for 1255 GMT 10 November 1985 showing a small cluster of deep convective cloud associated with a mesoscale cyclone under the central part of an upper-level cold low over of the Hudson Bay (from Parker *et al.*, 1991). (d) an infra-red satellite image for 1825 GMT 10 November 1985 showing a mature mesoscale cyclone approaching the eastern coast of Hudson Bay (from Parker *et al.*, 1991).

1.6.4 Merry-go-round signatures

When a large-scale, cold core upper-level cyclone is situated over the sea, polar lows/mesoscale cyclones may form in the central part of the low (i.e. the cold low-type described above), as well as in the region of cyclonic, upper-level flow along the rim of the cyclone, most likely due to the action of small-scale, upper-level troughs embedded within this flow. The cloud structure

34   1 Introduction

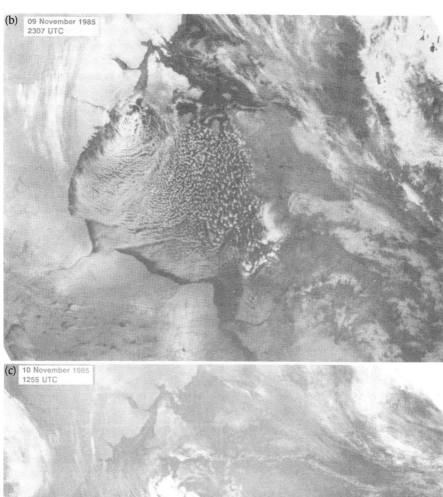

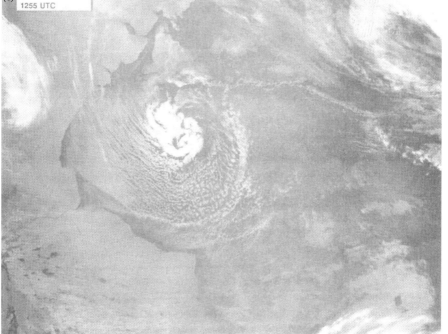

Figure 1.15 (cont.).

Figure 1.15 (cont.).

formed by the multiple mesoscale cyclones around a centre of rotation has led to the name 'merry-go-round' for this system of multiple polar lows (Forbes and Lottes, 1985). Northern Hemisphere examples are shown on Figures 1.1 and 1.17, and an example from the Southern Hemisphere in Figure 1.18. The large-scale, cold core systems causing the merry-go-rounds to develop may be part of, or originate within, the circumpolar vortex or be the end-product of an occluded synoptic-scale cyclone.

1.6.5 Instant occlusion developments

When mesoscale cyclones, particularly comma clouds, occur close to synoptic-scale cold fronts, an important development known as an 'instant occlusion' (Anderson *et al.*, 1969) may occur. In an instant occlusion, the comma cloud or polar low interacts with, and induces a wave upon, the frontal cloud band generally located to the east. On satellite imagery, the merging of the comma cloud and the induced frontal wave results in a system which, without going through the normal occlusion process, bears a striking resemblance to a mature occluded baroclinic wave (see Kurz in Bader *et al.*, 1995). Instant occlusions may result in strong systems, as discussed by Mullen (see Section 3.1.3).

36   1 Introduction

Figure 1.16. An infra-red satellite image showing a polar low with a spiraliform cloud signature south of Iceland at 1549 GMT 9 December 1993. (Image courtesy of the NERC Satellite Receiving Station, University of Dundee.)

Two examples of instant occlusions are illustrated on Figures 1.19 and 1.20. In the case from February 1979, illustrated on Figures 1.19a and b, the spiraliform polar low seen just southeast of Iceland (Figure 1.19a) approached the polar front and formed an instant occlusion, as shown on Figure 1.19b.

1.6 Cloud signatures   37

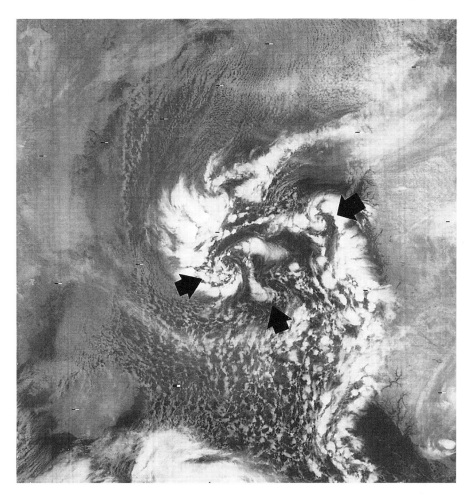

Figure 1.17. An infra-red satellite image showing a 'merry-go-round' polar low over the Greenland/Norwegian Seas. Several minor lows (marked by arrows) were rotating around a common centre. In contrast to the merry-go-round shown on Figure 1.1, no vortex had formed in the centre. The image was taken by NOAA 6 at 0920 GMT 11 December 1981. (Image courtesy of the University of Dundee.)

In the second case, from March 1997 (Figure 1.20) a comma cloud interacted with a developing wave (Figure 1.20a) forming the instant occlusion seen on Figure 1.20b. Other examples and a synoptic-climatological assessment of instant occlusions were presented in Carleton (1985a). The instant occlusion process over the Southern Hemisphere oceans was noted in early investigations of the utility of satellite imagery in synoptic meteorological analysis (Rutherford, 1969; Zillman and Price, 1972), and confirmed in a subsequent climatology (Carleton, 1981a, b). In the Southern Hemisphere, instant occlusions tend to occur particularly frequently to the west of areas favourable for blocking (see

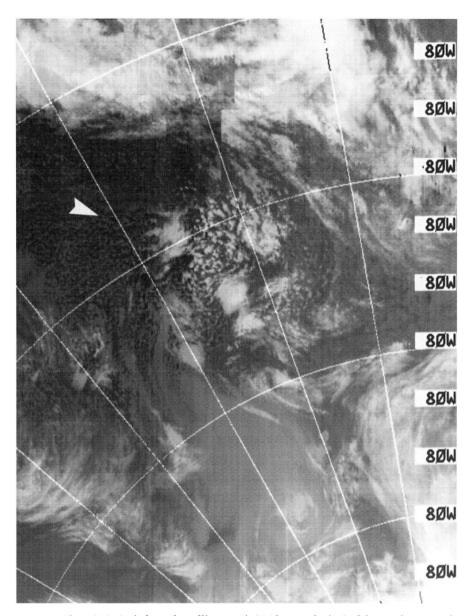

Figure 1.18. An infra-red satellite mosaic (5.4 km resolution) of the southeast South Pacific showing a 'merry-go-round' comprising multiple mesoscale cyclones in the incipient stage. The merry-go-round has formed within a decaying synoptic-scale cyclone (indicated by an arrow), centred at $c.$ 55° S, 102° W. The image was taken at 0249 GMT 3 August 1988. Produced from the USAF DMSP film transparencies archived for NOAA/NESDIS at the University of Colorado, CIRES/National Snow and Ice Data Center, Campus Box 449, Boulder, CO 80309, USA.

1.6 Cloud signatures

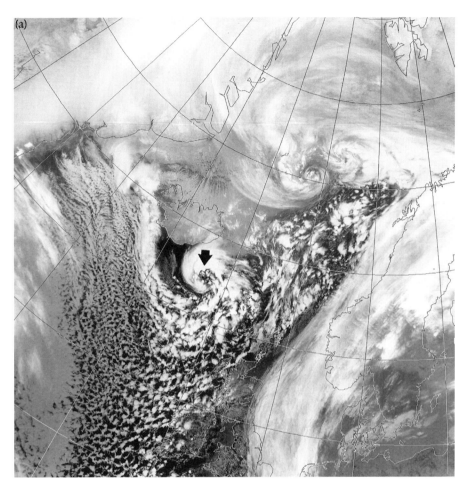

Figure 1.19. The development of an instant occlusion over the Norwegian and Greenland Seas as seen by satellite images for (a) 1427 GMT 28 February 1979 and (b) 1417 GMT 1 March 1979. The spiraliform polar low seen southeast of Iceland on (a) (indicated by an arrow) moved east towards the incipient wave on the polar front east of the United Kingdom to form the instant occlusion seen over the Norwegian and the Greenland Seas on the following day (indicated by an arrow). (Images courtesy of the NERC Satellite Receiving Station, University of Dundee.)

Section 2.2.2), since cold fronts that approach the blocking high are slowed, which increases the opportunity for an interaction between the cold-air comma cloud and the frontal wave.

1.6.6 Baroclinic waves

Mesoscale cyclones and polar lows may form through baroclinic instability along secondary baroclinic zones poleward of the main baroclinic

40   1 Introduction

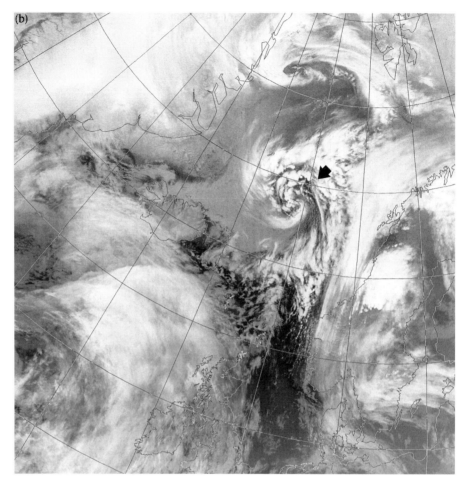

Figure 1.19 (cont.).

zones (the polar front), including shallow baroclinic zones along ice edges and/or ice/snow-covered coastlines. The cloud patterns associated with baroclinic mesoscale cyclones may, apart from the horizontal scale, be strikingly similar to those characterizing synoptic-scale extratropical cyclones. An example of this type of system is illustrated in Figures 1.21a–d, showing four stages of the development of a baroclinic polar low. In this case the polar low developed along a strong secondary, east–west oriented baroclinic zone south of Iceland, well north of the polar front. The satellite images show how a baroclinic leaf cloud (seen on Figures 1.21a and b) developed into a comma-like structure and finally into a tight vortex. Weldon (1979) described a baroclinic leaf as a cloud system associated with frontogenesis aloft within a westerly wind field, and pointed out that this cloud pattern is associated with the frontogenetic

1.6 Cloud signatures 41

Figure 1.20. The development of an instant occlusion between Iceland and the North Sea region as seen by satellite infra-red images for (a) 1721 GMT 2 March 1994. At this time a comma cloud west of the British Isles was approaching an incipient wave on the cloud band associated with the polar front further south. (b) 0534 GMT 3 March 1994. Twelve hours later the comma and the wave had merged to form the instant occlusion over the North Sea region and the sea northwest of Scotland. (Images courtesy of the NERC Satellite Receiving Station, University of Dundee.)

part of the development, whereas the comma pattern is the cyclogenetic phase (Figure 1.21c). In the final stage (Figure 1.21d), the baroclinic polar low occluded, at which time its cloud system, apart from the small horizontal scale, closely resembled that of a 'normal', large-scale occluded system.

The baroclinic polar low discussed above formed along an intense, deep baroclinic zone, and became quite strong (see the case of Rasmussen and Aakjær (1992) in Section 4.2, Baroclinic instability). On the other hand,

42   1 Introduction

Figure 1.20 (cont.).

mesoscale cyclones that develop along shallow, coastal baroclinic zones will often be quite weak, such as the system off the east coast of Greenland shown on Figure 1.22.

Most mesoscale cyclones develop as isolated vortices, but under certain conditions a series of lows can form as a wave train, as shown in Figure 1.3 (the 'Dannevig case'). Wave trains are quite common over the Greenland and Norwegian Seas where they form as baroclinic waves on so-called reversed shear zones. Two examples of reverse shear wave trains are shown on Figures 1.23a and b. The wave train in Figure 1.23a, consisted of three vortices that formed within a strong northeasterly surface flow along the boundary between a shallow cold air mass adjacent to the Greenland coast, and a warmer air mass over the Norwegian Sea further to the east. Between the wave train and the Norwegian coast an isolated mesoscale cyclone formed beneath an upper-level

1.6 Cloud signatures   43

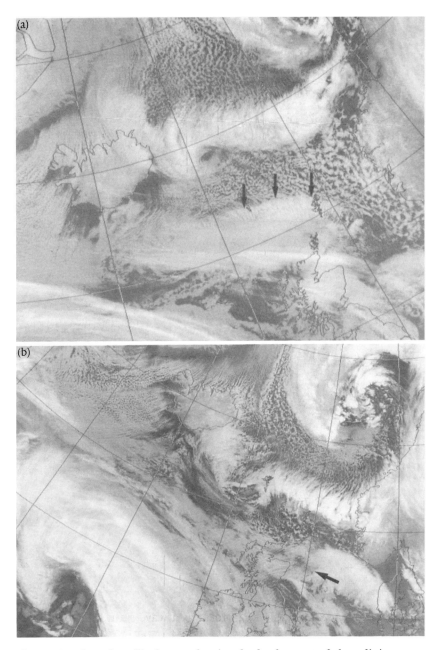

Figure 1.21. Infra-red satellite images showing the development of a baroclinic polar low (from Rasmussen and Aakjær, 1992). (a) 0818 GMT 28 March 1985. A baroclinic leaf cloud lay from south of Iceland to north of Scotland. The arrows indicate a striated cloud pattern resembling a horse's mane. Such striated clouds are often observed on satellite images, but the physical cause of their striations is not clear. (b) 1842 GMT 28 March 1985. The arrow shows where a 'slot' had begun to form in the baroclinic leaf cloud indicating the transition to the comma stage.

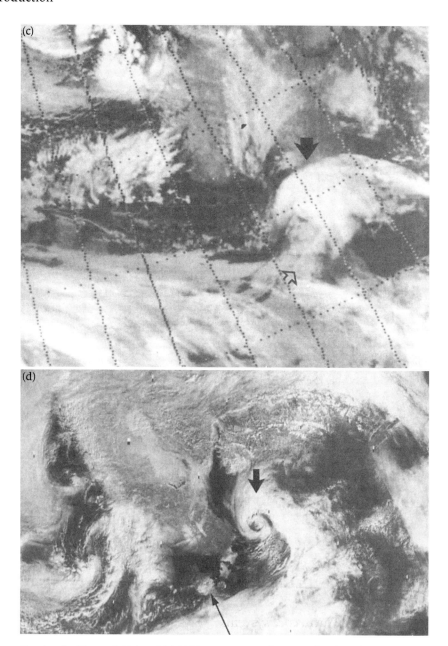

Caption for Figure 1.21 (*cont.*) (c) 0315 GMT 29 March 1985. The cloud field at this time resembled that of a large-scale extratropical cyclone, but of a much smaller horizontal scale. The open arrow indicates the position of a comma tail partly hidden by high level cirrus belonging to the main baroclinic zone further south. (d) 1128 GMT 29 March 1985. Showing the occluded polar low (indicated by a short arrow) over the Baltic. The long arrow shows the track of snow over Jutland left by the polar low.

1.6 Cloud signatures   45

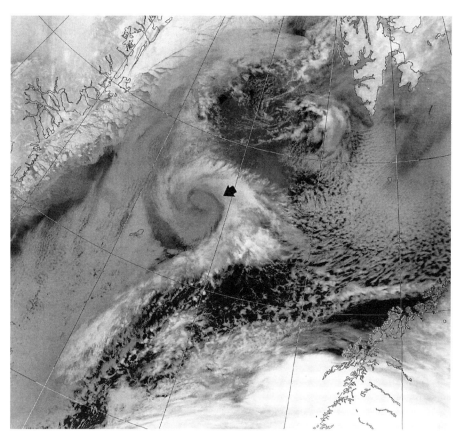

Figure 1.22. An infra-red satellite image showing a baroclinic mesoscale cyclone (indicated by an arrow) off the coast of northeast Greenland. The image was taken by NOAA 12 at 1739 GMT 11 December 1996. (Image courtesy of the NERC Satellite Receiving Station, University of Dundee.)

cold core low. Another wave train that formed in a similar situation is shown in Figure 1.23b. Examples of baroclinic mesocyclones in the Antarctic are shown in Figure 1.24.

1.6.7 Warm core systems

Within some polar lows the convection becomes so intense that, due to the release of latent heat, a warm core develops through a significant depth of the troposphere. Such systems are characterized by low-level cyclonic inflow and upper-level anticyclonic outflow, often indicated by a shield of high level cirrus cloud, capping the convective clouds (Figure 1.25). Another example of a system with a pronounced anticyclonic outflow is shown in Figure 3.50. This low, which developed over the Mediterranean, had a strong resemblance to a polar low.

Figure 1.23. Infra-red satellite images showing characteristic reverse shear systems over the Greenland and Norwegian Seas. (a) A reverse shear wave train of three waves (indicated by arrows) along a boundary between a shallow cold air mass adjacent to the Greenland coast and a warmer air mass further east. Taken by NOAA 11 at 0526 GMT 20 March 1994. (Image courtesy of the NERC Satellite Receiving Station, University of Dundee.) (b) An infra-red satellite image showing a reverse shear wave train of two vortices (indicated by arrows) between northern Norway and Iceland. Taken by NOAA 6 at 0840 GMT 20 March 1981. (Image courtesy of the NERC Satellite Receiving Station, University of Dundee.)

1.6.8 The evolution of cloud signatures

In a number of cases, the cloud associated with a mesoscale cyclone has been observed to change from comma-shaped to spiraliform or vice versa. For example, the mesocyclone examined by Bromwich *et al.* (1996), which

Figure 1.23 (cont.).

was initially observed to be spiraliform at mid-tropospheric levels, changed to a comma shape with middle and low-level cloud, before finally becoming a comma consisting mostly of low cloud. Of course there are some mesoscale cyclones than can first be observed with a single hook of cloud, but because of the rapid spin-up of the low, the cloud band can pass two or more times around the centre so that the system could be regarded as a spiraliform system. Care must therefore be taken not to infer too much about the processes involved in the development of a low from the cloud signature alone. If possible, a sequence of images should be used to follow the evolution of a particular vortex.

The cloud associated with the various types of mesoscale cyclone can be at any level in the atmosphere and the lows can be made up of stratiform, cumuliform or deep convective cells. Infra-red satellite imagery indicates that the

48   1 Introduction

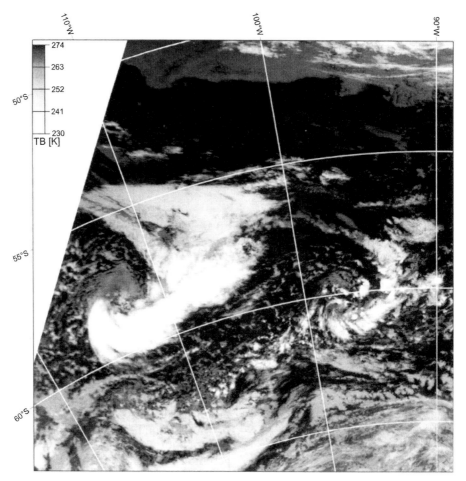

Figure 1.24. A thermal infra-red image of a pair of mesocyclones over the Amundsen/Bellingshausen Sea at 0200 GMT 12 January 1995.

cloud associated with mesoscale cyclones just north of the Antarctic coast is relatively low and vortices with higher cloud are usually found further north. However, some active mesoscale cyclones at relatively high latitude can contain convective cloud extending up to the mid-troposphere (Heinemann, 1996a).

Over the ice-free ocean there is a plentiful supply of moisture so that the mesoscale cyclones have well-defined cloud signatures. However, in the interior of the Antarctic the automatic weather station (AWS) observations have suggested that some lows have a surface pressure and wind perturbation but no cloud, as the atmosphere is very dry. This feature has been noted on the Ross Ice Shelf where there are many AWSs allowing mesoscale analyses to be prepared. In fact Bromwich (1991), in his study of two years of satellite and AWS data from

1.6 Cloud signatures 49

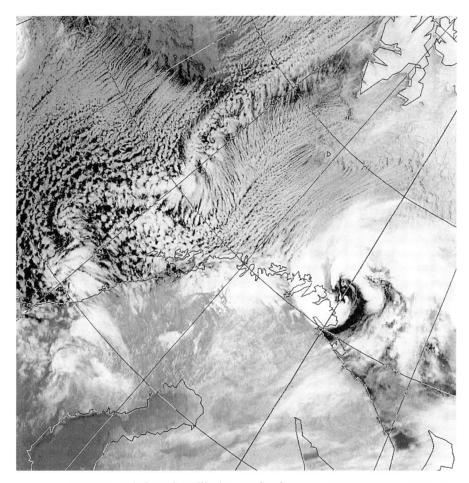

Figure 1.25. An infra-red satellite image taken by NOAA 6 at 0931 GMT 27 January 1982 showing a warm core polar low near North Cape. (Image courtesy of the NERC Satellite Receiving Station, University of Dundee.)

the Ross Ice Shelf area, found that most winter mesoscale cyclones initially had little or no cloud signature, so that the satellite imagery provided no help with the study of cyclogenesis events. However, when there was open water off the coast, surface fluxes of sensible and latent heat were more important and the mesoscale cyclones tended to have well-developed cloud signatures. The studies in this area have shown that if the wind direction changes to start drawing air from over the ice-free ocean, then a pre-existing mesocyclone with no cloud can develop a cloud band (Bromwich, 1991).

Well-developed mesoscale cyclones often have mid-level cloud while weak disturbances usually have much lower cloud tops (Figure 1.24). Well-developed mesoscale cyclones, such as the western low shown here, often have a cloud

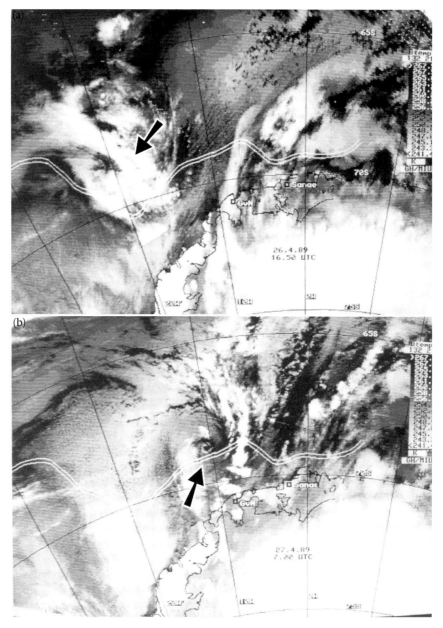

Figure 1.26. Infra-red satellite images of an Antarctic mesocyclone in the (a) incipient, 1650 GMT 26 April 1989, (b) mature, 0700 GMT 27 April 1989 and (c) dissipating, 1630 GMT 28 April 1989, states. The double white lines indicate the ice edge. After Heinemann (1996c).

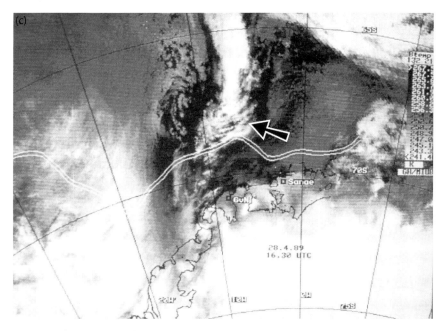

Figure 1.26 (cont.).

pattern that is very similar to that of a short baroclinic wave, frequently embedded in a field of cumulus cloud. When the system is active the comma of cloud can often take the form of a cold front separating continental, or cold air off the sea ice, from more moderate maritime air masses. The presence of the cold air can sometimes be diagnosed by the occurrence of convective clouds or cloud streets over the ocean while some mesoscale cyclones in deep cold air outbreaks can have shower bands consisting of deep convective cloud. With many cases examined, the cloud tops become higher as the mesoscale cyclone develops, before there is a general drop in cloud heights as the low declines.

Although the evolution and development of mesoscale cyclones is continuous the lows observed in infra-red satellite imagery can broadly be classified into three phases – the incipient, the developed (or mature) phase and the decaying stage. Examples of a low in these three states are shown in Figure 1.26. In the incipient phase there are only the first indications of rotation of the cloud elements and the low is dominated by the slightly curved linear cloud band. Once mature, the low has a well-developed cloud comma or spiral with sometimes a clear area or 'eye'-like feature at the centre. When declining, the cloud elements become much more disorganized and the frontal structure, if any, breaks down.

E. A. RASMUSSEN, K. NINOMIYA AND A. M. CARLETON

2

Climatology

2.1 The Arctic

2.1.1 Introduction

A large number of the most significant mesoscale vortices/polar lows form in northerly flows close to the Arctic coast or along the ice edges bordering the coast. For this reason, knowledge of the general weather and climatic conditions in the Arctic region is of major importance for an understanding of the formation of mesoscale cyclones, including polar lows.

As explained in Chapter 1, *polar lows* are a subclass of especially intense, maritime cyclones among the more general *mesoscale cyclones*. In Scandinavia and elsewhere in northwestern Europe, the main interest for many years has been focused upon the more intense systems, i.e. the polar lows. The term has been widely accepted throughout the meteorological community in the region, even for systems which do not, in a strict sense, fulfil the requirements of a wind speed around or above gale force. For this reason in this section we will generally use the term 'polar low' instead of the more general 'mesoscale cyclone' or 'mesocyclone'.

The Arctic region is dominated by the huge, generally sea ice-covered Arctic Ocean. It is approximately as large as the Antarctic continent, but apart from this, there are striking differences between the two regions (see Section 2.2). The Arctic Ocean surrounding the North Pole is bordered by Scandinavia, Siberia, Alaska, Canada and Greenland. It consists of a large basin, the Arctic Ocean, plus a number of marginal seas along the continental shelves. It is connected to the Atlantic Ocean via two other Arctic Seas, the Greenland Sea and the Norwegian Sea, and with the Pacific Ocean by the Bering Strait. The Arctic Ocean is the world's smallest ocean with an area of 14 million km^2. The marginal Arctic seas, the Beaufort Sea, the Chukchi Sea, the East Siberian Sea, the Laptev Sea, the Kara Sea, and the Barents Sea over the continental shelf occupy around 35% of the entire ocean area.

Ice coverage characteristics

The average area of ice coverage of the Arctic Seas changes dramatically with the seasons, shrinking from around 12 million km² in late winter to half this size in late summer (Figure 2.1).

Shallow atmospheric baroclinic zones, which are of great importance for the formation of mesoscale cyclones/polar lows, tend to form along or close to the ice edges especially during the winter.

Ocean currents

Ocean currents have a profound influence upon the climate of the Arctic region, and several studies have indicated the importance of locally high sea surface temperatures (SSTs) and strong SST gradient adjacent to the ice, for the development of polar lows (e.g. Rasmussen, 1979, 1985a). Figure 2.2 shows the surface currents within the North Atlantic, the Nordic Seas and in waters west and southwest of Greenland.

General climate conditions

According to Sater *et al.* (1971), four main climatic factors characterize the Arctic.

1. The surface weather systems associated with the large-scale, cold core circumpolar vortex (the planetary vortex) present in the free atmosphere over the area.
2. The distinctive regime of daylight and darkness which, together with low solar elevation, gives rise to a prolonged period of radiational loss from the Earth's surface in winter.
3. The surface cover of snow or ice for a significant part of the year.
4. The strong temperature inversion above the snow- or ice-covered surface resulting from strong radiational cooling.

In the following account, points 1, 3 and 4 in this list are discussed in some detail.

The circumpolar vortex

The upper-air flow patterns associated with the circumpolar vortex are apparent at 850 hPa but reach their greatest intensity near 200 hPa (Sater *et al.*, 1971). The circumpolar vortex displays a multi-wave pattern that reflects the influence of land, oceans and mountains on the Northern Hemisphere westerly circulation. During winter, three lows/troughs normally appear, one

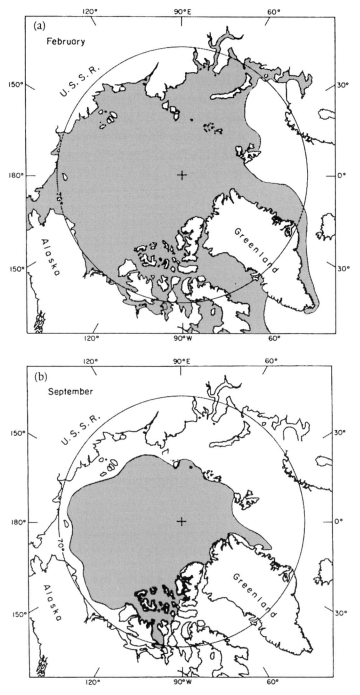

Figure 2.1. Arctic maximum (a) and minimum (b) sea ice extent based on 75% concentration limits (from Herman, 1989).

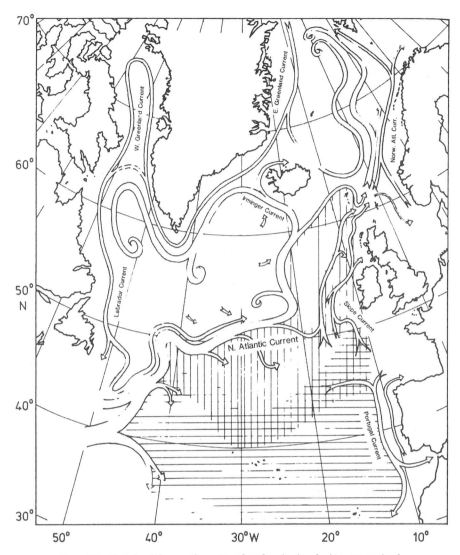

Figure 2.2. Sketch of the northern North Atlantic circulation. Water in the horizontally hatched area is influenced by the Subtropical Gyre, and water in the vertically hatched area by the Subpolar Gyre (from Ellett, 1993).

over the Canadian Archipelago and the Hudson Bay region, one near Novaya Zemlya, and one near Kamchatka and the Sea of Japan (see Figure 2.3). The corresponding Southern Hemisphere winter circulation shows much simpler circulation, as seen by comparing Figures 2.3a and b with their counterparts from the Southern Hemisphere shown in Figures 2.3c and d.

During spring and summer the circumpolar vortex contracts and weakens to a single centre approximately over the pole, and a more zonal flow replaces

56   2 Climatology

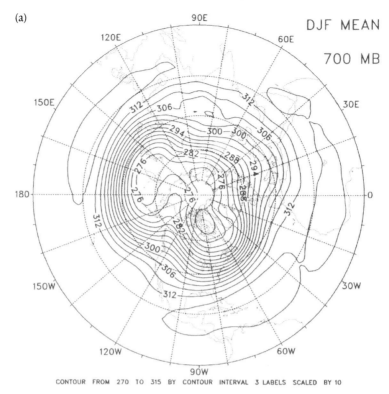

Figure 2.3. Mean winter season contour charts for the period 1979–88.
(a) Northern Hemisphere 700 hPa, (b) Northern Hemisphere 500 hPa, (c) Southern Hemisphere 700 hPa, (d) Southern Hemisphere 500 hPa (from Bluestein, 1993).

the much more meridional flow of winter. In autumn, as the Pole to Equator temperature gradient builds up again, the vortex expands and strengthens and the flow shifts to lower latitudes and becomes more meridional.

The strongest north–south temperature gradient associated with the vortex appears during the winter at mid-latitudes where the polar jet stream is also found. The jet 'steers' the large-scale synoptic cyclones and anticyclones in mid-latitudes and is itself associated with the formation and dissipation of these systems. Thus, areas south of the periphery of the vortex are generally characterized by a succession of eastward-moving frontal cyclones.

Further north, deep within the cold air and fairly close to the centre of the circumpolar vortex, persistent non-frontal, cold core lows resulting from the occlusion of frontal systems, can often be observed. Due to the influence of the strong stable surface layer over most of the Arctic, especially during the winter, the influence of these cold core vortices on the surface circulation over ice/snow-covered regions tends to be small. Smaller, upper-air cold core troughs or vortices, the origin of which is still uncertain, are regularly found as well. While hardly felt at the surface, over the snow/ice-covered regions these

(b)

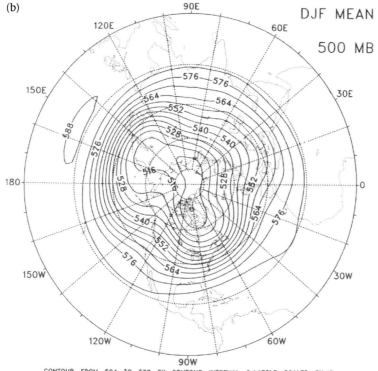

CONTOUR FROM 504 TO 588 BY CONTOUR INTERVAL 6 LABELS SCALED BY 10

(c)

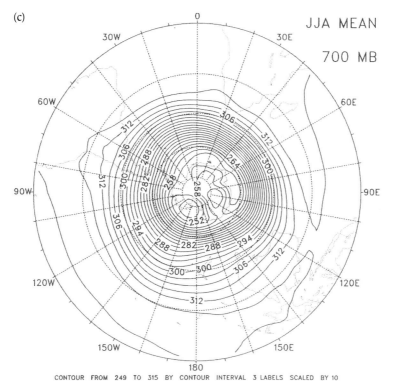

CONTOUR FROM 249 TO 315 BY CONTOUR INTERVAL 3 LABELS SCALED BY 10

Figure 2.3 (cont.).

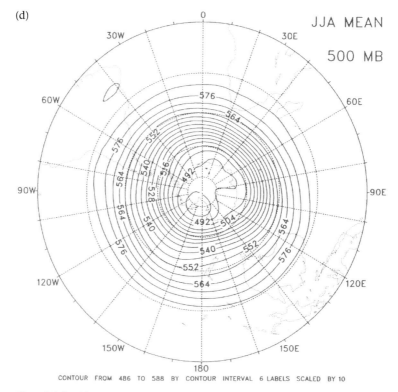

Figure 2.3 (cont.).

vortices, small as well as large, may, upon crossing the Arctic coast or ice edges, influence the lower layers and play an important role in the initial formation of polar lows.

On daily upper-level charts the closed circulation centres within the circumpolar vortex may move away from their mean positions (Figure 2.3), or may vary in number and strength. Based on 15 years of data, Bell and Bosart (1989) constructed a climatology of 500 hPa closed circulation centres between 24° N and 82° N. The frequency distribution of Northern Hemisphere 500 hPa closed cyclone systems for the winter season (December, January and February) found by them is displayed in Figure 2.4. The two centres over the Sea of Okhotsk and the Hudson Bay–Baffin Island region seen on the mean charts shown on Figure 2.3 are clearly reflected in the 500 hPa closed circulation centres distribution shown on Figure 2.4, while the Novaya Zemlya maximum is less well defined.

Most of the cyclone centres observed over the Nordic Seas region (i.e. the North Atlantic *east* of Greenland and north of 60° N, plus the North Sea, the Norwegian Sea, the Greenland Sea and the Barents Sea), and over the Arctic Sea are not *generated there*, as seen from Figure 2.5, which shows the total number of

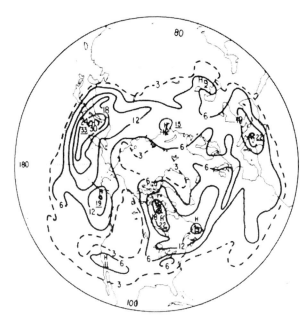

Figure 2.4. Total number of analysis periods (twice daily over 15 years) in which distinct closed circulation centres at 500 hPa were located in a given 2° × 5° latitude–longitude grid box during the winter period (December–February) (from Bell and Bosart, 1989).

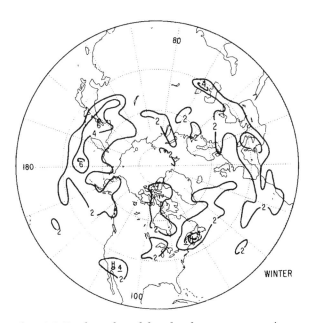

Figure 2.5. Total number of closed cyclone centre genesis events in any 2° × 5° latitude–longitude grid box for winter (December–February) (from Bell and Bosart, 1989).

closed cyclone centre genesis events for winter. Upper-level cold core vortices (closed 500 hPa cyclone centres) have often been observed to move south from high latitude regions triggering quite strong low-level circulations, but at the same time diminishing in strength at upper levels (see Bell and Bosart, 1989).

The surface cover of ice and snow

The third aspect pointed out by Sater *et al.* is the effect of the surface cover of ice or snow. Depending on the amount of ice and snow, the percentage of absorbed solar radiation can vary markedly. Up to 20% absorption occurs with a fresh snow surface compared with greater than 80% in most snow-free regions (Sater *et al.*, 1971). Over the regions covered by permanent ice and over the pack ice, surface temperatures may fall below $-30\,°C$ during winter, whereas during the brief summer, the temperature over ice/snow-covered areas is maintained around $0\,°C$ (Figure 2.6).

The Arctic inversion

The fourth distinctive aspect of the Arctic climate, according to Sater, is the spectacular temperature inversion unique to the Arctic and Antarctic (Figure 2.7). In extreme cases the temperature increases $20-30\,°C$ within the lowest 100 m. The inversions form because of a number of factors, including strong radiational cooling over ice/snow-covered surfaces, light surface winds, which preclude downward mixing of warmer air aloft, and clear skies, which maximize the radiational cooling. All these factors are optimized in the Arctic during the winter.

Serreze *et al.* (1992) studied the frequency and structure of low-level temperatures in the Eurasian Arctic. Some of their results, reproduced in Figure 2.8, show the frequency of all inversions in the region (surface-based and elevated), the frequency of elevated inversions, the depth and the temperature difference across the inversions. The depth (Figure 2.8) seems slightly smaller than the value in Sater *et al.* (1971), who gave the average depth of the inversion layer as 1200–1500 m. Serreze *et al.* concluded that inversion frequency, depth and temperature difference increases eastward from the Norwegian Sea towards the Laptev and East Siberian Seas, probably due to the effect of the proximity of open water, cyclonic activity and the effect of cloud cover. The deepest inversions, *c.* 1200 m, were found over the interior pack ice, while the strongest inversions (largest temperature difference across the inversion layer) were found over eastern Siberia.

During the extended cold season, and especially during mid-winter, pronounced low-level baroclinic zones may form between the extremely cold air over the ice- and snow-covered regions beneath the inversions, and the much milder air over adjacent seas. Short baroclinic waves may form along these

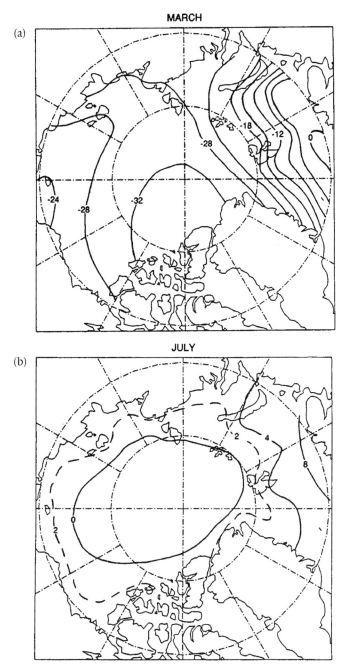

Figure 2.6. Mean surface air temperature (°C) for (a) March, (b) July (after Sechrist *et al.*, 1989. Figures provided courtesy of the Naval Research Laboratory, Marine Meteorology Division, Monterey, CA, USA).

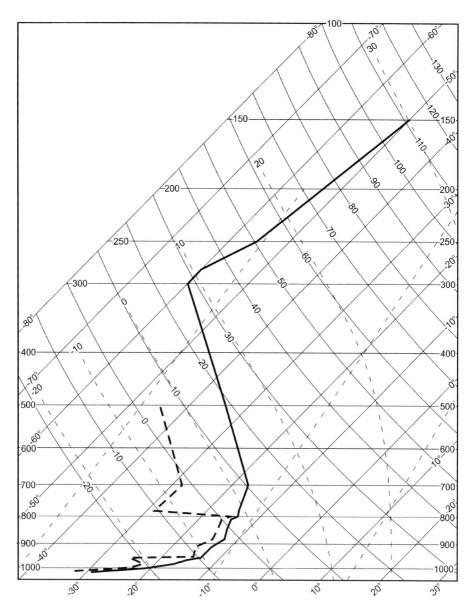

Figure 2.7. A typical sounding from a station in the Arctic. The ascent was made at Point Barrow at 1200 GMT 30 March 1989.

zones, especially when triggered by perturbations aloft. Some of these waves may develop into polar lows.

Air masses

The unique conditions over the Arctic, including the strong radiational heat losses over the snow/ice-covered surfaces, lead to the formation of air masses specific to the region.

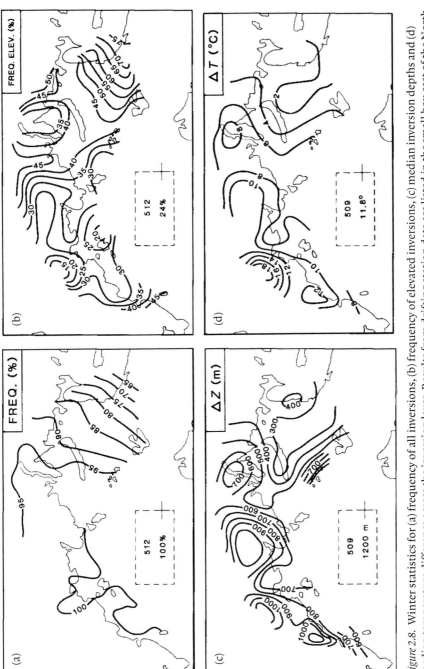

Figure 2.8. Winter statistics for (a) frequency of all inversions, (b) frequency of elevated inversions, (c) median inversion depths and (d) median temperature difference across the inversion layer. Results from drifting station data are listed in the small box east of the North Pole, which encloses all observations. In each figure, the top value in the box gives the number of cases, and the bottom number is the value of the appropriated statistics being examined (from Serreze *et al.*, 1992).

64 2 Climatology

Figure 2.9. Northern Hemisphere air masses in winter (from Barry and Chorley, 1992).

The Arctic Ocean has long been known as a principal source for Arctic air masses (Pettersen, 1950). The air mass source regions in winter for the Northern Hemisphere are shown in Figure 2.9.

The snow/ice-covered surfaces within the source regions lead to a marked cooling of the lower layers, causing pronounced low-level inversions to develop (see Figure 2.7). Also, above the low-level inversion, the stratification is generally stable throughout the troposphere, as illustrated in Figure 2.10 showing the average vertical temperature structure of air masses affecting North America.

In connection with the movement of large-scale synoptic systems, such as the major extratropical cyclones, air masses may be advected away from their region of origin so that they profoundly affect local weather conditions. The extent to which air masses are affected by the synoptic-scale cyclones is closely coupled to the path of these systems, i.e. to the *storm tracks* briefly discussed in the following section.

The storm tracks

Polar low developments generally take place in regions characterized by large amounts of low-level vorticity, such as synoptic-scale troughs, often observed west and southwest of large-scale cyclones, or within the central region of occluded cyclones. The storm tracks of the large-scale synoptic cyclones are

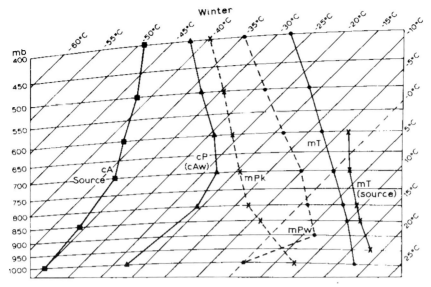

Figure 2.10. Average vertical temperature structure for selected air masses affecting North America at *c.* 45–50° N, recorded over their source areas or over North America in winter (from Barry and Chorley, 1992).

therefore an important climatological factor in explaining the genesis of polar lows and other mesoscale cyclones.

In winter, a number of depressions form along the North American east coast, an area which favours cyclogenesis because of the thermal contrast between the cold continental and warmer maritime air masses over the Gulf Stream. From the region around Newfoundland these storms, and others formed over the North American continent, generally migrate towards Cape Farewell and Iceland following the upper-air flow associated with the circumpolar vortex. A few of the lows enter the Davis Strait, moving north along the Greenland west coast. A number of these storms end as occluded systems in the climatological Icelandic low, but a significant number move further towards the northeast into the Norwegian/Barents Sea area from where the storm track continues eastwards, north of 70° N beyond Novaya Zemlya (Figure 2.11).

In the spring and autumn a primary storm track can be found from Newfoundland across Iceland to the Norwegian Sea and further on to the Barents Sea.

In the following we will consider the climate of the regions below, with special reference to the formation of polar lows.

1 The Nordic Seas (i.e. the northeast Atlantic east of Greenland and north of 60° N, the Greenland Sea, the Norwegian Sea, the Barents Sea and the North Sea).

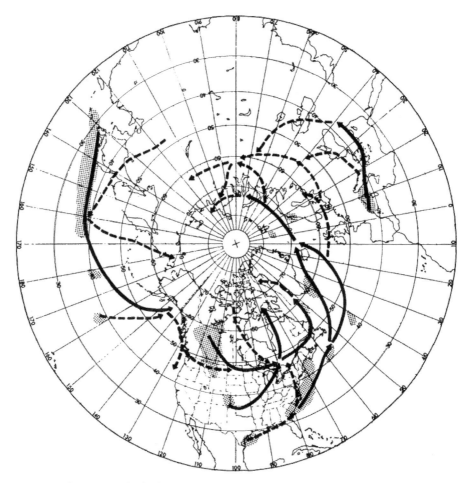

Figure 2.11. Principal northern hemisphere depression tracks in January. Solid lines show major tracks, broken lines secondary tracks which are less frequent and less well defined (from Whittaker and Horn, 1984).

2 The Labrador Sea, the Davis Strait, Baffin Bay and Hudson Bay.
3 The Beaufort Sea and the Chukchi Sea.
4 The Northeast Pacific and the Bering Sea.
5 The Northwest Pacific and the Sea of Japan.

2.1.2 The Nordic Seas

Physical characteristics – the seas

A number of significant currents are found in this region, including the East Greenland Current (Figure 2.2). Approximately 95% of the total sea ice expelled from the Arctic Ocean each year is carried southwards in this strong current. Sea ice can be found throughout the year, but is at a minimum in

September. A region of fast ice exists throughout the year along the northeast coast of Greenland (Figure 2.1).

Marginal ice zones (MIZs), defined as regions with intense air–sea interactions generated by the contrasts between a well-insulated, ice-covered surface and an ocean directly exposed to the air (Guest *et al.*, 1995), are found along the zone of fast ice off northeast Greenland and along other Arctic ice edges as well. Guest *et al.* considered two main MIZs in the Nordic Sea region, i.e. the Fram Strait–Greenland Sea MIZ (FMIZ) and the Barents Sea MIZ (BSMIZ) (see Figure 2.12). The ice edges, and the zones adjacent to the ice edges, change location seasonally. The median ice edge locations for April (lower thick line) and for November (upper thick line) shown on Figure 2.12 represent respectively the maximum and minimum ranges during a normal year.

The strong feedbacks and interaction of energy and moisture between the atmosphere, ocean and ice which take place in the MIZs profoundly affect the local weather and climate, including the formation of mesoscale vortices (see Section 3.1).

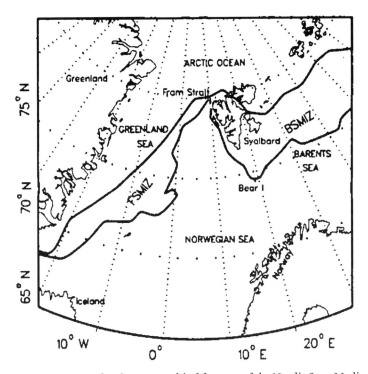

Figure 2.12. Map of major geographical features of the Nordic Seas. Median ice edge locations for the months of April (lower thick line) and November (upper thick line) represent maximum and minimum ranges during a normal year (after Guest *et al.*, 1995).

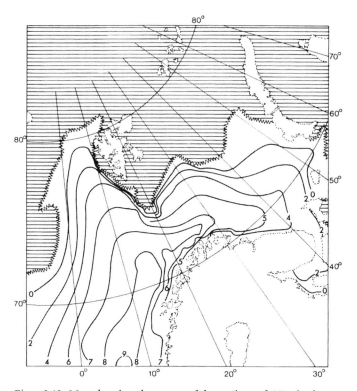

Figure 2.13. Map showing the extent of the sea ice and SSTs in the Norwegian and Barents Sea region on 10–12 December 1982. Areas wholly or partly covered by ice are hatched (from Rasmussen, 1985).

The Norwegian Coastal Current, a branch of the North Atlantic Current (see Figure 2.2), advects warm water northwards along the Norwegian coast and further into the Barents Sea. This results in high SSTs along the Norwegian coast of up to 7–8 °C even in mid-winter, as well as a very strong SST gradient in the region adjacent to the ice edge (Figure 2.13). The SST gradients are especially strong along the west coast of Spitsbergen and along the ice edge south of Bear Island and further east towards Novaya Zemlya. The SST gradient over the Greenland Sea is somewhat weaker, but nevertheless significantly affects the regional weather.

In the winter, the proximity between strongly cooled low-level Arctic air masses over the snow/ice-covered regions and the much warmer sea surface a short distance from the ice edge is of paramount importance for the development of polar lows in the region (see Section 3.1).

The Barents Sea is usually ice-free in its southwestern part where the SSTs are relatively high. In the winter the southern edge of the eastern MIZ stretches approximately west–east, roughly along 74–75° N to the region east of Novaya Zemlya, from where it runs approximately north–south along the coast line.

The Nordic Seas and especially the Norwegian Sea and the Barents Sea are the main genesis areas for polar lows in the region. However, in connection with the formation of major large-scale depressions, cold air masses may move far south over the eastern North Atlantic or even down over the Mediterranean. In such cases, small-scale lows resembling polar lows may be observed far to the south, outside of their usual genesis regions. A striking example of a vortex formation over the western Mediterranean, which in many ways resembled a polar low, was discussed by Rasmussen and Zick (1987) (see Section 3.1.8).

Physical characteristics – the land

The marine region of primary interest in this section is bounded towards the north by snow/ice-covered regions from northeast Greenland over Svalbard and Bear Island towards Novaya Zemlya.

Towards the west, the region is bounded by Greenland. Scorer (1988), in his presidential address to the Royal Meteorological Society, pointed out that Greenland, even in mid-summer, has a high albedo of close to unity, and acts as an atmospheric cold source for the Northern Hemisphere. Greenland is an impediment to many large scale cyclones and their frontal systems, and in many other ways is of far reaching importance for the weather and climate in the region.

Over the western part of the Greenland plateau there is, according to Scorer (and others), a permanent katabatic flow towards the west coast. Because the air has a very low absolute humidity the radiative exchange with the water vapour in the air above extends to a much greater height, so that the katabatic effect may extend to 500 m or more.

The katabatic flow on the east coast of Greenland is often perceptible on infra-red satellite images of the region around the fjord of Kangerdlugssuaq, which has a steep sloping hinterland (L. Rasmussen, 1989). A typical outflow situation is illustrated on Figure 2.14.

Another region with a steep sloping hinterland in the form of a pronounced valley rising steeply towards the ice cap and causing strong katabatic flows to develop is found around Angmagssalik on the east coast at a latitude around $60°$ N. Very strong winds, the so called Piteraqs, blowing out of the valley are occasionally observed in this region. Piteraqs is the East Greenland designation for a 'cold and strong wind and is directed out of the fjord and suddenly coming on' may become extremely destructive. During February 1970, northwesterly winds with gusts estimated to 80 m s^{-1} and $-20\,°C$ temperatures caused severe damage to Angmagssalik, the largest town in East Greenland (L. Rasmussen, 1989).

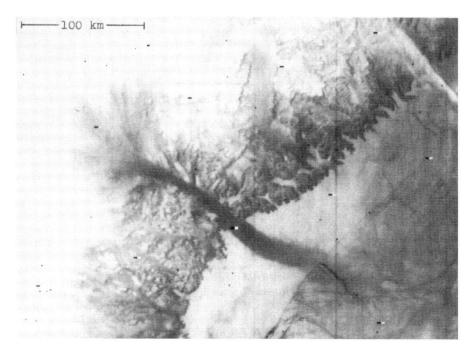

Figure 2.14. Black body temperature calibrated image (black/white = 0°–30°) showing the outflow through the Kangerdlugssuaq Fjord at the east coast of Greenland near 68° N (from L. Rasmussen, 1989).

Small-scale lows have occasionally been observed on satellite images of the Denmark Strait, east of Angmagssalik. Whether these lows form as the result of 'vortex stretching' of air descending from the Greenland Plateau remains to be investigated. The normal drainage flow is too shallow for significant vortex stretching to occur, but in synoptic situations with a strong northwesterly flow in the region, a deep atmospheric layer may be advected from the plateau towards the east and undergo significant vertical stretching as it descends towards the coast, leading to cyclogenesis.

Occasionally polar lows are observed to form further south along the southeast Greenland coast near Cape Farewell. These lee lows form in a synoptic-scale westerly or northwesterly flow as the air is forced to flow around or over, the southernmost part of the Greenland ice cap. So far little research has been carried out on the impact of Greenland on atmospheric flow, and cyclonic developments in the vicinity of Greenland and Iceland are often poorly captured by numerical models.

The impact of southern Greenland on a westerly flow is clearly illustrated by Figures 2.15a–c. On Figure 2.15a the centre of a large, synoptic-scale cyclone is

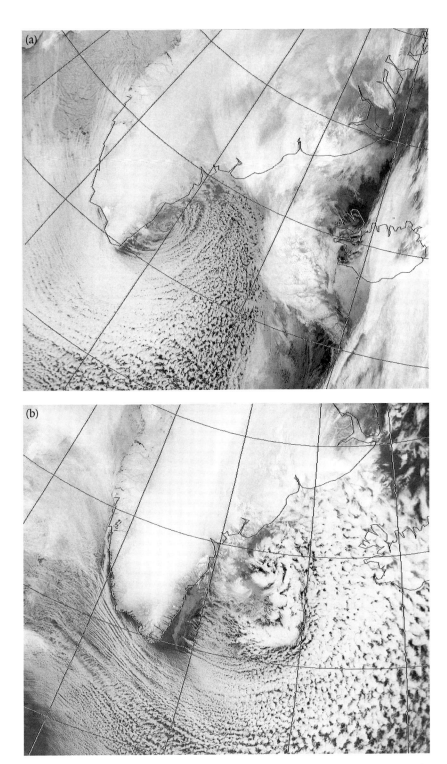

Figure 2.15. Infra-red satellite images showing the formation of a mesoscale vortex in the lee of South Greenland for (a) 1526 GMT 1 March 1983, (b) 1659 GMT 2 March 1983, (c) 1646 GMT 3 March 1983. (Images courtesy of the NERC Satellite Receiving Station, University of Dundee.)

Figure 2.15 (cont.).

situated close to the southeast coast of Greenland on 1 March 1983, not far from Cape Farewell. A satellite image from the following day (Figure 2.15b) shows that the cyclone at low levels remained 'anchored' to the southeast Greenland coast and that a mesoscale cyclone had begun to form within the central region of the large-scale circulation. Finally, Figure 2.15c illustrates the situation on 3 March 1983, when the low itself had moved east and was approaching Scandinavia. In its wake the large-scale system had left a mesoscale vortex over the Denmark Strait between Iceland and Greenland (the satellite image also shows the outflow from the Greenland ice cap through the valley near Angmagssalik).

A major large-scale meteorological feature associated with Greenland is the Icelandic low. This semi-permanent feature, which shows up on seasonal mean charts should, according to Scorer, be called the 'Greenland lee low' since it forms partly due to the lee effect of southern Greenland, and Iceland itself is not important in its formation.

The Icelandic low, also called the 'graveyard of cyclones', has the effect of creating a northerly wind along the east coast of Greenland in connection with the formation of a trough extending northward from the low. The low, together with the trough is, according to Putnins (see Orvig, 1970) 'one of the most important factors in the climatic features of Greenland'.

Svalbard is a group of islands including Spitsbergen, the largest of the islands, situated north of Norway from 77° N to 80° N (see map on opening endpaper). During the winter the western boundary of the eastern MIZ stretches from the southern tip of Spitsbergen south to Bear Island, a minor island at 74° N, 19° E. The low-level flow in the region is often northerly, i.e. along the western coast of Spitsbergen and the ice edge further south to Bear Island. In such circumstances a boundary layer front (BLF) is often observed to form between the cold air over the ice and snow and the warmer air over the sea that has been modified by a long fetch over a relatively warm sea surface. During cold outbreaks, the air originating east of Svalbard often has a slightly more easterly component than the air from the western side, and in such situations a shear/convergence line will form south of Svalbard (see Figures 1.8 and 3.10). The BLFs/shear lines may extend southwards along the Norwegian coast for hundreds of kilometres and occasionally reach as far south as the North Sea.

Novaya Zemlya consists of two main islands, oriented mainly north–south, in the southeastern part of the Barents Sea. During the cold season a MIZ exists near their western coast line along which BLFs may form in the same way as along the western coast of Spitsbergen.

Waves, with wavelengths ranging from the smallest observable by the satellite, up to hundreds of kilometres, develop along these BLFs and shear lines, and some of them may develop into polar lows provided a suitable upper-level forcing exists (for a further discussion see Section 3.1).

General synoptic features and climate/statistical studies – the circumpolar vortex

A factor of crucial importance for the formation of numerous polar lows in the region is the circumpolar cold core vortex. The vortex, which is strongest during the winter, displays a multi-wave pattern at this time of the year, and one, albeit the weakest, of the three centres or troughs seen on the mean upper-level charts is situated around Novaya Zemlya, close to the Barents and Norwegian Seas (Figure 2.3). The upper-air flow around the trough–ridge system is strongly meridional, and the surface cyclonic systems, steered by the upper-level flow, move in a northeasterly direction across the North Atlantic towards polar latitudes.

General synoptic features and climate/statistical studies – upper-level disturbances

Apart from its influence on the storm tracks, the changes in structure and position of the circumpolar vortex influence the formation of polar lows in a more direct way. One of the major results of the intensified polar low research in the early 1980s was the documentation of the importance of upper-level, short-wave cold troughs or vortices acting as triggers for polar low developments at the surface (e.g. Rasmussen, 1985a; Businger, 1985). These small-scale disturbances are generally found on the flanks of the major, synoptic-scale, cold upper-level lows or troughs, rotating around the central region of the vortex. As discussed above, one of the three centres within the circumpolar vortex is situated around Novaya Zemlya (Figure 2.3). Small-scale, upper-level disturbances embedded within the flow around this centre, may trigger significant polar low developments upon entering the Barents or Norwegian Sea (see Section 3.1). In addition, polar lows have frequently been observed to develop within the central regions of the polar vortex centres themselves in cases when these centres have moved away from their usual positions over snow/ice-covered regions.

General synoptic features and climate/statistical studies – air masses

The nature of Arctic air masses was briefly discussed in Section 2.1.1. However, polar lows form on the outskirts of the Arctic air mass source regions where the stratification (vertical temperature lapse rate) may be influenced by warmer air masses advected north or by strong heat transfer from an underlying warm sea. Depending upon the actual situation, the stratification within a continental Arctic air mass may be less stable than that illustrated by the average sounding shown on Figure 2.10 or the typical sounding shown on Figure 2.7.

Most polar lows are associated with convection that is rather deep, considering the latitude. In a study of 38 cases of gale-producing polar lows in the period 1977–82, Wilhelmsen (1986a) found a conditionally unstable lapse rate between the surface and the 500 hPa level in all cases. Therefore, continental, stably stratified Arctic air masses with a vertical structure similar to that indicated on Figures 2.7 and 2.10 necessarily must undergo a significant transformation before a polar low development. A characteristic transformation of a continental Arctic air mass upon crossing the ice-free ocean near Bear Island was described by Økland (1989) and illustrated on Figure 2.16. According to Økland, the sounding from Bear Island (shown as curve 1) is quite typical for cold (Arctic) air outbreaks in this particular region and may be considered as characteristic for the air mass in its source area over the ice-covered Arctic Ocean, although the stability above the low-level inversion varies

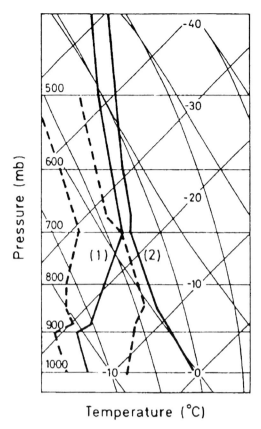

Figure 2.16. Soundings of temperature (solid lines) and dew point (broken lines) illustrating the transformation of a continental Arctic air mass upon passing over the ice-free ocean: (1) Bear Island (74° N, 19° E), 0000 GMT 21 November 1975, (2) Bodø (67° N 14° E), 1200 GMT 21 November 1975 (after Økland, 1989, in Twitchell *et al.*, 1989. © A. Deepak Publishing).

from case to case. When the air moved over the ice-free ocean the low-level inversion seen on the Bear Island sounding was gradually eroded away resulting in the much more unstable stratification observed at Bodø, 800 km further south, 12 h later (curve 2). The strongest transformation clearly had taken place within the originally stable layer adjacent to the surface, whereas the layers above were relatively unaffected. Occasionally, the transformation of Arctic air masses moving south may become very pronounced through a combination of upper-level cold air advection and strong low-level warming due to intense air–sea interactions over a warm sea surface. This is illustrated by the sounding shown on Figure 4.26. In such cases strong polar lows may develop as a result of latent heat release due to organized convection (see Sections 3.1 and 4.5).

Climatic and statistical studies concerned with the Nordic Seas and the North Atlantic

A number of climatic and statistical studies concerning the occurrence of polar lows and the large-scale environment in which they tend to form, have been carried out over the years in and around the Nordic Seas.

Businger (1985) studied the synoptic climatology of polar low outbreaks in the Norwegian and Barents Seas. He used the so-called superposed epoch-analysis technique computing a composite 500 hPa chart by averaging the 500 hPa fields for the starting days (key days) of all polar lows observed in his data set (about 10 years of data). He then computed an anomaly chart by subtracting the 500 hPa climatological mean height field for the season October–April from the composite chart (Figure 2.17).

Businger found significant negative height and temperature anomalies in the 500 hPa synoptic-scale fields on days when polar lows occurred over the area, indicating strong positive vorticity and low static stability over the area

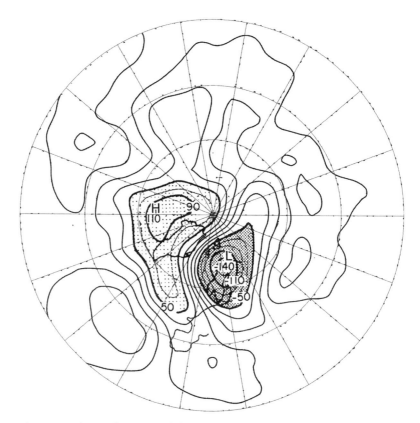

Figure 2.17. The 500 hPa anomaly height field for days with polar low developments (contour interval 20 m), 42 cases (from Businger, 1985).

on those days. Another interesting aspect noted by Businger was the fact that the composite surface pressure and 1000–500 hPa thickness field revealed only slight asymmetries in the respective contours, 'suggesting that the structure of the atmosphere is primarily equivalent barotropic over these storms on the larger synoptic-scale'.

Businger's work confirmed and quantified the generally accepted understanding that most polar low developments take place within deep outbreaks of polar/Arctic air masses. This work on the *large scale conditions* leading to the development of polar lows was extended and supplemented by a number of other climatological studies, many of them carried out as part of the Norwegian Polar Lows Project from 1982–85.

Ese *et al.* (1988), using a data set of observed polar lows covering January 1971 through January 1983 (i.e. basically the same data set as used by Businger), studied the large-scale (synoptic-scale) flow pattern during the formation of polar lows over the Norwegian, Greenland and Barents Seas. For most of the cases when polar lows were observed, a strong northerly flow was found over the Norwegian Sea, or a northeasterly flow over the Barents Sea.

The geographical distribution of the polar lows suggested a division of the events into two groups, a western group consisting of the lows first observed west of 5° E and an eastern group consisting of those first observed east of the same longitude. The mean values of the longitudes of the two groups are shown on Figure 2.18.

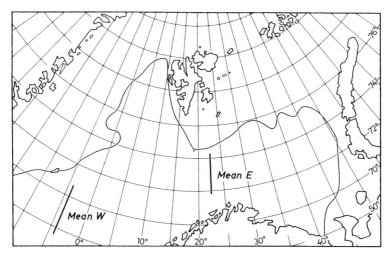

Figure 2.18. The average position of the ice edge in February for the years 1971–80. Mean W and Mean E, mean values of the longitude for the polar lows in the western and the eastern groups respectively (from Ese *et al.*, 1988).

78   2 Climatology

For days when the polar lows were observed in the western sector, the large-scale synoptic low pressure centre was generally situated over the northern part of the Norwegian Sea, while eastern sector polar lows were associated with large-scale low pressure centres further east over northern Scandinavia or the Barents Sea.

Composite and anomaly charts were computed for the key days for each of the two groups (eastern and western) for the winter season (December–March). The composite sea level (surface pressure) chart for the 19 polar low events that occurred in the western sector is shown on Figure 2.19a, and the corresponding chart for the eastern sector events on Figure 2.19b.

The 500 hPa anomaly patterns for the western and the eastern sectors are shown on Figures 2.20a and b, respectively. The structure of the upper-air anomaly fields is similar to those at the surface (not shown). The values, and especially those for the western sector, are large compared with those of Businger.

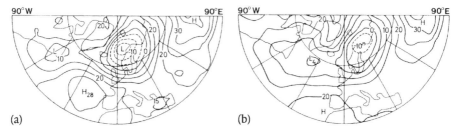

Figure 2.19. The composite sea level chart (contour interval 5 hPa) for the polar low events during the winter season (December–March, 1972–83). Dashed contours are used for values less than 1000 hPa; (a) western sector, (b) eastern sector (from Ese *et al.*, 1988).

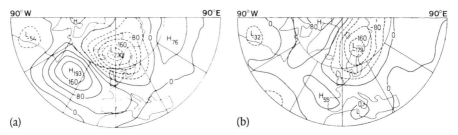

Figure 2.20. The 500 hPa level anomaly field (contour interval 40 gpm) for the polar low events during the winter season. Dashed contours are used for negative values; (a) western sector, (b) eastern sector (from Ese *et al.*, 1988).

According to Ese *et al.*, the reason could be that they used winter season cases only, and that they divided the polar low events into two geographically separated groups.

Wilhelmsen (1985) used mainly 3-hourly synoptic surface weather charts in a pioneering study searching for polar lows over the seas surrounding Norway during the period 1972–82. Wilhelmsen defined a polar low in a fairly general way as a 'mesoscale low or an irregularity in an otherwise uniform cold air stream north of and/or west of the polar front, the polar low having a horizontal scale of 100–500 km'. Only polar lows characterized by surface wind near gale force (15 m s^{-1}) or greater at Norwegian meteorological coastal stations, islands or weather ships near the Norwegian coasts were included in the study. These constraints reduced the total number of selected cases to 71. Heavy precipitation in the form of snow lasting for 3–6 hours was often observed in connection with the lows. The frequency distribution of the lows is shown on Figure 2.21.

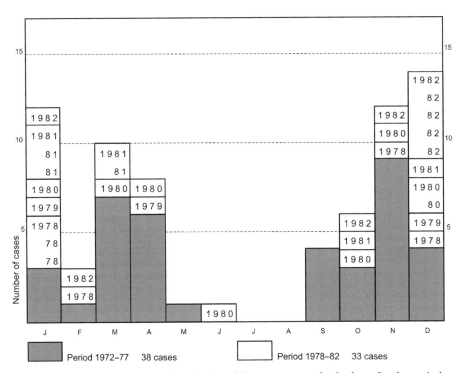

Figure 2.21. Frequency distribution of the occurrence of polar lows for the period 1972–82 (71 cases in 11 years) in the Norwegian Sea and the Barents Sea (after Wilhelmsen, 1985).

80   2 Climatology

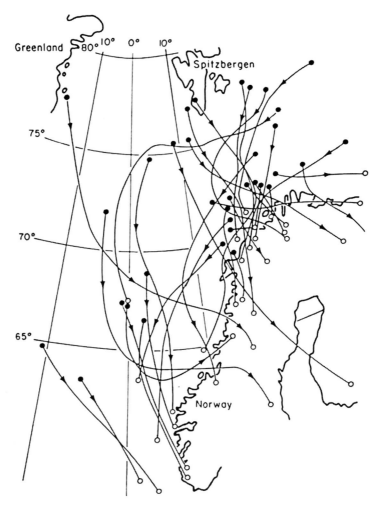

Figure 2.22. The tracks of polar lows for the period 1978–82 (after Wilhelmsen, 1985).

Figure 2.22 shows the points of origin of the polar lows and their tracks for the four years 1978–82. The figure shows that most of the polar lows in this period developed east or south of Bear Island.

One of the most important meteorological parameters of interest in connection with polar lows is their maximum wind speed. The distribution of maximum observed wind speed for the cases studied by Wilhelmsen is shown on Figure 2.23.

Wilhelmsen's study was extended by Hoem, Kristoffersen and Smits, their results being summarized by Smits (1985) and Lystad (1986). The same region was studied as considered by Wilhelmsen, and basically the same definitions

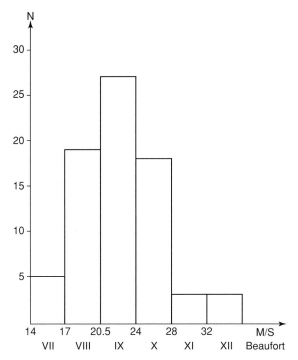

Figure 2.23. Frequency distribution of maximum observed mean wind in Beaufort classes (corresponding wind velocities in m s^{-1} are indicated) in the Norwegian and Barents Seas for the period 1972–82 (76 cases; from Lystad, 1986). (Figure courtesy of M. Lystad, Norwegian Meteorological Institute.)

and selection criteria were applied. However, in this study infra-red satellite images were also used extensively.

During the three winter seasons 1982/83, 1983/84 and 1984/85, a total of 83 polar lows were observed satisfying the 'gale force criterium' first applied by Wilhelmsen, i.e. *more* than the 71 cases found by Wilhelmsen in the much longer 12-year period considered by her from 1971 to 1982. The term 'winter season' or 'cold season' here was loosely defined as the period during which polar lows were observed, generally from October to May.

Even this fairly large number is bound to be an underestimation of the number of polar lows. During the winter season 1984/85, for example, a total of 44 cases of possible polar lows were detected by using satellite images and surface charts. However, for a number of cases, data were not available to determine whether the wind criterion (i.e. that the *observed* wind near the track exceeded 14 m s^{-1}) was fulfilled. For this reason only 28 of these 'candidates' were included in the final statistics.

The frequency distribution of polar lows for the three winter seasons studied by Smits was qualitatively similar to that found by Wilhelmsen

(Figure 2.21), apart from the fact that it comprised more cases. This includes the minimum in February, which was common for all three years studied by Smits. Also, the start positions reported by Smits were in accord with those found by Wilhelmsen (Figure 2.22) in that most polar lows formed in the region around and east of Bear Island. Another region favourable for polar low development was found around and east of Jan Mayen. To a large degree, the tendency for the start positions of the lows to cluster in certain regions can be explained by geographical and meteorological factors to be discussed in Section 3.1.

Grouping the polar lows according to the observed wind speeds (gale, storm, hurricane-force) gives the distribution shown on Figure 2.24. Only one occasion with hurricane force winds was observed, when the mean wind increased from 10 to 35 m s^{-1} in only 20 minutes. It should be noted that the distribution shows 'only' the 10-minute *mean* wind. The surface winds associated with polar lows will normally be very gusty and the instantaneous wind speed may therefore exceed hurricane force even if the mean wind is only in the storm category (24.5–32.7 m s^{-1}).

Following their formation, most polar lows drifted in a southerly direction. Most polar lows that developed near Bear Island had a relatively short track

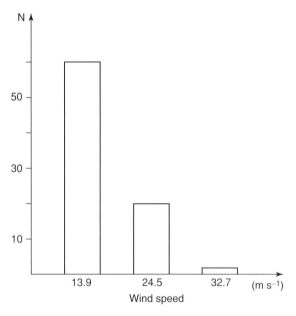

Figure 2.24. Frequency distribution of polar lows during 1982–85 as a function of maximum mean wind speed (gale, storm and hurricane) (after Lystad, 1986). (Figure courtesy of M. Lystad, Norwegian Meteorological Institute.)

before making landfall on the Norwegian coast (see Figure 2.22), and many persisted for only 12–18 hours. Others may exist for up to two or three days and traverse more than 1000 km before they dissipate, provided they do not make landfall in the meantime. Polar lows are maritime phenomena and generally lose their identity quickly after landfall, although there are noteworthy exceptions.

Forbes and Lottes (1985) studied 133 cases of mesoscale cloud vortices occurring in cold air masses over the northeastern Atlantic Ocean and adjacent seas between 1 December 1981 and 5 January 1982. The weather was unusually cold over southern Scandinavia during this period and cold air had pushed far southwards. Consequently most of the vortices were found at an unusually southerly position, the centre of the frequency distribution being placed over the North Sea.

Interestingly, and in contrast with the results of many other authors, Forbes and Lottes concluded that the initial development of the vortices was dominated by processes related to diabatic heating and CISK, but that later in the development, baroclinic processes became more important.

Aakjær (1992) studied the occurrence of polar lows (mesoscale vortices) over Denmark for ten polar low seasons (November–April), 1980–90. Polar lows do not occur every year over Denmark, but in some years several have been observed. During the ten year period, 18 vortices of different intensities were found, among which seven passed close to Denmark within only one week in December 1981, i.e. within the period also studied by Forbes and Lottes. The 18 polar lows observed crossing Denmark within the ten years period were traced back to their origin, mainly using satellite imagery. The positions of origin as well as the polar low tracks are shown on Figure 2.25.

All the places of origin of these polar lows were in a region with relatively high SSTs, favouring the formation of polar lows of predominantly convective nature.

Over the years, case studies as well as many climatological studies, have been focused on the occurrence of strong mesoscale vortices, i.e. polar lows. However, in addition to the more intense systems, numerous less developed vortices can be observed, especially on satellite imagery. Forbes and Lottes, in their study of mesoscale vortices in wintertime polar airstreams, included statistics of all 'organized cloud systems' showing signs of circulation, and they pointed out, that 'many of the cloud systems were associated with disturbances which were insignificant in terms of accompanying pressure and wind perturbations, and associated merely with enhanced showers of snow or small hail'. Fett et al. (1993) made the same point, noting that, while small cloud vortices on one hand 'have a tendency to develop in regions where cyclonic vorticity

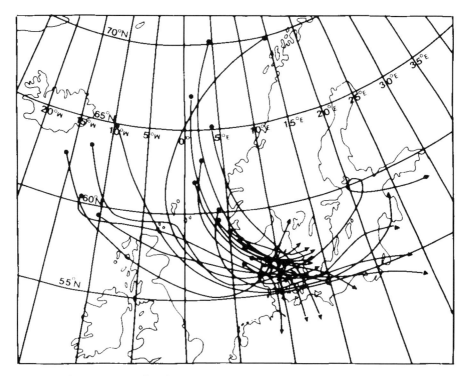

Figure 2.25. Map showing the origin and tracks of 18 polar lows passing Denmark within the period 1980–90 (after Aakjær, 1992).

is maximized', such as along trough or wind shift lines, then in such regions 'despite the cloud vortex, a closed wind circulation implying a pressure minimum does not exist'. They added: 'the term polar low seems inappropriate for the description of such features'.

Harold *et al.* (1999a, b) studied the spatial and temporal distribution of polar mesocyclones (mesoscale vortices) over the Northeast Atlantic and the Nordic Seas during a two-year period October 1993–September 1995, mesocyclones being defined as all cloud systems with a cyclonic curvature observed over the ocean and with a diameter less than 1000 km. An unexpectedly large number (4054) of mesocyclones were found, occurring throughout the year, although winter was the most active season. The study used a database of high latitude cyclonic cloud patterns produced from infra-red satellite images with a spatial resolution of approximately 6 km, and an average of five images per day. Satellite imagery was analysed for all cloud systems with a cyclonic curvature observed over the ocean, regardless of size. During the two-year study period a total of 7792 cyclonic cloud patterns were observed, many of which (64%) were observed only once (single-sighted). Most mesocyclones were observed in the northern part of the study area, near the ice edge, and were small, with 86% of the cyclones having diameters of less than 400 km. Most of the mesocyclones formed and decayed within one day and travelled only short distances.

The highest density of cyclone formation occurred in the north Norwegian Sea and in the Greenland Sea. Small lows were dominant in the northern regions, especially over the northern Greenland Sea/Fram Strait where the dominant cyclone size was less than 200 km and the corresponding lifetime (mean duration) only 8 hours.

Both single-sighted and multiple-sighted cyclones had a continuous spectrum of cyclone diameter, but with a distribution skewed towards smaller diameters. The multiple-sighted cyclones had a modal size around 100–150 km (Figure 2.26) compared to a much larger mean diameter (410 km).

The mean diameter of the multiple-sighted cyclones was much greater than that for single-sighted cyclones (160 km) because the spread of the distribution towards the larger sizes was greater. The majority of multiple-sighted cyclones (53%) were observed only twice. A rather small correlation ($r = 0.35$) between duration and size suggested that the lifetime of a mesocyclone is not strongly linked to its size.

The dominant characteristics of the majority of the mesocyclones were that they were small and lasted a relatively short time. However, a significant number of cyclones of all sizes were observed to persist for much longer periods.

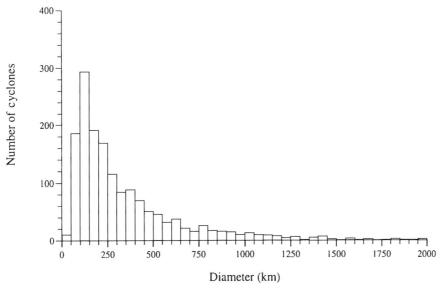

Figure 2.26. Distribution of mean lifetime diameter of multiple-sighted cyclones (after Harold *et al.*, 1999a).

The longest lived mesocyclone observed had a mean diameter of 700 km, and existed for 4.5 days.

The *shape* of cloud systems associated with the mesocyclones observed by Harold *et al.* was identified as either comma-shaped or spiraliform. Comma-shaped cloud patterns accounted for 55% of the sightings and the remainder were spiraliform. Only in the smallest and largest categories (less than 200, 200–400 and greater than 1000 km) was there a small but statistically significant tendency towards comma shapes.

The *cloud height* displayed a marked change across the size range reflecting the different nature of small and large cyclones. In general, the small-scale cyclones (less than 200 km) were associated with low-level clouds.

Figure 2.27 shows the proportions of cyclones occurring in the four seasons. The figure indicates that for most cyclones (except the 600–800 km group) the most active season is winter, confirming the perception that polar mesoscale cyclones are predominantly a winter phenomenon. In contrast, mesoscale cyclones, unlike 'real polar lows', occur in all seasons, including summer.

Carleton (1985a) studied the frequency of comma clouds, 'spiral polar lows' and 'instant occlusions' over the entire North Atlantic and North Pacific using infra-red satellite imagery mosaics derived from the USAF Defense Meteorological Satellite Program (DMSP) polar orbiting satellites for two years of mid-season months, January, April, July and October, 1977/78–1979. The occurrence of polar air cloud vortices was determined using a general (process-oriented) classification system developed earlier by Carleton and others (see Carleton, 1985b for details). Carleton found that polar lows (comma clouds and spiral polar lows taken together) decreased in frequency through the spring to a summer minimum. (Note that Carleton defined the term 'polar low' in a more general way than used elsewhere in this volume, by not applying any threshold value for the wind velocity.)

The data indicated a predominance of the incipient (newly developed) comma cloud formations in the North Pacific, but a greater frequency of the spiraliform type in the North Atlantic. Carleton interpreted this as an observational confirmation of the findings of Sardie and Warner (1983), that moist baroclinic processes dominate in polar low formation for the North Pacific, but that CISK is of additional importance for North Atlantic polar-air cyclogenesis. The extension of the open ocean to higher latitudes in the North Atlantic compared with the North Pacific is probably another crucial factor explaining the differences between the two regions.

In a review, 'Satellite climatological aspects of cold air mesocyclones in the Arctic and Antarctic', Carleton (1996) gave the average diameter of incipient Northern Hemisphere polar lows as 395 km and of mature polar lows as

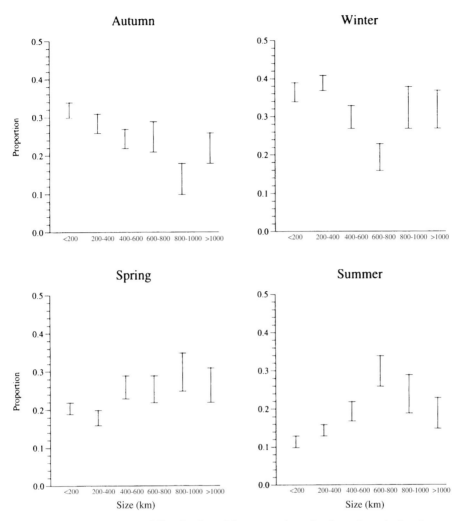

Figure 2.27. Seasonal distribution of the proportion of cyclones in each size class for multiple-sighted cyclones. Error bars indicate the 95% confidence limits (after Harold *et al.*, 1999a).

420 km. These values indicate that Northern Hemisphere mesocyclones, in the main, are larger than their Southern Hemisphere counterparts which have an average diameter of 375 km in all stages of development. Concerning the lifetime of Northern Hemisphere mesoscale vortices, an analysis of DMSP infrared imagery by Carleton showed that only 24% of incipient polar lows were also evident on the succeeding (12-hourly) satellite image, a value which decreased to only 12% for the mature polar low.

Concerning the frequencies of mesocyclones over the Northern Hemisphere, Carleton found a maximum in winter and a minimum in summer in accord with earlier results (e.g. Businger, 1985, 1987; Wilhelmsen, 1985; Lystad, 1986). The seasonal patterns of mesoscale cyclone occurrence in the Southern Hemisphere are quite different from those in the Northern Hemisphere as a result of differences in geography and the seasonal heating and cooling patterns between the two hemispheres (see Section 2.2).

Sardie and Warner (1983) used a quasi-geostrophic model together with wintertime climatological data to study correlations between growth rates and the preferred scales of disturbances, and local effects such as latitude and the climatology of the basic state. The model correctly predicted some of the preferred geographical regions for polar low development – in the Atlantic in the vicinity of Greenland/Iceland, and in the northern Pacific. Sardie and Warner pointed out that Pacific polar lows generally formed just north of the relatively strong baroclinic regions around 40° N, while the Atlantic polar lows generally formed far north of the mid-latitude baroclinic belts. Thus they concluded that there appear to be two types of polar lows: for the Pacific, the comma-cloud type polar lows, which benefit from the existence of strong, deep baroclinicity, while polar lows over the Atlantic, forming around 60° N near Iceland, are too far north to take advantage of any strong baroclinicity. Instead, the North Atlantic polar lows must utilize the CISK mechanism in conjunction with shallow baroclinicity from residual circulations or occluded systems.

Formation of Deep Sea (bottom) Water

Polar lows, and mesoscale vortices in general, are relatively small, short-lived, subsynoptic systems, but they may nevertheless play an important role in a climatological context.

The formation of *oceanic deep water* is of vital importance for the global thermohaline circulation. Oceanic deep convection is forced by a loss of buoyancy of the surface waters, a process that occurs through interactions between the ocean and the atmosphere. The loss of buoyancy occurs either as atmospherically-forced surface cooling from evaporation, or by brine rejection as sea ice forms. At this point, mixing in the vertical can be triggered by further buoyancy losses forced by the passage of weather systems (Renfrew *et al.*, 1999). Deep water formation through the cooling of surface water of high salinity in the Northern Hemisphere occurs only at a relatively few locations, including the Greenland/Norwegian Seas and the Labrador Sea, i.e. within regions where polar lows are especially abundant. For a detailed discussion of oceanic deep convection in the Labrador Sea the reader is referred to papers by The Lab Sea Group (1998) and Renfrew *et al.* (1999).

Polar lows are generally steered by the ambient atmospheric flow in which they are embedded. Because of this they seldom affect the same location for any prolonged period of time. Occasionally, however, polar lows may move very slowly, or hardly at all, as, for example, the so-called 'reverse shear-polar lows' (Section 3.1). Alternatively, several lows may affect the same region within short time intervals, a situation which may occur in a region affected by a so-called 'merry-go-round' system (Section 3.1). Owing to the strong winds and the large air–sea temperature differences associated with polar lows, these systems cause strong fluxes of latent and sensible heat from the sea into the atmosphere. In cases when one or more polar lows affect the same region for a prolonged period of time the integrated effect may be substantial.

Moore (see Rasmussen *et al.*, 1993) suggested that the cooling of the ocean mixed layer due to the strong surface fluxes associated with polar lows (occasionally exceeding 1000 W m^{-2}) may be large enough to initiate deep convection in the ocean. The integrated cooling effect may, as mentioned above, be substantial when several polar lows affect the same region over a period of several days. This occasionally happens when one of the centres of the Northern Hemisphere circumpolar vortex, normally situated around Novaya Zemlya, moves southwest in an outbreak of Arctic polar air to become situated over the northern part of the Norwegian Sea and the Greenland Sea. Such situations are characterized by 500 hPa (winter) temperatures below $-42\,°C$. Audunsdóttir (1999) found, over a period of 12 years (1979–93), 64 cases where the 500 hPa temperature was equal to or below $-42\,°C$ over a significant portion of the Nordic Seas. Each of these cases represented a deep outbreak of Arctic polar air masses and three of the most vigorous cases in terms of duration, convective activity and surface wind speed were chosen for closer inspection. According to Häkkinen (1995) ocean deep convection may occur in conditions with a prolonged cooling of 600 W m^{-2} or more for a period of 5 days within a region of surface temperatures around -0.6 to $-1.0\,°C$. Within the period studied by Audunsdóttir (1999), she found several locations within and along the MIZs of the east Greenland coast and east of Svalbard, which fulfilled the criteria mentioned above, *provided* that the strong winds characterizing the polar lows were correctly included in the objective surface wind speed analysis used for the calculations.

The results indicate that polar lows may in some cases be a crucial factor for the formation of deep water in the Greenland and Norwegian Seas.

The North Atlantic Oscillation: description

The North Atlantic Oscillation (NAO) is the dominant mode of interannual and inter-decadal variability of the atmospheric circulation and upper ocean temperatures in the Atlantic–European sector (Horel and Wallace, 1981; Wallace and Gutzler, 1981; Hurrell, 1995; Hurrell and van Loon, 1997).

In the atmosphere, the NAO involves a variation in the meridional exchanges of mass between the subtropics and subpolar latitudes. This is represented by an NAO index (NAOI) of the normalized monthly or seasonal anomaly differences of sea level pressure (SLP) within the Azores anticyclone, and those in the Icelandic time-averaged low (Rogers, 1981, 1984; Trenberth and Paolino, 1981). Whilst particularly evident in the spatial anomaly patterns of wintertime SLP, geopotential height, temperature and precipitation (e.g. van Loon and Rogers, 1978; Rogers and van Loon, 1979; Lamb and Peppler, 1987; Moses *et al.*, 1987; Hurrell and van Loon, 1997; Walsh and Portis, 1999), the NAO is also present in the other seasons (Barnston and Livezey, 1987). Thus, it contrasts with most other teleconnection patterns, which exhibit greater seasonal dependence. For example, the tropical El Niño–Southern Oscillation (ENSO) 'cycle' usually exhibits a marked weakening in the boreal springtime, limiting the climatic predictability of the Southern Oscillation Index (SOI) into that season. Geographically, the NAO appears to have some association with the circulation climate of the Arctic and North Pacific regions, via the so-called Arctic Oscillation (Thompson and Wallace, 1998). However, this influence does not, apparently, extend to the Eurasia/Siberia region (Rogers and Mosley-Thompson, 1995). The NAO is a significant contributor to the longer term variations in snow chemistry revealed in Greenland ice-core data (Barlow *et al.*, 1993).

In the ocean, the NAO is evident as large-scale anomalies of SST on inter-annual time scales that occur off the east coast of North America, extending northeastwards to the Greenland–Iceland area, and into the northeast Atlantic. Broadly similar anomalies in the upper ocean also appear to be important for North Atlantic climate variations and trends on inter-decadal time scales (Deser and Blackmon, 1993; Rogers *et al.*, 1998). These possibly also include the Great Salinity Anomaly (GSA) of the 1960s (Walsh and Chapman, 1990; Walsh and Portis, 1999).

There are two extreme modes of the NAOI: the high zonal index or positive phase, characterized by a deeper than normal Icelandic low and/or a stronger than normal subtropical high, and the low index or negative phase, having a weakened Icelandic low and/or a weaker Azores high. Through consideration of the geostrophic relationship, these contrasting extremes in the NAOI produce anomalies in the westerly winds, and associated temperature advection patterns, over the extratropical North Atlantic and Europe sectors. The westerlies are stronger than normal in the positive phase, and transport mild maritime air into western and northwest Europe. Conversely, the westerlies are weaker than normal in the negative phase, having an easterly anomaly that advects cold continental air westwards into western Europe during winter

(Rogers and van Loon, 1979; Moses *et al.*, 1987; Carleton, 1988a; Audunsdóttir, 1999).

The SLP anomaly variations associated with NAO are mostly equivalent barotropic: they are reflected as changes in the position and intensity of the Rossby wave pattern in the mid- to upper troposphere, primarily that connected with wavenumber two (Rogers and van Loon, 1979). During the negative phase of NAO, ridging occurs in the Iceland region with enhanced troughing over western Europe, and higher-latitude blocking is more commonly associated with this mode (Moses *et al.*, 1987). These circulation anomalies force colder than normal temperatures over Europe, but milder than normal conditions in the Greenland area. Indeed, the NAO was first described as a regional-scale phenomenon in winter, known as the Greenland–Europe temperature 'seesaw' (see van Loon and Rogers, 1978; Rogers and van Loon, 1979; Meehl and van Loon, 1979). The negative phase of NAO corresponds to the 'Greenland Above: GA' ('Europe Below') normal winter temperatures; the positive phase of NAO corresponds to the pattern 'Greenland Below: GB' ('Europe Above') normal winter temperatures. In other winters, temperatures in the Greenland area and northwest Europe may be of the same sign; either positive or negative (or 'Both Above', 'Both Below'). Rogers and van Loon (1979) and Meehl and van Loon (1979) documented the anomalies in upper-ocean climate associated with the extreme phases of the Greenland–Europe seesaw, particularly the sea ice extent variations and SSTs. During negative phases of the NAO, or GA, the greater frequency of milder southerly (colder northerly) winds in the Davis Strait and the Labrador Sea (the Norwegian and Baltic seas) induces less extensive (more severe) sea ice conditions in those areas. The opposite regional patterns of sea ice occur during the positive phase of NAO, or GB. There are internally consistent anomalies of SST in the higher latitudes associated with these sea ice anomalies.

There are evident multi-decadal trends in the atmospheric circulation of the North Atlantic sector, including the relative frequencies of positive and negative NAOI (e.g. Moses *et al.*, 1987; Hurrell, 1995). Particularly during the 1960s, the negative phase of NAO was relatively common, giving a greater incidence of cold winters in Europe (Moses *et al.*, 1987). The positive phase of NAO has been much more common in the period since the 1980s (Hurrell, 1996; Kushnir *et al.*, 1997; Walsh and Portis, 1999), giving generally milder and wetter winters in northwestern Europe, but colder conditions in the Greenland/Labrador area. This trend has coincided with an increase in the frequency of the negative phase of ENSO and its associated increase in SSTs in the central and eastern tropical Pacific. However, an expected negative association between the SOI and NAOI on interannual time scales is not evident;

at least, in the period through the early 1980s (Rogers, 1984). Rather, there was a tendency for cold winters in Europe, associated with the negative phase of NAO, to occur with negative values of the SOI, or ENSO warm events (Bjerknes, 1966; Rogers, 1984). This general association was mediated, over the North America sector, by the Pacific–North America teleconnection pattern (e.g. Dickson and Namias, 1976; Yarnal and Henderson, 1989b; Carleton *et al.*, 1990).

The recent trend to greater frequencies of the positive phase of NAO in winter has been accompanied by changes in the upper-ocean circulation of the North Atlantic sector that appear to be forced by combinations of the atmospheric circulation, particularly 'storminess' in the extratropics and the trade winds in the subtropics, and the time-dependent advection of anomalies of SST and upper-ocean heat content from lower latitudes occurring via the Gulf Stream/North Atlantic Drift (Deser and Blackmon, 1993; Sutton and Allen, 1997; Venzke *et al.*, 1999). Thus, there is increasing evidence for an interdecadal pattern of climate–upper ocean variability in the North Atlantic sector that resembles, in its grossest features, the NAO (Hurrell and van Loon, 1997; Hurrell *et al.*, 2001). This is akin to the Pacific Decadal Oscillation (PDO) and its resemblance to the ENSO.

NAO associations with cold-air mesocyclone activity

Because the transient and standing eddy activity in the atmosphere is superimposed upon, and partly derived from, the larger-scale extratropical teleconnection patterns, the extreme phases of the NAO and the Greenland–Europe temperature 'seesaw' have associated fluctuations in the frequency and geographically preferred areas of synoptic cyclone activity and the meridional transports of eddy sensible heat, as well as cold-air mesocyclones and polar lows (Carleton, 1988a; Rogers, 1997; Serreze *et al.*, 1997). During positive NAOI, the synoptic-scale storm track associated with the Icelandic low is strong. During negative excursions of the NAOI, the weakening of the Icelandic mean low is associated with reductions in the intensity of the storm track in this region, but increases in synoptic cyclone activity occur further south, generally located between Labrador and Portugal (Carleton, 1988a; Serreze *et al.*, 1997). Mesocyclones and polar lows may play a crucial role in the anomalies of salinity and upper-ocean temperature over higher latitudes of the North Atlantic. The extreme sea–air heat fluxes that accompany meso-cyclogenesis in the Labrador Sea and East Greenland Sea areas increase the production of cold deep water in the marginal ice zone (Rasmussen *et al.*, 1993). Moreover, decadal periods having enhanced versus weaker overturning of water masses in the North Atlantic have been suggested as being partly connected with the observed multi-decadal changes in the frequencies of intense tropical cyclone activity for the Caribbean Basin (Gray *et al.*, 1997).

Carleton (1985b) first studied the possible associations between the NAO and mesocyclone activity in the North Atlantic sector for two winters (December–February) representative of extremes in the Greenland–Europe temperature seesaw (GB 1974–75; GA 1976–77). He classified mesoscale and synoptic-scale cyclonic cloud vortices for these winters from twice-daily DMSP thermal infra-red imagery. Carleton (Carleton, 1985b, 1988a) found that, in GB winter 1974–75, representative of the positive NAO, more mesocyclones developed east of Labrador and Newfoundland than in GA winter 1976–77, or the negative phase of NAO, when more systems occurred off the east coast of the USA (Figure 2.28).

SST patterns in the two winters (see Carleton, 1996, his fig. 15a, b) were broadly representative of the longer term composite means for each contrasting 'seesaw' mode, as depicted in Meehl and van Loon (1979, their fig. 5). During the GB winter, positive anomalies of SST occurred east of North America and extended to the northwest coast of Europe. During the GA winter, negative anomalies of SST occurred in the European sector and the East Greenland sea, with associated increases in the regional sea ice-extent conditions (Carleton, 1985b). Carleton (1985b, 1996) noted a tendency for mesocyclones to be more frequent near the areas of enhanced SST anomaly gradients in both winters. There was more (less) extensive sea ice in Davis Strait and around southern Greenland in winter 1974–75 (1976–77). During GA winter 1976–77, mesocyclones forming in the Norwegian Sea exhibited a dominantly southward motion, associated with the long-wave pattern favouring cold-air advection into western and northwest Europe. Whilst mesocyclones occurred in the Mediterranean region in both winters, there was observed to be considerable inter-monthly variability.

A recent study of cold-air mesocyclone activity over the northeastern Atlantic region for the two years 1993–95 (Harold *et al.*, 1999a, b) documents the satellite-observed characteristics of the associated cloud vortices, and the larger-scale circulation teleconnection patterns, including NAO, and the sea ice extent conditions giving rise to mesocyclogenesis. These authors found that mesocyclones in the vortex diameter ranges 200–600 km showed the strongest association with the NAO. The frequencies of these mesocyclones over the Nordic Seas increased as the NAOI became more positive during the study period, in relation to the tendency for systems in this size range to develop behind major synoptic cyclones. Smaller-sized mesocyclones showed a closer association with formation at the sea ice margin, apparently in connection with pre-existing shallow boundary-layer fronts (Harold *et al.*, 1999b).

94 2 Climatology

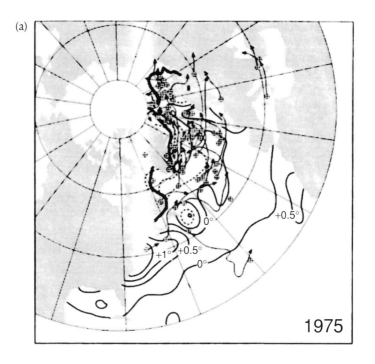

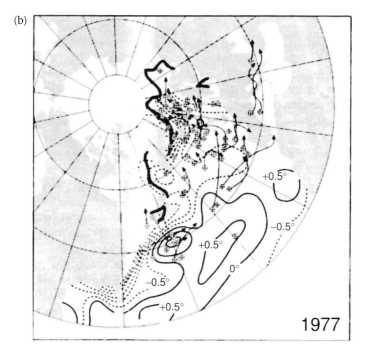

2.1.3 The Labrador Sea, the Davis Strait, Baffin Bay and the Hudson Bay (eastern Canadian waters)

In the early 1990s a comparatively southern genesis region for polar lows was found over the Labrador Sea between the east coast of Labrador and southwestern Greenland (Rasmussen, 1990; Rasmussen and Purdom, 1992). The Labrador Sea is a favoured place for outbreaks of extremely cold air masses originating over northern Canada. Polar lows also occasionally form in the Davis Strait and in Baffin Bay, as well as over the ice-free part of Hudson Bay (see Section 3.1).

In the preceding section we noted that an important factor contributing to the frequent formation of polar lows in the northern part of the Norwegian Sea and in the Barents Sea was the proximity of a semi-permanent centre of the upper-level circumpolar vortex situated near Novaya Zemlya. In an analogous way the east Canadian waters are located close to the mean position, near Baffin Island, of another of the semi-permanent centres of the circumpolar vortex. Movement of this vortex, or the short-wave troughs forming on its flanks, may cause the formation of polar lows just as observed in the Norwegian genesis region.

Using polar orbiting satellite images supplemented by geostationary GOES data, Hanley and Richards (1991) studied the occurrence of polar lows in eastern Canadian waters, covering the period 1977–89. Their study was later complemented by similar studies (see Parker, 1997 for details). The results of these studies are summarized in Figure 2.29, showing the occurrence of polar lows over eastern Canadian waters for the period 1977–93.

The figure shows that the main concentrations of events are found in the southern Davis Strait and in the Labrador Sea. The temporal distribution of the cases observed in the Labrador Sea and in the Davis Strait is shown on Figure 2.30.

Figure 2.28. Mesocyclone regimes (locations of cyclogenesis and cyclolysis, and system tracks) in the North Atlantic sector for (a) January 1975: GB mode, or positive NAOI, and (b) January 1977: GA mode, or negative NAOI, from the analysis of twice-daily thermal infra-red imagery from the DMSP. The spatial patterns in these two months are broadly similar to those evident in the other winter months (December, February) of their respective extremes in the NAO and Greenland-Europe temperature 'seesaw'. On the charts, crosses mark the locations of meso-cyclogenesis; arrows, the system tracks, and arrowheads, the locations of meso-cyclolysis. Also shown are the SST anomalies for each January with respect to the 1963–77 'long term' mean for January (solid: positive, warmer; broken: negative, colder). The extent of the Arctic sea ice in each month is given by the thick lines. (From Carleton, 1996, *The Global Atmosphere and Ocean System*, Taylor & Francis Ltd (http://www.tandf.co.uk/journals).)

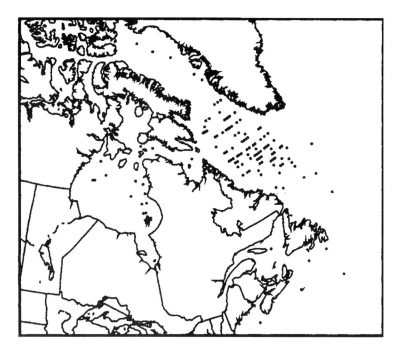

Figure 2.29. Cold-air vortices (mesoscale vortices) and polar lows over eastern Canadian waters for the period 1977–93 (after Parker, 1997).

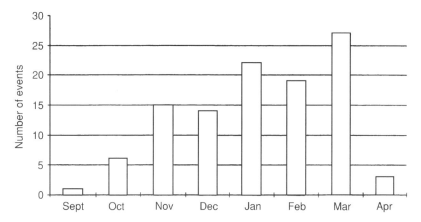

Figure 2.30. Frequency of cold-air vortices (mesoscale vortices) and polar lows in eastern Canadian waters for the period 1977–93. Events in Hudson Bay and Baffin Bay are not included (after Parker, 1997).

In the same period, 14 events were recorded in Hudson Bay, with four occurring in each of October and December and five in November.

Figures 2.29 and 2.30 show that mesoscale vortices and polar lows are not uncommon in eastern Canadian waters. However, they do not necessarily occur every season in the different regions as special meteorological conditions must be fulfilled. In some areas the polar lows have a limited season due to the expansion of the winter ice pack, the season ending in Hudson Bay in late December and within Baffin Bay by the end of October.

2.1.4 The Beaufort Sea and the Chukchi Sea

The Beaufort Sea, situated north of Alaska, is part of the Arctic Ocean. For much of the year the only open water areas in the Arctic Ocean consist of leads and polynyas which are too small to be important for the formation of polar lows, a condition that may change in the future due to an increasing greenhouse effect.

The surface Arctic water in the Beaufort Sea undergoes significant seasonal changes in temperature and salinity in response to freezing and thawing of the pack ice and of the rivers in the coastal area (Sechrist *et al.*, 1989). The waters of the southern Beaufort Sea are ice-free and relatively warm in summer and early autumn, and during this time the area is frequently affected by fairly small-scale but strong storms. Some of the storms seem to form or intensify *in situ* within the region and the question has been raised whether some of the storms should be classified as polar lows. Parker (1989) studied the frequency of polar lows in the southern Beaufort Sea by examining detailed surface weather charts during the open water seasons for the period 1976–85. During the period examined no instances of polar lows developing over the southern Beaufort Sea were found. One case developed over the Chukchi Sea further west and subsequently moved across northwestern Alaska into the Beaufort Sea region (see Sections 3.1.10 and 6.2.5). It was concluded that 'occurrences of true polar lows in the Canadian Beaufort Sea are rare events', but that 'Given optimum synoptic conditions during a year when ice cover is at a minimum it should be possible for a polar low to develop over the southeastern Beaufort Sea.'

Some reports, however, indicate that the formation of polar lows over the southern Beaufort Sea may be more frequent than shown by Parker's study. One reason for this apparent discrepancy may be that different definitions of a polar low have been applied (for a more detailed discussion, see Section 3.1.10).

2.1.5 The Gulf of Alaska and the Bering Sea

Based on a study of infra-red images from the polar orbiting satellites NOAA 4 to NOAA 7, Businger (1987) demonstrated that small-scale (less than 500 km diameter), spiral-shaped storms are often present over the northern Gulf of Alaska and the Bering Sea. Figure 2.31 shows a histogram of the number of days per month that polar lows were observed over the Gulf of Alaska or the Bering Sea during the period 1975–83.

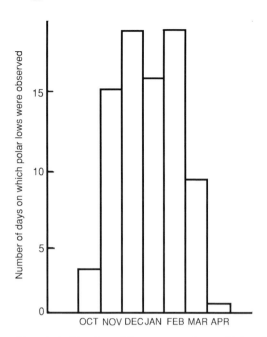

Figure 2.31. Number of days per month on which polar lows were observed in polar-orbiting satellite imagery over the Gulf of Alaska or the Bering Sea during the period 1975–83 (from Businger 1987).

In analysing the satellite imagery for Figure 2.31, Businger found that on c. 20% of the 'polar low days' more than one polar low was observed. On these days the polar lows tended to form in a series in close proximity to one another, analogous to the cyclone families described by Bjerknes and Solberg (1922), but on shortened time and space scales.

To develop a climatology of the synoptic environments conducive to the formation of polar lows, Businger composited 500 hPa heights, temperatures, 1000–500 hPa thickness and surface pressure data for days when mature polar lows were present. He made use of the same 'superposed-epoch method' as in his study of polar lows over the Norwegian Sea and the Barents Sea (Businger, 1985). Only vigorous polar lows exhibiting a well-developed spiral cloud structure and high cloud tops were considered. On days when well-developed polar lows occurred over the Gulf of Alaska the study revealed the presence of significant negative anomalies centred over the northern Gulf of Alaska in the 500 hPa temperature, height and thickness fields, and in the surface pressure field. The negative 500 hPa height and temperature anomalies, indicated enhanced positive vorticity and a potential for deep convection over the area. Comparing the results over the Gulf of Alaska with those over the Norwegian and Barents Seas, the environment conducive to the formation of polar lows appeared very similar in most of the variables studied for both regions.

2.1.6 The Northwest Pacific, the Japan Sea, the Yellow Sea and the East China Sea

Yarnal and Henderson (1989a, b) documented the distribution of polar lows in their early stage of development (the cyclogenetic stage) over the North Pacific Ocean based on the subjective interpretation of infra-red DMSP satellite images. Mosaics for seven 5-month winter seasons, November to March 1976–77 to 1982–83 were used. The classification used was based on the structure of the cloud patterns and only two morphological types of cyclogenetic polar lows were considered, the comma clouds and the spiraliform signature.

The charts in Figure 2.32 show the total number of comma clouds and spiraliform systems observed during the whole seven-year period, derived by

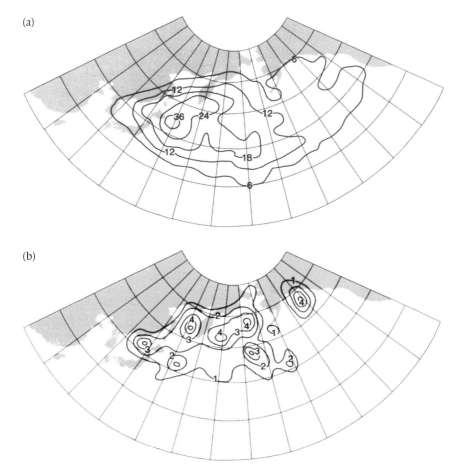

Figure 2.32. Observed number of (a) total comma clouds, (b) total spiraliform systems for the seven-year period 1976–83 (from Yarnel and Henderson 1989a).

Table 2.1. *Statistics of cyclogenetic polar lows, stratified by type*

	Comma clouds	Spiraliforms
(a) Frequencies and percentages of total		
Frequency	1391	128
Percentage	92	8
(b) Radius (degrees latitude)		
Mean	1.47	2.17
Standard deviation	0.40	0.76
(c) Tail length of comma clouds (degrees latitude)		
Mean	2.69	–
Standard deviation	0.88	–
(d) Cloud form (per cent)		
Cumuliform	89	90
Stratiform	11	10
(e) Cloud height (percent)		
Medium to high	77	50
Low	23	50

From Yarnal & Henderson (1989a).

plotting the locations of the first observation of the systems, i.e. during their cyclogenetic phase.

Comma clouds comprised the vast majority of polar lows observed in the North Pacific, being more than 92% of all the lows during the study, and with a peak area of cyclogenesis located close to Japan. Since the comma clouds were the most populous type of system in the North Pacific, the total distribution of *all* polar lows (comma clouds plus spiraliform) was almost the same. Comma clouds form most frequently just off the east Asian coast in a band that extends from northern Japan through the Kamchatka Peninsula and into the western Bering Sea. Another high frequency band extends from northern Japan due eastward into the North Pacific. The eastern North Pacific is only slightly active in terms of polar low cyclogenesis, with the lowest frequencies occurring along the west coast of North America. The pattern of formation of spiraliform systems is very different, mainly hugging the coasts of extratropical east Asia, the Aleutians and the south coast of Alaska.

The different patterns of distribution clearly reflect the different mechanisms responsible for the formation and development of the two types of polar lows; the comma clouds being associated with deep baroclinicity close to the polar front, and the spiraliform systems to processes deeper within the cold air masses closer to the ice edges. A similar result is found in the Antarctic (Section 2.2). Based on their measurements, Yarnal and Henderson presented a number of general statistics shown in Table 2.1. Yarnal and Henderson pointed out that the apparent difference between the mean radii of the spiraliform polar lows and of the comma clouds is a function of their measurement technique and should not be used to suggest that the former are larger on average than the latter. Both the comma clouds and the spiraliform systems appear to be dominated by cumuliform clouds. Considering the cloud heights, comma clouds tend to be composed of medium to high cloud, while spiraliform systems are much more likely to contain low clouds. Comma clouds usually form in the cold air masses trailing an advancing cold front over the open waters, and often develop as air passes from the cold continent or ice edge over the relatively warm water of the North Pacific or associated seas. Much less common is the spiraliform polar low, which usually forms as cold air moves off the cold land/ice edge and over warm water.

During any one winter season, the distribution of polar low cyclogenesis changes over space and time. The observed distribution of polar low cyclogenesis with a pronounced maximum during the winter months (see Yarnal and Henderson, 1989a, fig. 3) can be explained in terms of the seasonal cycle of the large-scale climatic system of the North Pacific Ocean (Figure 2.33). The figure shows the mean positions of the 5150 m and 5400 m contours of the 500 hPa surface and the 3 °C isotherm of the SST–air temperature differences for each winter month. The purpose of these charts is to portray the seasonal expansion of the circumpolar vortex and its interaction with areas of low-level instability.

The 5400 m contour was chosen as a rough approximation to the position of the polar front, while the 5150 m contour characterizes the 'cold core' of the trough over the North Pacific. The semi-permanent trough that forms during the winter months over the extreme east Asia region and the adjacent North Pacific pulls cold, continental polar or Arctic air over the relatively warm waters of the North Pacific and marginal seas, giving rise to localized convective instability and/or baroclinic instability along the coast. Polar air traversing the open waters of the western North Pacific experiences considerable SST–air temperature differences conducive to low-level convective instability. Passing the sharp thermal gradients associated with the Kuroshiu current may further affect the vertical stability of the polar air masses, and polar lows may be triggered provided that upper-level disturbances, such as short-wave, upper-level troughs are present. Thus polar lows should be expected to form along the east Asian coast and in the vicinity of the Kuroshio current in regions where the necessary upper-level forcing can be found, i.e. near the east Asian jet, as confirmed by the study of Yarnal and Henderson.

As in the North Atlantic sector, the large-scale environment controls the average intra-annual variability of polar low formation. In November the circumpolar vortex is relatively far north. The instability is small throughout

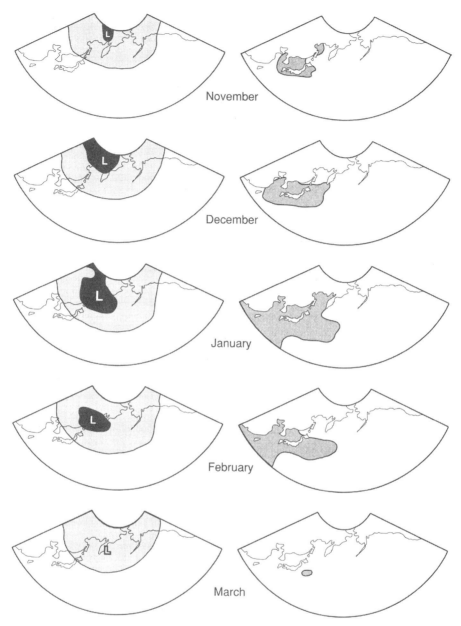

Figure 2.33. Representative mean 500 hPa heights (left) and mean SST minus air temperature (SST − T) patterns (right) through the seasonal cycle. The grey-shaded area on the 500 hPa charts is surrounded by the mean 5400 m contour, while the blackened area of lowest heights is contained by the 5150 m contour. For the SST − T charts, the 3 °C isotherm surrounds the area of greatest SST − T differences (from Yarnal and Henderson, 1989a).

the basin, and as a result only a small number of polar lows form along the colder coastal areas and in the vicinity of sea ice. During December, the circulation strengthens and the circumpolar vortex expands southwards. Polar air sweeping off the continent becomes colder, increasing the instability across large portions of the basin. Following this the polar low frequency increases with large numbers forming not only along the coasts but also over open water. By January there is a rapid increase in the frequency of polar low cyclogenesis as the circumpolar vortex overruns the Kuroshiu current, and the difference between the SST and air temperature reaches its maximum. This pattern starts to weaken in February, and in March the polar low formation drops to its lowest frequency of the winter.

Despite the fact that the intra-annual variability seems tied to the general climatic controls of the basin, in any given winter the regions of peak cyclogenesis can shift dramatically and may be located nearly anywhere in the extratropical North Pacific.

Polar lows in the vicinity of the Japanese Islands

As seen from Figure 2.34, polar lows are especially abundant over the Japan Sea and the adjacent seas. Such polar lows have for years been studied extensively by Japanese scientists, and in the following account we summarize some of the most important results from these many studies.

The genesis region for mesoscale cyclones/polar lows around Japan is situated at a relatively low latitude $c.$ 35°N to 50°N. Because of this, polar low developments in the Japan Sea and surrounding waters can be followed in detail using visible and infra-red satellite imagery from the Japanese GMS (Geostationary Meteorological Satellite). Further, the fine structure of the mesoscale vortices can be observed when they pass within the observation

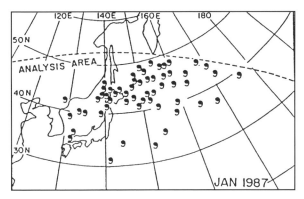

Figure 2.34. Locations of polar lows found over Japan Sea and adjacent seas at 0000 GMT and 1200 GMT in January 1987 (from Ninomiya, 1989).

range of the dense radar network of the Japan Meteorological Agency. Therefore, over the years, several Japan Sea polar lows have been studied extensively (e.g. Miyazawa, 1967; Asai and Miura, 1981; Ninomiya, 1989, 1991, 1994; Ninomiya and Hoshino, 1990; Ninomiya *et al.*, 1990, 1993, 1996; Tsuboki and Wakahama, 1992; Yamagishi *et al.*, 1992; Nagata, 1993). Further east, over the Northwest Pacific, data are much more sparse, apart from satellite images, and only two reports for systems in these areas (Ninomiya, 1989; Ninomiya *et al.*, 1996) have been published in Japan.

Although the scale and the development processes of polar lows over these regions vary widely from case to case, the following general features can be deduced from the papers mentioned above:

1. The polar lows over the region tend to be generated in the polar air streams over the relatively warm sea west of the large, synoptic-scale lows, especially under the influence of a cold vortex aloft or of a deep, cold westerly trough (cf. the results of the studies of Yarnal and Henderson discussed in the preceding section).
2. Many of the polar lows over the northwestern Pacific develop poleward of the main polar front and fall into the category of meso-α-scale lows (scale definition according to Orlanski, 1975: meso-α-scale 200–2000 km, meso-β-scale 20–200 km, meso-γ-scale <20 km). They are accompanied, in general, by comma-shaped cloud systems in their developing stage.
3. The polar lows that develop over the northeastern part of the Japan Sea, i.e. along the western coast of northeastern Japan, tend to develop in the northwestern quadrant of a parent large-scale low. Their horizontal scale is relatively small and, in some cases, falls within the meso-β-scale.
4. The horizontal scale of polar lows forming over the western part of the Japan Sea is also relatively small and, in some cases, falls into the meso-β-scale. A considerable number of polar lows over this area are generated in association with 'the Japan Sea polar air mass convergence zone (JPCZ)', which is a confluence-convergence zone in the polar air streams downstream of the coastal high mountain area of the Asian continent (Nagata, 1987, 1992).

Climatological conditions over East Asia and the northwestern Pacific favourable to the formation of polar lows

The locations of polar lows identified on the surface charts and nephanalysis charts for 0000 and 1200 GMT, January 1987 are presented in Figure 2.34. Almost the same distributions were obtained for December 1986 and February 1987. The generation of polar lows over the Yellow Sea and the western part of the Okhotsk Sea are rare events, even in mid-winter. The close

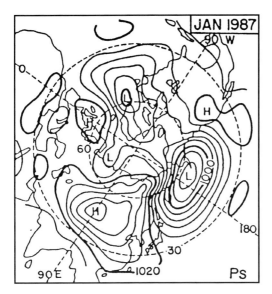

Figure 2.35. The one-month averaged sea level pressure field for January 1987 (after Ninomiya, 1989).

relationship between the polar lows, the main low and the upper-level cold core vortex or deep troughs is illustrated in Figures 2.34 to 2.36.

In many cases, polar lows over the Japan Sea form or develop 500–1000 km northwest of the main large-scale lows, which pass along the Pacific coast of Japan. In some cases, two or three polar lows develop successively within a parent large-scale low. In general, polar lows over the Japan Sea and the northwestern Pacific form and develop in the vicinity of the cold core in the upper cold vortex or deep westerly trough. In contrast, polar lows over the northwestern Pacific tend to develop 1000–2000 km west of the large-scale low, which attains its maximum intensity around the Aleutian Islands.

Polar lows tend to develop 1–2 days after the passage of the main large-scale low over the Japan Sea and the northwestern Pacific off the coast of the Japan Islands. Consequently, the period of the 'polar low burst' of frequent polar low development is $c.$ 5 days. The 12-hourly locations and the central sea level pressures of the large-scale main lows for January 1987 are shown on Figure 2.37. It is clear from Figures 2.34 and 2.37 that development of polar lows is never seen to the south of the track of the large-scale main lows.

As already pointed out, polar lows over east Asia and the northwestern Pacific develop at lower latitudes ($c.$ 35° N to 50° N), compared with other regions of frequent polar low generation such as the Norwegian Sea and the Barents Sea ($c.$ 60° N to 80° N). To explain this fact, features of the large-scale circulation of east Asia and the northwestern Pacific will be considered.

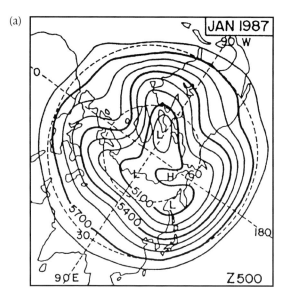

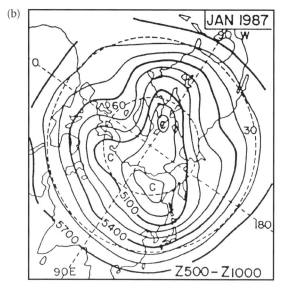

Figure 2.36. (a) The average 500 hPa geopotential height field for January 1987. (b) The average 1000–500 hPa thickness for January 1987 (after Ninomiya, 1989).

The large-scale lows travelling northeastwards over the northwestern Pacific usually reach their maximum intensity around the Aleutian Islands to form the deep 'mean' Aleutian low. This is much deeper than its counterpart in the Atlantic, i.e. the Iceland low. On the mean charts, the Siberian anticyclone predominates over the northeastern part of the Asian continent. Thus a very strong pressure gradient is maintained between the Aleutian low and the Siberian

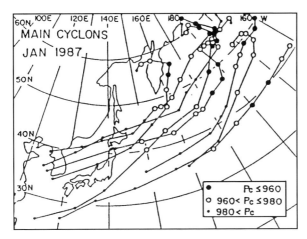

Figure 2.37. The 12-hourly locations and the central sea level pressures (P_c) of the large-scale main lows for January 1987 (after Ninomiya, 1989).

anticyclone (see Figure 2.35). Consequently, polar air outbreaks, on average, are very strong during winter. The lower tropospheric circulation over the Japan Sea and the northwestern Pacific is characterized by the cyclonic branch of the polar air outbreak, whereas the Yellow Sea and the northwestern part of the East China Sea are characterized by the anticyclonic branch of the polar air outbreak (Ninomiya, 1976). This is a major reason why polar lows seldom develop over the Yellow Sea and in the northwestern part of the East China Sea.

On the 500 hPa mean contour chart (Figure 2.36a), a deep trough predominates between 140° E and 160° E and the area bounded by the 5100 gpm height contour extends southward to $c.$ 50° N. On the 1000–500 hPa thickness chart (Figure 2.36b) the cold area bounded by the 5100 gpm thickness contour reaches southwards to $c.$ 45° N over the Japan Islands. The very strong thermal gradient of $10\text{–}15\,\text{K}\,(1000\,\text{km})^{-1}$ is maintained in the lower troposphere along $c.$ 40° N during winter. When a cold vortex within this stationary trough moves southwards towards the Japan Sea, the static stability in the lower troposphere decreases considerably, and at the same time the low-level baroclinicity and the cyclonic circulation increase. These features are favourable conditions for the genesis of polar lows.

Another important factor related to the frequent development of polar lows is the large air–sea temperature difference. The 10-day averaged SST distribution for 11–20 January 1987 is shown in Figure 2.38.

The area of sea ice during this period is also presented on this map. The SSTs over the Japan Sea and the northwestern Pacific are relatively high owing to the predominant warm sea currents. The strong horizontal gradient of SST along $c.$ 40° N over the Japan Sea is associated with the northern side of the Tushima

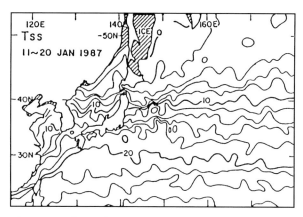

Figure 2.38. The 10-day averaged SST for 11–20 January 1987 (after Ninomiya, 1989).

current, whereas that over the northwestern Pacific is along the north side of the Kuroshio current.

On the other hand, surface air temperatures over the Asian continent are extremely low in winter resulting in the formation of sea ice around Sakhalin. The sea ice area extends to *c.* 45° N, which is the southernmost extension of sea ice in the Northern Hemisphere.

The air–sea temperature differences over the Japan Sea and the northwestern Pacific, around 40° N, are 10–15 °C and 3–10 °C, respectively. Owing to the large air–sea temperature difference and the strong winds, the transformation of polar air masses over these areas is very rapid. The large amount of sensible heat and the moisture supplied from the underlying sea surface destabilizes the stratification in the lower atmosphere and leads to the formation of convective clouds, which may play a role for CISK-related developments of polar lows. The aforementioned climatological features explain why polar lows in east Asia and the northwestern Pacific are generated at lower latitudes compared with other parts of the world.

2.2 The Antarctic

2.2.1 Introduction

The physical geography of the Antarctic and sub-Antarctic regions is very different from that of the Arctic. Whilst the Arctic comprises a polar ocean largely covered by multi-year sea ice and surrounded by land areas extending into middle and lower latitudes, the South Pole is located within a continent, beyond which lies a circumpolar ocean that is frozen for only part of the year, but which effectively doubles the size of the continent between March and September. The extreme continentality of the south polar region is compounded by the presence of an ice sheet that exceeds 4 km altitude in Greater (or East) Antarctica. These factors help to isolate the interior of Antarctica from atmospheric and oceanic circulations over lower latitudes for the greater part of the year. The much lower altitude of most of the north polar region, and the generally higher mean amplitude of the tropospheric wave pattern, facilitate the penetration of moderating air masses deep within that basin, even in winter (Barry, 1983). Moreover, because the middle and higher-middle latitudes of the Southern Hemisphere are essentially uninterrupted ocean, the patterns of SSTs and tropospheric circulation are much more zonal, on average, than in the Arctic: the mean patterns of both SST and sea ice distribution (Figure 2.39) manifest a zonal wavenumber one pattern (Anderssen, 1965; Rayner and Howarth, 1979). The zone of strongest SST gradient, associated

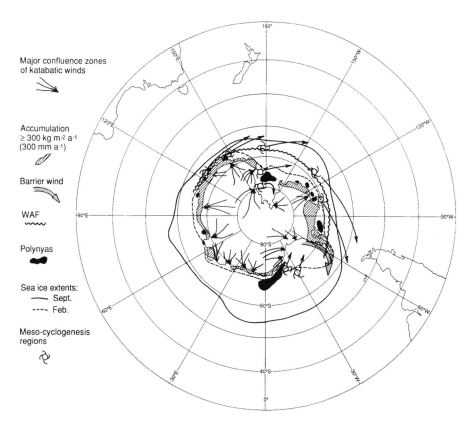

Figure 2.39. Schematic diagram showing the dominant features of the surface and near-surface climatology of the Antarctic region described in the text, compiled from various sources. The ACC (not shown) is displaced approximately 5°–10° of latitude equatorward of the mean position of the sea-ice edge at maximum extent (from Carleton, 1992).

with the Antarctic Circumpolar Current (ACC), is located further equatorward in the Antarctic region. The ice in the marginal ice zone (MIZ) is mostly less than one year old, is thinner, and also far less constrained in its patterns of drift and seasonal expansion and contraction compared with the Arctic. On average, the atmosphere tends to be more stable over high southern latitudes than it is at comparable latitudes in the Arctic, especially the northern North Atlantic in winter (Taljaard, 1969; Wendland and McDonald, 1986). Moreover, the phenomenon of a snow-covered land surface (Scandinavia in winter) that is located *equatorward* of a zone of strong SST gradient, and which influences strongly the development of mesoscale storm systems in the Norwegian Sea, does not have a counterpart in the Antarctic. Physical geographic differences, such as these between the two polar regions, influence the seasonal frequencies of mesocyclones, the relative frequencies of the different cloud vortex signature types that are observed on satellite imagery, and the physical processes of meso-cyclogenesis.

The Antarctic ice sheet is a sink of energy all year, but net radiation values are most negative in the winter half-year (e.g. Carroll, 1982). Figure 2.40 is an example of a radiosonde ascent for McMurdo station near the western Ross Sea at a time when a mesocyclone was developing just to the north (2 March 1988) (compare with Figure 2.7, which shows a typical Arctic sounding). Strong stability in the lower troposphere is associated with the marked radiation inversion in the temperature plot between the surface and 919 hPa. Evidence for subsidence of air from the interior high plateau is given by the inversion around 600 hPa and the very dry air at and above this level. This temperature profile is typical of coastal Antarctica in the early autumn, although the relative moistness of the lower to mid-troposphere, given by the dewpoint values, results from the cloud associated with the mesocyclone. Temperature soundings for the winter (summer) season would show deeper (shallower) and more intense (weaker) surface-based inversions and a lower (higher) tropopause, here located at 360 hPa.

In the Antarctic, much of the surface net radiation deficit must be made up by advection into the continent of warm, moist air associated with synoptic-scale cyclones originating over mid-latitudes (Andreas, 1985; Nakamura and Oort, 1988; Giovinetto *et al.*, 1997), mostly via West Antarctica, which is of lower altitude (Hogan, 1997). To a far lesser extent, the radiation deficit is also redressed by heat storage from within the snow (Carroll, 1982), and by the convective fluxes of heat from leads and polynyas in the MIZ, e.g. Figure 2.39 (Weller, 1980; Allison *et al.*, 1982); especially under conditions of southerly cold-air flow (Andreas and Markshtas, 1985). The boundary-layer cold-air that is generated by the net radiation deficit, drains off the continent as a strong

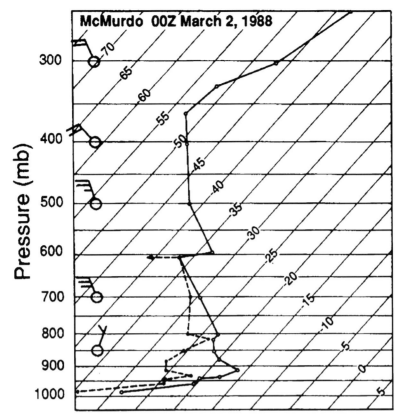

Figure 2.40. Radiosonde ascent for McMurdo Station, Antarctica (77.9° S, 166.7° E) at 0000 GMT 2 March 1988. The temperature sounding (solid curve) shows the strong stable stratification characteristic of stations in coastal East Antarctica. The relative moistness of the lower to mid-troposphere (dewpoint temperature: dashed curve) is less typical of climatology, being associated with a mesocyclone that had developed just northeast of the station. (From Carleton, 1995, *International Journal of Remote Sensing*, Taylor & Francis Ltd (http://www.tandf.co.uk/journals).)

katabatic wind which achieves remarkable directional constancy: Figure 2.39 (e.g. Parish and Bromwich, 1986; Wendler *et al.*, 1988; Streten, 1990), and which can also propagate considerable distances away from the continent's edge (Bromwich, 1989). Its only corollary in the Northern Hemisphere, occurring on a considerably smaller scale, is around Greenland. The katabatic wind is strongest where the ice-sheet orography is steepest: near the edge of the continent in East Antarctica and, in particular, where confluence occurs between katabatic flows (Parish, 1982; Wendler *et al.*, 1997c). These areas are favoured for the development of coastal polynyas Wendler *et al.*, 1997b), which supply heat and moisture to the boundary layer, especially in winter

(Adolphs and Wendler, 1995). Accordingly, coastal regions such as the western Ross Sea, the Siple Coast, the eastern Weddell Sea and the Adélie Coast, are sites of 'spin-up' of mesoscale cyclones: Figure 2.39 (Turner and Row, 1989; Bromwich, 1991; Carleton, 1992; Carrasco and Bromwich, 1996; Carrasco *et al.*, 1997a). The katabatic wind is also strongest when and where synoptic circulation patterns favour enhanced outflow of cold air; that is, on daily time scales when a synoptic frontal cyclone is located to the east or northeast of a given coastal location (Streten, 1968; Parish and Bromwich, 1998). The katabatic winds comprise the lowermost branch of a large-scale meridional circulation system that involves divergence of air away from the sloping ice sheets, sustained by a mean cold-cored cyclonic circulation in the free atmosphere above Antarctica (Figure 2.41), low-level convergence and ascent

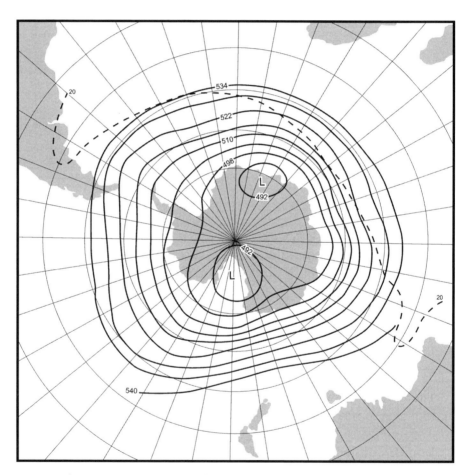

Figure 2.41. Mean geopotential height (dm) over high southern latitudes at 500 hPa in July. Broken lines enclose areas where the zonal component of the geostrophic (westerly) wind exceeds 20 m s^{-1} in the mean (from Carleton, 1987).

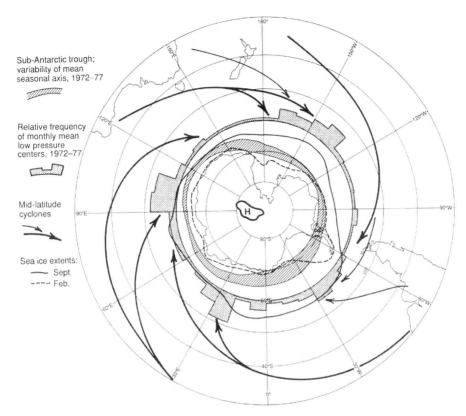

Figure 2.42. Schematic diagram showing the dominant features of the synoptic climatology of higher southern latitudes, compiled from various sources (from Carleton, 1992).

into the circumpolar, or sub-Antarctic trough of low pressure (the circumpolar trough) (Figure 2.42), and eddy convergence of momentum and removal of high vorticity air by incoming synoptic-scale cyclones (James, 1989). The majority of mesoscale cyclonic circulations that form in response to the low-level katabatic winds tend to have an equatorward component to their motion, at least in their earlier stages of development: Figure 2.39 (Fitch and Carleton, 1992; Carleton and Fitch, 1993; Carleton, 1996), and are, therefore, likely to be partly involved in the 'shedding' of cyclonic vorticity at the continental margin.

The varied physical geography of the Antarctic region also produces spatial differences in mesocyclone formation areas. On the broadest level, one can identify two zones having high frequencies of mesocyclone formation: those areas near the Antarctic coast and over the ice shelves, associated with the confluence areas of katabatic winds (e.g. Figure 2.43); and the sub-Antarctic ocean (both sea ice-covered and ice-free areas). The sea ice–ocean margin is particularly favoured for mesocyclogenesis (Carleton, 1981b; Heinemann, 1996a, c),

Figure 2.43. A thermal infra-red direct-readout image (1.1 km nominal resolution) of the western Ross Sea showing a mesoscale vortex near the ice front of the Ross Ice Shelf. The dark streaky signatures coming into the system from the east are associated with low cloud that is warmer than the underlying ice surface. Taken by NOAA on 13 March 1988.

because it marks a discontinuity in most or all of the following climate variables: the surface albedo; the net radiation; the fluxes of sensible and latent heat and resulting Bowen ratio; and the surface roughness (e.g. Weller, 1980; Allison *et al.*, 1982; Kottmeier and Hartig, 1990; Wendler *et al.*, 1997a). Mesocyclones forming in association with these baroclinic zones provide a focus for instability and convergence in the lower troposphere, when cold air is advected off-ice by synoptic-scale low pressure systems (located eastwards). Figure 2.44 shows a mesocyclone forming at the ice edge in the southeast Pacific Ocean.

With the exception of certain isolated island stations, most of the Southern Hemisphere located between the three lower-latitude continents and the Antarctic coastline lacks long-period conventional surface and free-troposphere observing records. This has impacted, in several major ways, the direction and progress of research on weather systems in general, and mesocyclones in particular, over the southern higher latitudes. These include:

1. A continuing heavy reliance on the observations from ships and also drifting buoys transiting the sub-Antarctic region (e.g. Radok *et al.*, 1998).
2. An early (mid-1960s onward) dependence on the interpretation and analysis of satellite polar orbiter visible and thermal infra-red images for identifying synoptic and subsynoptic-scale cyclonic circulations,

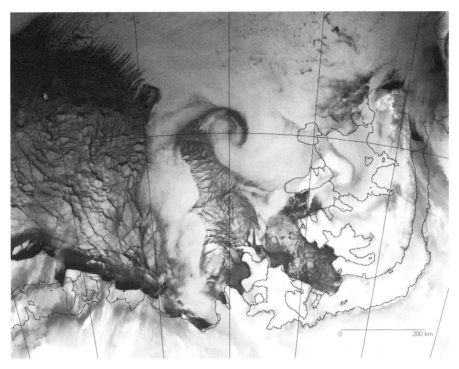

Figure 2.44. AVHRR thermal infra-red direct-readout image (1.1 km nominal resolution) of the Bellingshausen Sea, showing a mesocyclone developing near the sea ice edge. Taken at 0938 GMT 26 October 1998.

determined on the basis of their characteristic cloud vortices (e.g. Martin, 1968).

3 Consequently, a reasonable understanding of the synoptic climatology (i.e. the spatial distributions and tracks) of frontal cyclones and mesocyclones over the open ocean and the seasonal sea ice zone (SSIZ), and also their general associations with larger-scale climatic teleconnections (e.g. Streten and Troup, 1973; Carleton, 1979; Carleton and Carpenter, 1990; Carleton and Fitch, 1993; Carleton and Song, 1997).

4 However, there has been a delayed understanding of the detailed structure of individual mesocyclones in the Antarctic. This gap in mesocyclone research between the two hemispheres has only recently started to close, spurred largely by the deployment of automatic weather stations (AWSs) in some key regions of mesocyclone formation, particularly the western Ross Sea. Over the ice-free oceans, the greatest impetus has come from the application of newer satellite microwave data to determine the marine weather associated with mesocyclones (e.g. Carleton et al., 1995; McMurdie et al., 1997; Marshall and Turner, 1997a) (see also Section 3.2.4).

2.2.2 The synoptic-dynamic climatology of higher southern latitudes

The major physical geographic differences between the Arctic and Antarctic regions (e.g. their land–sea distributions; sea ice, snow cover and SST regimes; and extent of continental glaciation), are reflected in their respective large-scale upper-ocean and atmospheric circulations (Section 2.2.1). In turn, the differences in synoptic-dynamic climatology described below are also apparent in the characteristics and dominant regimes of mesoscale cyclonic circulations over southern higher latitudes (Sections 2.2.3 and 2.2.4).

The circumpolar trough

The Antarctic circumpolar trough (ACT) of low pressure is the dominant feature of the atmospheric circulation over the sub-Antarctic region on daily through climatological (monthly and longer) time scales (e.g. Jones and Simmonds, 1993; Sinclair, 1994, 1995). It marks the poleward end of 'storm tracks' originating primarily over lower-middle latitudes (Figure 2.42). The ACT is most clearly seen on mean sea level pressure charts, such as shown in Figure 2.45. In the Southern Hemisphere, a storm track may be defined (Trenberth, 1986, 1991) as a zone having maximum day-to-day variability in tropospheric heights, the v (south–north) scalar component of the vector wind, and vorticity (i.e. high alternation rates of pressure systems), but having a minimum variance in the u (west–east) wind component. Thus, the storm track is associated with individual cyclone tracks. The polar front jetstream (PFJ) is located on the equatorward side of the storm track, and exhibits marked variance of the u wind component related to the passage of speed maxima. In winter, the PFJ spirals in towards the Amundsen/Bellingshausen sector of Antarctic from a position over lower-middle latitudes much further westward, in the western South Atlantic (van Loon, 1966, his fig. 8b). This pattern reflects the importance of the cyclone 'graveyards' as centres of frontal cyclonic decay and stagnation (cyclolysis) in the Pacific sector, and as sources of snowfall for coastal West Antarctica (Figure 2.39). Also, in that sector during winter, the Subtropical jetstream (STJ) occurs over lower latitudes. Its intensity varies out of phase with that of the PFJ during the extreme modes of the El Niño–Southern Oscillation (ENSO) climatic teleconnection (see below).

Whilst the ACT is continuous around the sub-Antarctic, it comprises several centres of lower mean sea level pressure (SLP). Climatologically, these 'mean lows' include the Amundsen Sea Low (ASL), a low in the eastern Weddell Sea, and in the longitudes of Wilkes and Adélie Lands. On a daily basis, cyclonic systems move into the ACT from lower latitudes (mostly the synoptic frontal cyclones), and also form within it. Jones and Simmonds (1993) found that an

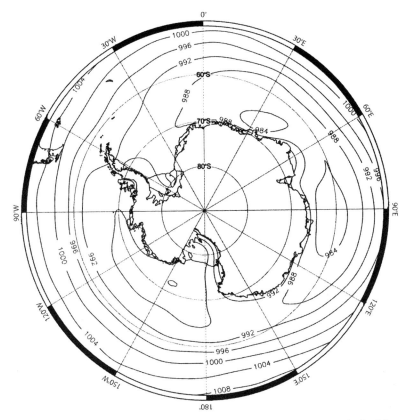

Figure 2.45. The mean winter PMSL field for the period 1971–2000 as derived from ECMWF re-analysis data.

ACT defined as a zone of most frequent cyclolysis, was located some 8°–10° latitude equatorward of the zone of lowest SLP associated with the cyclones. Accordingly, the ACT represents an integrated measure of cyclonic activity over the higher southern latitudes (cf. Jones and Simmonds, 1993; Sinclair, 1995). However, an aspect of the ACT notably absent from climatologies is its exact association with the regions of mesocyclogenesis over southern high latitudes; probably because of the relatively coarse resolutions of the operational analyses used to generate most cyclone climatologies. Thus, the association between the ACT and mesoscale cyclone activity has been far better deduced from the interpretation of high resolution satellite imagery (Section 2.2.3).

Although the ACT is located adjacent to the Antarctic continent, its distance from the coast varies interseasonally, comprising a strong semi-annual (twice-yearly) signal, as well as interannually (Streten, 1980; van Loon and Rogers, 1984). It is a zone of convergence between the very cold easterly winds occurring on its poleward side (at about 65° S), and the warmer maritime air of the westerlies occurring on its equatorward side (Wendland and McDonald, 1986). Variations in the ACT, particularly its semi-annual fluctuation in latitude location, induce wind stress anomalies in the upper ocean which, in turn, help influence the location and strength of the oceanic ACC (Large and van Loon, 1989). Similarly, the wind stress variations associated with the ACT are likely to be involved in the seasonally asymmetric advance/retreat patterns of the sea ice, and its interannual variations (Enomoto and Ohmura, 1990), the latter also in conjunction with differences in temperature advection by the atmosphere (Streten, 1983; Harangozo, 1997). The cyclonic centres comprising the ACT, particularly the ASL, also undergo interannual variations in intensity and longitude location in response to the ENSO phenomenon over the tropical Pacific: see below (Chen *et al.*, 1996; Cullather *et al.*, 1996). These teleconnections, in turn, influence the longitudes of most frequent cold air outbreaks in Antarctica and, accordingly also, the patterns of mesocyclone formation.

Several studies suggest decadal and possibly longer-term variations in the intensity and location of the principal cyclonic centres within the ACT, particularly the lows in the Weddell and Ross Seas. These seem unlikely to be the result of differences in data density or the analysis methods used between time periods (cf. van Loon, 1962; Streten, 1980; Mo and van Loon, 1984; LeMarshall *et al.*, 1985).

Meridionality and blocking

Despite the strong zonality displayed by the atmospheric circulation over southern higher latitudes on a time-averaged basis, marked meridionality may occur on synoptic time scales (i.e. daily to about one week) (Davidova, 1967) as the ACT is interrupted by high pressure ridges from both lower and higher latitudes (Streten, 1980). These episodes lead to anomalies of wind direction, temperature and sea ice conditions (Alvarez, 1958; Streten, 1977; Streten and Pike, 1980), particularly in the following longitude sectors, which are minima in the frequencies of cyclones (Figure 2.42): 10° W–10° E; 40° E–70° E; 120° E–160° E; and secondarily between 120° W and 30° W. Eastward of these preferred longitudes for ridging, southerly geostrophic wind anomalies, cold air advection, upward-directed convective heat fluxes, and possibly also, strong equatorward advances in the sea ice edge occur that combine to increase mesocyclone activity (Carleton and Carpenter, 1989b; Heinemann, 1990, 1996a; Fitch and Carleton, 1992; Carleton and Fitch, 1993; Carleton and Song, 1997, 2000; Carleton *et al.*, 1998). The opposite conditions, and a suppression of mesocyclonic activity, tend to occur to the west of ridges.

When persisting over weekly and longer time scales, blocking in the Southern Hemisphere tends to occur with a zonal wavenumber three pattern (e.g.

Trenberth and Mo, 1985; Sinclair *et al.*, 1997). Each of the three southern continents located in the subtropical/middle latitudes seems to participate in the development of anticyclonic blocking downstream (eastward) (van Loon, 1956; Alvarez, 1958; Streten, 1977; Coughlan, 1983). However, Antarctica may play an active role in the generation of split flow and blocking regimes near New Zealand (James, 1988) that can be characteristic of the lead-up to an ENSO warm episode, or El Niño event (Trenberth, 1980; Smith and Stearns, 1993). Accordingly, the propensity for more frequent blocking near New Zealand and enhanced troughing downstream in longitudes of the Ross Sea associated with ENSO, produces concomitant increases in the frequencies of cold air outbreaks over the southwest Pacific and, therefore, greater frequencies of mesocyclones (Section 2.2.4).

There are other consequences of the association between meridional atmospheric circulation patterns and mesocyclone development in the Antarctic and sub-Antarctic. For example, synoptic cyclones approaching a block over southern South America tend to dissipate and also to become quasi-stationary west of the Antarctic Peninsula. The superimposition of these large, cold core vortices over water that remains mostly ice-free for much of the year, results in relatively frequent occurrences of a mesocyclone type that is characterized by multiple vortices and known as a 'merry-go-round' signature (Turner and Thomas, 1994) (Sections 1.6.4 and 2.2.3).

The semi-annual oscillation (SAO)

Unlike in the Northern Hemisphere, where the circumpolar vortex has a dominantly annual variation in extent and intensity over middle and higher latitudes, the seasonal variability of the circumpolar vortex over higher latitudes of the Southern Hemisphere is dominantly twice-yearly (see Section 2.1.1). Equinoctial intensifications and solsticial relaxations in the gradient maxima of tropospheric temperature, SLP, tropospheric height, and wind speed over the sub-Antarctic comprise the semi-annual oscillation (SAO). This manifests itself as SLP mean maxima in the summer and winter, and minima in the transition seasons at coastal Antarctic stations. The SAO is driven primarily by differences in the seasonal cooling and heating rates between the middle and higher southern latitudes dominated, respectively, by ocean and the Antarctic continent (Schwerdtfeger, 1960; Mo and van Loon, 1984). The out-of-phase relationship in the annual march of SLP between middle and high southern latitudes that is connected with the SAO, results in two equatorward (in December and June) and two poleward (March, September) excursions of the ACT (van Loon, 1967). Climatologies of synoptic-scale cyclones derived from daily data also show a semi-annual movement in the latitude zone of their maximum frequency consistent with the movement of the time-averaged ACT (Carleton, 1981c; Howard, 1983). This matches the seasonal variation of precipitation at stations in the Antarctic and sub-Antarctic (e.g. Turner *et al.*, 1997). Moreover, the frequencies and rates of travel of mesocyclones over the sub-Antarctic also exhibit associations with the SAO (Sections 2.2.3 and 2.2.4).

Semi-annual variations in the longitudinal location and amplitude of the tropospheric waves seem connected with differences in the cooling and heating rates between land and sea *within* the middle latitudes; especially between Australia and the Indian and South Pacific oceans (van Loon, 1967, 1972). These changes in the wave in this sector are reflected in changes in the preferred longitudes of cold air outbreaks intraseasonally and, accordingly, the occurrences of mesocyclones (Carleton and Carpenter, 1990). On average, a progressive westward shift occurs in the maximum frequencies of mesocyclones: from the southwest Pacific in early winter, to the southern Indian Ocean in late winter. The amplification of the tropospheric waves between the March/April and May/July periods, and the resulting increase in the poleward advection of clouds and warm air into the Ross Sea region, have been connected with the 'coreless winter' displayed by station temperature regimes there (van Loon, 1967; Stearns and Wendler, 1988). Studies of climatic data from manned stations and AWSs elsewhere reveal the widespread character of the coreless winter in the Antarctic (e.g. Bromwich, 1988; Allison *et al.*, 1993).

The strength of the SAO varies interannually (van Loon and Rogers, 1984) and also decadally (Mo and van Loon, 1984). In particular, since about 1979 there has been a delay in the component of the SAO associated with the weakening of the circumpolar vortex in southern spring (Hurrell and van Loon, 1994), especially near Antarctica (Simmonds and Jones, 1998). This has shifted into November from an occurrence previously in late September/early October. It is believed to have resulted, at least in part, from a strengthening of the latitude gradient of pressure and temperature associated with a rise in the SSTs over lower latitudes and a decrease in SLP within the ACT, especially in the Pacific sector. These changes are also consistent with an intensification of the poleward-directed sensible heat fluxes carried out by the atmospheric eddies during the 1980s (van Loon and Kidson, 1993). However, the impact of these decade-scale changes in the SAO on cyclonic activity regimes, including those of mesocyclones, has yet to be documented.

The El Niño–Southern Oscillation (ENSO) and other teleconnections

On annual and interannual time scales, the dominant sources of circulation variability over southern higher latitudes comprise the eccentricity of zonal wavenumber one, and the teleconnections to ENSO. The signals of these low-frequency variations are particularly evident in the South Pacific sector. There also, the Antarctic coast from east of the Ross Sea to the western side of the Antarctic Peninsula communicates directly with middle and lower latitudes via incoming synoptic systems, the time-averaged ASL, and the South Pacific Cloud Band (e.g. Streten, 1975; Zillman and Johnson, 1985; Carleton, 1989; Turner *et al.*, 1997). Accordingly, the region west of the Peninsula is a sensitive indicator of interannual climate variations, including ENSO, and longer-term climate trends involving temperature–sea ice inter-relationships (King, 1994; Jacobs and Comiso, 1997). This region is in stark contrast to that east of the Peninsula. The Weddell Sea is ice-covered for the greater part of the year, and experiences a continental climate. There is also significant export of ice and cold air northeastwards into the South Atlantic Ocean (Schwerdtfeger, 1979). This advection is enhanced by a mean cyclonic circulation in the atmosphere and ocean, and a strong thermally-induced 'barrier wind' in the boundary layer on the eastern side of the Antarctic Peninsula (Schwerdtfeger, 1975).

The eccentricity of zonal wavenumber one manifests the tendency for SLP and 500 hPa height anomalies to be out of phase between Australia–New Zealand and the Antarctic Peninsula–South America regions on monthly and longer time scales, especially during the summer (Rogers and van Loon, 1982). The phase and amplitude of the wave can be depicted synoptically by a Trans-Polar Index (TPI); or the SLP anomaly difference between Hobart, Australia and Stanley, Falkland Islands (Pittock, 1980, 1984). Positive (negative) values of TPI are associated with the trough – and zone of strongest westerlies and coldest air – displaced towards the South America (Australia) sector. Variations of sea ice extent and concentration in the Antarctic south of Australia, in the Ross Sea, and also the Scotia Sea, show associations with the TPI (Rogers and van Loon, 1982; Streten, 1983; Carleton, 1989), as do the areas of most frequent meso-cyclones; at least in southern winter (Carleton and Carpenter, 1990). These associations arise because of coupled anomalies in the geostrophic wind direction and temperature advection (Section 2.2.4). There are decade-scale variations apparent in the correlation of SLP between Hobart and Stanley (Carleton, 1989; Villalba *et al.*, 1997), and these appear linked with lower-latitude teleconnections, possibly ENSO.

Over the southern oceans, the teleconnections to ENSO appear as alternating anomalies of SLP and zonal wind between low and middle/high latitudes (Mo and White, 1985). These are such that the trade winds are stronger in the cold phase, or La Niña (weaker: El Niño), and westerlies north (south) of about 45° S are stronger (weaker) (Trenberth, 1981). The development of the warm extreme, or El Niño, of ENSO is evident in the pressure and temperature anomaly fields over much of the Antarctic continent (Smith and Stearns, 1993), as well as in the sea-ice extent patterns of adjacent oceans; especially in the South Indian and South Pacific (Carleton, 1988b, 1989; Simmonds and Jacka, 1995; Gloersen, 1995). Chen *et al.* (1996) and Cullather *et al.* (1996) demonstrated links between ENSO and the intensity and location of the ASL. In the warm (cold) events of ENSO, a strengthened (weakened) STJ, and a weakened (strengthened) PFJ, accompany a weaker (stronger) ASL. The weaker ASL during the warm phase is the result of both a displacement of this centre of action westward into the Ross Sea sector, and a reduction in the numbers of synoptic cyclones moving into this region. Thus, the occurrence of extreme winter temperature anomalies along the Antarctic Peninsula shows some association with the phase of ENSO: these are colder (warmer) than normal when the ASL is weak (strong), as can occur in the El Niño (La Niña) modes (Marshall and King, 1998). The longitude variations in position of the ASL also dominate the fluxes of moisture into West Antarctica, and influence the frequencies of mesoscale cyclones occurring in the cold air outbreaks over the Amundsen/Bellingshausen Seas (Bromwich *et al.*, 1995; Carleton *et al.*, 1998). Specifically, westward displacements of the ASL in El Niño (eastward displacements in La Niña) tend to be associated with decreased (increased) frequencies of mesocyclones in that sector of Antarctica (Carleton and Song, 2000) (Section 2.2.4).

The intensity variations of the ASL that are a significant part of the ENSO signature in the Pacific sector of the Antarctic, themselves comprise a larger teleconnection known as the PSA: Pacific–South America pattern (Mo and Ghil, 1987). The PSA comprises the second and third most dominant modes of variation in the winter 500 hPa geopotential heights, expressed as a wave train of anomalies extending from the subtropical Pacific southeastward into the Weddell Sea (Kidson, 1988; Farrara *et al.*, 1989). The teleconnections to ENSO in the South Pacific sector involving the ASL have accompanying changes in the seasonal cycle of the trough in the Tasman Sea, which is enhanced in the lead-up to a warm ENSO event contrasted with a non-ENSO year (van Loon, 1984). Accordingly, changes in cyclonic activity in the New Zealand region, as manifest in the seasonally-averaged anomalies of SLP and tropospheric height, appear connected with atmospheric changes in the Indonesian region that comprise one 'pole' of the tropical Walker Circulation (Trenberth and Shea, 1987). These changes are evident as cyclonic activity variations over most of the Southern Hemisphere extratropics (Sinclair *et al.*, 1997). They include reduced (increased) numbers of wintertime cyclones during an El Niño over the Indian Ocean, Australasia and the Amundsen Sea (the circumpolar trough near Wilkes Land, and the subtropical eastern Pacific). The changes in the opposite extreme of ENSO are virtually mirror images of those occurring during an El Niño, suggesting a linear cyclone response to ENSO over the

Southern Hemisphere. These teleconnections also have a mesocyclone signature (Section 2.2.4).

2.2.3 Time-averaged patterns of mesocyclones in the Antarctic

Basic characteristics of mesocyclone vortices

The interpretation of image mosaics from the DMSP polar orbiters for cloud vortices of *all* types and sizes occurring poleward of 50° S, reveals some interesting properties of synoptic and subsynoptic-scale cyclonic systems (Carleton, 1995). By subjectively classifying vortices into two nominal classes of 'synoptic-scale cyclones' and 'mesoscale cyclones' it was found that the former (latter) have very different mean diameters of 1035 km (375 km). These differences are statistically significant, and are maintained even when the data are stratified seasonally. A similar analysis of the size characteristics of mesocyclones identified on both DMSP and AVHRR (Advanced Very High Resolution Radiometer) images of the southeast Pacific Ocean and Antarctic Peninsula area on selected dates in 1992 (Carleton and Song, 1997) seems to confirm this basic pattern (Table 2.2). Almost all classes of mesoscale cloud vortex have mean diameters less than 400 km, and in some cases are as small as 345–350 km (see also Turner and Thomas, 1994). Spiraliform mesocyclones are on average slightly larger than comma clouds, at least in their initial stages (Table 2.2a); however, mesocyclones of both types increase in size as they continue to evolve (Table 2.2b). The small differences in mean size observed between stratiform and convective mesocyclones (Table 2.2c) are unlikely to be significant statistically, given the relatively large standard deviations of the samples.

Figure 2.46 shows a lack of cyclones of either size class (i.e. synoptic-scale, mesoscale) in the diameter range 460–680 km. This result, which is supported by regional-scale climatologies of Antarctic mesocyclones (Turner and Thomas, 1994; Turner *et al.*, 1996a), may indicate a 'spectral gap' in cloud vortex circulations occurring over higher southern latitudes. Interestingly, a similar analysis conducted for the Northern Hemisphere (see Carleton, 1996) suggests no such gap in cloud vortex sizes there: these mesocyclones are slightly larger than their southern ocean counterparts (mean diameter = 410 km), and synoptic cyclones are smaller (mean diameter = 697 km). However, it should be remembered that these results are for all cyclones poleward of 20° N; it remains to be determined if mesocyclones (synoptic cyclones) might become appreciably smaller (larger) with increasing latitude in the Northern Hemisphere. Over higher latitudes of the Southern Hemisphere, at least, mesocyclones can be readily distinguished on the basis of size by an experienced satellite image analyst, as well as by the other descriptors of cloud vortex type (see Section 1.6) and the larger-scale cloud environment within which the mesocyclone develops.

Table 2.2. *Size characteristics of mesocyclone cloud vortices according to satellite-viewed characteristics, 1992: southeast Pacific region*

(a) Signature type[a]

	Comma clouds	Spiral systems	All vortices
$\bar{r}$ (° latitude)	1.57°	1.78°	1.63°
$\hat{s}$ (° latitude)	0.576°	0.441°	0.540°
$\bar{D}$ (km)	345	392	359
n	21	9	30

(b) Development stage[b]

	Incipient/developing	Mature	Dissipating	All vortices
$\bar{r}$ (° latitude)	1.61°	1.72°	1.94°	1.72°
$\hat{s}$ (° latitude)	0.502°	0.521°	0.417°	0.499°
$\bar{D}$ (km)	354	378	427	378
n	18	18	8	44

(c) Dominant cloud type[c]

	Stratiform clouds	Convective clouds	All vortices
$\bar{r}$ (° latitude)	1.69°	1.59°	1.63°
$\hat{s}$ (° latitude)	0.597°	0.507°	0.540°
$\bar{D}$ (km)	372	350	359
n	13	17	30

[a] The first observation of the vortex.
[b] The first observation (either comma or spiral) in that development stage.
[c] The first observation (any stage, either comma or spiral).

The positive association between the length scale, or system diameter, and the time scale, or duration, of meteorological motion systems (e.g. Barry, 1970) means that mesocyclones typically have life spans of hours, rather than the several days characteristic of the large frontal wave cyclones. Conversely, vertical motions in mesocyclones are likely to be greater, especially in association with the (relatively shallow) cumulus convection typical of many comma-cloud mesocyclones over the higher latitude southern oceans. Thus, mesocyclones might be the oceanic and cold-season equivalent of the Mesoscale Convective Systems (MCSs) which develop over heated land surfaces in summer, and which move large quantities of heat and moisture from the atmospheric boundary layer into the mid- and upper troposphere (see Carleton, 1996). Both types of system have strong upward vertical motions and cyclonic rotation, and develop in an environment characterized by weak static stability. In the case of MCSs, the instability results primarily from solar heating of the land surface, whereas

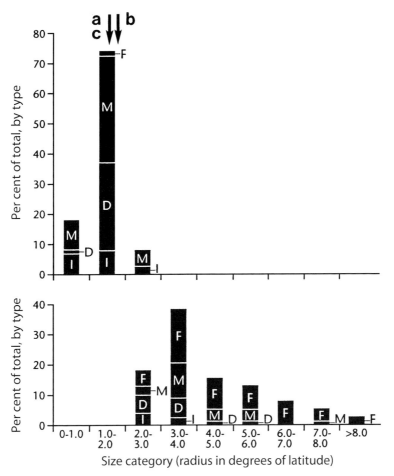

Figure 2.46. Distribution of cloud vortex sizes according to a nominal classification into 'mesocyclones' (upper panel) and 'synoptic cyclones' (lower panel), based on the interpretation of 20 days' DMSP imagery of the Southern Hemisphere during 1990. Vortices are classified by stage of development (I; incipient; D; developing; M; mature; F; dissipating/decaying), and grouped by radius class (in degrees latitude). Arrows in the top panel indicate the mean sizes of mesocyclones determined from analysis of AVHRR and DMSP images for a different year: 1992 (refer Table 2.2). (After Carleton, 1995, *International Journal of Remote Sensing*, Taylor & Francis Ltd (http://www.tandf.co.uk/journals).)

for mesocyclones it results from the strong vertical gradient of temperature as cold, dry air moves over a warmer sea surface.

The exact determination of mesocyclone mean durations is made difficult by the less than ideal temporal sampling offered by polar orbiting satellites, even over quite high latitudes where the successive orbital swaths overlap substantially. Thus, estimates of the mean lifetimes of Southern Ocean and Antarctic mesocyclones vary widely. Carleton and Fitch (1993) and Carleton

(1995) used twice-daily DMSP infra-red images to estimate the mean duration of mesocyclones. These varied from between about 13 h to almost 30 h according to month and signature type (comma cloud or spiraliform). More recently, the study of mesocyclone attributes for the southeast Pacific Ocean region during selected time periods of 1992 (Carleton and Song, 1997), used all available overpasses of both the AVHRR and DMSP systems. The mean duration of the 19 mesocyclones that could be identified and tracked was 10 h; however, the standard deviation was quite high at 7.25 h. Stratifying systems into comma clouds (spiraliform vortices) gave mean durations of 9.4 h (11.6 h), although these were unlikely to be significantly different, owing to the large standard deviations and small sample sizes involved. Putting these results into a climatological context, 452 mesocyclones were identified by Turner *et al*. (1996a) for the Bellingshausen/Amundsen seas in the period March 1993–February 1994. Over half of these systems had durations of 6 h or less; 15% lasted between 7 and 12 h, and only 11% exceeded 12 h in duration.

The different estimates of mesocyclone duration noted above show particularly the impact of temporal sampling of the satellite imagery. Another source of these differences is likely to be that related to the length of time required to develop a cloud signature that is detectable in the satellite data; particularly over the moisture-starved ice shelves and the sea ice close to the continent (Bromwich, 1991). The estimates of mesocyclone duration for ice-free ocean areas should improve in the near future with application of the wind data from satellite scatterometers (i.e. active microwave non-imagers).

Distributions and seasonality of Antarctic mesocyclone types

The use of satellite images to develop reliable climatologies of Antarctic mesocyclones is obviously superior to that of most Southern Hemisphere synoptic chart series; however, these data also have associated issues of representativeness. Mesocyclones developing close to or over Antarctica tend to lack deep clouds (e.g. Figure 2.43), and may lack clouds altogether in the incipient stages, even though a cyclonic signature may be present in the AWS wind field (Bromwich, 1991). Thus, mesocyclone climatologies developed using satellite imagery for these regions may include a bias associated with the inability to detect the accompanying low-level cloudiness against a high albedo and low temperature ice surface (i.e. their frequencies of occurrence are likely to be underestimated). However, mesocyclones occurring over ice-free ocean areas in the sub-Antarctic and middle latitudes, and over sea ice of lower concentration, tend to be deeper and more organized; therefore, they are more readily detected in the visible and infra-red imagery. In addition, there is the tendency for mesocyclones to develop near zones of strong temperature contrast, such as that between the MIZ and open ocean, and also that between the continent and the sea ice (Carleton, 1981b, 1995; Carleton and Carpenter, 1990; Carleton and Fitch, 1993; Turner and Thomas, 1994). This is a feature also noted for mesocyclones and polar lows in the Northern Hemisphere (e.g. Carleton, 1985a,b). For the transition and winter season months of 1988 and 1989, Carleton (1995) found that 147 out of a total of 250 mesocyclones (i.e. 59%) occurring poleward of 50° S developed cloud signatures within only ±2.5° latitude of these boundaries. The remaining 41% of mesocyclones occurred over much more extensive ocean or sea ice surfaces which tend to be more 'homogeneous' from an energy budget standpoint, with the exception of the ACC (cf. Viebrock, 1962; Zillman and Dingle, 1969; Taylor *et al.*, 1978; Weller, 1980). The tendency for mesocyclones to form near surface boundaries is also clearly evident in the charts of mesocyclone distributions (see below).

The ratio of multi-banded, or spiraliform, to single-banded, or comma cloud mesocyclones increases with latitude in the Southern Hemisphere (Carleton and Carpenter, 1990); from about 1:10 for the extratropics as a whole, to about 1:3 at higher latitudes in July. It should be noted that it is sometimes difficult to distinguish between comma and spiraliform cloud signatures and that the signatures also may change during the system's lifetime (see Section 1.6). Broadly similar results have been found for the Bellingshausen and Weddell Seas by Turner *et al.* (1996a); however, the latitudinal increase in the relative frequencies of spiraliform vortices that they reported was not as large. This possibly reflects differences in the time periods studied, and the nominal spatial resolutions of the imagery used. Examining only the latitudes poleward of about 50° S, and for the months of March, April, June, July, August and October in 1988, and June, July, and August of 1989, differences in the relative frequencies of spiraliform and comma cloud vortices were less evident (Carleton and Fitch, 1993). On average for this zone, the ratio of spirals to commas was about 0.8 (i.e. they are almost equal), however, there was considerable inter-monthly variability in these values (Figure 2.47). The ratio of spirals to commas decreased to about 1:2 in July, at least for the sector located between 100° E eastward to 50° W. Charts showing the spatial distributions of mesocyclones by type for this region often reveal no marked differences in the geographical regimes of spiraliform and comma cloud systems (Figure 2.48). However, in certain months these vortex types may be better differentiated. For example, in June 1988 (Figure 2.49) there was a clear tendency for comma cloud systems to be more evenly distributed across the sector compared with the spiraliform systems, which were mostly confined to the region westward of about 140° W. The data on mesocyclone frequencies

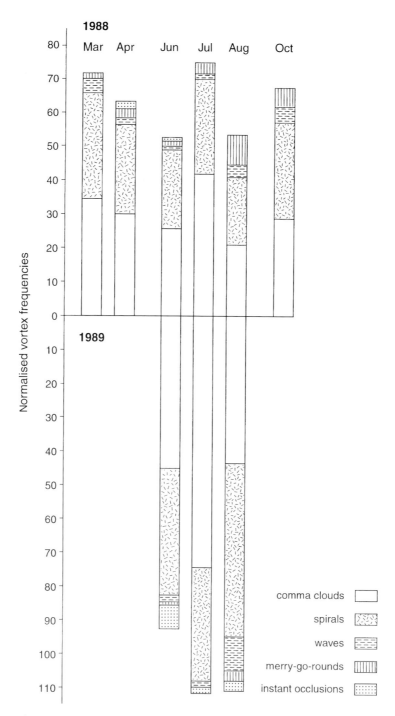

Figure 2.47. Normalized frequencies of mesocyclone cloud vortices according to five satellite-observed classes (comma clouds, spiraliform vortices, frontal waves, 'merry-go-round' systems, and instant occlusions), for 9 study months in 1988 and 1989 (region south of 50° S and between longitudes c. 100° E–60° W). Note particularly the strong differences between the two winter seasons of June, July and August.

(a)

(b)

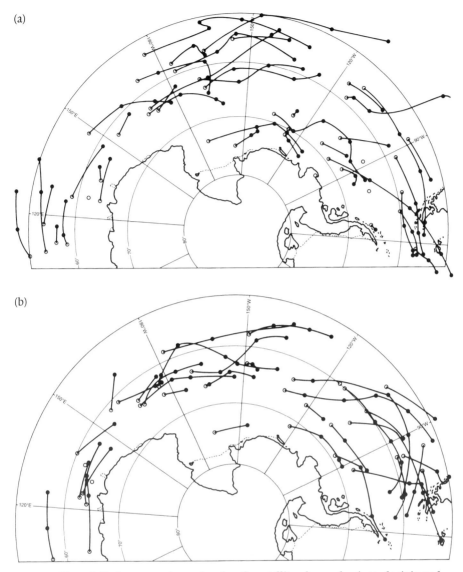

Figure 2.48. Summary charts showing the satellite-observed points of origin and tracks of mesocyclones during June 1989, separately for the two major classes of comma clouds (a), and spiraliform vortices (b). These are determined from the subjective interpretation of DMSP images. Once-only observation of a system is denoted by a single open circle.

for 1988 (Figure 2.47; also Table 2.3a) suggest a signal of the SAO, with maximum frequencies in the transition-season months (March, October), and again in mid-winter. There are also large between-winter differences in mesocyclone frequencies for this sector, where the 1989 winter was much more active than that of 1988. These differences are discussed more fully with respect

130   2 Climatology

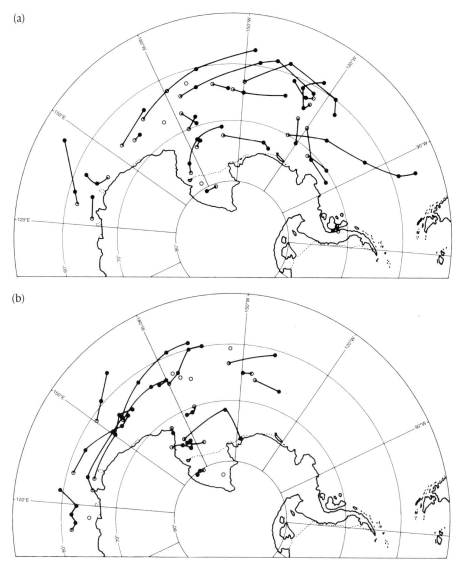

Figure 2.49. (a, b) Similar to Figure 2.48, except for June 1988.

to their association with ENSO, in Section 2.2.4. The dominance of the comma cloud and spiraliform mesocyclones over other types is clearly evident in Figure 2.47. The number of mesocyclones forming as mesoscale frontal waves or on boundary-layer fronts, as 'merry-go-rounds' or instant occlusions, is relatively small in most months.

The spatial distributions of the 'merry-go-round' vortices in the sector 100° E eastward to 50° W for the transition and winter seasons of 1988 and 1989 (Figure 2.50), reveal two areas favoured for these developments. These are the

Table 2.3. *Normalized frequencies of mesocyclones in the Southern Hemisphere, by year and region*

	Jan	Feb	Mar	Apr	May	Jun	Jul	Aug	Sep	Oct	Nov	Dec
(a) Large scale												
Sector 100° E eastwards to 60° W, South of 50° S (Fitch and Carleton, 1992; Carleton and Fitch, 1993)												
1988			71	61		46	75	47		67		
1989						83	113	93				
Full hemisphere, south of 30° S (Carleton and Carpenter, 1990)												
1977–83						150	164	143	41			
Sector 70° E eastwards to 150° W, all latitudes (Carleton and Song, 1997)												
1992			51	63		54	50	58		60	55	
(b) Regional scale												
Southwest Ross Sea (Bromwich, 1991; Carrasco and Bromwich, 1996)												
1984		8	6	1	3	4	2	5	4	5	1	4
1985	7	7	8	9	8	6	11	12	10	5	6	11
1991	10.8	12	16.4	6.4	7.2	10.4	9.2	10	6.4	8	12	10
Northwest Ross Ice Shelf (Carrasco and Bromwich, 1996)												
1991 (av.)	7.2	6	9.2	5.6	9.2	6.4	10.8	10	4.8	5.6	7.6	4.4
Marie Byrd Land (Carrasco and Bromwich, 1997a)												
1991	19	62	68	39	26	22	20	18	1	27	26	15
Bellingshausen Sea (Carrasco and Bromwich, 1997b)												
1991	34	33	37	33	26	21	7	26	24	26	29	40
Weddell Sea (Heinemann, 1990; Carrasco and Bromwich, 1997b)												
1983	17	9										
1984	26	28										
1985	14	19										
1987	30	19										
1988	29	39										
1991	24	19	17	15	11	16	12	11	10	23	22	53

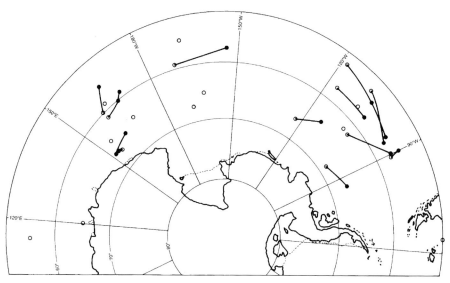

Figure 2.50. Spatial distribution of all 'merry-go-round' vortices (points of origin and tracks) observed for the nine study months of 1988 and 1989.

southeast Pacific between about 120° W and 90° W, and the Ross Sea sector between about 150° E and 160° W. Each of these areas corresponds to a region typically associated with high frequencies of decaying synoptic-scale cold-core depressions that have migrated into the ACT (e.g. Carleton, 1979).

The spatial distribution of mesocyclones identified as having formed on boundary-layer fronts, or on old synoptic-scale frontal bands (Figure 2.51) reveals the favoured areas for these developments to be the Ross Sea, and also near the sea ice–ocean margin. The former area is favoured given its location southeast of where many frontal cyclones and their frontal cloud bands typically stagnate, whilst the latter location is favoured for the occurrence of the low-level baroclinic zones associated with boundary-layer fronts.

On average in the winter and for the hemisphere as a whole, the instant occlusion system is most frequently observed in the latitude zone 45°–50° S (Carleton, 1981a). This zone is where frontal wave cyclones, having maximum frequencies between about 30° and 50° S, and comma cloud mesocyclones, with maximum frequencies between about 45° and 60° S (Carleton, 1979), are most likely to interact. The DMSP-based study of mesocyclones over the southern oceans for the nine study months of 1988 and 1989 (Carleton and Fitch, 1993) also identified occurrences of instant occlusions (see Figure 2.47). Whilst not particularly numerous in the study region south of 50° S ($n = 11$), all but one of these instant occlusions occurred in the 10° latitude zone of 50°–60° S, supporting the results of Carleton (1981b). Owing to the association of the instant occlusion type with the regions favoured for frequent frontal activity (or

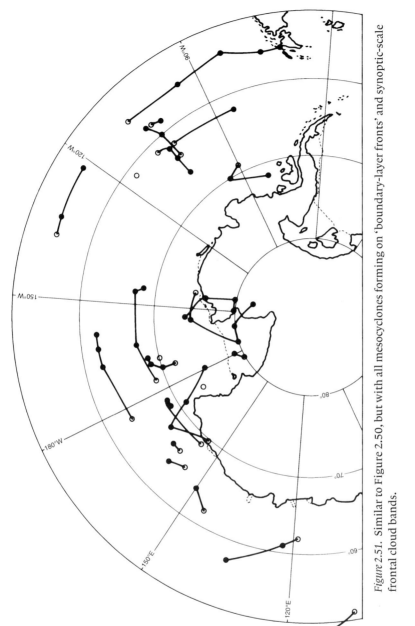

Figure 2.51. Similar to Figure 2.50, but with all mesocyclones forming on 'boundary-layer fronts' and synoptic-scale frontal cloud bands.

climatic fronts: Hattle, 1968) and cold-air outbreaks, these systems have the highest frequencies spatially in the southeast Indian Ocean to southwest of the Australian Bight (Carleton, 1981a). That the Australia–New Zealand area is favoured for high amplitude ridging and even blocking (e.g. van Loon, 1956; Streten, 1977) means also that the frontal cloud bands associated with synoptic cyclones are slowed as they approach from the west. This results in the relatively high frequencies of instant occlusions southwest of Australia. These synoptic associations also seem confirmed by the data on mesocyclone activity in the Southern Ocean sector between 70° E and 150° W for seven transition and winter-season months of 1992 (Carleton and Song, 1997). The locations of all instant occlusion vortices at the time of their first observation, and the associated mesocyclone tracks preceding each occurrence, are shown in Figure 2.52a. This distribution may be compared with that in Figure 2.52b, showing the mean monthly frequencies of mesoscale cyclogenesis (i.e. the points of origin of *all* mesocyclones) for the entire seven-month study period. Even though there is some correspondence between the occurrences of instant occlusions and the highest frequencies of mesocyclone formation (i.e. over the southeast Indian Ocean extending south of Australia), this does not appear to be the case further east (i.e. south and east of New Zealand).

Regional synoptic climatologies

Mesocyclones have been observed at some time over just about all longitudes within the Antarctic and sub-Antarctic. However, three sub-regions have been particularly focused upon in climatological and diagnostic studies; these are the western Ross Sea, the Weddell Sea, and the Bellingshausen/Amundsen seas. These areas mostly have been emphasized because of their proximity to some major Antarctic bases, and the resulting need to determine and understand mesocyclone climatology for operational and forecasting concerns. It is possible that these areas also have higher frequencies of mesocyclones compared with some other regions, owing to the conjunction of physical processes favourable for mesocyclogenesis (e.g. confluence of katabatic outflows, topography, sea–air interactions near coastal polynyas and the ice margin). However, this may partly reflect the greater attention that has been paid to these areas. For example, the coasts of Adélie and Wilkes Lands also seem to experience quite high frequencies of mesocyclones (Streten, 1990; Carleton and Fitch, 1993), yet these areas have been relatively little studied (cf. Callaghan and Betts, 1987).

The Ross Sea

The western Ross Sea, including Terra Nova Bay and Byrd Glacier, are sites both of frequent mesocyclogenesis and katabatic winds (e.g. Bromwich, 1989, 1991; Carrasco and Bromwich, 1993). The proximity to McMurdo base

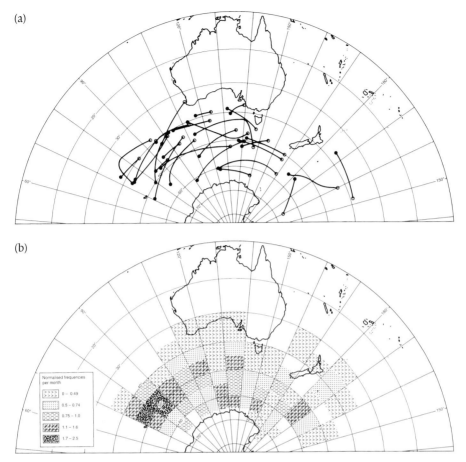

Figure 2.52. (a) Spatial distribution of instant occlusion vortices over the southern ocean sector between longitudes *c.* 70° E–150° W, for seven study months (March, April, June, July, August, October, November) of 1992. The point of 'instant cyclogenesis' is denoted by the open circle marking the end-point of the associated mesocyclone track. In (b) are shown the normalized frequencies per day (choroplethed by 5° latitude/longitude units) of all mesocyclogenesis events for this sector over the study period.

has focused attention on mesocyclones in this region; for example, many significant snowfall events at McMurdo are associated with mesocyclones (Rockey and Braaten, 1995). A three-year (1984, 1985, 1991) climatology of mesocyclones, developed from combinations of DMSP and AVHRR image analysis, synoptic charts and the mesoscale array of AWSs on the Ross Ice Shelf and western Ross Sea (Bromwich, 1991; Carrasco and Bromwich, 1993, 1996), shows high monthly frequencies of these systems (Table 2.3b). However, many mesocyclones lack deep or well-defined cloud signatures in the cloud imagery, especially early in their development.

A mechanism for mesocyclone development in the western Ross Sea was proposed by Carrasco and Bromwich (1993). They suggested that the export of cold continental air into this region by katabatic winds leads to the development of a baroclinic zone further east when milder maritime air from West Antarctica moves south and west over that boundary-layer air (see also Bromwich *et al.*, 1996). This scenario is activated by a synoptic-scale cyclone or short-wave passing through the northern Ross Sea. Column stretching of the air is almost certainly a contributing factor in the development of these mesocyclones (Carrasco and Bromwich, 1995), by which the descent of air downslope increases its cyclonic vorticity via the conservation of potential vorticity. This mechanism is also likely to be involved in mesocyclogenesis in Marie Byrd Land and along the Siple Coast (Carrasco *et al.*, 1997a). In that sub-region, the very high frequency of mesocyclones occurring in March 1991 (Table 2.3b) probably resulted from a strengthening of the boundary-layer baroclinic zone; possibly activated by the poleward movement of the ACT associated with the autumnal phase of the SAO, in a month when open water still extends to quite high latitudes.

The Weddell Sea

The presence of the British Antarctic Survey's Halley station on the coast of the eastern Weddell Sea, and the active summer programmes of German scientists researching ice–atmosphere interactions in the Weddell Sea, has helped focus attention on mesocyclones there (Turner and Row, 1989; Heinemann, 1990; Turner and Thomas, 1994). Seasonally, the greatest frequencies of mesocyclones seem to occur in summer (Table 2.3b), when the ice has retreated and more open water is exposed to the atmosphere (Turner *et al.*, 1996a; Carrasco *et al.*, 1997b). Significant mesocyclogenesis also takes place further equatorward near the sea ice–ocean margin (Turner and Thomas, 1994; Heinemann, 1996a, c), where strong low-level temperature gradients develop under conditions of off-ice airflow (Kottmeier and Hartig, 1990). In common with the western Ross Sea, mesocyclogenesis in this region tends to be favoured during periods when cold air is advected downslope, as occurs particularly when synoptic-scale low pressure systems are situated to the east (Heinemann, 1990).

The Bellingshausen and Amundsen Seas

Unlike both the Ross and Weddell Seas, the Bellingshausen and Amundsen Seas are ice-covered for a considerably shorter period of the year and, hence, are characterized by a more maritime climate (Section 2.2.2). Given the greater amount of open water in these seas, even in winter, the total frequencies of mesocyclones are greater, and the seasonal variation is less than that in the Weddell Sea (Carrasco *et al.*, 1997b, their table 2.2.2b). About one-third of mesocyclones originate in association with the upper 'cold pool' of a decaying synoptic-scale cyclone (i.e. the 'merry-go-round' type) in this region (Turner and Thomas, 1994). Many of the stronger mesocyclones that develop near the sea ice margin subsequently move northeastwards to affect the Drake Passage and Tierra del Fuego (Lyons, 1983; Carleton and Fitch, 1993; Turner *et al.*, 1996a; Carrasco *et al.*, 1997b; also Figures 2.49a and 2.50). However, Turner *et al.* (1995) found little evidence of a mesocyclone precipitation signature for two stations (Rothera, Faraday) located in the western Antarctic Peninsula, during a one-year (1992–93) period.

Mesocyclone associations with standard synoptic fields

Although an individual mesocyclone is relatively limited in its horizontal, vertical and temporal extents, the tendency for mesocyclones to occur as 'outbreaks' means that their temporal and spatial influence is extended to larger scales. Accordingly, much insight into mesocyclone regimes and formation processes has come from comparing synoptic meteorological and upper-ocean fields (e.g. the tropospheric layer thickness; SST–surface air temperature difference) with the occurrences of mesocyclones in particular sub-regions of the southern ocean (e.g. Heinemann, 1990; Turner and Thomas, 1994; Song and Carleton, 1997). The simplest approach to identifying these associations is to compare the time-averaged (e.g. monthly) patterns of mesocyclogenesis derived from satellite image interpretation, with the larger-scale pressure/height anomalies for that same period (Carleton, 1996). Two examples for the Australian sector of March and April of 1992 (Figure 2.53) illustrate the tendency for mesocyclones to be associated with negative anomalies of SLP (specifically, southwest of Australia and southeast of New Zealand in March: Figure 2.53a) and to 'avoid' the positive anomalies associated with time-averaged high pressure. When SLP anomalies are oriented more meridionally, as occurred particularly in the Tasman Sea/New Zealand region in April 1992 (Figure 2.53b), mesocyclone tracks tend also to be meridional, since they are influenced by the co-occurring anomalies of lower-tropospheric geostrophic and thermal winds.

The associations gleaned from examination of charts such as Figure 2.53 rely upon the assumption that mesocyclones occur frequently during the month in question. A more exact determination of the associations between mesocyclogenesis and the synoptic atmospheric environment is obtained by compositing, or averaging, the daily SLP and tropospheric height anomalies (i.e. the departures from longer-term monthly means) for only those days

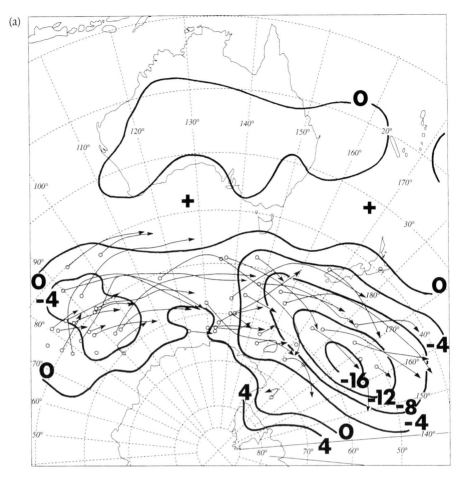

Figure 2.53. Charts showing tracks of all mesocyclones for the sector denoted in Figure 2.52, for (a) March 1992, (b) April 1992. Superimposed on each map are sea-level pressure (SLP) anomalies for each month. (Instant occlusions are indicated by squares.)

on which mesocyclones are present in a particular region (e.g. Heinemann, 1990; Fitch and Carleton, 1992, 1993; Turner and Thomas, 1994; Turner *et al.*, 1996a; Carrasco *et al.*, 1997b). The use of daily anomalies does not restrict this type of analysis to any one month, since the procedure removes the seasonal 'cycle' and other (e.g. SAO) effects, allowing the daily anomalies to be averaged together over different months. This method borrows from earlier mesocyclone studies undertaken in the Northern Hemisphere (e.g. Businger, 1985, 1987; Ese *et al.*, 1988; Ninomiya, 1989), and many of the relationships revealed by those studies appear broadly similar to those found for the Southern Hemisphere, though with some interesting differences as well. Thus, composite anomalies of the 1000–500 hPa layer thickness associated with mesocyclone

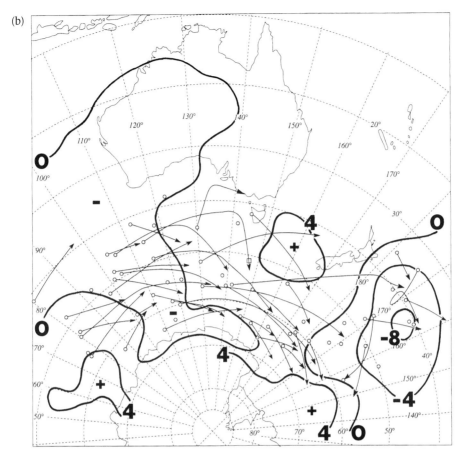

Figure 2.53 (cont.).

outbreaks in the Amundsen/Bellingshausen Seas (Carleton and Fitch, 1993, their fig. 11), showed that the majority of mesocyclones in this region develop *within* a synoptic-scale tropospheric cold-core anomaly. Moreover, compositing the pressure/height anomalies for successive days leading up to the outbreak day (or '*superposed epoch*' analysis) adds insight to this association: the cold-core anomaly is observed to migrate from the Pacific coast of West Antarctica northwards into the Amundsen/Bellingshausen seas. Associations such as these are reminiscent of the synoptic patterns connected with the development of 'polar low' mesocyclones in the North Pacific near Alaska (Businger, 1987).

In contrast, Fitch and Carleton (1992) showed that mesocyclone outbreaks in the Ross Sea are associated with 1000–500 hPa thickness anomalies that are negative (positive) to the eastward (westward), and where most mesocyclones develop in the cold southerly flow between these two anomalies. This suggests

(a)

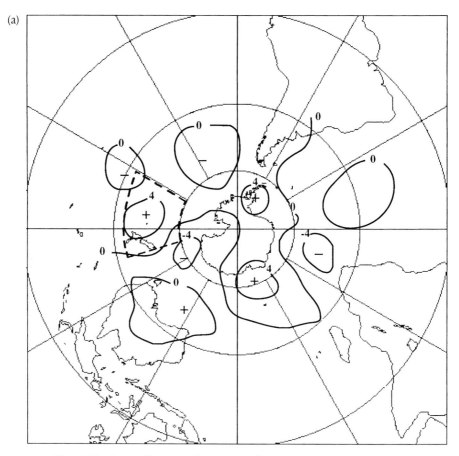

Figure 2.54. Composite anomaly patterns of SLP associated with (a) a sample of days during the 1992 study period on which mesocyclones were absent in the New Zealand region (area outlined), $n = 10$, (b) for all days during the same period when mesocyclones were observed in this region, $n = 7$. Isopleths in hPa.

the important role of katabatic outflows from the Ross Ice Shelf in mesocyclogenesis for this region (see also Carrasco and Bromwich, 1996), a result which appears confirmed if one also composites the synoptic-scale pressure/height anomalies for days on which the 'thermal plumes' associated with katabatic winds are observed on the infra-red imagery (Bromwich *et al.*, 1993): the pressure/height anomalies are broadly similar for the two sets of days.

Confirmation of the role of synoptic-scale forcing of Southern Ocean mesocyclone outbreaks can also be obtained by compositing the pressure/height anomalies for days on which *no* mesocyclones occur in a given region, and contrasting these patterns with those obtained for days on which mesocyclones *are* observed (Carrasco *et al.*, 1997a, b). For example, Figure 2.54 compares these composite SLP anomaly charts for the New Zealand/southwest

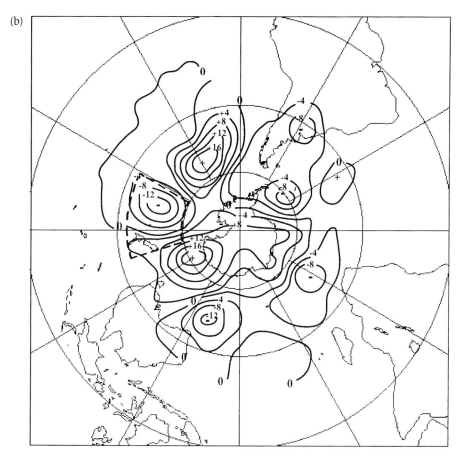

Figure 2.54 (cont.).

Pacific region during 1992. On days without mesocyclones (Figure 2.54a) positive (negative) anomalies are located to the eastward (westward) with corresponding northerly flow and warm advection, in sharp contrast to the patterns for days on which mesocyclone outbreaks occurred (Figure 2.54b).

2.2.4 Temporal variability of Antarctic mesocyclones

Daily to weekly time scales

A feature of Figure 2.54b is the co-occurrence of a couplet of pressure/height anomalies that should be favourable for mesocyclogenesis in the southeast Pacific/southern South America region, at around the same time that mesocyclones are observed near New Zealand. Interestingly, such a couplet was absent on days when mesocyclones occurred in a 'key' region south of Australia (Carleton and Song, 1997). The implication from Figure 2.54b is that mesocyclone outbreaks may be linked across long distances on daily time scales; presumably via the connectivity of the tropospheric waves, which give a preferred wavenumber three pattern over middle and higher latitudes of the South Pacific region on time scales of 10–50 days (Kidson, 1991). To evaluate the possibility that mesocyclone outbreaks may occur in the South America sector around the time when they also occur near New Zealand, Carleton and Song (1997) interpreted direct-readout DMSP and AVHRR imagery acquired by Palmer Station on the Antarctic Peninsula for mesocyclone vortices. It was found that, on days when mesocyclones occurred in the New Zealand area, the frequencies of such systems also increased in the southeast Pacific just west of Chile. Carleton and Song (2000) subsequently confirmed that large positive values of the sea–air temperature difference also occur in the two regions. However, on days when mesocyclones occurred south of Australia, mesocyclone frequencies decreased, on average, over southern South America but increased further south within the Amundsen and Bellingshausen Seas. On those same days, the frequencies of synoptic-scale cyclones also increased in the southeast Pacific. These results raise the possibility of two preferred modes of cyclonic activity in the southeast Pacific/Drake Passage region. In this scenario, one circulation mode is dominated by synoptic-scale (mesoscale) cyclones over higher-middle (sub-Antarctic) latitudes, whilst the other mode is characterized by mesoscale cyclones occurring over more middle latitudes of this sector. These opposing modes are reminiscent of those proposed by Carleton (1979) and Mayes (1985), and perhaps as also seen in the contrasting synoptic patterns associated with the openings and closings of leads within the sea ice on the eastern side of the Antarctic Peninsula (Schwerdtfeger, 1984; his fig. 3.21). Further studies are required to verify these possible preferred cyclone type and frequency modes, as well as the potential forecasting benefit of a long-distance association between mesocyclone outbreaks in the New Zealand area and those in the southeast Pacific.

Monthly to seasonal time scales

Large numbers of mesocyclones are observed in the Antarctic seas in summer as well as winter (Heinemann, 1990; Turner *et al.*, 1993a, 1996a; Turner and Thomas, 1994). These results come from the fact that the Antarctic continent is a source of cold air all year, unlike the Arctic where the land areas lose their snow cover and warm up in summer. Moreover, Southern Hemisphere cloud-vortex studies show that mesocyclones occur frequently in the transition-season months of March/April and September/October (Lyons, 1983; Carleton and Fitch, 1993). Since these are times when, on average, the ACT reaches its most poleward location, this suggests an association with the SAO: see also Table 2.3a. For the Australian sector of the southern oceans,

Carleton and Song (1997) found a possible SAO signal in the mean speeds of movement of mesocyclones for 1992: these were fastest in March, July and October, which are all extreme months in the oscillation (December and January were not studied). The attribution to the SAO seems reasonable given that mesocyclones tend to be superimposed upon the large-scale background flow, which itself exhibits a strong semi-annual signal over sub-Antarctic latitudes. Even when considering the southern oceans as a whole, the SAO dominates the *frequencies* of mesocyclones and frontal cyclones alike (Carleton, 1981c; Carleton and Song, 1997). Grouping mesocyclone frequencies according to their occurrence in either the transition-season months (March, April, October) of 1988 or the winter season (June, July, August: both 1988 and 1989) reveals more mesocyclones close to Antarctica (the 65°–70° S zone) in the transition season than in the winter (Figure 2.55). This pattern dominates despite the greater extent of the sea ice later in the season (i.e., the study months of June through October) or the large between-winter variability in mesocyclone frequencies for this region (Figure 2.47).

When considering only the *formation (genesis)* patterns of mesocyclones during the winter and early spring (June through September), a simpler pattern of equatorward movement suggests a stronger association with the advance of the MIZ over sub-Antarctic latitudes, rather than with the SAO. Figure 2.56 shows the spatial regimes of comma cloud systems for the months of April, August and October, 1988. These charts also may be compared with that for June 1988 (Figure 2.49a). They indicate a general equatorward, as well as eastward, movement of the mesocyclone points-of-origin and system tracks consistent with the advance of the sea ice during this period (see Carleton and Fitch, 1993; their fig. 10a). This pattern may again emphasize the importance of low-level baroclinicity induced by horizontal gradients of temperature across the

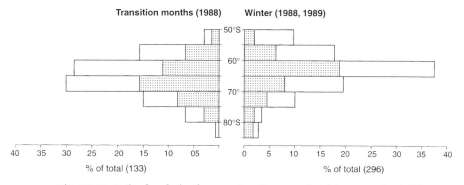

Figure 2.55. Latitude relative frequencies of comma cloud (open) and spiraliform (shaded) mesocyclones shown separately for the transition-season months of 1988 and winter months of 1988 and 1989.

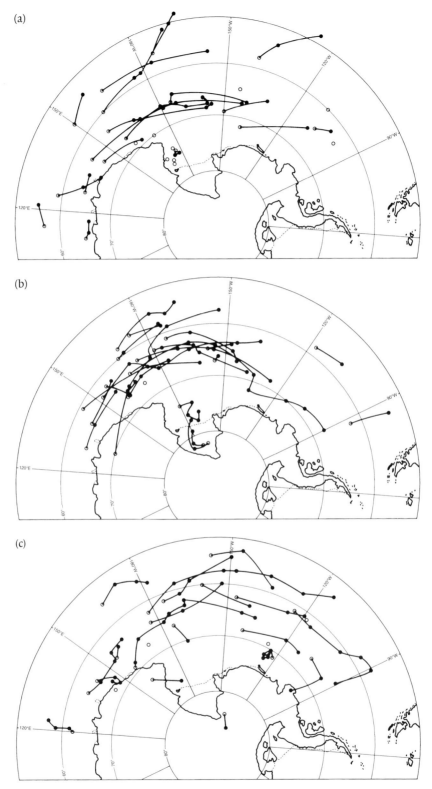

Figure 2.56. Similar to Figures 2.50 and 2.51, showing the comma cloud mesocyclones only for (a) April 1988, (b) August 1988, (c) October 1988.

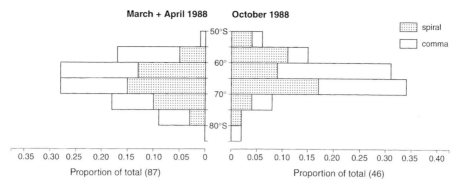

Figure 2.57. Similar to Figure 2.55, but shown separately for the autumn and spring transition-season months of 1988.

sea ice edge, for initiating many mesocyclone events. Separating out the latitude occurrences of mesocyclones according to dominant cloud signature, and comparing the distributions for March and April 1988 with those for October of the same year (Figure 2.57), suggests that the effect of the sea ice advance in these regions may be even more significant for the spiraliform vortices. These developed at somewhat lower latitudes, where they were also more frequent, in the spring transition-season month contrasted with the autumn; at least in 1988.

Interannual time scales

The seven-winter (1977–83) climatology of mesocyclones developed by Carleton and Carpenter (1990) showed marked interannual variations in the frequency and longitude of maximum occurrence of these systems over the southern oceans. This variability was related statistically to the eccentricity of wavenumber one, and also the ENSO. Thus, in a winter for which the seasonal cycle of the Tasman Sea trough was amplified (van Loon and Rogers, 1984), as during the developing El Niño of 1982, large numbers of mesocyclones were observed in the New Zealand area. A similar pattern was also observed during the 1991–92 El Niño event in this region (Carleton and Song, 1997). This contrasted with the situation in the same region for 1981 (i.e. year −1 of El Niño), when the annual cycle of the trough was suppressed, and the associated frequency of cold air outbreaks near New Zealand (and consequently, also, of mesocyclones) was reduced.

Interannual changes in the broad-scale atmospheric circulation associated with the teleconnections to ENSO over southern higher latitudes (Section 2.2.2), affect the frequency and spatial distribution of mesocyclones downstream of New Zealand in the Ross Sea sector. This change was clearly evident for the southern winters of 1988 and 1989 (Figure 2.47): considerably more

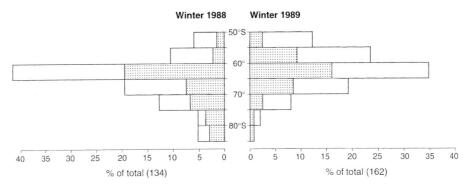

Figure 2.58. Similar to Figure 2.55, but shown separately for the winter-season months of 1988 and 1989.

mesocyclones occurred in 1989 contrasted with 1988. There were also latitudinal changes in the relative frequencies of mesocyclones for this region between the two winters (Figure 2.58), particularly over the sub-Antarctic where frequencies increased in winter 1989. The winter of 1989 was associated with a positive extreme in the Southern Oscillation Index (SOI), or La Niña event. This contrasted strongly with the winter of 1988, which followed a negative extreme in the SOI (El Niño) during 1987. Mesocyclones were more (less) frequent in longitudes of the Amundsen/Bellingshausen seas in 1989 (1988), in association with a strengthened (weakened) ASL in those winters (Carleton and Fitch, 1993). Cold air sweeping equatorwards east of the amplified ridge in the Ross Sea area in 1989 increased the frequency of mesocyclogenesis near the sea-ice edge and over the strong gradients of SST located north of the Amundsen and Bellingshausen Seas (Carleton and Song, 2000).

The role of differences in low-level temperature advection in these mesocyclone interannual differences for this sector is also evident at smaller scales. Figure 2.59 depicts the zonal, or east–west, gradient of the monthly mean departures of surface air temperature for two AWSs, Martha-2 and Manuela, that are located on the Ross Ice Shelf, for the winter ice-growth seasons of 1988 and 1989 (Carleton *et al.*, 1998). The gradient is positive (negative) for warm (cold) advection, occurring with 'thermal winds' directed from the north (south). These were more frequently northerly in 1988, when mesocyclone activity was also suppressed in longitudes of the ASL, but more consistently southerly in 1989, when mesocyclone frequencies increased. Additionally, the sea-ice extents in longitudes of the Ross Sea were reduced (greater) in 1988 (1989). Commensurate with the increases in mesocyclone activity in winter 1989 relative to winter 1988 were the SST anomaly patterns adjacent to the sea-ice edge at the end of each winter (Carleton and Song, 2000) (not shown). Areas of positive

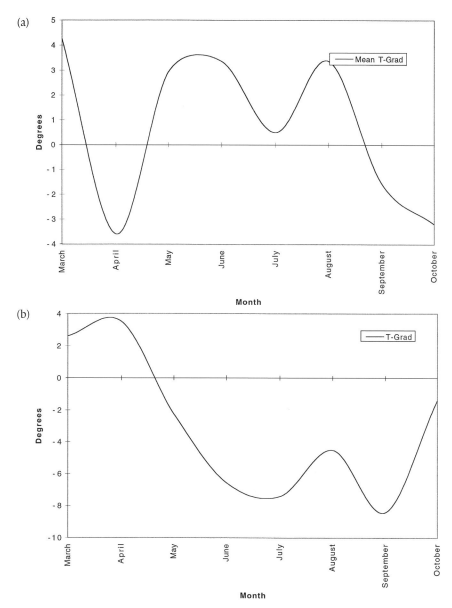

Figure 2.59. Smoothed time plots of differences in the mean daily departures of temperature from the 'long-term' station monthly means for AWSs 'Martha-2' and 'Manuela', for the ice-growth seasons of (a) 1988, and (b) 1989 (adapted from Carleton *et al.*, 1997).

(negative) SST anomaly in the Amundsen Sea area had developed by September 1988 (1989). These upper ocean anomalies are consistent with the contrasts in low-level thermal winds between years (Figure 2.59). The uptake of large amounts of heat from the upper ocean by the greater frequencies of southerly winds during winter 1989, is implicated in the increased mesocyclogenesis observed near the ice edge for that winter.

2.2.5 Conclusions and implications for mesocyclone processes

Studies of mesocyclones using satellite remotely-sensed and conventional surface and upper-air observations, as well as theoretical studies, suggest some similarities but also strong differences in the predominance of different mesocyclone types between polar regions, and even between higher latitude regions of the same hemisphere. For example, the greater frequency of deeply convective spiraliform 'polar lows' in the North Atlantic contrasts with the situation in the North Pacific, and particularly with that over the southern oceans (Carleton, 1996). This observation implies differences in the relative contributions of important physical processes to mesocyclogenesis (Businger and Reed, 1989b; Turner *et al.*, 1993a). The surface-atmosphere fluxes of heat, moisture and momentum associated with strong sea–air temperature gradients near the sea ice edge seem particularly important for the intensification of convective spiral systems (dominate in the Arctic), but are less important in the Antarctic where column stretching and the generation of cyclonic vorticity are involved in mesocyclogenesis close to the continent. Over the MIZ and ice-free ocean areas of the sub-Antarctic, upper-level baroclinicity in the form of a jet streak or short-wave trough becomes important, as evidenced by the high frequencies of comma clouds over a wide range of latitudes (Carleton *et al.*, 1995). When the upper-level forcing is superimposed on a lower-level baroclinic zone, such as a boundary layer front, occurring near the sea ice–ocean margin or at the edges of the ice sheet, the likelihood of mesocyclogenesis is also increased.

Confirmation of Antarctic mesocyclone processes is difficult in the absence of data having high spatial resolution and for several tropospheric levels. However, the application of satellite data in the microwave region to case studies using shipboard and aircraft observations, as well as AWS data, are helping to clarify these processes (Section 3.2). In addition, computer modelling of individual mesocyclones is illuminating the relative importance of sea–air interactions, baroclinicity and topography in the 'spin up' of Antarctic mesocyclones (Section 5.2).

Associated topics in the mesocyclone climatology of the Antarctic and sub-Antarctic that beg to be better explored include the following: a detailed satellite analysis of mesocyclones for additional sub-regions, particularly coastal

Adélie and Wilkes Lands; improved determination of the surface wind fields associated with Antarctic mesocyclones; a better understanding of the temporal and spatial variations of these mesocyclones, particularly as they may manifest the major modes of circulation variability, including possible decade-scale changes in the atmosphere and upper-ocean associated with the 'Antarctic Circumpolar Wave' (see White and Peterson, 1996). In all these research thrusts, satellite images in the traditional (visible and infra-red) wavelengths are likely to be increasingly supplemented by the data acquired by active and passive sensors in the microwave frequencies.

E. A. RASMUSSEN, J. TURNER,
K. NINOMIYA AND I. A. RENFREW

3

Observational studies

3.1 The Arctic

3.1.1 Introduction

For several decades observational investigations in the form of case studies have supplied an important part of the attempts to understand the structure and development of mesoscale vortices. Apart from obtaining a description of the individual cases, an underlying purpose has been, through a synthesis of the different cases, to gain sufficient knowledge to describe the basic properties of these systems, including their structure and dynamics. Present-day high resolution numerical models have proved to be very effective for simulating the structure and development of mesoscale systems, such as polar lows in data sparse regions, and case studies in the form of model simulations of polar low developments have yielded much important information about these systems. The results from these studies will be discussed separately in Chapter 5, but also, when relevant and where model studies have been coupled with observational investigations, in this chapter.

A very significant part of the polar low research over the last 30 years has been dedicated to the Nordic Seas (defined as the North Atlantic east of Greenland and north of 60°N, plus the North Sea, the Norwegian Sea, the Greenland and Barents Seas), which is a primary genesis region for polar lows. The following discussion will start therefore by presenting the results from research carried out in this region. This discussion will be followed by an overview of parallel work carried out in other parts of the Northern Hemisphere, including important results obtained by Japanese researchers. It should be noted though, that because of limitations of space it has been impossible to mention *all* studies of relevance that have been published over the last 20 years or more.

In a discussion of mesoscale features in polar air masses, Scorer (1986) noted that 'satellite pictures give a much better picture of the cause of this than any weather chart on which the patterns are smoothed and which contain an illusory simplicity'. This statement is very relevant for the study of mesoscale cyclones. Over the years satellite images have provided an invaluable tool for polar low studies and extensive use is made of these images throughout this chapter.

3.1.2 Early studies

The Harrold and Browning case

Harrold and Browning (1969) published the first major investigation of polar lows. Their work was based on data collected in connection with the passage of two polar lows over the United Kingdom as described by Stevenson (1968). Harrold and Browning's paper marked the starting point of a debate about the nature and structure of polar lows that went on for several years. According to them, the polar low was basically a baroclinic disturbance of short wavelength, i.e. the same type of system as the well-known extratropical cyclones seen on virtually all weather charts, except that the polar low was much smaller. In their work, they documented that at the time the polar lows crossed the UK the associated widespread precipitation was produced within a stream of air steadily ascending at about 10 cm s^{-1}, within, using a modern term, a 'warm conveyor belt'. The idea of the warm conveyor belt structure of the polar low during its passage of the British Isles was supported by the nature of the cloud field showing a characteristic comma-shape. Harrold and Browning hypothesized that polar lows developed within low-level baroclinic zones and that high-level disturbances might be a factor contributing to their development, although, according to them, it seemed unlikely that high-level pertubations were the primary cause of the low-level polar lows. This point of view was supported by Lyall (1972), who in a discussion of 'The polar low over England' concluded that 'For a long time, polar lows were thought to be purely convective in origin – possibly the merger of several shower systems. It is now clear that this supposition is untrue; polar lows are basically low-level baroclinic disturbances…'.

Reverse shear polar lows

Duncan (1978) in a small but influential paper investigated the structure and development of a 'polar air depression' that developed northeast and east of Scotland on 10 December 1976 in a 'baroclinic reverse shear flow' (see Section 4.2). Using a simple linear, quasi-geostrophic model, Duncan showed that under conditions similar to those that existed on 10 December, unstable baroclinic waves, resembling the observed 'polar air depression' in wavelength, phase speed and growth rate, might form. As will become clear from the following, a significant number of polar low developments in the Nordic Seas region are due to 'reverse shear baroclinic instability'.

The role of convection

Rasmussen (1977, 1979) and Økland (1977) challenged the point of view of Harrold and Browning and other British meteorologists by supporting the old idea that polar lows were basically driven by deep convection and that baroclinic instability played a minor role. Rasmussen and Økland made use of the so-called CISK (Conditional Instability of the Second Kind) theory developed by Charney and Eliasen (1964) and Ooyama (1964) according to which tropical cyclones grow through a positive feedback process between organized convection on the cloud scale and a balanced vortex of much larger horizontal scale, i.e. the tropical cyclone (or the polar low) (see Chapter 4). Results from a quasi-geostrophic model, including a simple CISK mechanism and using parameters observed during polar air outbreaks, showed somewhat surprisingly, that the CISK mechanism might work even over cold polar/Arctic waters. The sea surface temperatures here are far below the 26–27 °C found across the tropical seas over which tropical storms and hurricanes develop (Rasmussen, 1979). Rasmussen re-analysed the polar low case considered by Harrold and Browning and argued that the two polar lows did not form initially as waves on a baroclinic zone, but as lee-vortices south and southwest of Iceland[1]. Following the initial formation of the polar lows, the further development until they crossed southwestern England was, according to Rasmussen, mainly due to CISK. Polar lows have often been observed to change their structure during their lifetime as they move from one region to another acquiring different characteristics. It is plausible therefore that the systems initially grew due to CISK and later, while crossing Great Britain, acquired the characteristics of a shallow baroclinic disturbance, as observed by Harrold and Browning.

Another polar low development near the Norwegian coast and weather ship *Polar Front* (situated in the Norwegian Sea around 66° N, 2° E) on 13 October 1971 in which baroclinic instability seemed to have played a minor role was discussed in the same paper (Rasmussen, 1979). A radiosonde ascent from the weather ship, together with coastal measurements, indicated significant CAPE (Convective Available Potential Energy; see Section 4.5, Figures 4.26 and 4.27), and widespread, deep convection was observed within the region in which the polar low developed. The development, which resembled the type of developments envisaged by Dannevig (see Section 1.2), led to the formation of a

---

[1] Polar lows occasionally develop from vortices forming in the lee of major orographic barriers, such as Iceland or, more rarely, southern Norway.

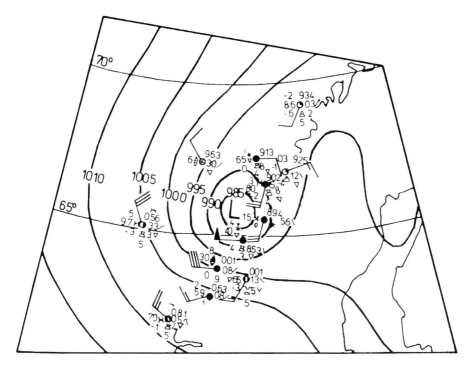

Figure 3.1. A surface analysis showing a polar low off the Norwegian coast at 0000 GMT 13 October 1971 (from Rasmussen, 1979).

symmetric, warm core vortex with a closed circulation with a diameter of a few hundred kilometres and mean surface winds reaching 25 m s^{-1} as illustrated on Figure 3.1. The low formed in a region close to three radiosonde stations allowing a detailed upper-air analysis to be carried out. This analysis for the first time documented that *some* polar lows had a vertically aligned structure similar to tropical systems with a warm core throughout the troposphere. The development of a small, but intense symmetric vortex within a non-frontal environment characterized by significant CAPE, led to the conclusion that the polar low development on 13 October 1971 was *a convective, warm core 'CISK-driven' system and as such the first analysed example of a polar low which in important ways resembled a tropical storm (hurricane)*. However, the October 1971 development, which took place at a relatively southerly position and in an environment characterized by significant CAPE, should not be considered typical for either the comma clouds discussed in the following section, nor for the type of polar lows that develop further north.

As discussed throughout this book, most polar lows, like the October 1971 case considered above, are associated with convection promoted by the relatively warm sea surfaces in the regions where the lows form, as well as the high humidity over the sea. If making 'landfall', i.e. being steered across a coastline, this type of polar low will invariably become situated over a cold, often snow/ice-covered surface. Therefore, upon making landfall the lowest layer above the surface will cool and dry out, causing the convection to weaken and die out with a resulting rapid decay of the low within a short period of time. A typical example of the rapid demise of a polar low was observed in connection with the 'most beautiful polar low in the world' which, upon crossing the Norwegian coastline near North Cape in February 1987, quickly disintegrated, losing its hurricane-like cloud structure and well-defined eye (see Section 5.1.2, 'Nordeng and Rasmussen's case of 26 February 1987').

3.1.3 Comma clouds

Reed (1979) studied cyclogenesis that occurred in polar airstreams poleward, but rather close to major frontal bands (see also Section 1.6.2). Reed, referring to British sources, called these systems 'polar lows'. Satellite images revealed that the systems investigated by Reed in their mature stage were characterized by a comma-shaped cloud pattern, giving rise to the name 'comma clouds'. Prior to Reed's work, Anderson *et al.* (1969) had pointed out that the comma-shaped cloud patterns revealed by satellite images from the North Pacific were found downstream of an upper-tropospheric vorticity maximum in the region of 500 hPa positive vorticity advection (PVA), a point of view supported by Reed.

In the cases studied by Reed, which were all from the Pacific region, the comma-shaped cloud patterns evolved from regions of enhanced convection. The cloud patterns, with a typical dimension of 1000 km in the direction normal to the polar front, were found in regions of appreciable, deep baroclinicity, located on the cold side of the baroclinic zone where they had an ample supply of potential energy available for their growth. The lapse rate was conditionally unstable through a considerable part of the lower troposphere resulting in a potential for deep convection. The commas extended from roughly the position of the jet axis (belonging to the main baroclinic zone) to the strong shear zone north of the jet stream. A surface trough was found under the trailing edge of the comma tail, and in more intense cases, a surface pressure centre under the comma head. The temperature at 500 hPa in the region of the cloud system ranged from about -30 to $-40\,°C$.

While Harrold and Browning ascribed the formation of polar lows to shallow baroclinicity, the systems studied by Reed were associated with baroclinicity of appreciable depth and the presence of an upper-tropospheric jet stream.

Mullen (1979, 1982, 1983) in a series of papers extended Reed's work in different ways. Based on 22 composite cases Mullen (1979) described the large-scale environment within which small-scale cyclones (comma clouds) developed over the wintertime North Pacific behind or poleward of major frontal bands. Based on the results of the composites, Mullen concluded that the oceanic polar air cyclones were baroclinic instability phenomena whose small scale related to the low values of the Richardson number near the surface and whose large upper-level amplitudes related to the effect of latent heat release on baroclinic development.

In his 1983 paper, Mullen discussed the important subject of the 'instant occlusion', a phenomenon closely related to polar lows. As first pointed out by Anderson *et al.* (1969), the small vortices in the polar air masses may excite waves on the main frontal band if they approach sufficiently close to the front. Such waves appear to occlude rapidly, jumping from an open wave to a fully occluded stage without going through the intermediate stages of development described by the Norwegian frontal model (see Figures 1.19 and 1.20). In forecast practice, this process was named 'instant occlusion'. As shown by Mullen, some polar air stream cyclones (polar lows) and frontal waves induced by the approach of the cold air comma cloud (i.e. the instant occlusion process) exhibit deepening rapid enough to meet the 'bomb-criterion' as defined by Sanders and Gyakum (1980).

Reed and Blier (1986a, b) continued Reed's 1979 work through studies of two comma cloud developments over the eastern Pacific. The results from the two companion papers to a large extent confirmed the results from Reed's 1979 paper. The forcing was not strong in comparison to typical mid-latitude storms, but the effect of relatively modest PVA aloft was enhanced by the relatively small static stabilities in the region. Latent heat release in cumulus convection was another likely significant factor for the growth. Low-level sensible heating was not an immediate cause of the development but was nevertheless important because of its effect of decreasing the vertical stability, allowing deep convection to proceed. In the early stages of development the thermal advection was weak and likely to be of little consequence. Simultaneously with the formation of the comma head out of a cluster of convective clouds, a surface low centre was observed to form. Figure 3.2 shows a schematic diagram of a comma cloud development according to Reed and Blier (1986a).

Both comma clouds studied by Reed and Blier started initially as regions of enhanced, organized convection that eventually evolved into systems which, in their mature stage, were mainly composed of stratiform clouds with only small regions of embedded convection. In their second paper, Reed and Blier considered especially the physical processes responsible for the organized growth of deep convection, which plays an important part in the genesis of comma clouds. They found that the rapid growth of the comma cloud and the associated appearance of a surface cyclone took place when the disturbance

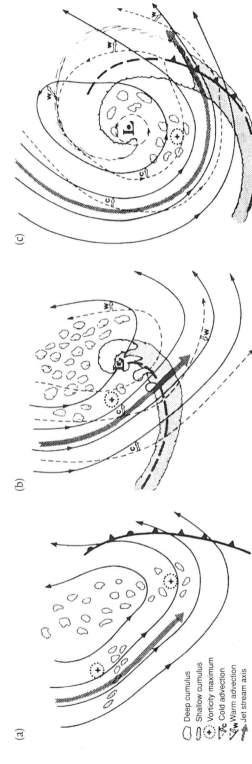

Figure 3.2. A schematic diagram of a comma cloud development. (a) Incipient stage, (b) intensifying stage, (c) mature stage. Solid lines are 500 hPa contours, broken lines are surface isobars. Other symbols are explained on the figure (from Reed and Blier, 1986a).

migrated over increasingly warm ocean water and they concluded that this southward migration was a principal factor for the long and sustained growth that characterized the comma cloud. A similar 'migration effect' is bound to also be significant for the convective type of polar low which forms at higher latitudes. Low-level warming and moistening produced by strong fluxes of heat and moisture from the underlying 'warm' ocean served to maintain and enhance an unstable environment in which the comma cloud could develop. However, a potential for deep convection alone is not sufficient to produce a comma cloud. An initiating or organizing element is also required. Reed and Blier suggested that this element was the advance of a pre-existing short-wave trough, or secondary vorticity centre, that either formed within the large-scale basic flow, or migrated into the region from upstream (concerning the role of an initiating element in the form of an upper-level, pre-existing short-wave trough organizing the convection, see also Zick, 1983). The two comma clouds studied by Reed and Blier both developed in north to northwesterly flows, whereas most comma clouds in the Pacific move in a more zonal direction. The question of whether there are fundamental differences in the development of southward and eastward moving systems, i.e. comma clouds, were answered with 'we think not'.

Comma clouds are often associated with rather weak surface systems in the form of a trough along the trailing edge of the comma tail or a weak surface low. Occasionally, however, comma clouds may develop strong surface circulations, i.e. polar lows. An example of a comma cloud over the North Atlantic, which developed into a strong surface polar low was shown on Figure 1.4.

3.1.4 The 'Polar Low Spectrum'

During the 1980s it was realized that polar lows appear in many forms, including ideal types of (almost) 'pure convective systems' and 'pure baroclinic systems'. This is not surprising considering that some polar lows form close to ice edges along the shallow baroclinic zones characterizing these regions. Some others form deep within nearly equivalent barotropic, but vertically unstable air masses, and some lows, further south, close to the main baroclinic zone. This means that a variety of forcing mechanisms are active, leading to a whole 'spectrum' of polar lows. Although the idea of a spectrum of polar lows has been generally accepted for a long time (the term 'polar low spectrum' was probably first introduced by Kerry Emanuel at the International Conference on Polar Lows in Oslo, 1986), no systematic study has so far been carried out to specify which types of polar low actually constitute this so-called spectrum. The published case studies, however interesting they may be, are of limited value in this respect. The fact that the cases have been selected in a highly subjective way reflecting the special interests of the respective authors and other factors as well, means that these cases do not necessarily represent the true population of polar lows.

Businger and Reed (1989a, b), noting that 'As yet no widely accepted method exists for classifying polar lows', differentiated between 'three elementary types of polar low development based on associated distinctive synoptic patterns': (1) the short-wave/jet-streak type, characterized by a secondary vorticity maxima and PVA aloft; deep, moderate baroclinicity, and modest surface fluxes; (2) the arctic-front type, associated with ice boundaries and characterized by shallow baroclinicity and strong surface fluxes; and (3) the cold low-type, characterized by weak baroclinicity, strong surface fluxes, and deep convection.

Rasmussen (unpublished data) carried out a pilot study in order to extend, and if possible improve, the classification of polar lows suggested by Businger and Reed. As the basis for this study he selected the population of polar lows defined by the cases considered by Wilhelmsen (1985). Wilhelmsen studied the occurrence of gale-producing polar lows near Norway in the period 1978–82, using mainly conventional meteorological data, such as synoptic surface charts and other surface observations in the Norwegian area (in the following we will refer to the Wilhelmsen cases (the 'Wilhelmsen file') in the form WF 1978–1, to indicate the first polar low in 1978 listed in her 1985 study, table 1). Only a few satellite images were used by Wilhelmsen and only polar lows that gave near gale force winds (15 m s^{-1}) or more at Norwegian meteorological stations or at weather ships near the Norwegian coast were included in her study. For the period 1978–82, 33 cases of gale-producing polar lows were identified and listed by Wilhelmsen. Apart from Wilhelmsen's table, the data for the study by Rasmussen were routine meteorological charts, although extensive use was made of satellite images available from the NERC Satellite Receiving Station, University of Dundee and other sources as well. Except for one of the cases for which satellite data were not available, it was possible to place each of the cases within the seven categories or 'types' shown in Table 3.1.

Like the Businger and Reed classification, the scheme defined by Table 3.1 has been based partly on distinctive synoptic patterns associated with the occurence of polar lows, and partly on the basic mechanism for their formation. The categories in Table 3.1 as such form a *pragmatic classification* system, based on a combination of synoptic features and physical considerations, for polar lows in the Nordic Seas. The scheme is fairly general so that practically all polar lows observed in the region can be placed within one of the groups. However, since polar low developments are strongly dependent on local climatological factors, the scheme may not be applicable for other genesis areas. It

Table 3.1. *Polar lows found in the Wilhelmsen file*

Group	Type of polar low	Number of cases
1	Reverse shear-systems	8
2	Trough systems	5
3	Boundary layer fronts	5
4	Cold lows (including merry-go-round systems)	9
5	Comma clouds	1
6	Baroclinic wave–forward shear	2
7	Orographic polar lows	2

should be noted that the Wilhelmsen file contains no examples of polar lows forming along occlusions. Such developments, which are quite frequently observed over the Nordic Seas, may often be classified as 'reverse shear systems'), or, alternatively, added as an extra group.

In the following account, the different groups will be considered and some characteristic examples, mostly from the Wilhelmsen file, will be discussed in some detail.

Group 1: Reverse shear systems

Six of the reverse shear cases (for a definition of a 'reverse shear baroclinic development' see Section 4.2) listed in Table 3.1 initially developed over the Barents Sea, the remaining two further west over the Norwegian Sea (see Figure 3.3). Inspection of the relevant upper-air charts indicates that all eight developments might have been triggered by upper-level disturbances, in the form of upper-level, cold, short-wave troughs. The observations available suggest that both baroclinic instability as well as deep convection contributed to the development of the lows.

Reverse shear type polar lows often develop in the region between Iceland and Spitsbergen (see Figure 1.23) within a low-level northerly to northeasterly flow on the western flank of a large-scale low centred further east. West of this flow, cold temperatures prevail over the Fram Strait and the pack ice region along the east coast of Greenland, while relatively warm air is present further east and southeast over the Norwegian Sea. Less well known, but documented by the several cases in the Wilhelmsen file, reverse shear baroclinic developments also seem to be quite frequent over the Barents Sea along the roughly west–east oriented ice edge near $75\,°$N. In the following we will illustrate two reverse shear developments over the Barents Sea and the northern part of the Norwegian Sea.

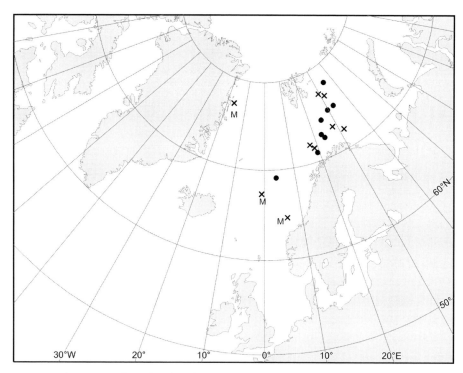

Figure 3.3. A chart showing the initial development locations of two major types of polar lows from the "Wilhelmsen file" (see Table 3.1). Reverse shear polar lows (group 1, denoted by dots); polar lows of the cold low type (group 4, denoted by crosses). The cold low type development far to the west near the Greenland coast, and the two developments far to the south over the southern part of the Norwegian Sea (marked M) were all associated with developments within cold core vortices formed as the result of an occlusion of a large-scale extratropical cyclone.

Case: 14–15 January 1981 (WF 1981–3)

The situation leading to the polar low formation in this case is illustrated by the synoptic surface chart from 0600 GMT 14 January 1981 (Figure 3.4) showing a broad northeasterly surface flow over the Norwegian Sea, the Greenland Sea and further south beyond Iceland. On Figure 3.5 is shown a corresponding satellite image from 0935 GMT. The thickness chart for the layer 1000–500 hPa (not shown) indicated that the direction of the thermal wind for this layer was opposite to the surface flow in the region of interest. A wave train formed by the two disturbances, one over the Norwegian Sea west of Lofoten near the position 70° N, 3° E and another just northeast of Iceland was situated within the northeasterly surface flow. The synoptic charts, as well as the structure of the cloud field, indicate that the disturbances formed due to reverse shear baroclinic instability.

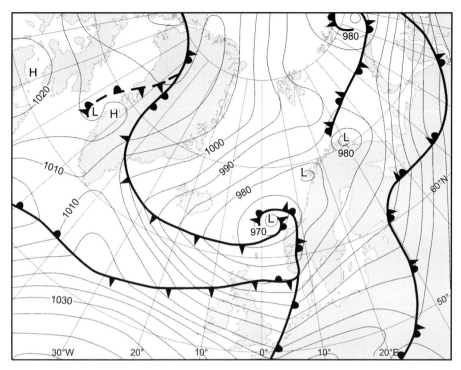

Figure 3.4. Synoptic surface analysis for 0600 GMT 14 January 1981 (based on the Berliner Wetterkarte, Freien Universität Berlin).

The wave west of Lofoten developed further, as illustrated by Figure 3.6 showing a satellite image of the system at 1745 GMT on the same day. The low-level vortex, as well as the cirrus outflow from the widespread deep convection is clearly seen.

Grønås *et al.* (1987a) used a limited area high resolution model to study a number of polar low developments in the Norwegian Sea. In all of their simulations baroclinic development took place in a reverse shear flow leading to a synoptic situation favourable to further development through CISK.

Case: 3–4 December 1981 (WF 1981–7)

Two of the cases studied by Grønås *et al.* were chosen from the Wilhelmsen file, including the development on 3–4 December 1981. During the morning of 4 December 1981 a reverse shear disturbance first observed on 3 December had developed into the strong vortex seen over the Norwegian Sea on Figure 3.7. The cloud pattern at this time resembled a swan. It had formed through a merging of the cloud systems belonging to the parent synoptic-scale cyclone and the reverse shear-generated polar low. Reverse shear developments in the region close to the Norwegian coast often take place

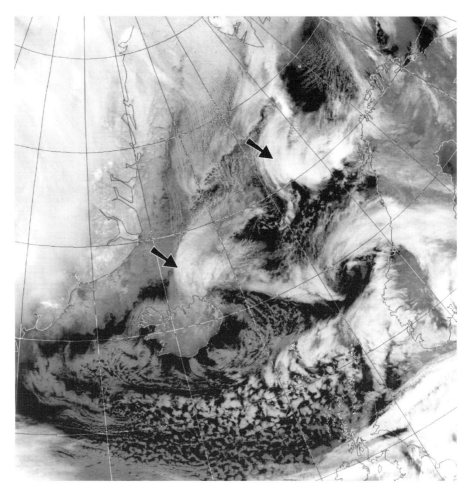

Figure 3.5. An infra-red satellite image for 0935 GMT 14 January 1981 showing a wave train of two baroclinic waves (indicated by arrows) that formed in a region of reverse shear northeast of Iceland. The clouds in the lower right corner were associated with the polar front. (Image courtesy of the NERC Satellite Receiving Station, University of Dundee.)

rather close to centres of major cyclones and 'swan-like' cloud signatures can frequently be observed (see Figure 5.13).

This reverse shear development was also characterized by deep convection, as seen from the satellite image shown on Figure 3.7. The cloud system here also resembles an instant occlusion. However, the events resulting in the formation of the 'swan' were different from those normally associated with the formation of instant occlusions.

Reverse shear systems forming further west, closer to Iceland and the Greenland coast, seldom form swan-like cloud signatures but quite often

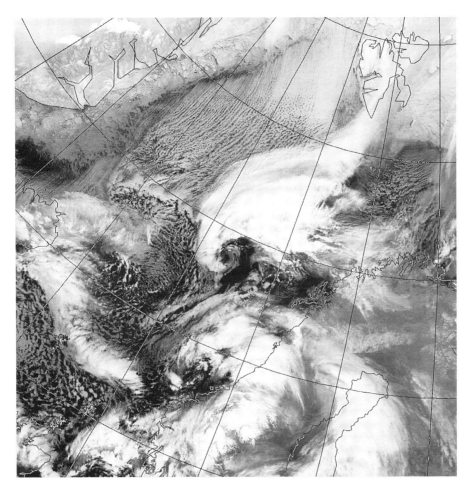

Figure 3.6. An infra-red satellite image for 1745 GMT 14 January 1981 showing a polar low over the Norwegian Sea west of Lofoten. The surface centre near 69° N, 3° E is indicated by an 'eye' surrounded by a mixture of high- and low-level clouds. (Image courtesy of the NERC Satellite Receiving Station, University of Dundee.)

impressive wave trains. For examples of this kind of reverse shear system, see Figure 1.23.

Group 2: Trough systems

During the winter, a number of large-scale cyclones follow a primary cyclone track over the northern part of the Norwegian Sea into the Barents Sea (see Figure 2.11). In this situation polar lows may form in east–west oriented 'trailing troughs' characterized by large values of low-level vorticity on the western flank of the synoptic-scale lows. Polar lows forming in this synoptic situation have been placed in the group 'trough systems'. These polar lows

164   3 Observational studies

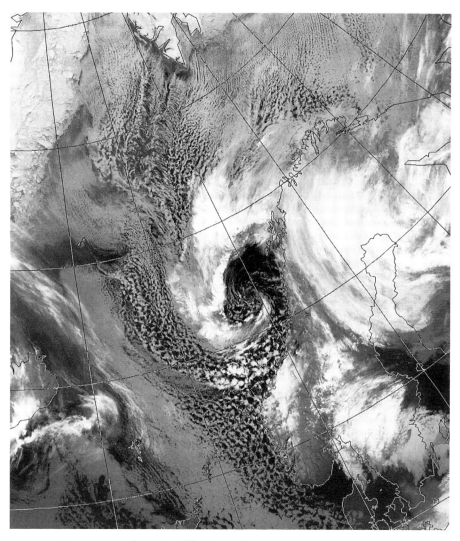

Figure 3.7. An infra-red satellite image for 0841 GMT 4 December 1981 showing a polar low over the Norwegian Sea southwest of Lofoten. The cloud pattern, formed by the clouds associated with the polar low and the large cloud shield further east belonging to a baroclinic wave on the polar front, resembling a swan, is typical of many reverse shear developments in this particular region. (Image courtesy of the NERC Satellite Receiving Station, University of Dundee.)

are found in more or less the same region as the reverse shear-generated polar lows discussed above.

The upper-air charts indicate that polar lows that form within synoptic-scale surface troughs are generally triggered by upper-level, short-wave cold troughs. With the approach of the upper-level short-wave trough (or jet streak), the low-level vorticity will, owing to the general ascent associated with the upper-level PVA, be increased locally (the 'spin-up-effect') causing a low-level vortex to form. At the same time, deep convection will be promoted due to the combined effects of general ascent within the region and destabilization via upper-level cold advection. Once formed, such lows may develop further due to the effect of released latent heat, as discussed in detail in Chapter 4.

An example of a polar low with a pronounced spiral structure that formed within a trailing trough over the Norwegian/Barents Sea behind a large, synoptic-scale cyclone was shown on Figure 1.8 in Section 1.6, Cloud signatures.

Polar lows developing within 'trailing troughs' north of 70° N constitute an important group of polar lows, but such systems may form within synoptic-scale troughs in other regions as well, provided the right conditions for their formation are present. An example of such a development on 13 October 1971 near the Norwegian coast was briefly discussed earlier.

On the synoptic scale, vorticity is maximized along trough axes, which for this reason may be considered as regions especially prone to polar low developments. An example of 'multiple mesocyclones' forming a wave train within a sharp synoptic-scale surface trough, such as illustrated on Figure 3.8, is shown

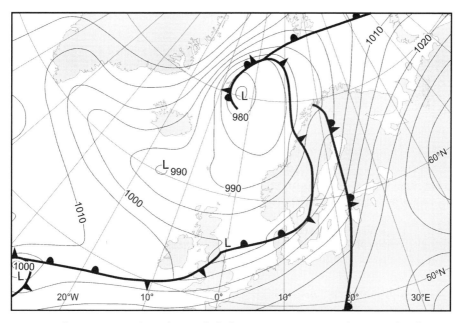

Figure 3.8. Synoptic surface analysis for 1200 GMT 18 November 1981 showing a sharp synoptic-scale trough over the North Atlantic along 60° N between South Greenland and the sea north of Scotland (based on the European Meteorological Bulletin, Deutschen Wetterdienstes).

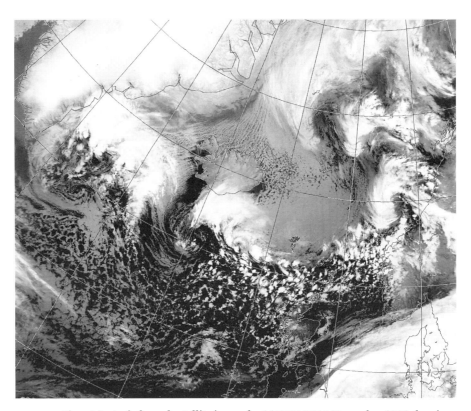

Figure 3.9. An infra-red satellite image for 1423 GMT 18 November 1981 showing four mesoscale vortices along the trough axis seen on Figure 3.8. (Image courtesy of the NERC Satellite Receiving Station, University of Dundee.)

on Figure 3.9. The sharp trough on Figure 3.8 over the sea along 60° N, from southern Greenland to southern Norway, represents a region of strong confluence between a northerly flow of Arctic air from the Greenland Sea and a westerly flow of a warmer, maritime air mass. As illustrated by Figure 3.9, showing a satellite image from approximately the same time as the surface chart on Figure 3.8, several mesoscale vortices had formed along the trough axis. The dynamic mechanisms involved in the formation of the vortices are not known, but judging from the synoptic situation in general and the structure of the vortices as seen on the satellite imagery, baroclinic instability most likely played a significant role.

Ralph *et al.* (1994), noting that 'observations suggest that multiple mesocyclones may not be uncommon components of synoptic-scale maritime cyclones', distinguished between two major classes:

1. mesocyclones that formed due to 'frontal instability' and characterized by *quasi-simultaneous growth* of more than one cyclone, and

2 mesocyclones that formed due to 'downstream development', which requires the presence of an initial cyclone, followed by the *sequential development* of new cyclones.

In the case discussed above, satellite imagery indicated that the mesocyclones formed more or less simultaneously, i.e. due to frontal instability (see also 'Polar lows along occlusions', later in this section).

Occasionally polar lows may develop along minor mesoscale troughs within Arctic air masses, away from the main (synoptic-scale) trough axis. A striking development of this type took place in January 1982 (see below, Group 4: Cold lows), resulting in an unusually strong polar low.

Group 3: Boundary layer front type

Numerous satellite images such as Figure 1.8 and others used throughout this book have documented the frequent occurrence of boundary layer fronts (BLFs) along and south of the west coast of Svalbard in situations with a northerly large-scale flow. BLFs with their origin near the ice edges separate shallow, low-level Arctic air masses from warmer, maritime polar air masses over the sea and as such are examples of Arctic fronts[2]. The three-dimensional structure of these shallow fronts was studied by Shapiro and Fedor (1989) and Shapiro *et al.* (1989) as part of the Arctic Cyclone Expedition. A numerical study of the formation of one of the BLFs observed during the Arctic Cyclone Expedition was carried out by Thompson and Burk (1991).

The orientation of the cloud streets as seen on satellite images often indicates a strong, low-level horizontal wind shear across the BLF which may have a cross-front length scale of only a few tens of kilometres. Numerous small-scale vortices can be seen along these fronts. The individual vortices generally have a short lifetime and remain small-scale, rather insignificant systems. Occasionally, however, when influenced by upper-level forcing, generally in the form of a short-wave trough, such a vortex may intensify and eventually develop into a polar low.

Similar systems have been observed at many other places, including the coastal region west of Novaya Zemlya and in the vicinity of the Japan islands. The role of baroclinic instability for developments along these BLFs is still an open question. Nagata (1993) pointed out, that when a large amount of vorticity is concentrated within narrow shear zones across a width of tens of kilometres, the development of relatively small meso-β vortices may be

---

[2] The term Arctic front has sometimes been used quite generally to designate any front situated poleward of the polar front (the main baroclinic zone), or, sometimes more specifically, to designate a front between unmodified polar/Arctic air originating over an ice/snow-covered surface and another, warmer air mass. In the present context this 'warm air mass' will nearly always be of maritime origin.

expected to be mainly the result of *barotropic instability* (see Section 4.3). The importance of BLFs for polar low formation was noted by Fett (1989a, b) who pointed out that BLFs are likely to be important factors in polar low evolution in the northern Greenland, Norwegian and Barents Sea regions, and that 'vortex generation is an integral part in the life cycle of a BLF'.

Example: 1–2 March 1981 (WF 1981–4)

On 1 March 1981 a generally northerly flow prevailed over the Greenland Sea and further east. A weak north-south oriented trough was situated along the west coast of Svalbard and a satellite image (not shown) indicated that a BLF had formed within the trough. Only small, insignificant vortices could be seen at this time. The following morning, 2 March, another satellite image showed a number of more developed vortices along the BLF. The largest vortex was situated at the southernmost extension of the BLF (Figure 3.10).

The upper-level wind at 500 hPa around the time of formation of the vortices seen on Figure 3.10 was unusually strong, exceeding 50 m s^{-1} at Bear Island. The formation of the major vortex northwest of the Norwegian coast was most likely triggered by forcing associated with an approaching upper-level trough

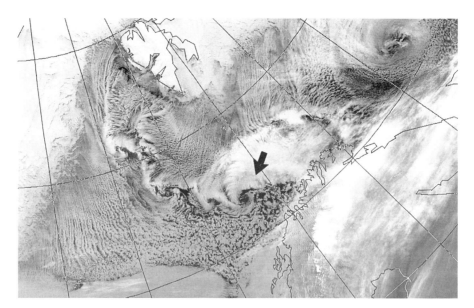

Figure 3.10. An infra-red satellite image for 0844 GMT 2 March 1981 showing a polar low (indicated by an arrow) under development at the southern end of a Spitsbergen boundary layer front (BLF) along which several other vortices have formed. The BLF which originally formed close to the ice edge west of Spitsbergen had, at the time of the picture, become detached from its place of origin and was drifting away in a southwesterly direction. (Image courtesy of the NERC Satellite Receiving Station, University of Dundee.)

west of Svalbard. Later on 2 March the vortex developed further into a strong polar low near the Norwegian coast.

The event illustrates how a BLF, which initially formed close to the Spitsbergen coastline may become detached from the coast and move out over open water, driven by the synoptic-scale flow. The leading edge, i.e. an Arctic front, in these cases can often be seen on satellite images as a cloud band along which vortices have formed, as illustrated by Figure 3.10. Occasionally, when the vortices are very small, the cloud band has the character of a 'rope cloud' (Figure 3.11). These cold air outbreaks of shallow Arctic air masses (termed 'surges' by Fett, 1989a), resemble gravity currents characterized by an ageostrophic flow which implies that the leading edge of the surge (the BLF) is not necessarily associated with a conventional frontal trough. Being shallow systems they are not normally reflected in the 1000–500 hPa thickness field. Although seldom analysed on the conventional synoptic charts, such BLFs nevertheless can quite often be followed on satellite images for a substantial time. After they have moved away from their place of origin, many BLFs disappear within a relatively short time without undergoing any significant development while others can be traced for days and act as seats of significant polar low developments. An example of this is illustrated in Figures. 3.11 to 3.13 and described below.

Example: 1–2 March 1980 (WF 1980–2)

On 29 February 1980 a satellite image from 0852 GMT (not shown) indicated a typical Spitsbergen BLF over the sea from west of Spitsbergen down to North Cape along which several minor vortices had formed. One day later, on 1 March, the northern part of the front could still be seen west and southwest of Svalbard, while the southern part had moved southwest leading an Arctic air outbreak (Figure 3.11). A disturbance discernible the day before as a small hook of cloud had developed further forming the vortex on the western flank of the BLF seen on Figure 3.11.

Within the next 24 h this vortex developed into the polar low seen on Figure 3.12 at the Norwegian coast near Ålesund. Strong winds exceeding gale force were reported within the region of enhanced surface pressure gradient on the western flank of the low. The synoptic situation at this time is illustrated by Figure 3.13, which shows the synoptic surface chart for 1200 GMT 2 March 1980. The situation, with the polar low (indicated by a cross) situated close to the coast on the western flank of a large synoptic-scale low centred over Finland, is quite typical for many polar low developments over the Norwegian Sea.

170   3 Observational studies

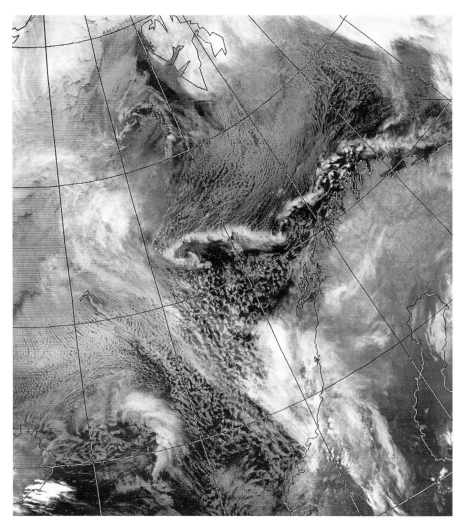

Figure 3.11. An infra-red satellite image for 0831 GMT 1 March 1980 showing a BLF along *c.* 71° N, detached from its region of origin. During the following day the small vortex seen on the western flank of the BLF developed into the strong polar low shown on Figure 3.12. (Image courtesy of the NERC Satellite Receiving Station, University of Dundee.)

The development illustrated by Figures 3.11 to 3.13 resembles in a striking way a reverse shear baroclinic development from February/March 1984 studied by Rabbe (1987) and by Grønås *et al.* (1987a).

Only a few studies have been published about the nature and structure of Arctic fronts. Shapiro and Fedor (1989), in a study for which a number of dropsonde data were available, discussed the structure of a frontal zone south of Spitsbergen during the Norwegian Polar Lows Project on 14 February 1989.

3.1 The Arctic   171

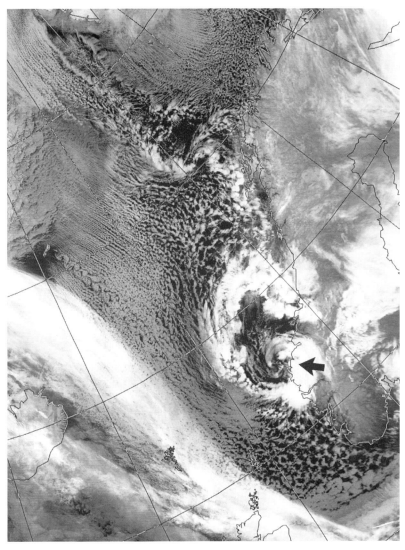

Figure 3.12. An infra-red satellite image for 0949 GMT 2 March 1980 showing a polar low (indicated by an arrow) near the Norwegian coast near Ålesund. (Image courtesy of the NERC Satellite Receiving Station, University of Dundee.)

However, during this period the main baroclinic zone with strong frontal activity was situated at a very northerly position, including the northern part of the Norwegian Sea and the Barents Sea. The rather deep frontal structure found by Shapiro and Fedor for that occasion (see their fig. 4) therefore cannot be taken as representing the structure of a local ice-edge generated BLF.

Another study by Shapiro *et al.* (1989) of a BLF within the same region around Spitsbergen a few days later on 18 February 1984, showed a more

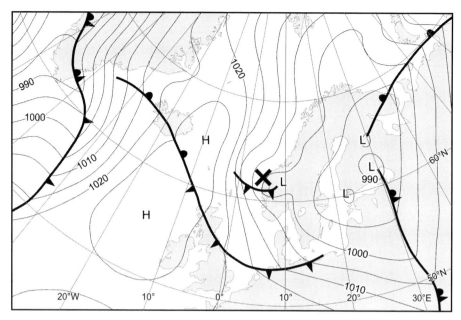

Figure 3.13. Synoptic surface analysis for 1200 GMT 2 March 1980 with the location of the polar low (shown on Figure 3.12) at 0949 GMT near Ålesund, Norway indicated by a cross (based on the European Meteorological Bulletin, Deutschen Wetterdienstes).

shallow baroclinic zone only a few hundred metres deep, capped by a rather deep and very stable layer (Figure 3.14). The figure shows the Arctic front as a tilting continuation of the Arctic inversion down to the surface south of the ice edge. The horizontal temperature gradient associated with the Arctic front was strong but the baroclinic layer was, as evident from Figure 3.14, very shallow and capped by a rather deep layer of very stable Arctic air. When ice-edge BLFs are driven out to sea by changes in the synoptic flow, some of them may, as documented by the example from 1–2 March 1980, maintain their identity in spite of the modifying effect of the diabatic heating by the warm underlying sea (Shapiro and Fedor, 1989).

Group 4: Cold low types

Small spiral, or occasionally comma-shaped cloud patterns of convective character are sometimes observed to flare up within the inner cores of old occlusions or cold lows, without any obvious association with upper-level short waves or low-level baroclinic features (Businger and Reed, 1989a, b). This type of polar low, called a 'cold low type', constitutes the largest of the seven groups (group 4) within Table 3.1, and was among the three elementary types considered by Businger and Reed in their 1989 classification. The large-scale cold lows

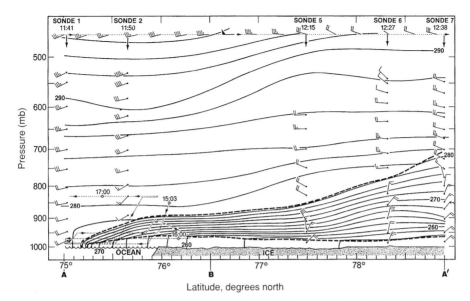

Figure 3.14. A cross-section analysis of potential temperature (K, thin solid lines) for an Arctic front south of Spitsbergen on 18 February 1984 (after Shapiro *et al*, 1989a, in Twitchell *et al.*, 1989. © A. Deepak Publishing).

within which the polar lows form have, as implied by their name, a cold core, because of which the intensity of their circulation increases with height. The cold low type polar lows are often accompanied by impressive cloud spirals (see Figures 1.15a and 1.16), and can as such be identified with the type of polar lows that occasionally (and especially in the early phases of polar low research in the 1970–1980s) were called 'real' or 'true' polar lows in order to distinguish them from the comma cloud type.

Several polar low developments have been noted to occur in the Arctic when an upper cold low moves out over an open body of water. Once over water, the low-level circulation may increase, causing a vortex and eventually a polar low to develop (for a more detailed discussion of the means of formation of the polar lows see the following case studies).

The origin of the cold core lows found within the circumpolar vortex have been ascribed to synoptic-scale baroclinic waves, most of which have occluded before they reach the high Arctic. After reaching the Arctic, these non-frontal systems move in the 'sluggish cold core of the vortex' (Sater *et al.*, 1971).

The initial positions of development of the cold low type polar lows found in the Wilhelmsen file are shown on Figure 3.3 (marked by crosses). The chart shows a group of six cases at high latitudes over the easternmost parts of the Norwegian Sea and the Barents Sea. These lows formed within cold lows originating within the circumpolar planetary vortex. The remaining three cases (marked M), two of which developed at fairly southerly latitudes, were associated with so-called 'merry-go-round systems' (see Section 1.6.4), formed through the occlusion process of large-scale baroclinic waves.

A few polar lows within this category have been observed to form during the *warm season* whereas practically all other polar low developments (with a few exceptions) are confined to the winter (see Section 2.1). Actually a case chosen by Fett (1989a) to illustrate this type of development was from early June 1985. In this case the upper-level cold low, as well as a surface low, were evident near the North Pole, three days prior to the polar low formation. As the upper cold low and the accompanying surface low moved southwards from the sea ice to a position over the Barents Sea, a small-scale polar low developed within the parent circulation of the cold low system. However, most upper-level cold lows are *not* accompanied by a noteworthy surface circulation as long as they remain over snow/ice-covered regions because of the presence of an extremely stable boundary layer. Only after moving from the snow/ice-covered regions out over the open ocean, the upper-level vortices may penetrate to the surface as the Rossby penetration depth increases over the sea owing to the destabilization of the boundary layer (see Chapter 4). Quite often low-level vortices formed in this way become quite strong and develop into polar lows. It is noteworthy that this type of development takes place within the central part of the upper-level cold core low. Here, as pointed out by Businger and Reed (1989a, b), the lowest temperatures are generally observed, while the dynamic forcing associated with upper-level PVA may be quite small or non-existent.

The low upper-level temperatures associated with cold core vortices are conducive to deep convection leading to the release of latent heat. However, since the heating is occurring in anomalously cold air (the heating being negatively correlated with temperature) the 'large-scale' cold core vortex will, in such situations, decay with time. Generally a low-level vortex must, once formed, achieve a warm core in order to intensify into a strong polar low. The dynamics of polar lows within a large-scale, cold-core environment will be considered in detail in Section 4.5.

Example: 14–18 October 1980 (WF 1980–5)

In this case the centre of an upper-level cold low situated just north of Spitsbergen early on 14 October 1980, was moving slowly south. A satellite image for 1344 GMT 14 October (not shown) indicated that the vortex was accompanied by high-level stratified clouds. At the surface the vortex was discernible only as a weak trough over the snow/ice-covered region on the eastern flank of a large high centred over northeast Greenland. On the following day,

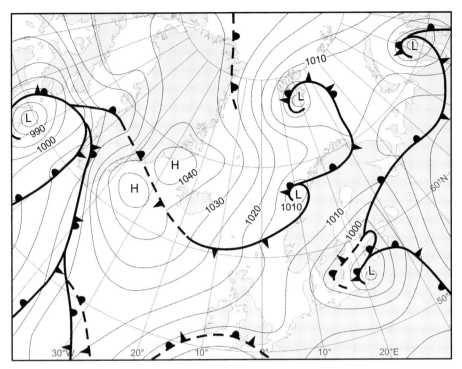

Figure 3.15. Synoptic surface analysis for 0600 GMT 15 October 1980 showing a polar low (indicated by L) near Spitsbergen. Note that an occluded front was erroneously drawn into the polar low (based on the Berliner Wetterkarte, Freien Universität Berlin).

15 October, when the upper-level cold low was situated south of Spitsbergen, a surface low had formed (see Figure 3.15) directly beneath the upper-level centre and accompanied by the tight cloud spiral shown on Figure 3.16. The cyclonic circulation at this time stretched from the surface through the whole troposphere, and even into the stratosphere. Within the troposphere the vortex, still having a cold core, intensified with height. The clouds around the centre of the cold core low were mainly convective, converging into the central region of the low forming the cloud spiral. Further east, on the flank of the cold core vortex, parts of a comma cloud could be seen. In contrast to the clouds within the central region of the vortex, the cloud deck belonging to the comma further east seemed more stratified, most likely caused by stable ascent due to differential vorticity advection associated with an upper-level trough on the eastern flank of the vortex. No significant surface system accompanied the eastern comma. After its formation, the surface low drifted southwards without any noteworthy intensification, until on the following day it formed an instant occlusion with a wave on a frontal zone further east. Still later, on

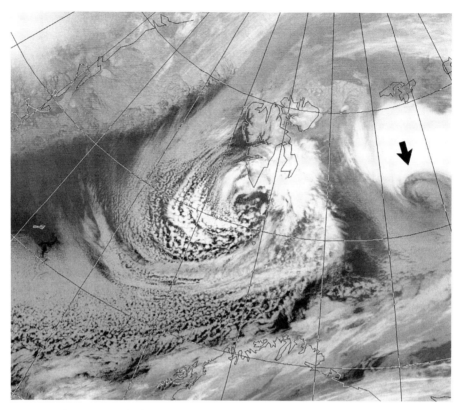

Figure 3.16. An infra-red satellite image for 1333 GMT 15 October 1980 showing the cloud spiral associated with a polar low that formed within the central part of a cold core vortex centred near southern Spitsbergen. The position of the polar low at 0600 GMT was shown on Figure 3.15. Further to the east, parts of a comma cloud (indicated by an arrow) are seen. (Image courtesy of the NERC Satellite Receiving Station, University of Dundee.)

18 October a new polar low formed within the central region of the circulation associated with the instant occlusion. The newly formed low passed the southern part of Novaya Zemlya between 18 and 19 October and finally decayed over the Kara Sea. The combined lifetime of the two polar lows, the original system and the rejuvenated low within the instant occlusion circulation was unusually long at about five days. The satellite images during most of the period documented the existence of a surface circulation characterized by bands of rather shallow convective cloud spiralling towards the centre of the vortex. The fact that deep convection was only sporadically observed is in accordance with the fact that the polar low, as judged from the upper-level charts, failed to develop any significant warm core and that the surface circulation did not intensify.

Another example of an unusually strongly developed low of the cold low type, an 'extratropical winter hurricane' (Rasmussen, 1981), well documented by numerous satellite images, is discussed in the following.

Example: 25–27 January 1982 (WF 1982–1)

The genesis of this polar low development can be fixed precisely from satellite imagery to the morning of 26 January 1982. On the day before, an upper-level, cold core low moved south from the region north of Svalbard over the sea around Bear Island. The cloud signature of the upper-level low, as shown on Figure 1.12, is characteristic for this kind of low when situated over an snow/ice-covered surface.

Upon crossing the sea south and southeast of Svalbard the low-level circulation associated with the cold low intensified, resulting in the formation of a synoptic-scale low. On the surface chart from 1200 GMT 26 January (not shown) the centre of this low was seen northwest of the Kanin Peninsula (near 70° N, 40° E) with a trough trailing in a westerly direction along the Norwegian coast and further west. The 500 hPa chart showed that one of the centres of the circumpolar vortex was situated between Svalbard and Novaya Zemlya from which a minor trough was stretching south over the Barents Sea roughly along 35° E (Figure 3.17).

The situation on the morning of 26 January, prior to the formation of the polar low, is illustrated by the satellite image shown on Figure 3.18. A shear line, emanating from the centre of the large-scale low and separating a northerly flow of unmodified Arctic air and a more easterly flow of modified, less cold air, had started to 'roll up' causing a number of minor vortices to form. In the following hours the incipient vortex marked 'V' at the northern part of the shear line developed into a hook of deep convective clouds and subsequently into a major polar low as illustrated on Figures 3.19 and 3.20 showing respectively the surface synoptic chart from 1200 GMT 27 January and a corresponding satellite image. Note that already at 0000 GMT 26 January (Figure 3.17) the upper-level trough axis was situated *east* of the location of the surface development, rendering any effect of upper-level PVA on the development unlikely.

The synoptic chart (Figure 3.19) shows that an impressive polar low had formed northwest of North Cape. A ship observation close to the centre of the system indicated that the surface pressure at this time was close to or even below 960 hPa. The low central pressure was later confirmed by a number of coastal observations. A satellite image from the same time as the surface chart (Figure 3.20), showed a well-developed cloud spiral characteristic of this kind of a 'real' polar low. The structure of the cloud field showed how a tongue of

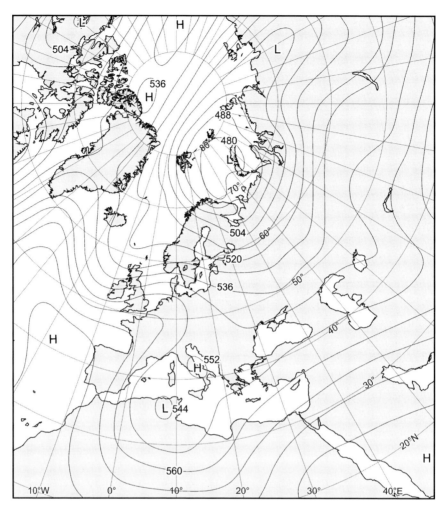

Figure 3.17. The 500 hPa contour chart for 0000 GMT 26 January 1982, showing one of the centres of the circumpolar vortex situated between Svalbard and Novaya Zemlya. A minor trough below which the polar low formed is seen emanating from the centre in a southerly direction towards the Kola Peninsula (based on the European Meteorological Bulletin, Deutscher Wetterdienst).

cold air spiralled into the central part of the vortex, secluding a region of warm air. Within the shallow, fresh Arctic air mass outbreak to the west, numerous low-level cloud streets could be seen. The satellite image illustrates how deep convection was triggered in regions where relatively warm and moist, modified maritime air was forced to ascend over a shallow wedge of cold, unmodified Arctic air. Away from this region of forced ascent little deep convection was seen apart from close to the coastal region where air was forced to ascend along the coastal mountains.

Figure 3.18. An infra-red satellite image for 0814 GMT 26 January 1982 showing a shear line (indicated by arrows) within a synoptic-scale trough behind the low (marked L) centred over the Barents Sea. Along the shear line a 'rope cloud' had formed along with several minor vortices can be seen including the disturbance marked by V which subsequently developed into the polar low shown on Figures 3.19 and 3.20.

An intriguing feature with this polar low was its unusual strength. While most well-developed polar lows are characterized by surface pressure perturbations of the order of 10 hPa, the 27 January 1982 case was associated with a pressure perturbation of no less than $c.$ 25 hPa. The unusually strong pressure perturbation characterizing this polar low was associated with a pronounced cold air outbreak causing extremely low tropospheric temperatures within the region of interest. The vertical structure of the atmosphere within which the polar low formed is illustrated by Figure 3.21, showing the radiosonde ascent from nearby Bear Island at 1200 GMT 26 January 1982, i.e. at the time when the hook cloud was first observed. Temperatures along the coast and from the ship

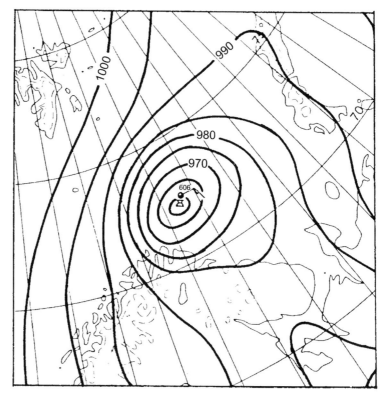

Figure 3.19. Schematic synoptic surface analysis for 1200 GMT January 27 1982 showing an 'extratropical winter hurricane' near North Cape (from Rasmussen, 1983).

situated near the centre of the polar low indicated that air parcels ascending in convective clouds within the central region of the low, would have approximately followed a moist adiabat, as indicated on Figure 3.21. Provided that the convection was vigorous enough and lasted for a sufficient time, the central core of the polar low would have ultimately reached a temperature more or less corresponding to this moist adiabat after which the convection would have become less vigorous and the deepening would have stopped. Using Eqn. (3.1) (see Rasmussen and Zick, 1987),

$$\delta p_0 = -gHp_0(\delta \overline{T})/(R\,\overline{T^2}) \tag{3.1}$$

the surface pressure decrement δp_0, due to the formation of a warm core corresponding to a mean temperature increment $(\delta \overline{T})$ within the layer of depth H, can be found (the other symbols have their usual meaning and the overbar denotes the mean value within the layer of depth H). Assuming that the cold air within the core was heated to a temperature corresponding to that of the moist

Figure 3.20. An infra-red satellite image for 1233 GMT 27 January 1982 showing a cloud spiral of mainly convective clouds associated with a fully developed polar low centred northwest of North Cape. Patches of high level cirrus indicate an anticyclonic outflow. (Image courtesy of the NERC Satellite Receiving Station, University of Dundee.)

adiabat, we find $\delta p_0 \simeq 30$ hPa, a value not far from the one actually observed. Note that the character of the radiosonde ascent shown in Figure 3.21 makes it very unlikely that significant CAPE would have been created due to heating of the layers adjacent to the 'warm' sea surface.

Immediately after landfall on the Norwegian coast, late on 27 January, the polar low started to fill and the impressive cloud spiral disintegrated into two.

The sequence of satellite images suggest that the polar low started as a barotropically-generated vortex along a zone of high horizontal wind shear. Upon formation, deep convection was triggered as warm air was forced to

182 3 Observational studies

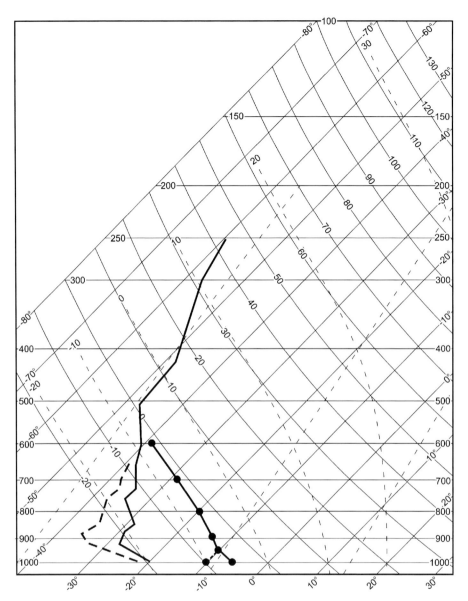

Figure 3.21. The radiosonde ascent from Bear Island taken at 1200 GMT 26 January 1982. The thick black line with dots shows the temperature traced out by an ascending parcel with surface properties as observed from a ship near the centre of the low.

ascend along the shallow wedge of fresh Arctic air. This type of forcing is unique to Arctic regions and may be a key factor in explaining the problem of why these disturbances develop much more frequently at high latitudes near the ice edges than at more southerly locations. The development discussed in the preceding and other cases as well (see the discussion of the Bear Island polar low development in December 1982 later in this chapter) indicate that this type of forced ascent due to shallow Arctic air masses being drawn into the circulation of a developing low, may be more effective in promoting deep convection than the frictionally-induced inflow traditionally considered in 'classical' CISK models. In order to distinguish between the two types of CISK, the latter may be designated *Arctic CISK*.

Polar lows of the cold-low type often develop within cold core vortices which have formed as the result of the occlusion of a large-scale extratropical cyclone. In these cases, where the occluded low may or may not be situated far north, a pronounced surface cyclonic circulation will generally be present prior to the formation of the polar low. An example of this is the formation of 'The most beautiful polar low in the World', discussed in Section 5.1.

Another and quite typical example of a polar low development within the core of an occluded cyclone was discussed by Rasmussen and Aakjær (1992). The satellite image in Figure 3.22 shows the polar low off the Norwegian coast (indicated by an arrow), south of a band of stratified clouds associated with the parent occluded cyclone within which the polar low formed. The 1000–500 hPa thickness chart for 1200 GMT on the following day, 13 March (not shown), at a time when the polar low was situated near Bergen, shows that the polar low had developed a warm core.

The formation of a polar low within the parent circulation of a synoptic-scale occluded cyclone generally leads to an intensification of the low-level cyclonic flow at a time when it, according to the Norwegian cyclone model, should be weakening. This revival of occluding cyclones was already noted and discussed by Bergeron (1954). In a review, 'The problem of tropical hurricanes', Bergeron included a discussion of what he called 'extra tropical hurricanes', defining these as disturbances of small horizontal extent and of strong intensity. Bergeron noted that the distribution of cloud and temperature was rather symmetric, and that there was much evidence which supported the assumption that extratropical hurricanes form as secondaries within a low no longer containing any fronts. Bergeron considered five cases of typical extratropical hurricane developments in the Baltic region and concluded that the basic difference between the two classes of hurricanes, i.e. the tropical and the extratropical, 'seems mainly conditioned by the difference in the Coriolis force'.

In the next section we will consider a group of polar lows commonly observed over the Nordic Seas, but not represented in the Wilhelmsen file, and not necessarily connected with upper-air cold lows. The discussion of these lows has, for practical reasons, been placed here, following the discussion of other polar lows associated with large-scale occluded systems.

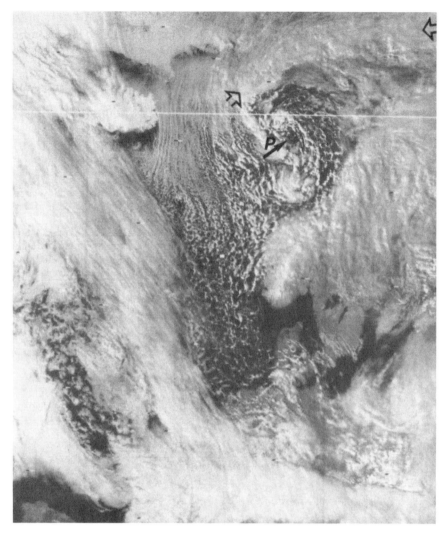

Figure 3.22. A visible satellite image for 1438 GMT 12 March 1988 showing a polar low, P, (indicated by an arrow) that formed within the core of an occluded cyclone. Clouds seen north of the polar low and belonging to the occluded cyclone are indicated by open arrows. (Image courtesy of the NERC Satellite Receiving Station, University of Dundee.)

Polar lows along occlusions

In addition to polar lows that form in the central region of the cyclonic wind field associated with the occluded system, polar lows may form *along* an occluded front as illustrated by Figures 3.23 and 3.24. As noted earlier, no such cases were found in the Wilhelmsen file; nevertheless, such developments are quite common in the Nordic Seas region. Figure 3.23 shows

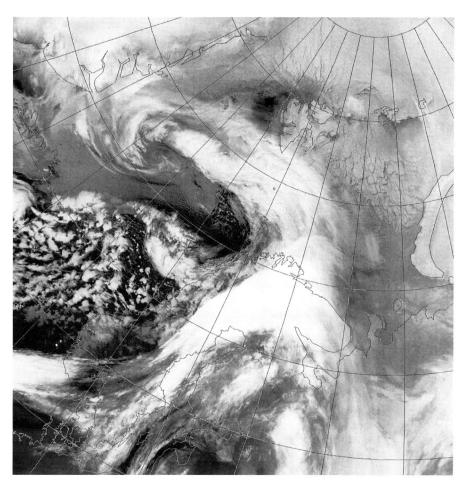

Figure 3.23. An infra-red satellite image for 1139 GMT 26 February 1998 showing an occluded extratropical cyclone over northwestern Europe. (Image courtesy of the NERC Satellite Receiving Station, University of Dundee.)

an occlusion associated with a baroclinic wave over northern Scandinavia on 26 January 1998 stretching from the Barents Sea to east Greenland. On the following day a number of small-scale vortices had formed along the occlusion (Figure 3.24). The formation of vortices of this kind was first discussed by Bond and Shapiro (1991) in a study of polar low formation over the Gulf of Alaska (see Section 3.1.11), and explained by them as a result of confluence and differential cold air advection between the polar air stream to the west of the synoptic low and the relatively warm air secluded near the core of the low (see Section 3.1.11).

Ralph *et al.* (1994) discussed the formation of mesoscale vortices along occlusions within the general framework of 'mesoscale cyclones within synoptic-scale systems'. Ralph *et al.* studied four events in which multiple mesocyclones

186   3 Observational studies

Figure 3.24. An infra-red satellite image for 1308 GMT 27 February 1998 showing a vortex train (indicated by arrows) over the Norwegian Sea. These formed along an occlusion seen on the satellite image from the previous day (Figure 3.23). (Image courtesy of the NERC Satellite Receiving Station, University of Dundee.)

were observed within synoptic-scale cylones. For the cases considered, they found that the events appeared to fall into two major classes, i.e. frontal instability and downstream development. However, according to Ralph *et al.*, besides the two major cases found by them, other important cases of multiple mesocyclones exist, such as multiple polar lows. The life cycles of the two major classes of multiple mesocyclones are summarized in Figure 3.25. The most fundamental difference between the two classes is that downstream development requires an initial cyclone, followed by the sequential development of new cyclones downstream, whereas pure frontal instability is characterized by quasi-simultaneous growth of more than one cyclone. Based on the relatively

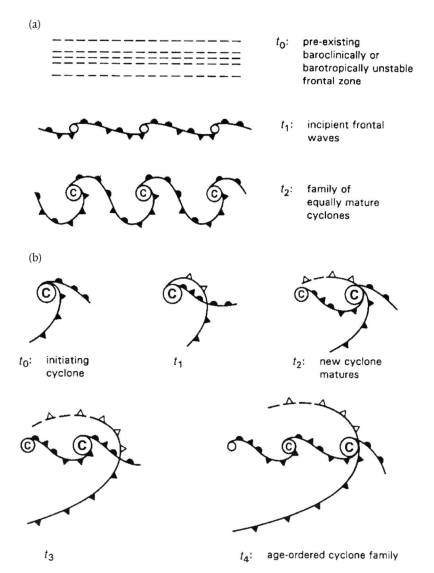

Figure 3.25. Summary of the life cycles of two classes of multiple mesocyclones observed within synoptic-scale cyclones. (a) Pure frontal instability, (b) mesoscale dowstream development shown in five stages over *c.* 24 h (from Ralph *et al.*, 1994).

few cases considered in their study, Ralph *et al.* concluded that the average scale of the downstream development was consistent with the scale of baroclinic instability found in a number of polar low studies. As an example of downstream development, Ralph *et al.* pointed towards the reverse shear development on 29–30 January 1986 over the Greenland/Norwegian Sea. This event, studied by Reed and Duncan (1987) and Moore and Peltier (1989), was characterized

188   3 Observational studies

Figure 3.26. An infra-red satellite image for 1551 GMT 7 February 1988 showing a number of mesoscale vortices along the occlusion from southern Greenland to the sea north of Scotland. (Image courtesy of the NERC Satellite Receiving Station, University of Dundee.)

by the sequential development of four separate cyclones, an evolution characteristic of downstream development rather than of pure baroclinic instability.

Figure 3.26 shows an example of mesoscale cyclone formation along an occlusion over the North Atlantic. The rather large horizontal wavelength, as defined by the two most developed vortices close to Cape Farewell (only one of the two vortices can be seen on the figure) and the fact that the mesocyclones are in a different stage of development, indicate that the formation of the mesoscale cyclones was the result of a downstream development. The small wavelength characterizing the mesoscale vortices at the eastern end of the cloud band point more in the direction of a frontal instability development.

Merry-go-round systems

'Merry-go-round systems' form a special group of lows within the cold low-type of polar lows. Forbes and Lottes (1985) introduced the descriptive name 'merry-go-round' for a vortex consisting of several small lows revolving about a common centre. Within the period studied by them, this category occurred quite often, i.e. three times.

The basic feature of a merry-go-round is a large, cold-core synoptic-scale vortex, typically formed as the end product of an occlusion process. Alternatively, a merry-go-round may form from a migrating upper-level cold low originating over the Arctic which, upon moving over the open sea, causes a cyclonic surface circulation to develop. Quite often a small-scale vortex may form within the central region of the upper-level cold core low. The multiple vortices giving their name to the merry-go-round generally form in the low-level cyclonic flow around, but some distance away from, the centre of circulation. Example of merry-go-round systems were shown on Figures 1.1, 1.7 and 1.18 in the section on cloud signatures.

Group 5: Comma clouds

Comma clouds were discussed earlier in this chapter (Section 3.1.3) and examples of such systems were shown in Figures 1.4 to 1.6. Most of the polar low developments listed in the Wilhelmsen file took place at rather northerly latitudes away from the main baroclinic zone and for the total period of five years covered by the file only one (and rather dubious) comma-cloud case was found and, as expected, at a rather southerly latitude.

Group 6: Baroclinic waves

Baroclinic instability has been considered as a primary mechanism for the formation of polar lows ever since Harrold and Browning's 1979 paper, 'The polar low as a baroclinic disturbance'. Baroclinic development in its pure form is characterized by stable ascent leading to the formation of characteristic cloud patterns of stratified clouds. These baroclinic 'cloud patterns' or 'cloud signatures' include leaf-like structures as seen on Figure 1.21a and b showing the early stages of a baroclinic polar low development. In connection with reverse shear baroclinic instability, wave trains may often form as illustrated on Figure 1.23. Also, most comma clouds, of which several examples have been shown above, albeit where convection was often present, are examples of baroclinic systems. Therefore, to *some extent* (but certainly not always), it is possible to decide whether a disturbance is of baroclinic nature or not, from satellite images alone.

Examples: 10–12 June 1980 (WF 1980–4) and 11–12 October 1982 (WF 1982–3)

Two examples of 'normal' (forward-shear) baroclinic polar low developments were found in the Wilhelmsen file. One case, in the *summer* from June 1980, and another case from October 1980. The two cases were surprisingly similar, considering that one of them was an example of a rare summer polar low development.

During both of the periods considered the main baroclinic zone (the polar front), was situated far south, and the polar lows formed as small-scale waves on secondary baroclinic zones far north. The frontal zones in these cases were deep, and as such resembled the polar front rather than shallow ice edge-generated BLFs.

The 1980 summer polar low developed as the result of an occlusion process along a baroclinic zone situated far north. Following the formation of the initial vortex, convection broke out leading to the formation of a polar low resembling a winter type system.

Group 7: Orographic polar lows

Two of the cases found within the Wilhelmsen file could best be characterized as 'orographic polar lows', i.e. polar lows that developed because of influences primarily from the orography. Orography is known to play an important role in polar low developments at many locations. For example, during periods with a northerly flow, vortices formed in the lee of Iceland have been observed to develop into polar lows. Another region where orographic influences may contribute significantly to the formation of polar lows is the sea east and northeast of Cape Farewell, at the southern tip of Greenland. Here vortices may form when a westerly or northwesterly flow is forced to ascend over or flow around the mountains in connection with outbreaks of polar/Arctic air from the North American continent (see Sections 2.1.2 and 3.1.7).

Around Scandinavia, polar lows can form as secondary circulations in the lee of the Norwegian mountains due to orographic effects as a parent circulation moves eastwards. An example of this was found in the Wilhelmsen file (2–3 January 1978 (WF 1978–1), not shown), when a fairly large polar low was left over the Norwegian Sea behind the main cyclone centre as this passed the Scandinavian peninsula on an eastbound track.

At the end of December 1978 a rare type of orographic development led to the formation of a polar low that significantly affected the weather in the North Sea and The Netherlands for a period of several days. At the end of the month a strong *easterly* flow was established over southern Scandinavia north of an west–east oriented frontal zone situated further south over Germany. Within this flow, around 28 December, a lee trough formed in the lee of the mountains of southern Norway. Following the formation near the Norwegian coast, the trough/vortex moved westwards. As the large-scale flow changed, the polar low was steered southwards into the North Sea. In the afternoon of 2 January the polar low hit the Dutch coast accompanied by strong northwesterly winds, which at several locations reached storm force.

3.1.5 Cases from the Norwegian Polar Lows Project (NPLP)

During the Norwegian Polar Lows Project (1983–1985) a number of studies concerning the structure and development of polar lows, ranging from the first research aircraft measurements within a polar low (Shapiro *et al.*, 1987) to a number of minor studies based on routine meteorological data, were carried out. Several of these were published in a special issue of *Tellus* (Series A, Vol. 39A, August 1987). In the following the observational case studies published in this volume, together with studies published in two other special *Tellus* volumes on polar lows (Series A, Vol. 37A, October 1985 and Vol. 44A, March 1992) will be discussed. In addition, some papers published later, but relevant to the topics raised in the *Tellus* papers, will also be discussed.

The 13–14 December 1982 Bear Island polar low development

Several studies of this development have been published (Rasmussen, 1985a; Emanuel and Rotunno, 1989; Fett, 1989a, b; Rasmussen *et al.*, 1992). The development began as an upper-level, cold core vortex on a southward track crossed the ice edge boundary near Bear Island on 12 December 1982 (Figure 3.27). The upper-level, small-scale vortex was situated within the central region of one of the centres of the circumpolar vortex over the Barents Sea. The small-scale vortex passed very near to Bear Island on 12 December. A radiosonde ascent from Bear Island showed that the vortex extended from the top of a relatively shallow, cold and stable boundary layer with a depth of around 100 hPa, throughout the troposphere. Upper-air wind measurements from Bear Island from 0000 GMT 12 December (shown in Rasmussen 1985a, his fig. 9) immediately prior to the passage of the upper-level low, as well as from the time when the centre of the upper-level vortex was situated just west of Bear Island (1200 GMT 12 December 1982), indicate that the atmosphere at both times was nearly equivalent barotropic from the top of the boundary layer and throughout the troposphere. This is in accordance with a result from Businger (1985), who in connection with a study of the synoptic climatology of polar low outbreaks concluded that the structure of the composite surface pressure field and 1000–500 hPa thickness fields, for a number of polar low developments, suggested that 'the structure of the atmosphere is primarily equivalent barotropic over these storms on the larger synoptic-scale'.

192   3 Observational studies

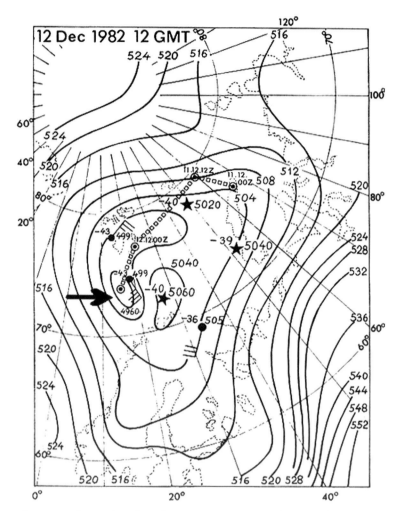

Figure 3.27. The 500 hPa chart for 1200 GMT 12 December 1982 showing a small-scale, upper-level cold vortex (indicated by an arrow), embedded within the core of the circumpolar vortex (from Rasmussen, 1985a).

The fact that the atmosphere in some cases of polar low development is nearly equivalent barotropic above the boundary layer has important dynamic implications because baroclinic disturbances, except from shallow developments in the boundary layer, will be precluded.

Figure 3.28 shows the vortex at the time when it was situated just west of Bear Island. Although high-level clouds partly obscure the lower layers, the satellite image shows how the vortex at the time had penetrated to the surface; this penetration being the result of a destabilization of the boundary layer over the ice-free sea. Cloud streets were converging into the region of the centre of

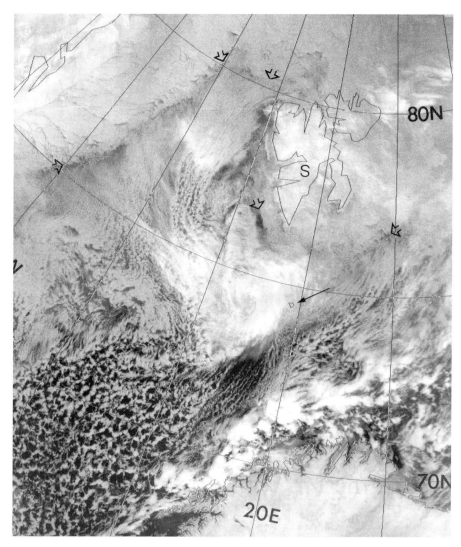

Figure 3.28. An infra-red satellite image for 1257 GMT 12 December 1982 showing a cloud vortex west of Bear Island (indicated by an arrow), associated with the upper-level vortex shown on Figure 3.27. Ice edges are indicated by open arrows. (Image courtesy of the NERC Satellite Receiving Station, University of Dundee.)

the low around 100 km west of Bear Island, including the cloud band associated with the Spitsbergen BLF. With the establishment of a surface circulation the vortex at this time formed a deep, nearly vertically aligned cyclonic system stretching throughout the troposphere. The radiosonde ascent (shown as fig. 9 in Rasmussen, 1985a) from nearby Bear Island, at the time of the satellite image shown on Figure 3.28, verifies that the cloud spiral west of Bear Island had formed within a stably stratified atmosphere.

194   3 Observational studies

Figure 3.29. An infra-red satellite image for 0250 GMT 13 December 1982 showing a major cloud spiral (associated with the upper-level vortex shown on Figure 3.27), and some minor vortices to the east. The cloud deck seen at the upper right corner of the picture was associated with a synoptic-scale cyclone south of Novaya Zemlya. One of the minor vortices subsequently developed into the stronger vortex as seen on Figure 3.37b. (Image courtesy of the NERC Satellite Receiving Station, University of Dundee.)

As the cold core vortex drifted south over a sea with steadily increasing surface temperatures the surface circulation gradually increased and an impressive cloud spiral with a well-defined centre developed, as seen on Figure 3.29. Near the centre, a pocket of warm air had been secluded forming the dark central eye seen on Figure 3.29. Notwithstanding the impressive cloud spiral, the surface circulation at this time was still rather weak.

As explained in Section 1.6 (Cloud signatures), a vortex superimposed upon a low-level baroclinic field will eventually form zones of increased temperature gradient corresponding to warm and cold fronts within the typical mid-latitude cyclone. A mesoscale cold front of this type can be seen on Figure 3.29, the front being situated to the south of the cloud-free slot leading into the centre of the low. East of the major vortex, two minor vortices had formed. These grew, but never developed into major circulation centres.

On a satellite image from 1740 GMT (not shown) the centre of the cloud spiral could still be identified near weather ship AMI. The low-level Arctic air at that time had penetrated further south and east so that the low at the surface was completely surrounded by a shallow Arctic air mass. The low-level circulation around the vortex had increased and surface winds close to gale force were measured at weather ship AMI.

During the following hours a second phase, in the form of a marked development, took place resulting in the intensification of the polar low and the formation of a 'hurricane-like' core (Reed, 1987). A dramatic change in the cloud field structure was seen at the time when strong pressure falls in the region around ship AMI resulted in the formation of the polar low shown on Figure 3.30, at a position somewhat east of the former centre. A satellite image from 0418 GMT 14 December (Figure 3.31) reflects the strong development seen on the surface charts.

The development of the intense polar low seen on Figures 3.30 and 3.31 on the evening of 13 December at a position east of the upper-level centre, took place close to the centre of a region of enhanced CAPE (convective available potential energy; see Section 4.5.1). The CAPE here had increased within the hours before the intense development, most likely due to a combination of

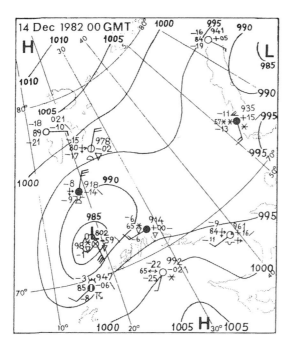

Figure 3.30. Synoptic surface analysis for 0000 GMT 14 December 1982 showing a mature polar low west of North Cape (from Rasmussen, 1985a).

Figure 3.31. An infra-red satellite image for 0418 GMT 14 December 1982 showing a rejuvenated polar low near North Cape. The centre of the polar low (marked A) at this time was characterized by a small cluster of convective clouds at the eastern flank of a band of deep convective clouds. The cloud spiral seen on the satellite images from the previous day (see Figure 3.29) is no longer discernible. The position of vortex B (also seen on Figure 3.37b) is also indicated. (Image courtesy of the NERC Satellite Receiving Station, University of Dundee.)

several factors, including upper-level cold advection, as well as low-level heating/moistening. Shortly after its formation, the centre of the polar low passed almost directly over weather ship AMI. Surface winds and air and sea temperatures from this station around the time of the passage of the polar low are shown on Figure 3.32, together with data from Bear Island. Surface air parcels ascending within the central region of the low, with the high values of temperature and dew point as indicated by these data, will be highly buoyant and as such support the notion that CISK may be of importance during this phase of development (see Section 4.5). Following its formation near AMI, the polar low moved east and later northeast to decay towards the end of the day over the sea north of North Cape.

In contrast to the cloud field associated with the 'parent vortex', which showed an eye-like feature (Figure 3.29), the cloud field associated with the strong 'hurricane-like core' polar low which formed during the evening of 13 December had the form of a small, compact cloud cluster (Figure 3.31). According to Økland and Schyberg (1987) the formation of an intense, small-scale

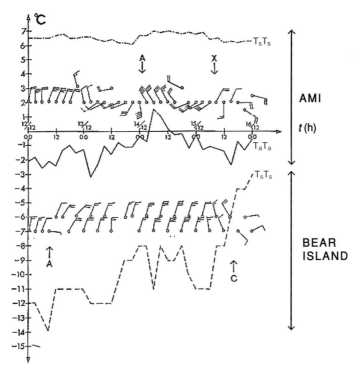

Figure 3.32. Surface winds (in knots and north upwards), air temperature ($T_a\,T_a$) and sea temperature ($T_s\,T_s$) from ship AMI (71.5° N, 19° E) and surface winds and temperatures from Bear Island from 0000 GMT 12 December to 0000 GMT 16 December 1982 (from Rasmussen, 1985a).

vortex within the central part of a parent low, accompanied by a cloud signature in the form of a cloud cluster *without an eye*, indicates heating proportional to the relative (low-level) vorticity. This, according to Økland and Shyberg, points to CISK as the governing mechanism for the development of the low.

The horizontal scale of the polar low that formed on 13–14 December was quite small. The tendency for convective systems to seek small horizontal scales became even more evident on the following day when a series of small-scale vortices associated with deep convection were observed to form along the BLF separating an outbreak of fresh Arctic air to the west from a modified (warmer) air mass to the east (Figure 3.33). A surface chart corresponding to the satellite image shown on Figure 3.33 is reproduced on Figure 3.34.

The satellite image (Figure 3.33) shows three cloud clusters L1, L2 and C that formed within the modified air mass just east of the outbreak of fresh Arctic air around Bear Island, the southern cloud cluster marked C being associated with a decaying vortex. The cloud cluster marked L2 was associated with a mature vortex and L1 with a vortex under development (see Rasmussen *et al.*,

198 3 Observational studies

Figure 3.33. An infra-red satellite image for 0406 GMT 15 December 1982. C, L2 and L1 are three cloud clusters. (Image courtesy of the NERC Satellite Receiving Station, University of Dundee.)

1992). The vortices, forming a 'mesoscale low family', are recognizable on the synoptic analysis on Figure 3.34 as, respectively, a minor wave corresponding to the southern cluster C and two minor lows along an Arctic front. The bottom part of the figure showing 3-hourly synoptic observations from AMI illustrates the very strong wind shear across the Arctic front which passed AMI between 0600 and 0900 GMT on 15 December.

The centre of one of these very small scale lows (low L2 on Figure 3.34) fortuitously passed over Bear Island and records of surface pressure and wind velocity during the passage are shown on Figure 3.35 (from Rasmussen *et al.*, 1992). A scale is indicated on the figure, showing the very small horizontal diameter of this low. Although this low and the polar low which formed near weather ship AMI early on 14 December formed close to each other there were significant differences between the two systems. While the air temperature at AMI, apart from a small increase as the centre of the low passed over the ship, was almost the same before and after the low had passed, the air temperature at Bear Island increased around 5 °C as the low passed, reflecting the fact that the latter low formed on a shallow but sharp low-level frontal zone.

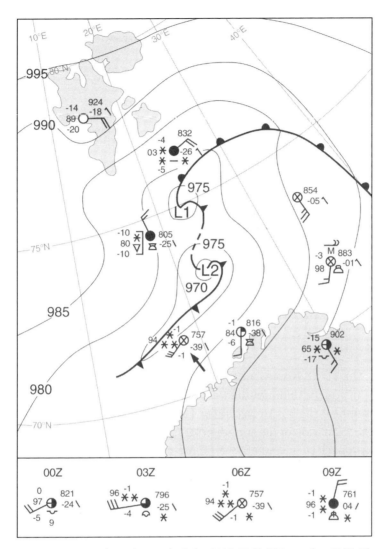

Figure 3.34. Synoptic surface analysis for 0600 GMT 15 December 1982. The plotted observations in the inset were taken at weather ship AMI (indicated by an arrow). Features L1 and L2 from Figure 3.33 are indicated (from Bader *et al.*, 1995a).

Apart from a small difference in horizontal scale, vortex L2, seen on Figures 3.33 and 3.34 and further illustrated by Figure 3.35, has a striking similarity to a strong vortex that formed over the Bering Sea along the east coast of Kamchatka on 6 March 1977 in a similar synoptic situation. In a study by Businger and Baik (1991) this vortex was referred to as an *Arctic hurricane* (see Section 3.1.12).

200   3 Observational studies

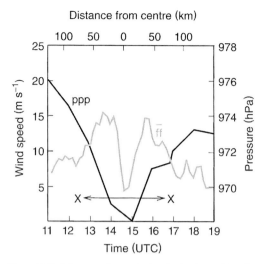

Figure 3.35. Mean surface wind velocity (m s^{-1}) and surface pressure from Bear Island for 1100 to 1900 GMT 15 December 1982 during the passage of the small-scale polar low L2 shown on Figures 3.33 and 3.34 (from Rasmussen *et al.*, 1992).

The 21–22 November 1983 Bear Island polar low development

Businger (1985) studied a case from November 1983 that in many ways resembled the 13–14 December 1982 case discussed above. Businger ascribed the formation of the surface polar low to baroclinic instability triggered by the approach of an upper-level, short-wave trough. The possible role of a BLF was not discussed. The same case was studied by Fett (1989a), who documented the existence of a well-developed Svalbard BLF prior to the development of the polar low. A comparison of the 500 hPa chart (Figure 3.36) from 1200 GMT 21 November 1983 showing a small-scale, upper-level cold low centred just north of Bear Island (with 500 hPa temperatures as low as −49 °C), and Figure 3.27, illustrates the similarity of the upper-level flows prior to the two developments.

A satellite image from 0844 GMT 21 November (not shown) indicates a region of high-level cloud associated with the upper-level vortex quite similar to Figure 3.28. Figure 3.37a, a satellite image from 0328 GMT 22 November, shows two polar lows, a minor system to the west and the major system to the east. The two systems resemble in a striking manner the systems from December 1982 shown on Figure 3.37b for comparison (the only difference being that the position of the stronger vortex is now west of the minor one). Both polar low developments from, respectively, 13–14 December 1982 and 21–22 November 1983, should be classified as 'cold low types'.

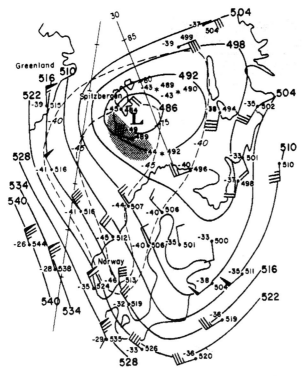

Figure 3.36. The 500 hPa chart for 1200 GMT 21 November 1983 showing a cold upper-level vortex between Svalbard and Bear Island prior to the surface development of the polar low. Solid contours indicate heights (60 m interval). Broken contours indicate temperature (5 °C interval). The shaded region indicates the approximate extent of the cloud cover derived from satellite imagery (after Businger, 1985).

When two polar lows coexist in the same area, binary interactions may occur that are similar to the vortex motions first noted by Fujiwhara (1923, 1931). These binary interactions consist primarily of a cyclonic co-rotation of the polar lows around their centroid, due to mutual advection. During the polar low outbreak 13–15 December 1982 over the Norwegian-Barents Sea more than one polar low was observed simultaneously in the region around Bear Island. Renfrew *et al.* (1997) studied two occasions of binary interaction during this period. The first example was between two vortices A and B, shown on Figures 3.37b and 3.31. On Figure 3.29, before the vortex interaction began, vortex A was associated with the well-developed cloud spiral seen in the central part of the image. Vortex B, associated with a minor vortex further east at this time, was still quite weak. Later at 0250 GMT 13 December 1982 vortex B had intensified and appeared nearly as significant as vortex A (Figure 3.37b). As demonstrated by Renfrew *et al.*, the two lows 'captured' each

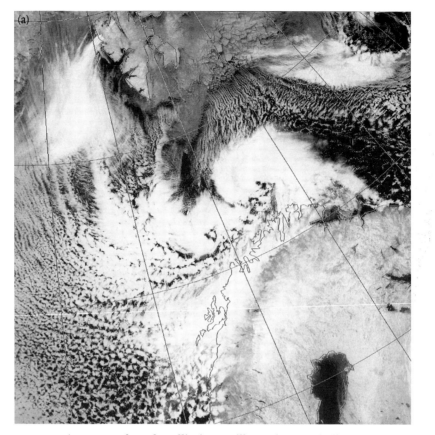

Figure 3.37 Infra-red satellite images illustrating two similar polar low developments over the sea in the vicinity of North Cape at (a) 0328 GMT 22 November 1983, (b) 1244 GMT 13 December 1982. (Images courtesy of the NERC Satellite Receiving Station, University of Dundee.)

other around this time, and subsequently they co-rotated to the morning of 14 December.

Following the decay of the lows A and B, other lows developed in the same region as illustrated by Figures 3.33 and 3.34 showing two lows L1 and L2 which formed along an Arctic front. These two lows also underwent a cyclonic co-rotation as well as a general system translation (for details see Renfrew *et al.*, 1997).

Co-rotation can, as demonstrated by Renfrew *et al.*, drastically alter the path and translation speed of polar lows, and their study demonstrated that binary interactions should be an important consideration in forecasting the tracks of polar lows when two or more of these systems exist in the same area.

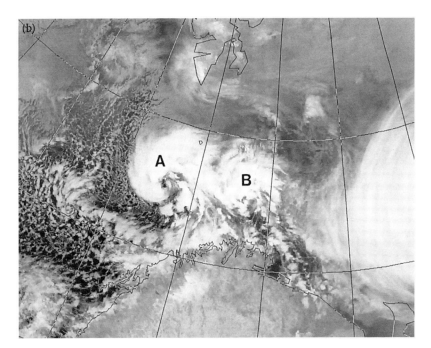

Figure 3.37 (cont.).

The polar low on 27 February 1984 (the 'ACE polar low')

During January and February 1984, the Arctic Cyclone Expedition (ACE) flew research flights over the North Atlantic, Norwegian and Barents Seas. The flight into this polar low southeast of Jan Mayen on 27 February 1984 represented the first documentation of the three-dimensional structure of a mature polar low by a research aircraft. The results of the flight were published by Shapiro *et al.* (1987) and were discussed by Fett (1989b). A numerical modelling study of this development by Aarhus and Raustein is discussed in Section 5.1.

The 'ACE polar low' developed within the confluence zone between a warm low-level southwesterly flow over the Norwegian Sea and a cold northeasterly flow in the region around Jan Mayen. Shapiro *et al.* (1987) pointed out that this polar low developed as the dominant circulation within a family of five polar lows that had individual life cycles as short as 6 h. Ralph *et al.* (1994) described the development of the ACE system as an example of a downstream development. On Figure 3.38, showing the situation a few hours before the ACE polar low was most developed, a system of multiple mesoscale cyclones is seen with the ACE polar low as the dominant system situated furthest east. Figure 3.39 shows a satellite image of the polar low around the time when it

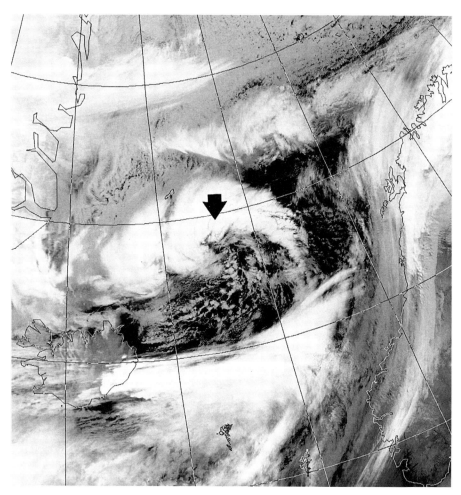

Figure 3.38. An infra-red satellite image for 0830 GMT 27 February 1984 showing a system of multiple mesoscale cyclones just south of 70° N. Clouds associated with the Polar Front are seen east and south of the mesoscale cyclones. The mesocyclone indicated by an arrow developed into the 'ACE polar low' shown on Figure 3.39. (Image courtesy of the NERC Satellite Receiving Station, University of Dundee.)

was most developed. This illustrates how the polar low formed along an Arctic front, north of a major frontal cloud band.

Only very few cases similar to the ACE polar low have so far been reported or discussed in the meteorological literature. That this kind of polar low may be occurring more frequently than generally presumed is indicated by the satellite image shown on Figure 3.40. This shows some resemblance to Figure 3.38 and illustrates a system of multiple mesoscale cyclones that formed along an Arctic front between an outbreak of a shallow Arctic air mass to the north and a maritime polar air mass further south.

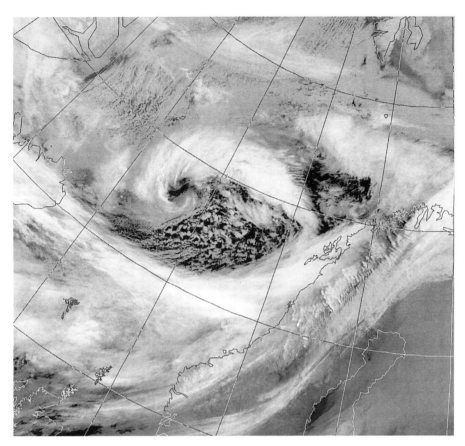

Figure 3.39. An infra-red satellite image for 1340 GMT 27 February 1984 showing the 'ACE polar low' close to the Polar Front. (Image courtesy of the NERC Satellite Receiving Station, University of Dundee.)

The ACE polar low formed close to a major baroclinic zone within an upper-level southwesterly flow. Fett (1989a), in his analysis of the synoptic-scale background leading to the formation of the ACE polar low, stressed the importance of the upper-level flow patterns and jet streaks. In particular, he drew attention to the superposition of the 500 hPa trough and the position of a left exit region of an upper-level jet streak over the region where the development took place.

Woetmann Nielsen (1998) discussed the formation of a polar low development on 17–18 January 1998 over the Norwegian Sea (Figure 3.41) which showed some similarities to the 27 February 1984 case. The polar low was correctly forecast an exceptionally long time ahead by the ECMWF model, i.e. more than five days, as illustrated by Figure 3.41a showing a 132 hour prognosis for the surface pressure, valid at 0000 GMT 18 January 1998. The low formed, as illustrated by Figure 3.42, on the flank of a cloud cluster situated

206   3 Observational studies

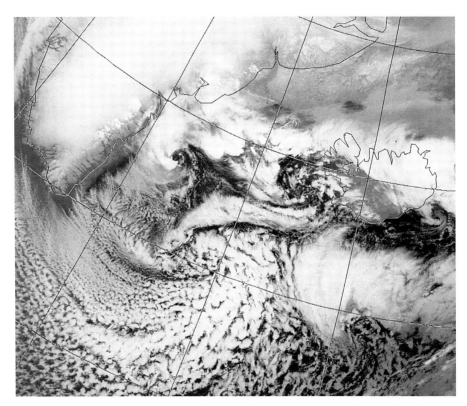

Figure 3.40. An infra-red satellite image for 1548 GMT 2 March 1979 showing a system of multiple mesoscale vortices near 65° N along an Arctic front between a cold outbreak of Arctic air to the north and a maritime polar air mass further south. (Image courtesy of the NERC Satellite Receiving Station, University of Dundee.)

within a trough in a vast cold air outbreak between Greenland and Scandinavia. The cloud field at this stage showed a striking resemblance to the cloud field associated with the ACE low (Figure 3.39). The development took place on the downstream side of a 500 hPa trough, below a region of high-level divergence associated with an upper-level jet streak. The divergence associated with the left exit region of an upper-level jet streak situated north of Iceland was, according to Woetmann Nielsen, an important factor for the establishment of the low-level convergence/upper-level divergence system which eventually led to the formation of the polar low. The couplet of low-level convergence/upper-level divergence at the time of formation of the polar low is shown on Figure 3.43.

3.1.6 Polar lows around the British Isles

Although being situated at a relatively southern latitude, the British Isles have often been affected by polar lows as documented by numerous

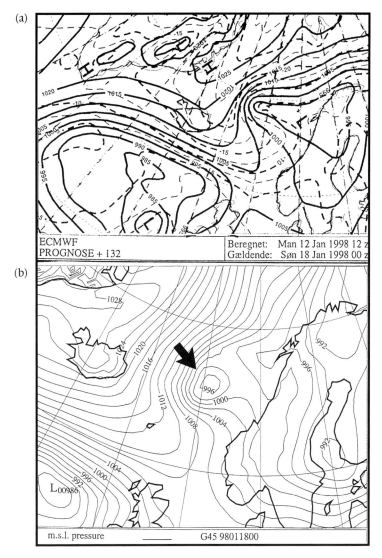

Figure 3.41. Charts from the European Centre for Medium-range Weather Forecasts showing the prediction of a polar low over the Norwegian Sea. (a) The 132 h forecast of MSLP (solid line) valid at 0000 GMT 18 January 1998, (b) the MSLP analysis at 0000 GMT 18 January 1998. The polar low is indicated by an arrow (from Woetmann, 1998).

studies, some of which were mentioned in Section 1.2. A comprehensive study by Harrold and Browning (1969) of two polar lows crossing the British Isles in December 1967 was briefly discussed in Section 3.1.2. Recently, two more studies of polar lows affecting the British Isles, which supplement these earlier studies, have been published, i.e. a study of the evolution and mesoscale

Figure 3.42. An infra-red satellite image for 1727 GMT 17 January 1998 showing a polar low over the Norwegian Sea. (Image courtesy of the NERC Satellite Receiving Station, University of Dundee.)

structure of a polar low outbreak by Hewson *et al.* (2000), and a study by Browning and Dicks (2001) on the mesoscale structure of a polar low that crossed the British Isles in the spring of 1998.

In the study by Hewson *et al.*, an outbreak of four polar lows was described, which all formed near southeast Iceland within a broad north to northwesterly current on the southwestern flank of a large filling depression. The lows displayed different structures and evolutions. The first was a 'convective low' associated with severe weather which, upon crossing north of Scotland, gave rise to heavy showers of snow, rain and hail, and strong winds in excess of

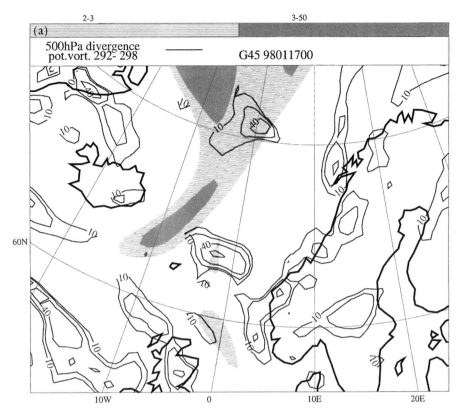

Figure 3.43. Analyses for 0000 GMT 17 January 1998. (a) 500 hPa divergence, (b) 925 hPa convergence. Contour intervals are shown every 10^{-6} s^{-1}. Hatched areas indicate potential vorticity between 2–3 PV units in the layer $\theta = 292$–298 K, while regions with PV >3 are indicated in grey (from Woetmann, 1998).

50 kts. The second low was described as 'a relatively weak surface signature of an upper-level feature'. The third low appeared as a wave train with the last one of the series which developed into an intense disturbance. Because of the high resolution of surface data, it was possible to identify unambiguously, mesoscale frontal structures within the polar lows. A close examination was made of the final intense low which, in the absence of perceptible upper-level forcing, went through an evolution and showed a frontal structure which closely matched those found in standard conceptual models of rapidly developing maritime cyclones. The horizontal and vertical structure of this low around its time of maximum intensity, i.e. 1400 GMT 14 March 1992, is illustrated on Figures 3.44a–c.

Figure 3.44a shows the large-scale situation on 1200 GMT 14 March 1992. The low formed, as seen from the figure, within a trough on the southwestern flank of a large, synoptic-scale low centred over northern Scandinavia. An

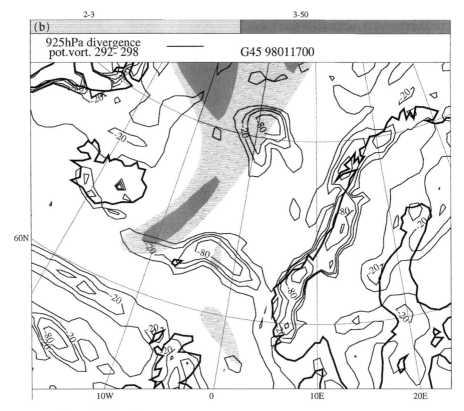

Figure 3.43 (cont.).

enlargement of the analysis for the region around the polar low and with surface observations added is shown on Figure 3.44b. The main features shown by the analysis include: well-defined frontal zones showing a 'T-bone frontal structure', high winds to the west of the low, and regions of heavy precipitation associated with the warm front. The widespread precipitation, mostly in the form of snow or snow showers, fell from a large 'cloud head feature' which had developed northeast of the low. The precipitating cloud head was, accordingly to Hewson *et al.*, comparable to the region of uniform slantwise ascent found by Harrold and Browning (1969) in their analysis of a polar low. Low D was, as illustrated on Figure 3.44c, trapped beneath a large-scale frontal zone associated with a synoptic-scale cyclone further west. Within the cloud head shown on Figure 3.44c, the ascent was slantwise corresponding to a region with stratiform precipitation rather than convective.

Browning and Dicks (2001) studied a polar low that crossed the British Isles on 14 April 1998 and which showed some of the same features as the low studied by Hewson *et al.* According to Browning and Dicks, the low had a structure

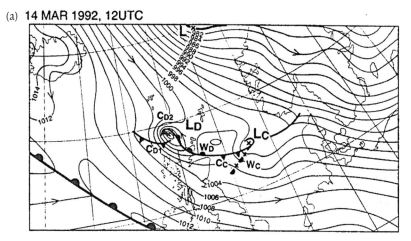

Figure 3.44. (a) Re-analysed surface chart for 1200 GMT 14 March 1992. (b) Enlargement of part of the surface analysis for 1200 GMT 14 March 1992 shown in (a), with surface observations added. Dash-dot line denotes position of the cross-section shown in (c). (c) a Cross section, position shown on (b), through cloud features associated with polar low D. Radiosonde data for Stornoway and Lerwick are shown at the top of the figure. F and H respectively indicate a warm front and a cloud head (from Hewson *et al.*, 2000).

which was a mesoscale version of the larger 'archetypical extratropical cyclone structure' as described by Bader *et al.* (1995). A narrow band of precipitation was seen along a surface cold front associated with the low, but most of the precipitation, mainly snow, fell from the cloud head in a region of slantwise ascent on the poleward side of a surface warm front. The polar low had a compact vortex centred near 3 km elevation, a cold core below this level and a warm core above. An analysis of data from a research radar and several other data sources, including the limited area and mesoscale versions of the UK Meteorological Office's operational numerical model, showed that the polar low was dominated by upper-level forcing. The low was affected by a stratospheric intrusion that brought air from near tropopause level down to 3 km.

3.1.7 Polar lows around Iceland and in the Denmark Strait

Besides the major genesis regions for polar lows at high latitudes, such as the northern parts of the Nordic Seas, some secondary genesis regions have been identified at lower latitudes. An example of this is the sea around Iceland. Several factors, including geographical as well as meteorological, contribute to the frequent formation of polar lows in this particular region.

The southern part of Greenland, forming the western boundary of the region, in many significant ways effects the meteorological conditions in the

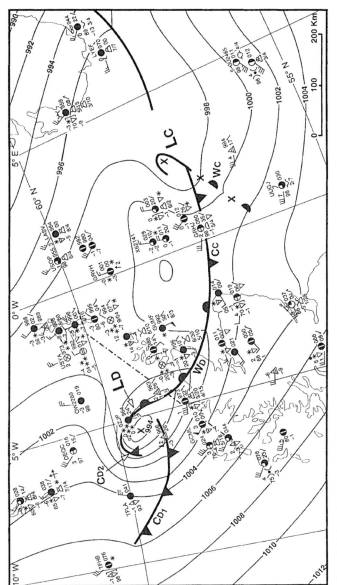

Figure 3.44 (cont.).

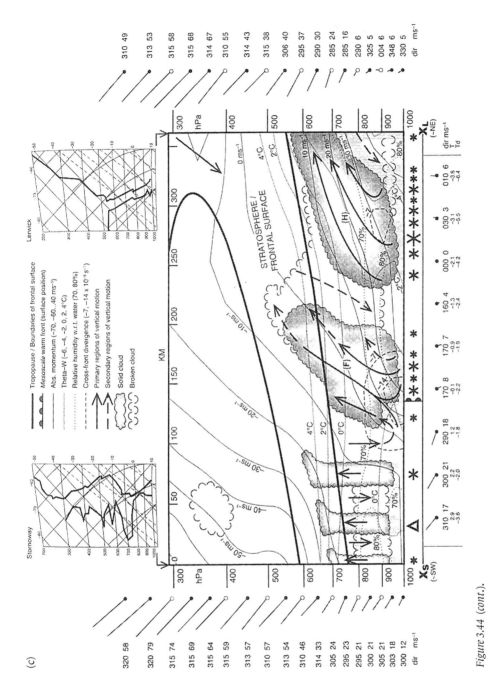

Figure 3.44 (cont.).

region. First, the existence and formation of the Icelandic low, a semi-permanent feature which shows up clearly, even on the monthly and yearly mean charts, is closely connected to the presence of southern Greenland, which blocks the westerly flow from 60° N and northwards. The Icelandic low may serve as a vorticity source for polar low developments as shown by Rasmussen (1981).

Orographic forcing on a more local scale also plays an important role in the formation of polar lows in the region. Small vortices are occasionally observed over the Denmark Strait near Angmagssalik, most likely forming due to vortex stretching as air is forced to descend from the Greenland plateau by the passage of a major cyclone further east. Such lows may subsequently move east and effect the weather in Iceland. Further south, other lee-lows have frequently been observed east of South Greenland near Cape Farewell (Kap Farvel) in situations with a westerly large-scale flow (see Section 2.1.2, Figure 2.15), and south of Iceland in situations with a northerly basic flow.

In Section 3.1.4 (Group 4: Cold low types), some examples of polar low developments within upper-level cold core systems were presented. In all these cases the upper-level cold core system had its origin around the centre of the circumpolar vortex normally found around Novaya Zemlya. Also, Iceland may occasionally be affected by upper-level cold core systems. However, in such cases the upper-level lows will often have their origin within the circumpolar vortex centre situated around Baffin Island. In the following, a brief account is given of a cold low-type polar low development around Iceland.

On 7 December 1993 a well-defined cold core vortex over the Davis Strait could be seen on the 500 hPa chart. The following day the vortex moved southeast until, on 9 December 1993, it was situated southwest of Iceland, as illustrated on Figure 3.45 showing the 500 hPa chart from 0000 GMT 9 December. The surface chart (Figure 3.46) showed how a polar low had formed below the upper-level cold vortex. A satellite image of the polar low at 1732 GMT 9 December (Figure 1.16) shows the spiral cloud system characteristic of many polar lows of the cold low-type. Already at 0000 GMT 9 December the 1000–500 hPa thickness field had indicated the beginning of the formation of a warm core around the low. This feature was much more clearly indicated on the following day by the presence of a pronounced warm core at the centre of the low, which at this time was centred just south of Iceland (Figure 3.47).

3.1.8 Mediterranean systems

Polar lows are, as implied by their name, phenomena associated with the polar regions. However, cyclones akin to polar lows are occasionally observed in other regions. Hurricane-like cyclones are known to form over

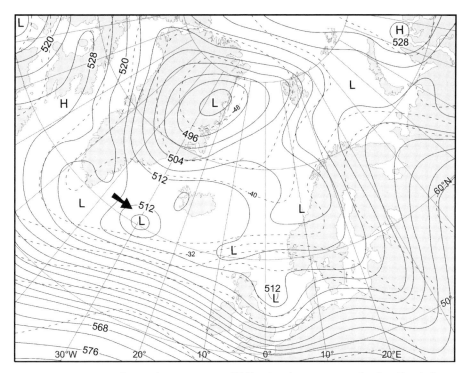

Figure 3.45. The 500 hPa contour (solid lines) and temperature (broken lines) chart for 0000 GMT 9 December 1993 showing a cold core vortex (indicated by an arrow) southwest of Iceland (based on the Berliner Wetterkarte, Freien Universität Berlin).

the Mediterranean and have been discussed by several authors, including Ernst and Matson (1983), Billing *et al.* (1983), Rasmussen and Zick (1987), Pytharoulis *et al.* (1999), and Reed *et al.* (2001). As demonstrated, for example through the case study of Rasmussen and Zick, these subsynoptic vortices resemble in many ways the convective type of polar low. Convection is, as discussed in the preceding, a basic mechanism for polar low developments. On the other hand, shallow baroclinic zones are almost invariably present in the regions where polar lows form, and it is sometimes difficult to assess the relative importance of convection versus baroclinic instability. However, in the case studied by Rasmussen and Zick of a subsynoptic vortex development over the Mediterranean, baroclinic instability hardly played any part at all.

The Mediterranean system initially formed, within the central region of a deep, cold, cut-off low (Figure 3.48).

The formation of an intense surface subsynoptic vortex over the warm Mediteranean Sea (sea surface temperature 23–24 °C) just northeast of Carthage was caused by a rapid spin-up of pre-existing vorticity associated with

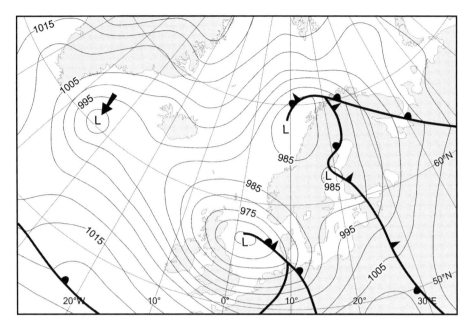

Figure 3.46. Synoptic surface analysis for 0000 GMT 9 December 1993 showing a polar low (indicated by an arrow) southwest of Iceland below the upper-level cold core vortex seen on Figure 3.45 (based on the Berliner Wetterkarte, Freien Universität Berlin).

a synoptic-scale low. The spin-up process in its turn was triggered by deep convection in the central part of the cold core low, facilitated by the combination of a warm sea surface and cold air aloft. The synoptic chart (Figure 3.49) shows the newly formed small-scale vortex embedded within a larger, synoptic-scale flow.

During the following five days the vortex followed an almost circular, anticyclonic track in the southwestern Mediterranean until it finally made landfall and dissipated near its place of origin. Before the crossing of Corsica on 30 September, the vortex showed the characteristics of a vertically aligned warm core system with low-level cyclonic inflow and upper-level anticyclonic outflow. Using half-hourly satellite-derived winds it was possible to determine dynamic quantities such as the low-level convergence and upper-level divergence, i.e. quantities normally not available for polar low studies. A NOAA 7 satellite image from 1418 GMT 29 September (Figure 3.50) shows clearly the clouds associated with the cyclonic low-level inflow as well as the clouds associated with upper-level anticyclonic outflow characteristic of a warm core vortex. Radiosonde ascents from the region indicated that substantial amounts of CAPE were available during the first stage of development until the crossing of Corsica, reaching values exceeding 1200 J kg^{-1}. As the result of released latent

3.1 The Arctic   217

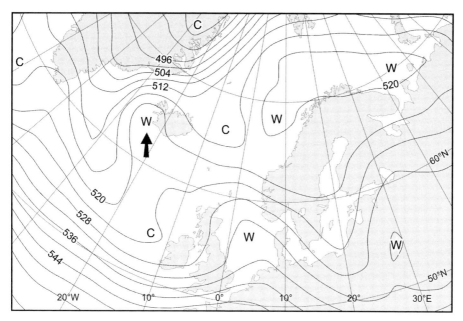

Figure 3.47. The 1000–500 hPa thickness chart for 1200 GMT 10 December 1993 showing a warm ridge/core (indicated by an arrow) southwest of Iceland, associated with the polar low (from 9 December) seen on Figure 3.46 (based on the European Meteorological Bulletin, Deutschen Wetterdienstes).

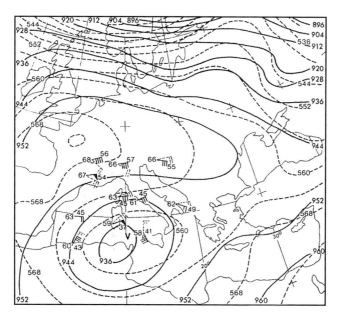

Figure 3.48. The 300 hPa height (solid contours) and 1000–500 hPa thickness (broken lines) analyses for 0000 GMT 28 September 1983 showing a cold cut-off cyclone over the western Mediterranean (from Rasmussen and Zick, 1987).

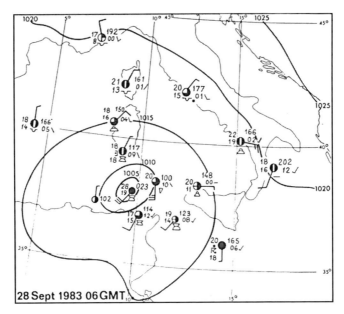

Figure 3.49. A synoptic surface analysis for 0600 GMT 28 September 1983 showing a subsynoptic vortex within a larger-scale circulation over the western Mediterranean (from Rasmussen and Zick, 1987).

heat through deep convection within the central part of the vortex, a warm core was formed. A thickness analysis from 0000 GMT 1 October, when the low was situated over the sea between Corsica and Italy, showed the warm core embedded within the large-scale cold trough (Figure 3.51). This case demonstrates that an intense, low-level vortex, in many ways similar to a polar low, can form rapidly within a cold core system, and in a situation where baroclinic instability can be ruled out as an important mechanism.

A recent study (Pytharoulis *et al.*, 1999) of a hurricane-like Mediterranean cyclone (see Figure 3.52), using observational data and the United Kingdom Meteorological Office (UKMO) Unified Model, confirmed most of the results obtained by Rasmussen and Zick concerning the structure and development of these systems. The formation of the Mediterranean cyclone took place during the morning of 15 January 1995 over the sea between Greece and Sicily in a region of strong convective activity, near the centre of a synoptic-scale low associated with a deep cold core vortex cut off from the basic westerly flow. The investigation of the cyclone's structure gave strong evidence to support the assertion that it was similar to a tropical cyclone and some polar lows. The presence of an 'eye' and a warm core together with available surface observations showed that the Mediterranean cyclone exhibited hurricane-like features.

A numerical experiment showed that the vortex did not develop in the absence of surface heat and moisture fluxes. Another experiment showed that

3.1 The Arctic   219

Figure 3.50. An infra-red satellite image for 1418 GMT 29 September 1983 showing a polar low-like vortex over the western Mediterranean. (Image courtesy of the NERC Satellite Receiving Station, University of Dundee.)

sensible and latent heat fluxes were equally important in its development, but baroclinic instability did not seem particularly important.

Reed *et al.* (2001) carried out a numerical study of a case from January 1982 also studied by Billing *et al.* (1983) and Ernst and Matson (1983), and for which winds near the centre approached hurricane force. They found that a considerable degree of instability to moist uplifting in the air near the centre was evident, and concluded 'that the event was mainly of convective origin'.

3.1.9 Labrador Sea and East Canadian Waters polar lows

Labrador Sea polar lows

During the 1980s most polar low research was focused on the Nordic Seas, but it soon became apparent that polar lows occurred in other regions as well. An active, southerly genesis region for polar lows was identified over the Labrador Sea, a favourite place for outbreaks of extremely cold air masses

220   3 Observational studies

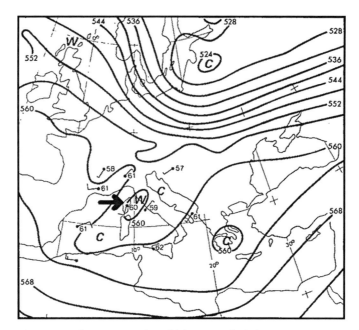

Figure 3.51. The 500–1000 hPa thickness analysis for 0000 GMT 1 October 1983 showing the warm core (indicated by an arrow) of the vortex situated between Corsica and Italy embedded within a large-scale cold trough. The plotted winds are the thermal winds for the layer (from Rasmussen and Zick, 1987).

originating over the northern part of Canada and the ice-covered Arctic Sea (Rasmussen, 1990; Rasmussen and Purdom, 1992). The temporal and spatial distribution of the lows were shown on Figures 2.29 and 2.30.

Labrador Sea polar lows are quite frequently observed and often show a remarkable resemblance to each other, with a typical asymmetric structure including a 'tail' extending from the centre of the low towards the east or southeast (Figure 3.53). Because of the paucity of surface observations little is known about the strength of these systems, but a typical surface wind is believed to be around 25 m s^{-1}. They are rather short-lived phenomena, lasting generally only one or two days. After their formation most of the lows move eastwards towards Cape Farewell, beyond which they usually rapidly lose their organization and decay.

A number of Labrador Sea polar low developments between 1984 and 1995 were studied by Rasmussen *et al.* (1996), and in particular a development from 11 January 1989, which was investigated utilizing observations from several meteorological satellites. A satellite image showing this polar low, which may be considered as typical for these systems, is shown on Figure 3.53. The impressive cloud shield seen on this afternoon NOAA satellite image, including

Figure 3.52. An infra-red satellite image for 0907 GMT 16 January 1995 showing a hurricane/polar low-like cyclone over the Mediterranean. (Image courtesy of the NERC Satellite Receiving Station, University of Dundee.)

an 'eye-like' feature suggesting an intense system accompanied by strong surface winds. For the 11 January Labrador Sea polar low the surface wind, an important parameter for forecasters, was determined in two different ways using remote sensing techniques. One method, based on the application of DSMP SSM/I (Special Sensor Microwave/Imager) measurements, provided the strength of the surface wind. The second method was based on tracking cloud elements using half-hourly GOES images. Using these techniques, winds around 20 m s^{-1} were observed south and southeast of the central region of the low.

The observational results discussed in Rasmussen *et al.* (1996) and in Moore *et al.* (1996) showed a number of characteristics that seem to be common for most Labrador Sea polar lows. These include the above-mentioned asymmetric

222   3 Observational studies

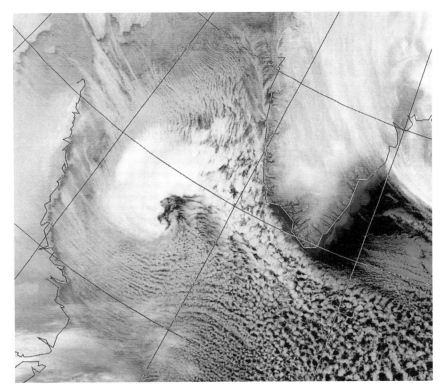

Figure 3.53. An infra-red satellite image for 1527 GMT 11 January 1989 showing a Labrador Sea polar low. (Image courtesy of the NERC Satellite Receiving Station, University of Dundee.)

cloud shield with a cloud band in the form of a 'tail' of convective clouds extending from the centre in an easterly direction. Small vortices, most likely caused by barotropic instability associated with horizontal wind shear, can often be observed to form along this tail, like the vortices seen along the Spitsbergen BLF.

The initial phase of a typical Labrador Sea polar low development typically starts as an eastward-moving upper-level disturbance, in the form of a cold trough/potential vorticity (PV) anomaly crossing the coastline. From a PV point of view, the initial development can be described as a downward extension of an upper-level PV anomaly as the Rossby depth (see Section 4.4) increases owing to decreasing static stability over the sea, forming the initial surface circulation. The west–east oriented tail, which forms more or less simultaneously with the low, separates cold, unmodified Arctic air advancing from the ice-covered continent, from modified 'warm' air masses to the north and northeast.

The interpretation of satellite observations in this region where *in situ* data are scarce may be supplemented by numerical simulations. Mailhot *et al.* (1996)

3.1 The Arctic   223

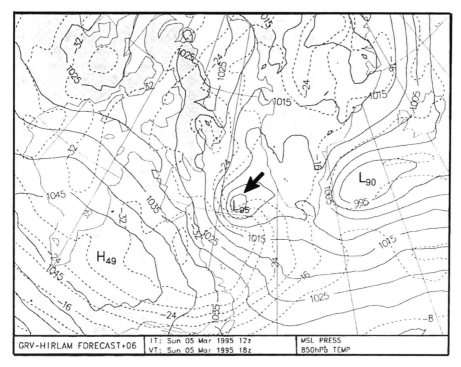

Figure 3.54. A 6-hour HIRLAM forecast of PMSL (thin solid lines) and 850 hPa temperature (broken lines) valid at 1800 GMT 5 March 1995 showing a polar low (indicated by an arrow) at the northern flank of an outbreak of Arctic air from the continent (from Rasmussen *et al.*, 1996, The Global Atmosphere and Ocean System, Taylor & Francis Ltd (http://www.tandf.co.uk/journals).)

made a successful simulation of the 11 January 1989 Labrador Sea polar low case which shed much light over this development (see Section 5.1.2, 'The Mailhot *et al.* case').

Figure 3.54 shows results from another numerical model run of a typical Labrador Sea polar low development on 5 March 1995 (see Rasmussen *et al.*, 1996) which adds to the understanding of the development of these systems. As seen from Figure 3.54, the polar low (which was correctly forecast by the 'HIRLAM' Danish numerical model) formed on the *northern flank* of an outbreak of Arctic air from Labrador out over the Labrador Sea at a position where the easterly thermal wind was parallel but of *opposite* direction to the westerly low-level flow; i.e. a situation conducive to a reverse shear type-development.

Hudson Bay polar lows

In addition to developments over the Labrador Sea, polar lows have been observed to form over other sea areas near Canada. In *The Polar Low Handbook for Canadian Meteorologists* (Parker *et al.*, 1991) and in a second edition of the Handbook, *Cold Air Vortices and Polar Lows: Handbook for Canadian Meteorologists* (Parker et al., 1997), the authors give examples of polar lows over Hudson Bay, Hudson Strait, the Labrador Sea, Davis Strait, the Great Lakes (Lake Superior) and the Gulf of Alaska.

Parker et al. (1991, 1997) described a Hudson Bay polar low (cold air vortex) development on 9–10 November 1985 when the air was flowing over an open (ice-free) Hudson Bay. In an 'overview' of this case they described the situation leading to the development as follows:

> At the surface, a northwesterly flow with temperatures in the minus 20s (°C) developed across the open waters of Hudson Bay on 9 November 1985. Streamers formed across Hudson Bay in response to the low level instability. An upper centre with an estimated cold core temperature of −45 °C very soon afterwards moved southeast across the bay allowing the instability layer to expand to mid and upper levels. By 1835 GMT 9 November a circulation was evolving over northwestern Hudson Bay and by 0820 GMT 10 November 1985, a well developed circulation was evident over central Hudson Bay. Thus, the polar low event of 10 November 1985 resulted from a coupling of low and upper level instability.

Although no reports from vessels operating in the Hudson Bay during the event were available, the development was well documented by a number of satellite images (three of which are shown in Chapter 1, Figures 1.15b–d). Prior to the development the low-level flow was remarkably uniform as illustrated by satellite images (not shown). The uniform flow was interrupted on 9 November, as shown by the satellite image in Figure 1.15b, showing the incipient surface circulation below the centre of the upper-level cold core vortex entering the northwestern part of Hudson Bay. Around noon a small cluster of deep convective clouds, spiralling into the centre over central Hudson Bay, had formed (Figure 1.15c). Finally, Figure 1.15d shows the vortex as it approached the east coast of Hudson Bay.

The sequence of satellite images in Figures 1.15b–d illustrate in a striking way how an upper-level cold core vortex affects its environment. After being decoupled from the strong Arctic surface inversion over land the upper-level system was able to penetrate into and affect the low-level boundary layer where the system of streamers (cloud streets) over the western Hudson Bay indicated the low-level circulation (Figure 1.15b). Deep convection formed *only* in the central region of the system where the lowest upper-level temperatures, near −45 °C at 500 hPa, were found, and consequently, where the vertical stability was a minimum (Figure 1.15c). Away from the centre where the upper-level temperatures, relative to the central region of the low, were higher, the vertical stability was more pronounced and the convection was unable to develop beyond the boundary layer (see also Section 4.6.1, Figure 4.62).

Satellite images, such as Figures 1.15c and d, indicate that a closed circulation formed under the upper centre. However, as noted by Parker *et al.* (1997), a 'tight' circulation never developed, for which reason in the 1997 handbook they describe the system as a 'cold air vortex' rather than a polar low. Despite the fact that the system never developed into an intense system the development in a striking way illustrates some of the basic factors of importance for the formation of cold low-type polar lows (see Section 3.1.4).

A detailed account of a successful numerical simulation of a Hudson Bay development on 8–9 December 1988, in many respects similar to the one discussed above, is presented in Section 5.1.2 (Albright *et al.*'s case). In this case, which was also discussed by Parker *et al.* (1997), most of Hudson Bay was covered or nearly covered by ice, except for a small region to the east. Nevertheless, a small, intense surface low developed over the eastern part of the bay.

Hudson Strait polar lows

A *Hudson Strait polar low* discussed by Parker *et al.* formed as a pre-existing weak surface low over the western part of the Hudson Strait and redeveloped in response to upper-level forcing and enhanced convection associated with an upper-level cold vortex. The low subsequently moved east, through the Hudson Strait, and developed a small tight spiral of clouds. During the passage of a coastguard vessel the pressure was observed to fall 5.2 hPa during a 3-hour period, and winds increased to 45 kts.

Davis Strait polar lows

Polar lows, often associated with upper-level cold core vortices (polar lows of the cold low type) have frequently been observed in the Davis Strait on satellite imagery. A satellite image showing a Davis Strait polar low was shown on Figure 1.15a. However, surface observations from these cases are rare.

Parker *et al.* presented a noteworthy case of a Davis Strait polar low development for which surface data were available from two vessels in the region of the low. The small scale of the polar low was reflected by the observations from the two vessels. One of the vessels, *Louis St Laurent*, experienced storm force *northwesterlies* (at 1500 GMT 4 November), and only 6 hours later, at about 125 km north-northwest of the 1500 GMT position, storm force *southeasterlies*. On 4 November at 1800 GMT, when the two vessels were only 75 km apart, one of the vessels reported a moderate southeast wind while the other was reporting gale force northwesterlies. During the time of development, a very cold vortex was located west of Baffin Island while a trough of cold air rotated northeastward across Davis Strait and Baffin Bay. Parker *et al.* (1997) assume that the cold trough caused an existing small vortex or weak baroclinic wave to deepen into the polar low.

Parker *et al.* (1997) point out that 'Conditions for reverse shear flow are not that uncommon in the northern Labrador Sea and Davis Strait. When synoptic scale lows pass south of Greenland they often develop a surface trough northward along the west Greenland coast. During the winter months this can produce a situation where the surface flow is closely oriented to the ice surface which aids in the formation of a low-level baroclinic zone along the ice edge.' In the 1991 and 1997 handbooks they presented an example of a polar low development over the southern Davis Strait-Labrador Sea from 9 to 12 February 1989 which was diagnosed by them as a reverse shear event (see also the section on Labrador Sea polar lows).

Great Lakes mesoscale vortices and polar lows

Forbes and Merritt (1984) documented the occasional occurrence of *mesoscale vortices* over the Great lakes during the winter months (October–April) during the years 1978–1982 over which period fourteen lake vortices were identified. The vortices were readily discernible in satellite imagery, in which they took one of three forms: a miniature comma cloud, a swirl of cloud bands (resembling a miniature tropical storm), or a swirl of cloud streets. The vortices, which were rather innocuous, were accompanied by a slightly lowered surface pressure, a weak low-level wind circulation and brief snow showers. Shoreline-parallel cloud bands occurred over the lakes during many of the developments and some of the vortices were initiated as small disturbances along these bands. Forbes and Merritt, however, concluded that vorticity production associated with diabatic heating over the lakes was more important for the lake vortex developments than the vorticity-generation processes operating on the scale of the shoreline-parallel cloud band. Synoptic-scale forcing appeared rather weak and unimportant for the development of the vortices, and upper-level positive vorticity advection or positive (warm) thermal advection, were not normally present.

While the vortices found by Forbes and Merritt were all rather weak, the same authors referred to reports of more extreme cases in which lake vortices appear to have produced strong winds and up to 20 cm of snowfall.

According to Parker *et al.* (1991) vortices over the Great Lakes typically have a diameter of 50–120 km, a duration of less than one day, winds less than 20 knots (with some gusts) and accompanying weather in the form of moderate snow flurries. They further state that *polar lows* are rare over the Great Lakes, but can occur. In December 1989 a polar low with a diameter of around

250 km and wind speeds generally around 20–25 kt (with one report of 35 kt) formed as a pool of cold air moved over Lake Superior. According to Parker *et al.* the combination of cold temperatures and cyclonic curvature aloft, and a lake which was still open with surface water temperatures of $+4\,°C$ led to the development.

Lake Superior, the largest of the Great Lakes is, according to Parker *et al.*, 'deemed to be the favoured lake for polar low development' partly because of its size and partly because it is closer to the track of very cold air than the other Great Lakes.

'Hurricane Huron'

Miner *et al.* (2000) documented that vortices, larger and much stronger than the kind of mesoscale vortices studied by Forbes and Merritt (see previous section), may form or redevelop over the Great Lakes.

The system dubbed 'Hurricane Huron' developed within a cutoff, cold core low over the Great Lakes during the period 11–15 September 1996. As the low deepened the cold core low evolved into a warm core system and eventually developed an eye and spiral bands of convective showers. In addition, the cyclone briefly produced storm force winds and excessive rain that caused flooding. From a satellite perspective, the system bore a striking resemblance to a hurricane. The term 'Hurricane Huron' was inspired by this resemblance and was not meant to imply that the system was of tropical origin.

Because the initially cold core cyclone moved slowly across the Great Lakes when they were near their climatological peak temperature, heat fluxes, particularly latent heat fluxes, were unusually large, exceeding 700 W m^{-2} during the low's development. Static stability analyses depicted the formation of a deep layer of conditionally unstable air over the lakes that extended to the tropopause. The thick layer of conditional instability ultimately assured that convective processes would extend the lakes' diabatic heating throughout the troposphere. Corresponding to this quite high values of CAPE were found, one sounding near Lake Erie indicating a value near 2300 J kg^{-1}. For these reasons it was hypothesized that the lakes, especially Lake Huron, played an integral role in the system's development, which was described by Miner *et al.* as 'akin to the cold-low class of polar lows'.

As pointed out by Miner *et al.*, the 'Hurricane Huron' development shows a striking similarity to some polar low developments such as the development over the Norwegian Sea on 13 October 1971 (see Section 3.1.2, 'The role of convection'), or the polar low-like development over the Mediterranean discussed in Section 3.1.8.

3.1.10 Polar lows over the East Siberian/Chukchi Sea/Beaufort Sea region.

In the following we will briefly describe two examples of developments in the Beaufort Sea region that both seem to qualify as polar lows.

The first case dates back to September 1970, i.e. outside the period covered by Parker's study (Section 2.1.4). In a report (Wilson, 1971) describing the formation of no fewer than 'three separate hurricanes' that formed in the southern Beaufort Sea over 14–15 September 1970, Wilson noted that the storms showed three unusual features:

1. three separate occurrences of winds of hurricane strength,
2. diabatic and barotropic deepening as well as baroclinic development, and
3. a storm tide that caused loss of life and property damage.

Wilson pointed out that one of the 'hurricanes', which may have had winds greater than $50\,\mathrm{m\,s^{-1}}$, was the worst storm recorded since 1944 in the Mackenzie Delta area. The surface and 500 hPa charts included in Wilson's report show fairly conclusively that at least one of these 'hurricane developments', which all occurred over the southern part of the Beaufort Sea, was a polar low within the central part of a cold dome.

Another case of an unusually rapid development in the Beaufort Sea was described by Black (1982). This storm, which produced strong winds of up to $30\,\mathrm{m\,s^{-1}}$ and waves up to 5 m over a drilling area in the Beaufort Sea, initially developed as a trough which 'intensified dramatically in the MacKenzie Bay' on 27 July 1982. For a discussion of polar low developments over the East Siberian Sea and the Chukchi Sea, see Section 6.2.5, ' Forecasting polar low developments in the East Siberian/Chukchi/Beaufort Seas'.

3.1.11 Polar lows over the Gulf of Alaska

As discussed in Section 2.1, the Gulf of Alaska is one of the important genesis regions for polar lows.

As part of his study of the climatology of polar low outbreaks over the Gulf of Alaska and the Bering Sea Businger (1987) discussed two outbreaks of polar lows, one over the Bering Sea and another over the northern Gulf of Alaska. Several polar lows formed during these outbreaks when winds in a deep layer advected cold Arctic air over open water from the ice or snow-covered regions. The two cases studied by Businger indicated that the environment conducive to the development of strong polar lows included a deep outflow of Arctic air over open water and a cold core, closed vortex aloft. Additionally, the formation of the surface lows was associated with forcing from small-scale vortices aloft. Businger also pointed out, that when synoptic conditions were favourable for the formation of polar lows, a series of systems often developed in close proximity to each other.

Douglas *et al.* (1991) analysed observations collected from two research aircraft flights on successive days into a polar low that developed over the northern Gulf of Alaska on 4–5 March 1987. The vortex, with a horizontal scale of *c.*300 km, developed along a shear zone between a north-northeasterly flow originating over Alaska and a southwesterly flow from the Gulf of Alaska. The vorticity associated with the polar low was largest near the surface and decreased rapidly with height, with only a weak circulation evident by 700 hPa. Coldest 500 hPa temperatures and lowest static stabilities were found directly above the surface low. On the second day the vortex was more intense at mid-tropospheric levels and convective clouds extended to higher levels.

Bond and Shapiro (1991) considered the formation of polar lows over the Gulf of Alaska in conditions of reverse shear along an occluded front. At low levels, as a result of confluence and differential cold air advection between the polar air stream to the west of the synoptic low and the relatively warm air secluded near the core of the low, conditions were brought about that were conducive for the development of the lows. Satellite imagery suggested that the formation of the polar lows was associated with mesoscale waves along the occluded front, probably due to reverse shear baroclinic instability enhanced by latent heating (Figure 3.55).

3.1.12 A Kamchatka polar low development

Businger and Baik (1991) and Bresch *et al.* (1997) (referred to as BB and BRA in the following account), carried out detailed observational/numerical simulation studies of a polar low development over the Bering Sea near the Kamchatka coast on 7–8 March 1977. The system studied by them had many similarities to developments observed over the Norwegian Sea and the Barents Sea. The work of BRA, being mainly a numerical model-based study is discussed also in Section 5.1.

The polar low developed over the western Bering Sea on 7 March 1977 and travelled eastward parallel to the ice edge. Surface winds in excess of 30 m s^{-1} were observed and satellite imagery revealed a spiral cloud band of unusual symmetry with a clear eye. The symmetric cloud signature of the polar low, the thermodynamic structure of the low, and the dominant role of the surface fluxes for the maintenance of the mature storm, led BB to adopt the term *Arctic hurricane* for the Kamchatka polar low. Reed (1992; see Section 1.4) objected to

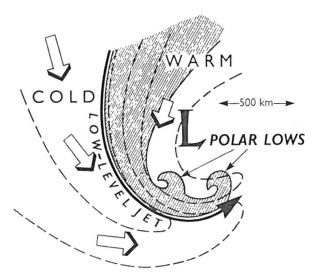

Figure 3.55. Schematic representation of polar lows forming along an occlusion with cold air advection west and south of the occlusion, and warm air advection to the east. Isotherms are indicated by dashed lines and clouds by hatching (from Bond and Shapiro, 1991).

the use of this term, arguing that polar lows and similar systems are neither true hurricanes nor indigenous to the Arctic.

The track of the low pressure centre passed over St Paul Island where time series of surface data showed a pronounced pressure drop at the core of the low (see Figure 5.26, p. 441). Synoptic analyses showed that the 'Arctic hurricane' formed at the leading edge of an outflow of Arctic air that originated over the ice and passed over the open water of the western Bering Sea.

As mentioned in the preceding, and discussed in Section 4.3, low-level shear lines with pronounced horizontal shear may act as precursors for some polar low developments, indicating that barotropic rather than baroclinic instability may be the decisive factor during the initial stages of these developments. The high resolution numerical model experiments reported in BRA seem to confirm the important role of these shear lines for the initial development of some 'convective' polar lows (Figure 3.56). According to BRA, a prominent feature of their full-physics simulation was the presence of several narrow bands of extremely high low-level vorticity (shear lines), and the Kamchtka low which developed in their simulation initially formed from one of the waves on Band 2 seen on Figure 3.56. As seen from Figure 3.56b, the baroclinicity was rather weak across the bands as demonstrated by numerical experiments, and baroclinic growth was able to proceed only when the atmosphere possessed reduced stability at low levels due to strong surface heat fluxes.

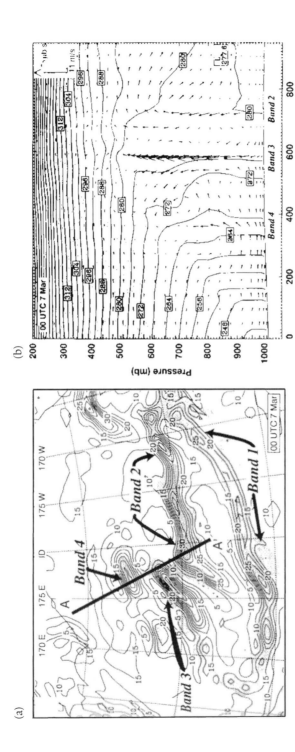

Figure 3.56. (a) Absolute vorticity at the lowest model layer in the numerical simulation of the Kamchatka polar low. (b) Cross-section along the line A–A' on (a), showing θ_e (solid lines) and winds in the plane of the cross-section (from Bresch *et al.*, 1997).

232   3 Observational studies

3.1.13 Observations of polar lows/mesoscale cyclones over the northwestern Pacific and the Japan Sea

Simultaneously with the studies of polar lows over the Atlantic sector of the Arctic and the northeast Pacific, a number of observational studies of polar lows were carried out over the northwestern Pacific and the Japan Sea by Japanese scientists. While polar lows over the Japan Sea have been investigated extensively via the dense observational network, polar lows over the northwestern Pacific have not yet been fully studied, since the observational data are scarce, except for GMS satellite images. The results of two studies (Ninomiya, 1989; Ninomiya *et al.*, 1996) on this subject as well as others will be briefly summarized in the following text.

The formation of two polar lows on 14 January 1987

Two polar lows formed on 14 January 1987 over the northwestern Pacific, off the coast of Japan within the cyclonic circulation of a synoptic-scale low, when this low was in its mature phase over the Aleutian Islands. The environmental conditions within which this pair of polar lows developed is illustrated by a number of charts shown in Figure 3.57.

The surface map illustrating the two polar lows is shown on Figure 3.57a, the cloud systems associated with the two lows being hatched. The main baroclinic zone associated with the synoptic-scale low was identified as the region of maximum temperature gradient ($c.$ 2 K $(100\,\mathrm{km})^{-1}$) on the 850 and 500 hPa surfaces. At 500 hPa the main baroclinic zone extended from $c.$ 30° N, 140° E to $c.$ 40° N, 170° E (Figure 3.57b). A secondary zone of strong temperature gradient ($c.$ 1.5 K $(100\,\mathrm{km})^{-1}$) is seen at the surface and 850 hPa levels, extending from $c.$ 40° N, 130° E to $c.$ 50° N, 160° E about 1500 km poleward from the major polar frontal zone. This low-level baroclinic zone formed, apparently, along the boundary between the cold air mass over the Asian continent and the Okhotsk Sea and the relatively warm air masses over the northwestern Pacific.

The two polar lows formed in this low-level baroclinic zone under the influence of mesoscale cold core vortices/troughs aloft (Figure 3.57c). The mesoscale cold vortices appear as maxima of the relative vorticity at the 500 hPa level associated with upward motion at 700 hPa (Figure 3.57d). During winter such pairs of upper-level, mesoscale cold vortices within a large-scale cold vortex/trough over the eastern part of the Asian continent have often been observed prior to the generation of polar lows.

The important role of the upper-level mesoscale cold vortices (short-wave troughs) has been pointed out for many cases of polar low genesis (e.g. Ninomiya, 1991, 1994; Ninomiya and Hoshino, 1990; Ninomiya *et al.*, 1993, 1996; Fujiyoshi *et al.*, 1996). This indicates that the mechanism for the genesis

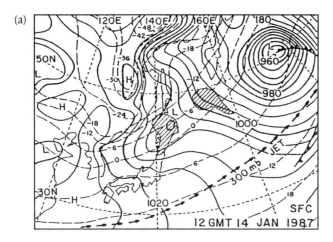

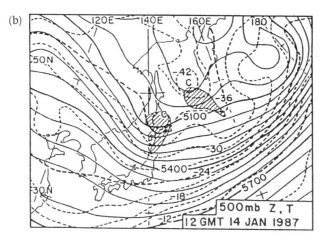

Figure 3.57. (a) Surface chart for 1200 GMT 14 January 1987. The isobars (4 hPa interval) and isotherms (6 °C interval) are given by solid and broken lines, respectively. The cloud systems of the polar lows are shown by the hatched areas. (b) The 500 hPa chart for 1200 GMT 14 January. The geopotential height contours (60 gpm interval) and isotherms (3 °C interval) are given by solid and broken lines. (c) The 500 hPa height and relative vorticity field for 1200 GMT 14 January. The height contours (60 gpm interval) and the isopleth of the vorticity (20×10^{-6} s^{-1} intervals) are given by thick and thin solid lines, respectively. Hatched areas indicate positive vorticity. The cloud systems of the polar lows are indicated by the blacked areas. (d) The 850 hPa temperature and 700 hPa vertical velocity (ω) fields for 1200 GMT 14 January. Isotherms (3 °C interval) and isopleths of ω (4 hPa h^{-1} interval) are given by the thick and thin solid lines, respectively. Hatched areas indicate areas of upward motion (negative ω) (from Ninomiya, 1989).

234 3 Observational studies

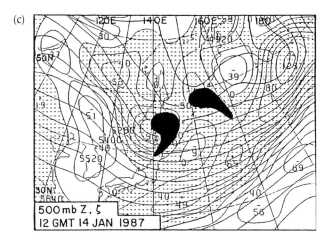

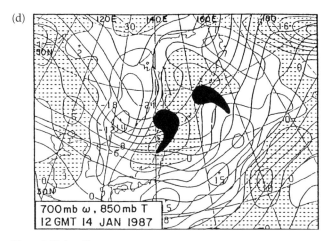

Figure 3.57 (cont.).

of the mesoscale circulation systems is dependent upon the conditions in the mid-troposphere, as well as on the presence of a low-level baroclinic zone. In brief, this case indicates that the cyclonic circulation of the parent large-scale low, the low-level baroclinic zone formed from the continent–ocean thermal contrast, and the mesoscale cold vortices aloft are all important for the polar low genesis.

Comma clouds

Just as in the Atlantic region, some of the mesoscale polar air disturbances over the Pacific and the Japan Sea are accompanied by 'comma clouds'. For a discussion of these systems, the reader is referred to Section 3.1.3.

Observations of polar lows and mesoscale vortices over the
northeastern Japan Sea

The northeastern Japan Sea, west of Hokkaido, has been known for some time as a region of frequent genesis of polar lows and mesoscale cyclones. These lows occasionally give very strong winds and intense snowfall over the western coast of Hokkaido. Although many reports about their features on radar and satellite imagery have been published (e.g. Motoki, 1974; Kibe, 1988), the evolution of the systems and the environmental condition associated with their development was not fully studied until the late 1980s. Below the structure and development of polar lows and mesoscale cyclones over this region will be discussed based on the results of observational studies.

Polar low genesis in the western part of synoptic-scale lows

One of the most characteristic features of polar lows in this region is their development within the cyclonic circulation of parent large-scale lows. Detailed observational studies of two cases on 9–11 December 1985 (Ninomiya, 1994) and 13–14 January 1986 (Ninomiya *et al.*, 1993) indicate similar evolutionary processes as well as similar large-scale environmental conditions. The 13–14 January 1986 development is summarized below.

The 13–14 January 1986 development

The surface chart for 0000 GMT 14 January 1986 (Figure 3.58) shows a polar low about 500 km west-northwest of the large-scale low in its developing stage. The 500 hPa analysis for 0000 GMT 14 January (Figure 3.59a) indicates that the polar low genesis was accompanied by a deep cold westerly trough aloft. Figure 3.59b shows the 850 hPa height field for the same time and indicates that a low-level trough extended northwestwards from the centre of the large-scale low. The 850 hPa chart shows pronounced warm air advection from the relatively warm ocean to the north, and cold air advection to the south of the centre of the large-scale low. The 273 K isentropic surface analysis at 0000 GMT 15 January (Figure 3.59c) demonstrates clearly how a wedge of continental cold air had moved towards the southwestern part of the large-scale low, as well as the seclusion-like intrusion of relatively warm air north of the centre of the large-scale low.

The advection of warm/cold air respectively north/south of the centre of the large-scale low leading to local frontogenesis near the east coast of the Asian continent has been schematically illustrated on Figure 3.60. On Figure 3.60a is shown a zone of strong thermal gradient in the lower troposphere over the eastern coast of the Asian continent, situated along the boundary between a continental cold air mass and relatively warm oceanic air. When the cyclonic

236   3 Observational studies

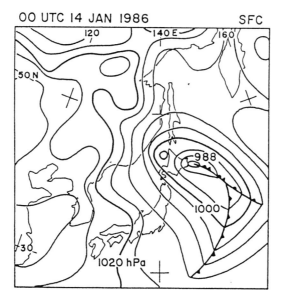

Figure 3.58. The surface chart for 0000 GMT 14 January 1986. The large-scale low in its developing stage is located off the east coast of Hokkaido Island. The polar low is seen over the sea, west of the Hokkaido coast (from Ninomiya *et al.*, 1993).

circulation of the large-scale low predominates over the coastal area of the Asian continent, the thermal field will change significantly, as illustrated in Figure 3.60b showing frontogenesis due to the shear flow. In this situation, an west–east elongated trough line (shear line) will form in the western part of the parent low. A polar low may form along the shear line under the influence of a cold vortex aloft.

The relationship between a mesoscale cold vortex aloft and the polar low is illustrated by Figure 3.61, showing the distribution of relative vorticity at 500 hPa and the vertical velocity (ω) at 700 hPa for 0000 GMT 14 January. The polar low developed in the area of strong vorticity advection, where 700 hPa upward motion was induced.

Another important feature for the development of the polar low was the significant decrease of the static stability under the cold vortex aloft as illustrated by Figure 3.62, showing the vertical stability ($-\partial\theta/\partial p$) in the 500–850 hPa layer at 0000 GMT 14 January.

The features of the polar low observed at the surface as the result of the genesis process discussed above can be seen in Figure 3.63 showing the time sequence of sea level pressure P_s, the air temperature T, the dew-point temperature T_d, precipitation PR, wind direction (indicated by arrows), maximum gust speed G, and 10 minute mean wind speeds V at Rumoi, located near the western coast of Hokkaido. The centre of the polar low passed over

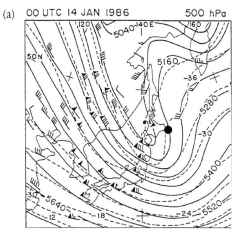
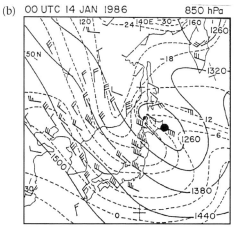
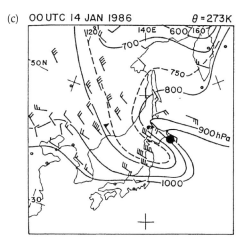

Figure 3.59. Charts for 0000 GMT 14 January 1986. The black dot and the comma indicate locations of the large-scale low and the polar low, respectively. (a) The 500 hPa heights and temperatures; (b) the 850 hPa heights and temperatures; (c) Isentropic chart for the potential temperature 273 K (from Ninomiya *et al.*, 1993).

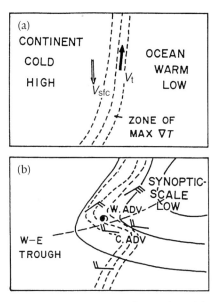

Figure 3.60. (a) Schematic illustration of the zone of strong thermal gradient in the lower troposphere along the boundary between the continental cold air mass and the relatively warm oceanic air mass in 'the undisturbed period'. V_{sfc} and V_t indicate schematically the surface wind and the thermal wind in the lower troposphere. (b) Schematic illustration of change in the thermal gradient under the influence of the cyclonic circulation of the large-scale parent low. Warm air advection is present in the northwestern quadrant of the parent low; cold air advection is present in the southwestern quadrant. Under this situation, a west–east elongated trough line (shear line) forms in the western part of the parent low (from Ninomiya, 1991).

this station at *c.* 0430 GMT 14 January and it can be seen that the polar low had a warm core structure in the lower atmosphere. The sharp drop of surface pressure around its centre and very strong winds indicate that the polar low had a very tight circulation concentrated within a few tens of kilometres. The mesoscale surface analysis (not shown) also indicated this small, strong cyclonic circulation of a few tens of kilometres radius associated with the warm core. The features mentioned above are commonly found associated with polar lows and mesoscale cyclones over the Japan Sea and should be compared with similar observations from other locations such as Bear Island (Figure 3.35) and St Paul Island (Figure 5.26).

A mesoscale low family forming over the Japan Sea and the northwestern Pacific

In some cases a sequence of mesoscale cyclogenesis events has been observed in the western portion of a developing large-scale low forming a

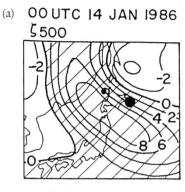

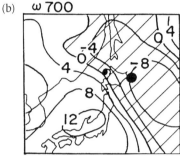

Figure 3.61. Distribution of (a) 500 hPa relative vorticity (units 10^{-5} s^{-1}) and (b) vertical velocity (ω) at 700 hPa (units hPa h^{-1}) for 0000 GMT 14 January 1986, reproduced from the objective analyses chart of the Japanese Meteorological Agency. The black dot and the comma indicate, as on Figure 3.59, locations of the large-scale low and the polar low respectively (from Ninomiya *et al.*, 1993).

'mesoscale low family'. A typical example from 3–4 January 1987 (Ninomiya, 1994) is briefly described as follows. As a large-scale low passed the eastern coast of the Japanese Islands, a polar low formed over the eastern Japan Sea around 0000 GMT 3 January 1987. The 12-hourly locations of the large-scale low and the polar low are shown on the surface chart for 0000 GMT 4 January (Figure 3.64). On this chart cloudy areas seen on the GMS infra-red image are indicated by the hatched area. A cloud zone is seen expanding westward from the centre of the polar low, and a meso-β-scale low had formed behind the polar low.

A surface analysis (not shown) utilizing the dense observation network, including the data from a few islands along the west coast of Hokkaido, revealed the genesis of two meso-β-scale lows in the cloud zone. The horizontal spacing between them was *c.* 200 km, while the distance between the eastern meso-β-scale low and the polar low was *c.* 400 km. The 500 hPa chart for 0000 GMT 4 January (Figure 3.65a) shows that both the polar low and the mesoscale lows

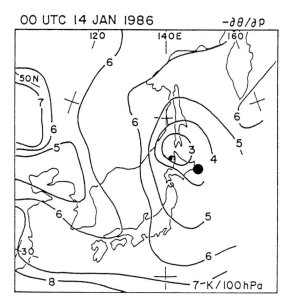

Figure 3.62. Distribution of the vertical stability ($-\partial\theta/\partial p$, units in K/100 hPa) in the 500–850 hPa layer for 0000 GMT 14 January 1986. The black dot and the comma indicate, as on Figure 3.59, the location of the large-scale low and the polar low respectively (from Ninomiya *et al.*, 1993).

developed in the vicinity of a deep westerly trough. The 850 hPa analysis for 0000 GMT 4 January (Figure 3.65b) indicates the formation of a shear line (trough) extending westward from the centre of the polar low. Warm air advection by easterly winds took place to the north of the shear line, whereas cold air advection by the westerly winds predominated to the south of the shear line, indicating frontogenesis within the shear flow. As seen from the figure, the mesoscale cyclogenesis occurred along the shear line behind the polar low, a feature similar to polar low genesis behind a parent large-scale low. In this case, however, the polar low played the role of the parent circulation for the genesis of the meso-β-scale low. The 500 hPa vorticity field and the 700 hPa vertical velocity ω field for 0000 GMT 4 January (not shown) indicate that the polar low and the meso-β-scale lows together formed within the region of upward motion associated with positive vorticity advection in the middle troposphere. The distribution of the static stability ($-\partial\theta/\partial p$) in the 500–850 hPa layer for 0000 GMT 4 January revealed the association of the polar low genesis and meso-β-scale cyclogenesis with the decrease in the static stability due to the passage of the cold core aloft.

Ninomiya (1994) examined whether the observed meteorological parameters for this case satisfied the conditions of barotropic or baroclinic instability as obtained by a linear instability analysis (e.g. Gill, 1982). The dense

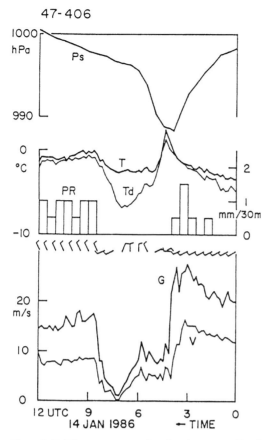

Figure 3.63. Time sequence of sea level pressure P_s, air temperature T, dewpoint temperature T_d, precipitation PR, wind direction (indicated by arrows), maximum 10 min gust speed (G) and 10 min mean wind speed (V) at Rumoi meteorological station (station number 47406) near the western coast of Hokkaido Island (from Ninomiya *et al.*, 1993).

surface observations over Hokkaido showed a strong wind shear of 5–10 m s^{-1} (10 km)$^{-1}$ over a zone 20–30 km wide. For widths of 20 km and 30 km, the maximum growth rate of barotropic instability occurs at wave-lengths of c. 160 km and c. 240 km, respectively. For lateral wind shears of 5 m s^{-1} (10 km)$^{-1}$ and 10 m s^{-1} (10 km)$^{-1}$, e-folding times of c. 3.2 h and 1.6 h were obtained, respectively, at the wave-length of maximum growth. This result indicates that barotropic instability in the shear flow may be one of the dominant mechanisms of meso-β-scale cyclogenesis in the present case.

It is well known from baroclinic instability theory that baroclinic disturbances of short wave-length are restricted to shallow systems or associated with

242   3 Observational studies

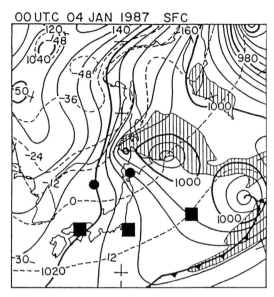

Figure 3.64. The surface chart at 0000 GMT 4 January 1987. The centre of the large-scale low (black squares) and the polar low (black circles) are shown at 12 h intervals. A mesoscale low is seen to the west of the polar low. Isobars (solid lines) and isotherms (broken lines) are given at 4 hPa and 12 °C intervals, respectively. Hatched areas indicate the cloud area seen on a GMS IR image (from Ninomiya, 1994).

small vertical stability. It has already been shown that small vertical stability in the 500–850 hPa layer appears under a cold core aloft. The upper-air observations near the meso-β-scale low family, revealed very small vertical stability in the layer between 980 and 850 hPa. The wave-length of the maximum growth rate for this condition was estimated to be *c*. 180 km, which is close to the observed wave-length of the present case. For the thermal gradient of 1 K $(100 \text{ km})^{-1}$, which was obtained from the regular upper-air observation network over Hokkaido, an e-folding time of *c*. 10 h was estimated at the wavelength of maximum growth. This e-folding time is not sufficient to explain the rapid growth of the development of the meso-β-scale lows. It should be mentioned, however, that the spacing of the regular upper-air observations is too wide to resolve the thermal gradient, which might have been concentrated within a narrow baroclinic zone. Although, the generation mechanism cannot be determined conclusively at this time, this case is important as a typical example of a multi-scale feature in the form of a mesoscale cyclone family behind a large-scale low under the influence of a cold core aloft. The case also indicates 'down scale influence' of the large-scale low toward the meso-β-scale through the generation of the polar low.

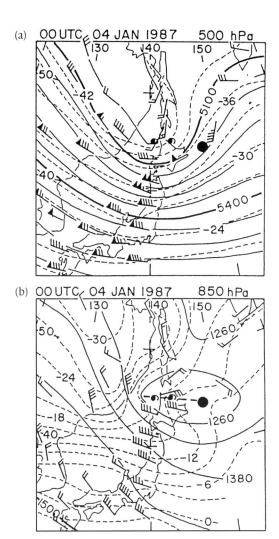

Figure 3.65. (a) The 500 and (b) 850 hPa charts for 0000 GMT 4 January 1987. Locations of the polar low and two mesoscale lows are indicated by the black circle and the small black commas, respectively. The shear line at 850 hPa (trough) westward from the centre of the polar low is indicated by a thick broken line on (b) (from Ninomiya, 1994).

Observations of polar lows and meso-β-scale lows over the western Japan Sea

The western part of the Japanese Sea is known as a region of frequent mesoscale cyclogenesis in polar air streams. These mesoscale lows occasionally give very strong winds and intense snowfall over the Japan Sea coast. Although several reports have been produced about these mesoscale disturbances as seen on radar and satellite images (e.g. Miyazawa, 1967; Asai and Miura,

1981), the evolution processes and the environmental conditions associated with their development had not been studied fully until the end of the 1980s. Recent studies have revealed that the combined influences of a cold vortex aloft, the large-scale parent low and the JPCZ (Japan Sea Polar Air mass Convergence Zone; a convergence/confluence zone of polar air streams forming on the lee-side of the coastal high mountain area of the continent (Nagata, 1987, 1992) occasionally result in the formation of mesoscale lows. Typical cases of the formation of meso-β-scale lows are illustrated below by a number of case studies.

Meso-β-scale vortices within a polar low forming over the Japan Sea

The genesis of a polar low beneath a meso-α-scale cold vortex aloft was observed over the western Japan Sea on 11–12 December 1985. As seen on the surface chart (Figure 3.66), the centre of the low appeared under the cloud vortex identified on a GMS IR image (not shown). The 850 hPa height analysis, and the height and temperature anomaly (not shown) revealed a meso-α-scale low with a low-level warm core structure. Radar observations of the polar low revealed interesting details. Six successive radar echo charts obtained at 1-hour intervals for 2200 GMT 11 December to 0300 GMT 12 December showed well-organized spiral mesoscale precipitation bands. Since the radius of the echo system was only approximately 30 km it was classified as a meso-β-scale vortex. The lifetime was estimated as several hours, which was obviously shorter than the lifetime of the polar low. A mesoscale surface analysis around the radar echo vortex revealed a cyclonic circulation, strongest around 30 km from the centre. The surface observations also indicated a surface pressure drop of c. 4 hPa

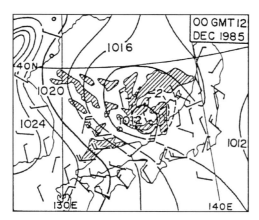

Figure 3.66. Surface analysis for 0000 GMT 12 December 1985. Solid and broken lines indicate isobars (2 hPa interval) and the isopleth of the pressure fall (in hPa) during the preceding 6 h, respectively. Cloud areas as seen on a GMS infra-red image are indicated by hatched areas (from Ninomiya et al., 1990).

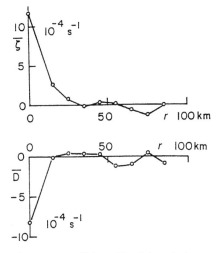

Figure 3.67. Radial profile of the relative vorticity (ζ) and divergence (D) of the meso-α-scale low evaluated by the dense surface observation data for the mature stage (from Ninomiya and Hoshino, 1990).

at the centre of the vortex and maximum wind gust of c. 25 m s^{-1} around the vortex centre. Since the meso-β-scale low passed over the dense surface observation network, it was possible to evaluate the tangential and radial wind velocity of the low using a composite analysis procedure, although the surface wind field showed significant large temporal and spatial variations. The radial distribution of the relative vorticity and divergence estimated from the composited wind velocity for the mature stage of the meso-β-scale low are presented in Figure 3.67. The large cyclonic vorticity and convergence were restricted to within c. 25 km of the centre of the meso-β-scale low. It is important to note that the significant pressure dip and strong cyclonic circulation were not directly associated with the polar low but with the meso-β-scale low generated within the polar low. The cyclonic flow and the deep layer with a moist quasi-neutral stratification within the parent polar low provided the favourable conditions for the genesis of the meso-β-scale systems.

It should be stressed that 'multi-scale structures' have been observed in connection with several atmospheric circulation systems. Businger and Reed (1989a, b) stated that 'In some cases, extreme winds are associated with small, hurricane-like vortices embedded within a large polar low or synoptic-scale low. It is not clear at this stage whether such systems should be regarded as a distinct class of polar low or as embedded substructure within a large system'. However, we do not conclude that a meso-β-scale low is always nurtured by a polar low. In some cases, they may develop under the influence of disturbances other than the polar low, and sometimes, even without the influence of a parent circulation.

246   3 Observational studies

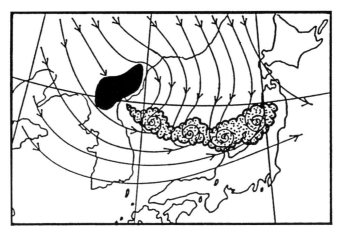

Figure 3.68. Schematic illustration of the flow field around four mesoscale vortices along the JPCZ (from Kuroda, 1992). The black area indicates the high mountain area.

Meso-β-scale lows aligned along the JPCZ

Meso-β-scale vortices aligned along the JPCZ are also frequently seen on GMS cloud images and radar refractive charts in winter. Kuroda (1992) described what he considered as a typical example of a meso-β-scale cyclogenesis along the JPCZ on 24 January 1990. A GMS infra-red satellite image from 1800 JST (0900 GMT) 24 January 1990 (not shown) indicated the long cloud zone of the JPCZ and three cloud vortices aligned along it. A stream-line analysis utilizing the movement of low clouds on successive GMS cloud images, indicated a flow field around the meso-β-scale vortices along the JPCZ as illustrated by Figure 3.68.

This case was further studied through a number of mesoscale numerical simulations (Nagata, 1993; see Section 4.3). Using a triple-nested (6, 18 and 54 km grid) primitive equations model, Nagata was able to simulate the generation of the JPCZ and the chain of meso-β-scale lows. Based on an energy conversion analysis of the simulated disturbances Nagata concluded that barotropic shear instability was the dominant mechanism for the development of this type of meso-β-scale vortex (see Section 4.3).

3.2 The Antarctic

3.2.1 Introduction

Observational studies of mesoscale weather systems over the Southern Ocean and the Antarctic continent have lagged behind comparable investigations in the Arctic because of the difficulties in assembling comprehensive data sets that can provide sufficient detail on the structure of the lows. In the Antarctic the research stations are widely spaced and clearly cannot provide information on the form of mesoscale phenomena, except when a system passes directly over a station. Mesocyclones and polar lows do occasionally pass over drifting buoys embedded in the sea ice or on the ice-free ocean, although such observing systems provide only surface data. However, many of the new forms of data becoming available can be of value to studies on the mesoscale and, when interpreted carefully, can give indications of the structure of these systems and the physical and dynamical processes involved. During some campaigns, such as the special observing periods of the Antarctic First Regional Observing Study of the Troposphere (FROST) project (Turner *et al.*, 1996b), additional *in situ* observations were collected and all the available model, satellite and *in situ* data were brought together in data centres so aiding case studies of individual mesocyclones (Lieder and Heinemann, 1999).

In this section we will review the insight that we have gained into the form, structure and environment in which Antarctic mesocyclones form through the use of a wide variety of data, collected both *in situ* and via polar orbiting satellites. In all the observational studies of Antarctic mesocyclones more than one form of data have been used so we will attempt to synthesize the various forms of data to gain insight into the processes involved.

3.2.2 The environment in which mesocyclogenesis takes place

Visible and infra-red satellite imagery provided the first indications that mesocyclones were a common occurrence over the Antarctic coastal region and the Southern Ocean (Streten and Troup, 1973; Auer, 1986) and Sinclair and Cong (1992) have claimed that 'It is difficult to find a satellite image of the midlatitudes of the Southern Hemisphere that does not include some post-frontal cold-air development.' This imagery proved valuable in allowing the first climatologies to be assembled on the occurrence of such systems (Carleton and Carpenter, 1989a, 1990) (see Section 2.2). An infra-red satellite image of the Bellingshausen Sea showing an example of an active mesocyclone and a less well defined system to the east is shown in Figure 3.69. Both the lows in this figure show the characteristic hook or comma of cloud, although the nature of the cloud is quite different in the two cases, with the western system having a continuous band of high cloud, while the cloud associated with the low to the east is much more broken.

The early satellite images were also useful for indicating where the lows developed and interpretation of the imagery allowed the broad-scale synoptic environment to be inferred at a time when the operational meteorological analyses were poor around the Antarctic. The imagery used in such early

248   3 Observational studies

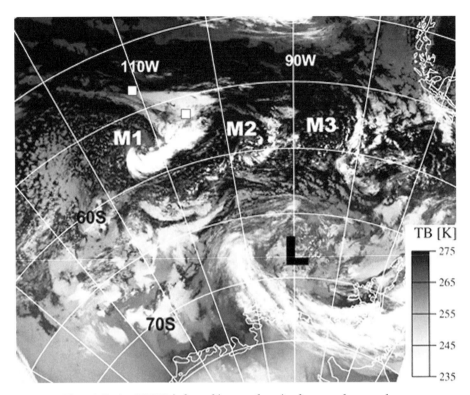

Figure 3.69. An AVHRR infra-red image of a pair of mesocyclones and a synoptic-scale low over the South Pacific at 1400 GMT 11 January 1995. M1–M3 refer to systems discussed in Section 5.2.4.

investigations was often of a relatively poor quality, consisting of hard-copy prints with few grey scales and a fairly coarse horizontal resolution. Whilst it was adequate to allow mesocyclones to be observed over the ocean areas, where there was a reasonable contrast in temperature and reflectivity between the cloud and sea surface, mesocyclones over the continental ice sheet were very difficult to observe and remained largely undiscovered until high quality, digital imagery became available in the 1980s.

Examination of the imagery shows that many of the lows form in outbreaks of cold air over the Southern Ocean, often to the west of synoptic-scale low pressure systems. These outbreaks can be recognized in visible and infra-red satellite imagery by the large number of convective clouds, and in passive microwave data by the comparatively dry air with low integrated water vapour content (Claud *et al.*, 1996). The satellite image in Figure 3.69 shows a typical situation of a large synoptic-scale low (marked L) and a cold airstream to the west in which there is convective cloud and at least three mesocyclones developing. The cold air has been drawn around the northern side of the low and the three mesocyclones are in an essentially westerly, recurved cold polar air mass.

Some mesocyclone developments have been observed to take place when an existing mesoscale cloud mass merges with a frontal band to form an 'instant occlusion' type of system (Carleton, 1981a). Often the outcome of this process is a larger synoptic-scale low with the comma-shaped cloud band of the mesocyclone merging with the frontal wave to create a new low with the appearance of a mature mid-latitude cyclone (Carleton and Carpenter, 1989a) (see Sections 3.1 and 1.6.5). Mesocyclones can also often merge with synoptic-scale lows or other mesoscale disturbances (Carrasco and Bromwich, 1991; Bromwich, 1991) to give one or more new systems which can be more vigorous than the original lows. In the study of Bromwich (1987) a mesocyclone system was observed to merge with a synoptic-scale low over the southwestern Ross Sea resulting in a more intense low.

During cold air outbreaks there can be large fluxes of heat and moisture into the lowest layers of the atmosphere over the open ocean and Heinemann (1996a) has reported fluxes of sensible heat of 300 W m^{-2} north of the sea ice edge in the Weddell Sea during winter. However, the fluxes are smaller than many of those measured in the Arctic since the ocean circulation of the Southern Hemisphere is much more zonal than in the north and warm water masses are not carried far south. In a study by Heinemann (1996c) it was found that in the Weddell Sea a vigorous mesocyclone development took place when the sea surface temperature was +2 °C and the air temperature −15 to −20 °C. Such conditions produce very unstable stratification at low levels but the inversion above often prevents the development of deep convection. For this case the fluxes of sensible heat were computed to be 280 and 360 W m^{-2} for air temperatures of −15 and -20 °C. Heinemann and Rose (1990) estimated that the flux of latent heat was about one-third of the flux of sensible heat and limited to about 180 W m^{-2} for the typical sea surface temperatures found near the Antarctic. Over ice-covered regions the surface fluxes are much smaller than in the open ocean and Weller (1980) has reported a sensible heat flux during winter of 89 W m^{-2} in the inner sea ice zone where the concentration was greater than 85%. However, as discussed in Section 2.2, many mesocyclones still form over the sea ice. The difference in temperature between the lower atmosphere over the sea ice and open ocean is greatest when the wind direction is parallel to the ice edge and under such conditions a belt of cloud is often observed near this boundary (Heinemann, 1996a), which is known as a BLF. Many studies have shown that the ice edge and the area immediately to the north are regions of frequent mesocyclogenesis (Carleton *et al.*, 1995) and a number of case studies have been carried out of developments in these regions, e.g. Claud *et al.* (1996). Some BLFs can be very extensive and Heinemann (1996a) shows an example that extended for more than 1500 km along the ice edge over the northern Weddell Sea. It has been found that if the wind direction changes from an along-ice-edge to an off-ice flow then the front can be advected over the open ocean where the ensuing strong sensible heat flux can induce mesocyclogenesis. When over the open ocean, the BLF often marks the leading edge of the cold air outbreak with the region behind often exhibiting the signs associated with very cold air over the warm ocean, such as extensive convective cloud and cloud streets. When mesocyclones develop on a BLF a number of centres have been observed to form, although one centre usually becomes dominant (Carleton *et al.*, 1995).

Many mesocyclogenesis events have been observed to take place over the area of the eastern Weddell Sea (see Section 2.2) but the air–sea temperature differences here rarely exceed 10 K and are typically around 5 K (Heinemann, 1996b). In one case examined by Heinemann (1996b), the surface flux of sensible heat was estimated at less than 25 W m^{-2}, suggesting that other factors are important here in the development of mesocyclones. The area is characterized by offshore flow of cold air, so relatively cold air masses are found over the ocean area. However, the extensive sea ice in winter limits the surface fluxes that can develop since the ice provides an effective cap on the ocean. Nevertheless, the offshore flow is important in the development of mesocyclones in this area (Heinemann, 1990) where the winds play a part in the establishment of low-level baroclinicity and minor frontal bands when the cold continental air meets more maritime air masses (Heinemann, 1996b).

Mesocyclogenesis has been observed to occur on low-level baroclinic zones in many locations from the inland ice sheets to the open ocean. Inland, on the Ross Ice Shelf the downslope katabatic winds flow down the glacial valleys (Parish and Bromwich, 1987), bringing cold air into contact with the more maritime air masses on the ice shelf and establishing significant horizontal gradients of temperature. Figure 3.70 shows such a development taking place via NOAA 10 infra-red satellite imagery. The black streaks emerging from the valleys indicate strong winds that break down the surface inversion and bring warm air down to the surface. Figure 3.70 shows a mesocyclone development taking place on the ice shelf with the dark katabatic wind signatures spiralling into a mesocyclone south of automatic weather station 15. The AWS data showed that a surface trough was present prior to mesocyclogenesis taking place and that in the presence of the katabatic wind-induced baroclinic zone the trough developed into the low.

As discussed above, the existence of low-level baroclinicity also appears to be important in the formation of many mesocyclones over the eastern Weddell

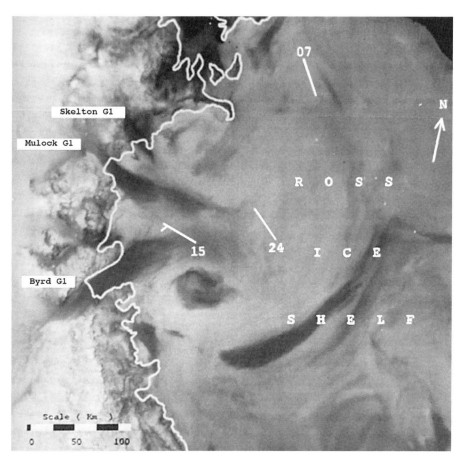

Figure 3.70. A NOAA 10 infra-red image for 1723 GMT 17 February 1988 showing the Ross Ice Shelf at a time when a mesocyclone was developing (from Bromwich, 1991).

Sea, the cold air for the establishment of this gradient coming from the interior of the continent. The interaction of mesoscale disturbances with the katabatic flow in this area also appears to be important, although we have little information at present about the detailed air flow of the region. The various cases examined to date for this area show that the cold air outbreaks can be either very deep or quite shallow and that an enhancement of the low-level baroclinicity by cold katabatic winds can lead to cyclogenesis. With the vigorous low examined by Turner *et al.* (1993b) the baroclinicity off the coast close to Halley Station formed as warm air was advected down the eastern Weddell Sea and onto the Ronne Ice Shelf. At the same time a cold trough moved down the coastal slope towards Halley giving a strong thermal gradient over the area as can be seen in the 1000–500 hPa thickness field derived from satellite sounder data and

252 3 Observational studies

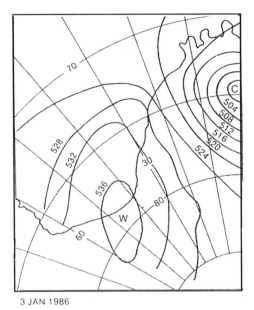

Figure 3.71. A TOVS 1000–500 hPa analysis of the eastern Weddell Sea for 0000 GMT 3 January 1986 (from Turner et al., 1993b).

shown in Figure 3.71. As the cold air descended towards Halley it undercut the warm air, as can be seen from the 1200 GMT 3 January 1986 radiosonde ascent from the base shown in Figure 3.72, which shows cold air up to a height of 930 hPa. In this area of the eastern Weddell Sea the steep orography just inland of the coast plays a further role in cyclogenesis as a result of the convergence of katabatic winds and the channelling of cold air into certain parts of the coastal region enhancing the baroclinicity. This zone of enhanced low-level baroclinicity and convergence aids vortex formation through a combination of baroclinic and barotropic instability (Münzenberg-St. Denis, 1994). The existence of a warm trough along the coast in this area has also been shown to be important in some mesocyclone developments since it will provide both warm air to enhance the baroclinicity and promotion of vertical motion through the presence of the trough (Heinemann, 1996b).

Moderate low-level baroclinicity is also found close to the edge of the sea ice and, as discussed above, a number of mesocyclones have been observed to form in such a location. Observational studies have shown that the lows in these regions are often quite shallow but can have well-established fronts and associated low-level jets (Heinemann, 1996a). The low studied by Heinemann (1996c) developed on an offshore cloud band and resembled the 'Arctic front' type of development observed in the Northern Hemisphere (see Section 3.1). This low was triggered by a 500 hPa short-wave trough, which interacted with a BLF

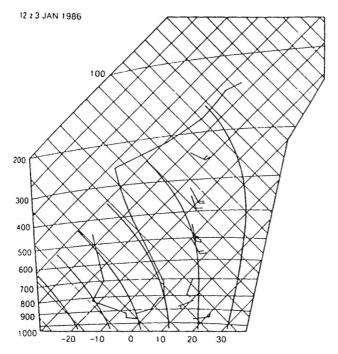

Figure 3.72. Halley radiosonde ascent for 1200 GMT 3 January 1986 (from Turner *et al.*, 1993b).

lying parallel to the edge of the sea ice. The formation of the low was characterized by a cold air outbreak combined with strong fluxes of sensible heat and low-level baroclinicity. This development was typical of many studied in the Antarctic in that a combination of two or more factors was important in the formation of the low.

Satellite sounder data can also indicate the presence of baroclinicity, since the satellite measurements can be used to derive thickness values for certain layers of the atmosphere. Sounder data were used in the study by Claud *et al.* (1996) where two lows were examined in detail. They found that one mesocyclone showed the presence of weak, low-level baroclinicity via the TIROS Operational Vertical Sounder (TOVS) data, while the second case did not have any indications of strong thermal gradient. Data from the first case, which occurred on 4 October 1988 over the northern Amundsen Sea, are shown in Figure 3.73. This mesocyclone developed close to 57° S, 105° W (indicated as a cross on Figure 3.73), just to the south of a strong thermal gradient indicated by the 1000–500 hPa thickness values, suggesting that this development was a predominantly baroclinic event. Other mesocyclone developments have been noted within the main cold air mass rather than in the baroclinic zones toward the edges of the polar air outbreaks.

254   3 Observational studies

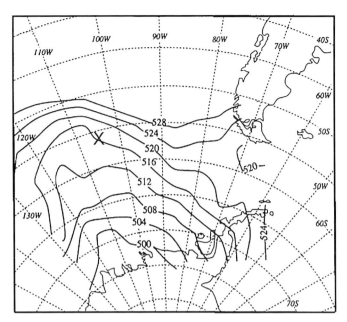

Figure 3.73. The 1000–500 hPa thickness field over the South Pacific produced from TOVS data covering the period 0240–0354 GMT 4 October 1988. The cross indicates the location of the mesocyclone. (After Claud *et al.*, 1996, The Global Atmosphere and Ocean System, Taylor & Francis Ltd (http://www.tandf.co.uk/journals).)

In the study by Claud *et al.* (1996) it was found that the baroclinic zones in which the systems developed were weaker than those in the Northern Hemisphere. This result is somewhat surprising since, overall, baroclinicity is stronger in the Southern Hemisphere, which accounts for the more active synoptic-scale lows that are found compared with the Northern Hemisphere. However, as so few cases have been investigated in depth in the Southern Hemisphere it is probably too early to draw general conclusions about the strength of the baroclinicity. In this study Claud *et al.* also found that the 1000–850 hPa thickness gradient was strongest in the incipient phase and became weaker at later stages of the development of the lows.

Observational studies have shown that mesocyclones can also develop on pre-existing frontal or non-frontal cloud bands in a variety of situations. Sometimes the fronts can be embedded in the cold air flow and the systems that form have the characteristics of the 'Arctic front' type of disturbance found in the Northern Hemisphere. On other occasions mesocyclones can form on the fronts associated with synoptic-scale lows, e.g. the two mesocyclones that developed on the front of a dissipating synoptic-scale low in the Drake Passage observed by Marshall and Turner (1997a). With the case examined by Turner *et al.* (1993b), a vigorous mesocyclone developed on a cold front at the leading edge of a cold air outbreak descending from the high interior plateau to the eastern Weddell Sea and this example shows that the fronts found around the Antarctic do not always conform to the Norwegian model and that the steep topography can result in unusual frontal structures in the coastal region.

The large number of AWSs installed in the Antarctic in recent years has allowed the production of high resolution regional analyses that have shed further light on mesocyclogenesis. In particular, the area around Terra Nova Bay has been shown to have many developments taking place in the baroclinic zone created where the cold katabatic winds descending from the interior (Bromwich, 1986) meet the more maritime air masses over the ocean. Factors favouring developments here are strong baroclinicity, enhanced katabatic flow, more open water than usual in the Ross Sea and greater synoptic-scale cold air advection from the Ross Ice Shelf onto the Ross Sea (Carrasco and Bromwich, 1996). Mesocyclones often form when the katabatic winds are strong but can also form in relatively weak downslope flow if the synoptic-scale flow is conducive to mesocyclogenesis (Bromwich and Parish, 1988). Figure 3.74 shows

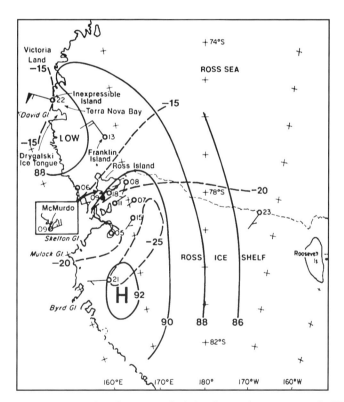

Figure 3.74. A regional MSLP analysis for the southern Ross Ice Shelf for 1200 GMT 20 February 1984 (from Bromwich, 1987).

a detailed analysis for the southwestern Ross Sea/Ross Ice Shelf area when a mesoscale low was developing. The system can be observed just to the south of the Drygalski Ice Tongue and the low centre is marked within a trough of low pressure extending from the northwest. Bromwich (1991) has described the mesoscale developments in the Terra Nova Bay and Byrd Glacier areas as being a result of a dry baroclinic process with marked baroclinicity being present, along with weak cyclonic vorticity in the boundary layer. In his study of two years of data there was little consistent upper-air support for the developments, although upper-level forcing often played a role in the subsequent development of the lows. On the other hand, synoptic forcing was found to play a significant role in the developments via troughs ahead of maritime cyclones located to the northwest of the Ross Sea. Overall, in the Terra Nova Bay area mesocyclogenesis appears to take place largely independently of the synoptic-scale pattern and is primarily driven by the katabatic winds, although synoptic activity can be important in the subsequent development of the lows.

This area of the southwest Ross Sea and Ross Ice Shelf area has been studied in depth for a number of years and we now have a great deal of information on mesocyclogenesis here through the use of *in situ* and satellite data. One major initiative (Bromwich, 1991) was to examine the general regional conditions that gave rise to the frequent mesocyclogenesis episodes by the preparation of daily and mean fields for the whole of 1985, including sea level pressure, potential temperature and vector average surface wind. The mean field of surface pressure on Figure 3.75 indicates the decrease of pressure from the Transantarctic Mountains towards the northeastern side of the Ross Ice Shelf and a trough extending from north of Franklin Island across McMurdo station to Minna Bluff. In addition, it also shows that intense katabatic winds blow into Terra Nova Bay, resulting in marked baroclinicity between this air mass and the surrounding environment. The figure also suggests that katabatic winds flow from Byrd Glacier, but these are less strong than at Terra Nova Bay; and that the two areas of frequent mesocyclogenesis at Terra Nova Bay and Byrd Glacier are characterized by persistent katabatic winds with negative buoyancy and close proximity to persistent surface pressure troughs.

Many mesocyclogenesis events occur in association with a trough either at low levels or in the mid- to upper troposphere. For mesocyclogenesis events over the eastern Weddell Sea, the existence of a trough over the area seems to be a common occurrence, with it often being associated with a synoptic-scale low north of the Antarctic coast, close to the Greenwich Meridian. This situation gives cold, offshore flow between Halley and Georg von Neumeyer stations and can generate cyclonic barotropic vorticity by vortex stretching, which was the case with the system on 8 February 1990 examined by Heinemann (1996b).

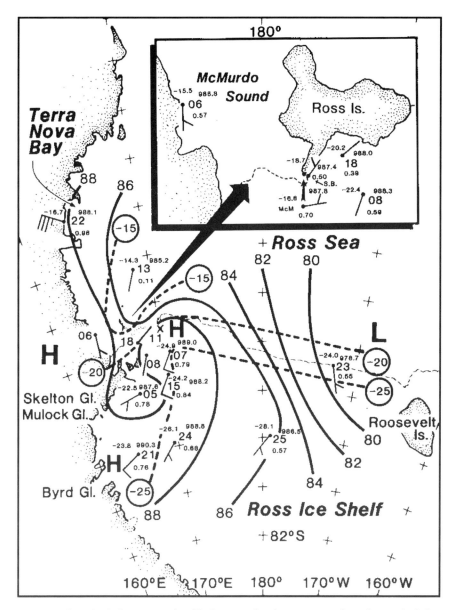

Figure 3.75. Average sea level isobars, surface isentropes and resultant winds for 1985 for the southwestern Ross Sea. Stations 13 and 05 are respectively Franklin Island and Minna Bluff (from Bromwich, 1991).

Over the eastern Weddell Sea the synoptic-scale flow can aid the development of mesocyclones by supporting the drainage flow off the continent (Heinemann, 1990) or inhibit the flow if the wind is onshore.

Vortex stretching can be important in other parts of the Antarctic; for example, Carrasco and Bromwich (1995) examined the case of a small low that descended from Victoria Land into the semi-permanent baroclinic zone over the southwestern corner of the Ross Sea and developed into a frontal system that existed for five days before declining over Marie Byrd Land. This study concluded that the low had developed because of vortex stretching as it descended the 2000 m from the interior plateau down to sea level, but was also influenced by the influx of cold katabatic air into the low.

Many observational studies have indicated that an upper-level trough is present when mesocyclogenesis takes place (Bromwich, 1991; Turner *et al.*, 1993b); or when an existing mesocyclone intensifies (Carrasco and Bromwich, 1991). New developments often take place to the east of the upper trough in the zone of enhanced vertical motion where synoptic-scale lows also frequently form. The trough is often important in instigating cyclonic vorticity advection at mid- to upper levels as, for example, in the case examined by Heinemann (1996a). Here a mesocyclone developed when a 500 hPa trough with upper-level cyclonic vorticity advection and a region of upward motion at 700 hPa moved over a BLF at the ice edge. In addition, further forcing came from the mesocyclone being beneath the exit region of a weak 500 hPa jet. However, strong upper-air cyclonic vorticity advection is not always present, e.g. the case of Heinemann (1996c) where there was little cyclonic vorticity advection but strong advection of cold air at the surface. The support of a synoptic-scale upper trough can also encourage the development of a pre-existing low so that relatively weak mesocyclones can develop rapidly if an upper-level trough moves over the system. Lieder and Heinemann (1999) have noted that the strength of the upper trough appears to be a factor in how the lows develop.

In one case examined by Bromwich *et al.* (1996), a mid-tropospheric mesocyclone developed within an upper-level trough at the edge of the Ross Ice Shelf. When first observed, this low had only a spiral of thin cloud at approximately the 500 hPa level but the low developed further within the short-wave trough as the lower atmospheric conditions became more unstable. In this case the thermal gradient across the region became more pronounced as warm air advection took place in association with a low pressure centre and an associated frontal band moved onto the ice shelf from Marie Byrd Land. Although there is often moderate baroclinicity in the vicinity of the southern Ross Sea the particular synoptic-scale pattern at this time resulted in a very unusual situation, with warmer air to the south and colder air to the north. Further development of the mesocyclone took place as the remains of the old synoptic low merged with the mesocyclone to give a single, vigorous mesoscale low. During this time the mid-level component of the low declined (possibly because of advection of anticyclonic vorticity at this level), leaving an active low-level vortex.

With many mesoscale developments around the Antarctic it is difficult to determine the role of upper-level forcing because of the poor quality of the upper-level fields. This was the case with the system examined by Bromwich (1987), when there were no upper-air observations available in the area and the upper-air charts could not be relied upon. However, as upper-air analyses for the Southern Ocean improve, further insight should be obtained on the role of upper forcing in the development of mesocyclones.

The role of different forcing mechanisms close to the coast has been investigated by Heinemann (1996b), who has suggested that developments take place because of two different processes: (1) as a result of baroclinic instability in association with upper-level forcing. Such mesocyclones often resemble short baroclinic waves (e.g. Turner *et al.*, 1993b) and often have deep vertical extent and lifetimes exceeding 24 h in some cases; (2) lows that develop in association with orographic forcing by katabatic cold air outflow (e.g. Carrasco and Bromwich, 1993). These lows, which are common over the eastern Weddell Sea, have more shallow vertical extent and a lifetime of usually less than 24 h.

Pools of cold air at mid- to upper levels have been shown to be important in the development of some mesocyclones through the analysis of TOVS sounder data (Heinemann, 1996c). This was the case with the lows over the eastern Weddell Sea examined by Heinemann (1990), where the presence of the cold air was thought to be essential for the development of the mesocyclones. Other studies involving TOVS sounder data to examine such developments have indicated that the multi-banded, 'merry-go-round' type of mesocyclone may occur in such a barotropic environment.

Synoptic-scale or large mesoscale lows can play an important part in triggering the development of new mesocyclones both over the ocean and in the Antarctic coastal region. Bromwich and Parish (1988) examined a case in which a large mesocyclone that had developed over the South Pacific moved into the area of Terra Nova Bay and triggered the development of two mesocyclones, one over the immediate coastal area and a second low over the interior plateau to the west of Terra Nova Bay. This second low formed within a strong baroclinic zone between warm, maritime air advected onto the plateau by the low in the coastal area and a cold, katabatic air mass blowing southwards along the mountains. The development of this low over the interior disrupted the surface wind regime, which is usually downslope and briefly gave marked upslope winds in a region of usually persistent katabatic flow.

When the conditions are favourable for mesocyclogenesis it is often observed that a series of lows form in a particular area over a period of several days (Bromwich, 1987, 1991). A case of five mesocyclones developing simultaneously over the Ross Ice Shelf was noted by Carrasco and Bromwich (1995), with the systems consisting of a mix of comma clouds and 'merry-go-round' vortices. In such a situation there is often a merging of mesocyclones and the emergence of one or more dominant vortices. In the case examined by Bromwich *et al.* (1996), a mesocyclone merged with an old, decaying synoptic-scale low and satellite imagery of the resulting active mesocyclones showed indications of the two centres within the new vortex.

Serial development of mesocyclones on a front has also been observed with a family of lows forming; for example, the two lows developing on a frontal cloud band associated with a declining depression examined by Marshall and Turner (1997a). Groups of several mesocyclones have also been observed to form over the eastern Weddell Sea during cold air outbreaks (Heinemann, 1990).

Although most investigations into Southern Hemisphere mesocyclones and polar lows have been concerned with systems in the Antarctic coastal region, early studies such as that of Auer (1986) examined mesocyclones at the relatively northerly latitude of Australia. More recently, Zick (1994) considered polar lows in the Tasman Sea, between Australia and New Zealand. Here we will consider briefly the additional insight that this work gave us of polar lows in the New Zealand sector via an examination of polar lows that occurred during the period 21–27 August 1990. Zick ascribes the frequent development of polar lows in the Tasman Sea area to the many extratropical cyclones and associated polar lows developing because of the occurrence of the two climatological jet streams at 30° and 60° S, along with the large-scale confluent motion of subpolar and subtropical air masses in the wake of the southeast Australian Alps. The Zick (1994) study was concerned with a series of polar low developments, which he called the 'James' series of vortices, since one of the vortices crossed the New Zealand ship *James Cook*. These consisted of two symmetric/convective vortices and three major comma clouds. The vortex named 'James' had a tight, 400 km diameter cloud signature, and developed over the Tasman Sea on 21 August 1990 and moved southeastwards towards the west coast of South Island. When close to the *James Cook* on 25 August the vortex had 45 kt winds, indicating that this system was an active polar low. The whole sequence of events took place within a large-scale, cold trough that extended northwards across the Tasman Sea. The 'James' vortex developed out of the head of a large comma-shaped area of convective cloud in the centre of a large, cold barotropic trough and a satellite image of the system about 12 h after emerging from the comma head in shown in Figure 3.76. During the subsequent 24 h the vortex unfolded as a strongly rotating subsynoptic-scale vortex, independent of the triggering comma cloud. On 23 August 'James' was located in the very centre of a 500 hPa cut-off low, but there are indications that the vortex was positioned

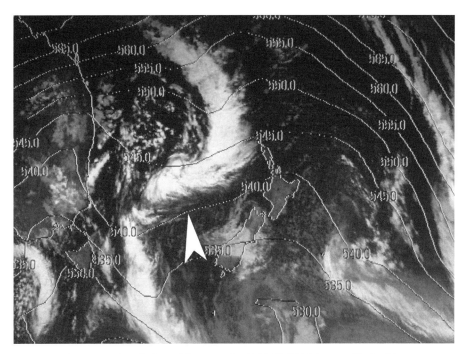

Figure 3.76. GMS infra-red imagery for 1200 GMT 21 August 1990 with superimposed 500–1000 hPa thickness field. The arrow indicates the initial pattern of 'James'.

in the centre of a warm region associated with convective activity. There are indications that the vortex was convectively driven and that it had a warm core during parts of its life cycle. Analysis of model fields showed that the low was associated with a low-level vorticity maximum on 23 August, with the vorticity decreasing up to the 300 hPa level. Evidence for this comes from the fact that 'James' had its lowest central mean sea level pressure (MSLP) at the time of the warmest core and through an objective cloud displacement analysis. This showed that on 25 August, at the western edge of the vortex cloud, there was a strong vertical shear from anticyclonic cirrus motion to cyclonic low-level cloud motion. In conclusion, there is evidence for a warm core system where CISK or ASII could have played a role.

Overall, the observational studies carried so far suggest that most mesocyclones occurring in the Antarctic form initially as a result of baroclinic instability, with support from an upper-level trough often being present. Surface fluxes of heat and moisture into the developing lows are usually much lower than in the Arctic but can be quite large at relatively northerly latitudes or in coastal polynyas. The importance of the various forcing mechanisms can vary considerably with different developments, which seems to account for the wide variety of cloud signatures observed, as discussed below.

3.2.3 The structure of Antarctic mesocyclones

Thermal structure

Some information on the thermal structure of mesocyclones can be obtained when the lows cross stations that launch radiosondes, but such events are rare and the personnel are not usually instructed to make additional ascents in such circumstances. Therefore the most common means of investigating the structure of mesocyclones is using the temperature profiles derived from satellite sounder data, such as the TOVS. An example of the 1000–500 hPa thickness data derived from TOVS measurements overlaid on AVHRR infra-red imagery is shown in Figure 3.77. This figure shows that the TOVS data were able to resolve the cold air behind the cold front that had descended from the interior of the continent and the warm pool over the Ronne Ice Shelf. A cross-section of potential temperature through the low, as indicated by the line A–B on Figure 3.77, is shown in Figure 3.78. This figure also has the locations of the cloud bands and the frontal surface indicated. To the west of the research station the isentropes below 600 hPa were almost horizontal with only a very slight slope towards the front. However, to the east of Halley there was a marked upward slope at all levels up to 300 hPa. The cross-section also shows that the low-level cloud bands were located ahead of the main lower-tropospheric thermal gradient and the surface cold front. Because temperature profiles derived from TOVS data are smoother

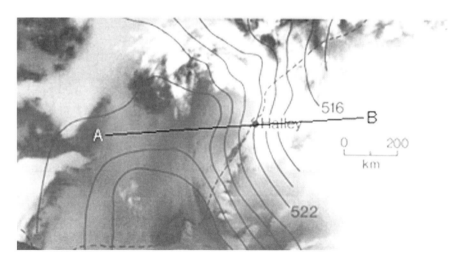

Figure 3.77. Overlaid AVHRR infra-red imagery and TOVS 1000–500 hPa thickness over the area around Halley Station at 2349 GMT 3 January 1986.

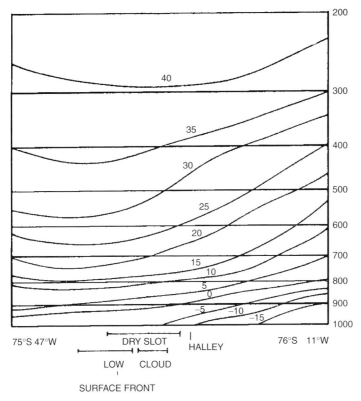

Figure 3.78. A cross-section of potential temperature through a mesocyclone off Halley at 0000 GMT 4 January 1986 (from Turner *et al.*, 1993b).

than comparable radiosonde ascents the cross-section will be smoother than in reality.

When mesocyclones are more mobile the rotation of the system, as occurs with synoptic-scale cyclones, distorts the thermal field with cold air often being pulled northwards behind the comma of cloud, which has the characteristics of a cold front. Although the existence of this cold air can be inferred from the presence of shower clouds in the unstable air mass the TOVS thickness analyses provide a very valuable means of monitoring the thermal evolution of the lows, with satellite data being available every few hours.

As discussed above in Section 3.1, there has been considerable discussion of the warm cores noted at the heart of a few Northern Hemisphere polar lows. In the Antarctic the vortices are not usually characterized by deep convective cloud and there is rarely the release of large amounts of latent heat from cumulonimbus cloud. It is therefore believed that any area of warmer tropospheric temperatures at the centre of a mesocyclone is present because of a seclusion process whereby warmer air becomes isolated in the centre of the low through being drawn into the centre of the circulation. Observational data show that the centres of mesocyclones have been associated with both warm and cold low-level anomalies (Heinemann, 1996b), adding to the debate over the origins of warm anomalies at the heart of these lows. At present the satellite sounder data do not have the horizontal or vertical resolution to easily resolve any small areas of warmer air associated with mesocyclones; however, Claud *et al.* (1996) examined a case where there was no warm core but a warm ridge at the 500–300 hPa layer was present, which extended into the centre of the mesocyclone. This ridge was more significant in the 500–300 hPa layer than at 1000–500 hPa, and generally more pronounced in the developed and mature stages of the system (Figure 3.79). In this case it was suggested that the warm core could be a result of adiabatic descent in the cloud-free region at the centre of the low or the result of a seclusion process.

The TOVS data can indicate the outbreaks of cold air associated with mesocyclone developments and show strong thermal gradients on their borders. The analysis of satellite imagery in conjunction with 1000–850 hPa thickness fields derived from satellite sounder data has suggested that in addition to cold air being present behind the main frontal band of a mesocyclone there are indications of a weak warm sector to the east of the front. The

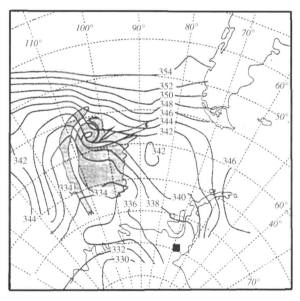

Figure 3.79. TOVS 500–300 hPa thickness over the South Pacific for 1152–1332 GMT 3 October 1988. Shaded areas indicate cloud diagnosed from satellite imagery. (From Claud *et al.*, 1996, *The Global Atmosphere and Ocean System*, Taylor & Francis Ltd (http://www.tandf.co.uk/journals).)

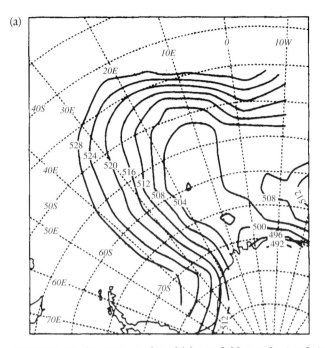

Figure 3.80. (a) The 1000–500 hPa thickness field over the South Atlantic produced from TOVS data covering the period 1856–2036 GMT 14 July 1988. (b) The 500–300 hPa thickness field for the same time. Shaded areas indicate cloud diagnosed from satellite imagery. (From Claud *et al.*, 1996, *The Global Atmosphere and Ocean System*, Taylor & Francis Ltd (http://www.tandf.co.uk/journals).)

dissipation of mesocyclones has also been monitored using TOVS data, which has shown a rapid increase in 1000–500 hPa thickness values as the lows decline.

When the spiraliform mesocyclone shown in Figure 1.14 was investigated using TOVS data the 1000–500 and 500–300 hPa thickness fields shown in Figures 3.80a and b respectively were obtained. The marked outbreak of cold polar air with 1000–500 hPa thickness values of less than 504 dm can be seen clearly in Figure 3.80a, and Figure 3.80b also shows that the development took place under a secondary cold pool in the mid- to upper troposphere.

Claud *et al.* (1996) computed a TLS (temperature of the lower stratosphere) quantity using a combination of TOVS channels in order to try and determine whether there was evidence in the satellite data of warm anomalies at the tropopause level in association of mesocyclones. Warm air at this level would be associated with a lower tropopause and indicate an upper-level potential vorticity maximum. In the two Southern Hemisphere cases they examined there was no evidence of any warming in the TLS product. This was probably

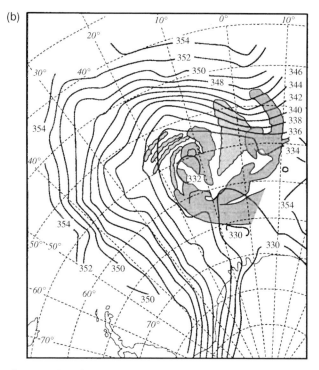

Figure 3.80 (cont.).

an indication that Southern Hemisphere mesocyclones are weaker than their northern counterparts.

The first direct aircraft observations of an Antarctic mesocyclone were made by Heinemann (1996b) over the eastern Weddell Sea during February 1990. In this investigation a flight was made on 8 February 1990 that crossed a mesocyclone twice, once at low level and once at mid-tropospheric level. During the passage at the 700 hPa level dropsondes were deployed in order to gain information on the three-dimensional temperature, moisture and wind fields. His temperature cross-section derived from the dropsonde data is shown in Figure 3.81a, with the satellite imagery and superimposed aircraft track shown in Figure 3.81b. In this case the flight at the 700 hPa level was everywhere above the cloud tops of the mesocyclone, indicating that the vortex was a relatively low-level feature. From the cross section it can be seen that neutral or unstable stratification of the atmosphere was found only in the lowest 60 hPa of the atmosphere, while in the cloud layers slightly stable conditions prevailed.

The moisture structure and fronts

Mesocyclones develop in a wide range of locations from over the very dry interior of the Antarctic continent to relatively northerly locations over the

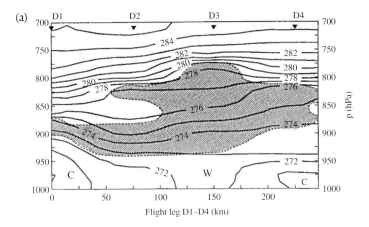

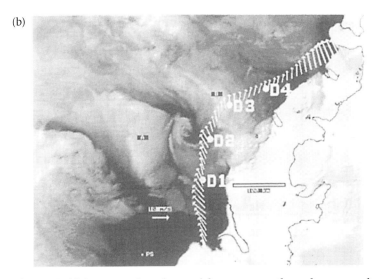

Figure 3.81. (a) A cross-section of potential temperature through a mesocyclone over the eastern Weddell Sea. Areas exceeding 80% humidity are shaded.
(b) AVHRR visible imagery of a mesocyclone over the eastern Weddell Sea at 1520 GMT 8 February 1990 with overlaid 700 hPa wind vectors from an aircraft overflight. (From Heinemann, 1996b, *The Global Atmosphere and Ocean System*, Taylor & Francis Ltd (http://www.tandf.co.uk/journals).)

Southern Ocean where moisture levels are comparable to mid-latitude values. To date, most of our knowledge of the moisture field of mesocyclones comes from passive microwave data since isolated radiosonde data cannot provide information on the spatial variability of moisture variables. One limitation of such data, though, is that moisture variables can only be derived over the ice-free ocean. However, the data that can be determined are very valuable and, for example, Katsaros *et al.* (1989) have shown that the integrated water vapour (IWV) field is a good indicator of frontal activity.

268   3 Observational studies

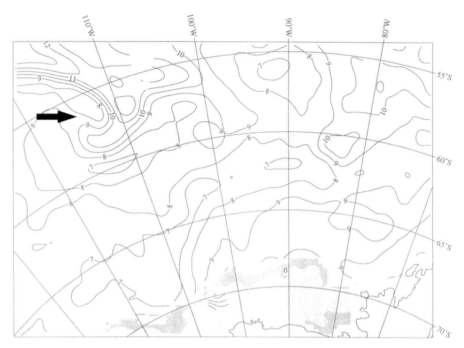

Figure 3.82. Integrated water vapour (isolines every 1 kg m^{-2}) over the ice-free ocean for 1000–1300 GMT 11 January 1995. The arrow indicates the mesocyclones and the shading the extent of the sea ice.

Recently, the IWV field as determined from SSM/I data was used by Lieder and Heinemann (1999) to examine the moisture structure of a mesocyclone over the Bellingshausen Sea. Their results, which are shown in Figure 3.82, indicate that the mesocyclone (indicated by an arrow) had levels of IWV with values up to 13 kg m^{-2} close to the low's frontal band, which compares to mid-latitude values as great as 48 kg m^{-2} found in comparable investigations of synoptic-scale cyclones (McMurdie and Katsaros, 1991). In this case the IWV values around the most vigorous mesocyclone showed high gradients ahead of the frontal bands separating the dry polar air with values of 7–8 kg m^{-2} from the moister mid-latitude air with values of 11–12 kg m^{-2}. These values are comparable with those found by Claud *et al.* (1996), who determined IWV values of up to 10–11 kg m^{-2} in the two Southern Hemisphere cases they examined.

Carleton *et al.* (1995) examined three examples of Antarctic mesocyclones using a combination of passive microwave data and conventional visible and infra-red imagery. They found that the systems, although appearing different in the imagery, were all characterized by low IWV. The water vapour field was found to be relatively featureless around the mesocyclone since the vortex occurred in the polar air mass. The IWV values were quite low and in the range 3–5 kg m^{-2}, indicating development in dry air of Antarctic origin. They found that the lowest values of IWV were located southeast of the low. In another case, values of IWV of 7–9 kg m^{-2} were found in the early stages of development of the low, which rose to 10–11 kg m^{-2} in a later satellite pass, possibly as a result of ingestion of warmer, more moist air or local convergence of moisture as a result of vertical motion.

The two cases examined by Claud *et al.* (1996) had IWV values in the range 2–9 kg m^{-2} during the incipient phase, with values increasing during the mature stage by about 50% to 4–11 kg m^{-2}. However, the algorithms used to derive the IWV values were not specifically tuned for the polar regions and may have underestimated the moisture values. The same systems had integrated cloud liquid water content values in the range 0.15–0.35 kg m^{-2} during the mature phase of the lows, when the frontal structures were most developed. Cloud liquid water (CLW) can also be derived from passive microwave data and fields of this quantity have been used by several workers to examine mesocyclones over the ice-free ocean. With the three cases examined by Carleton *et al.* (1995) it was found that all the lows were characterized by low to moderate values of CLW in the range 0.1–0.3 kg m^{-2} indicating the cloud features were not as well developed as those associated with mid-latitude disturbances. BLFs on which mesocyclones were observed to develop were barely detectable in the CLW field, indicating the weak signals that can be present in this quantity, although the front was easily detected in conventional infra-red imagery. Similarly, the mesocyclones themselves can often be difficult to detect in CLW data alone, although nearby synoptic-scale disturbances can often be quite easily identified (Carleton *et al.*, 1995).

The frontal and moisture structure of a mesocyclone over Halley Station was examined by Turner *et al.* (1993b) using a combination of radiosonde data from the station, conventional satellite imagery and the 6.7 μm water vapour imagery from the HIRS sounder instrument used as coarse resolution water vapour imagery. These data were used to construct conceptual models of the airflow, cloud and moisture fields at two stages in the development of the lows and figures illustrating these models are shown in Figure 3.83. These figures show that there were two main regions of flow associated with the system: (1) easterly descending air inland of Halley, which has been diagnosed by the dry slot in the water vapour imagery; (2) an area of low-level, slantwise ascent along the developing cloud on the cold front, which transported moisture into the system. At an early stage in the development of the low at 0000 GMT 4 January 1986 (Figure 3.83a) a region of low humidity could be detected in the water vapour imagery extending from over the continent into the centre of the

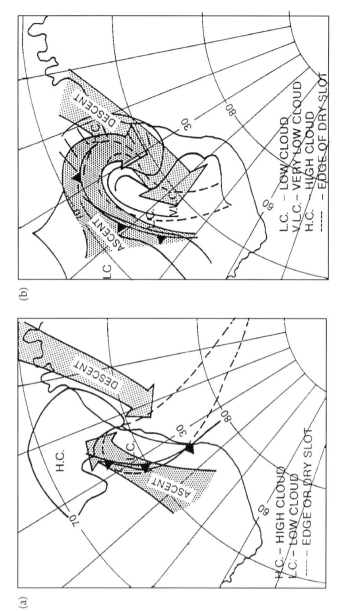

Figure 3.83. Conceptual models of a mesocyclone over Halley Station. (a) 0000 GMT 4 January 1986, (b) 1700 GMT 5 January 1986 (from Turner *et al.*, 1993b).

vortex. This air was behind the cold front on which the low developed and, from our knowledge of the characteristics of the particular channel, can be placed close to the 600 hPa level. An isentropic analysis carried out using model data showed that the air to the east of the dry slot was descending, bringing dry air down from the upper troposphere. A similar analysis within the warm air over the Weddell Sea indicated ascent ahead of the developing front. As the cold air pushed out into the Weddell Sea and undercut the more maritime air the comma of cloud on the cold front thickened but the cloud tops rose. The analysis indicated slantwise ascent up the frontal surface. At the later time of 1700 GMT 5 January (Figure 3.83b) the surface cold front and mid-level dry slot were over the Weddell Sea with a broad area of descending air poleward of these features. The descent was within about the lowest 400 hPa of the atmosphere and the ascent up the frontal surface was up to about 500 hPa.

In other cases examined over the southwestern Ross Sea and the immediate inland areas a dry slot has been observed to be present because of the intrusion of a dry katabatic airstream into a mesocyclone. Conceptual models of the frontal structure and air masses associated with a major mesocyclone over the Ross Ice Shelf were constructed by Carrasco and Bromwich (1995) using a combination of satellite and *in situ* data. Figure 3.84a shows the model for 0639 GMT 10 January 1988 when the vortex was well developed and had a cold front analysed to the north of the low centre. The analysis indicated relatively warm air being drawn into the circulation from the north and katabatic winds from Terra Nova Bay providing cold air from the west and north. The two air masses detected were found to spiral into the centre of the low. The more sophisticated model of Browning (1990) was used to incorporate the concept of conveyor belts into the model of this mesocyclone and the schematic development is shown in Figure 3.84b. In this figure the solid arrow indicates a warm conveyor belt with forward-sloping ascent relative to the movement of the front and Browning (1990) has suggested that such movement often occurs ahead of a diffluent upper-level trough. The dotted arrow indicates a cold conveyor belt, which Browning (1990) described as originating in the anticyclonic low-level flow to the southeast of the subsynoptic-scale cyclone. In this case the cold air moved westwards in relation to the cyclone's direction of travel and undercut the warm conveyor belt. In Figure 3.84b the dashed arrow indicates cold, dry boundary layer air (called in this study a boundary layer conveyor belt) that spiralled in behind the low. Carrasco and Bromwich (1995) summarize the flow in this case as the warm conveyor belt moving cyclonically southwards while being lifted by the boundary layer conveyor belt and then turning anticyclonically and ascending over the cold conveyor belt.

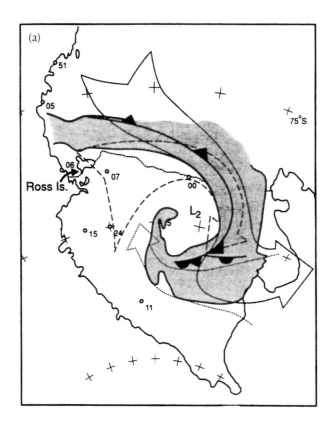

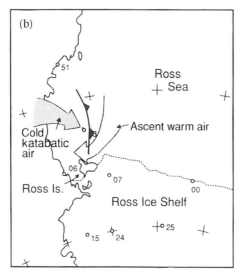

Figure 3.84. (a) Conceptual model of the airflows associated with a subsynoptic-scale vortex at 2100 GMT 7 January 1988. (b) Schematic illustration of the warm and cold airflows associated with the cloud organization at 0639 GMT 10 January. The solid arrow indicates the warm conveyor belt, the dotted arrow the cold conveyor belt and the dashed arrow the boundary layer conveyor belt (from Carrasco and Bromwich, 1995).

The passage of the front associated with a mesocyclone across a research station and observed via radiosonde data has been documented by Heinemann (1996c). In this case the cloud band took the form of a line of convective cloud and had the characteristics of a cold front with a cooling of about 5 K in the 500–700 hPa layer, but with a warming at the surface. The passage of the front therefore resulted in a destabilization of the lower troposphere.

The upper air wind field

Understanding the upper-air flow above mesocyclones is important since the lows generally move in response to the mid-tropospheric flow. Most mesocyclones are found in data-sparse areas where there are no radiosonde ascents and the quality of the analyses from numerical models can be suspect. However, some upper-air winds can be determined using conventional satellite imagery. For example, some upper-air wind vector could be determined for the Halley mesocyclone discussed earlier by tracking selected cloud features in a sequence of infra-red satellite images. These vectors are shown in Figure 3.85 superimposed on the infra-red image for 1700 GMT 5 January 1986. The vectors indicate the wind at the level of the cloud tops, which varied from 800 to 500 hPa. The vectors indicate two main flow regimes around the front to the west of the vortex centre. Behind the front the winds were from the south or southeast and quite light at around 5 m s^{-1}. Within the extensive low cloud in the warm air the winds were from the west in the ascending flow up the frontal surface. These winds were stronger than in the cold air, but everywhere less than 10 m s^{-1}.

Radiosonde ascents through Antarctic mesocyclones are fairly rare but the study by Heinemann (1996b) was fortunate in that some ascents were available from the research vessel *Polarstern*, which was close to the vortex. In the case of the low over the eastern Weddell Sea on 19 January 1990, the profile of wind vectors indicated a decrease of wind speed with height from values of about 9 m s^{-1} near the surface to less than 2 m s^{-1} at 800 hPa, the height of the inversion base. Above that level the wind speed increased with height to about 15 m s^{-1} at 500 hPa. The wind direction also showed a shear of about 100° at the base of the inversion. This low had a two-layer vertical structure of the wind field, with off-ice flow near the surface and coast-parallel flow above the boundary layer.

The surface pressure

We can gain information on the pressure perturbations associated with mesocyclones from occasions when the lows cross AWSs, research stations or drifting buoys. Lieder and Heinemann (1999) reported that a mesocyclone passed over a drifting buoy in the Bellingshausen Sea and gave a drop in surface

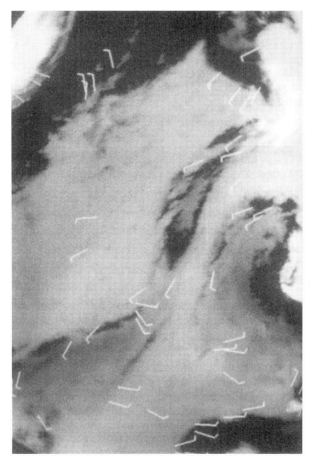

Figure 3.85. Cloud track winds determined from AVHRR imagery of a mesocyclone at 1700 GMT 5 January 1986 (from Turner *et al.*, 1993b).

pressure of 6 hPa, which is consistent with the relatively modest pressure signatures associated with most Antarctic mesocyclones. Similarly, small pressure perturbations of 3–5 hPa were found by Bromwich (1991) in his study of mesocyclones over the southwestern Ross Sea. Carrasco and Bromwich (1996) found, in their study of mesocyclones around the Ross Sea/Ross Ice Shelf area, that 77% of the mesocyclones had an intensity of 2 hPa or less, showing that few vigorous vortices develop or pass over this area. Heinemann (1990) also discusses cases of mesocyclones over the eastern Weddell Sea that had relatively small surface pressure perturbations of around 2 hPa, although we know that some much more active lows develop there.

Preparing surface pressure analyses for Antarctic mesocyclones is extremely difficult because of the very few *in situ* observations and the small horizontal scale of the lows. However, Heinemann (1996b) used the available *in situ* data

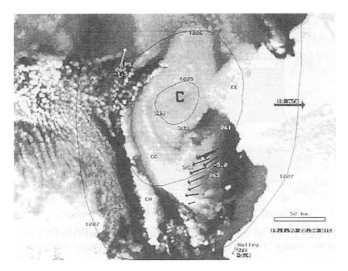

Figure 3.86. AVHRR visible imagery of a mesocyclone at 2250 GMT 19 January 1990 overlaid with wind vectors and MSLP. (From Heinemann, 1996b, *The Global Atmosphere and Ocean System*, Taylor & Francis Ltd (http://www.tandf.co.uk/journals).)

and measurements from dropsondes to construct an MSLP analysis for a mesocyclone over the eastern Weddell Sea on 19 January 1990. This analysis superimposed on AVHRR imagery is shown in Figure 3.86. This particular low had a surface pressure anomaly of about 2 hPa, which again is very weak in comparison to many of the polar lows observed in the Arctic.

As would be expected from the above, the scatterometer surface wind vectors discussed below suggest that most mesocyclones around the Antarctic have only moderate troughs in the surface wind field rather than a closed centre. Nevertheless, scatterometer surface wind vectors are valuable in that they can show the surface troughing corresponding to the frontal cloud bands that are present with many mesocyclones.

The large number of AWSs on the Ross Ice Shelf have allowed the investigation of the mesoscale pressure field across the area and have revealed the surface pressure anomalies associated with many of the lows. However, the data have shown that many of the mesocyclones are weak, with no pressure signal that could be resolved (Carrasco and Bromwich, 1995).

The operational numerical weather prediction systems have a horizontal resolution of around 50 km or more and are designed primarily to analyse synoptic-scale disturbances. Mesocyclones, and especially the smaller systems observed, are therefore difficult to represent in the current generation of models. However, Bromwich (1991) found that 33% of the mesocyclones around the southwestern Ross Sea were represented in the Australian operational surface analyses. Such a high figure probably occured because of the relatively large amount of *in situ* data in this area that were incorporated into the numerical analyses, compared with the much smaller quantities of data over the Southern Ocean. Of the mesocyclones with winds of at least 7.5 m s^{-1} the number of mesocyclones represented in the analyses rose to 77%. However, for the area of the Byrd Glacier where many mesocyclones have also been noted, only 18% of the lows detected with satellite and *in situ* data were represented in the Australian analyses.

The surface wind

Since the research stations in the Antarctic are very widely spaced, their observations can only provide information on the strength of the winds associated with mesocyclones when the vortices cross the stations. Some reported cases have indicated that the vortices affecting the stations or AWSs had gale force winds (Turner *et al.*, 1993b; Heinemann, 1996a, b, c; Adams, 1997), although these cases are exceptional and many other cases had winds in the range 10–15 m s^{-1} (Heinemann, 1990) and some developments, such as on the Ross Ice Shelf, had very low wind speeds (Carrasco and Bromwich, 1991). In their study of two years of mesocyclone activity in the southwestern Ross Sea, Bromwich (1991) defined a mesocyclone as being 'significant' if it had a surface wind speed of at least 7.5 m s^{-1}. This is only half the speed that is used to define a polar low, but does indicate the vortices that have a reasonable strength and surface signature. Of the vortices identified in the study, 43% were found to have winds greater than this threshold, or were enclosed by two 2 hPa contours if no wind data were available. In the study of Carrasco and Bromwich (1996), examining mesocyclones during 1991, it was found that 36% (13%) of the vortices at Byrd Glacier (Terra Nova Bay) were 'significant'.

Data from passive microwave radiometers, such as the SSM/I instrument (Hollinger *et al.*, 1987), can also be used to estimate the surface wind speed over the ice-free ocean (Goodberlet *et al.*, 1989), although the wind direction cannot be determined using these data alone. Such observations were used by Carleton *et al.* (1995) to investigate a mesocyclone over the Southern Ocean with very strong surface winds of 22 m s^{-1}, which would make the mesocyclone a polar low according to the accepted definition (see Section 1.3). With the case examined, the strongest winds were on the equatorward side of the low, which is what would be expected if we considered the vortex as a vorticity anomaly superimposed on a uniform westerly flow (Marshall and Turner, 1997). Similarly when the vortex develops within a cold southerly flow the winds on the western side will be stronger than those to the east of the vortex centre.

Despite the value of passive microwave data, there are some limitations to its use. In particular, the speed cannot be estimated when there is significant rain falling or when there are large values for cloud liquid water. Since we would expect strong winds to be present in the vicinity of fronts the speeds derived from passive microwave data may not reflect the strongest winds in the vicinity of a low because of frontal precipitation. However, radar altimeter data are able to show the strongest winds associated with mesocyclones in the cloud bands (Carleton *et al.*, 1995). These data have been useful in estimating the intensity of mesocyclones over the ice-free ocean. Carleton *et al.* (1995) used SSM/I data and Geosat altimeter measurements to examine the wind speed associated with three Antarctic mesocyclones. Two of the systems they examined had a couplet of strong winds to the west of the vortex centre and lower wind speeds to the east, giving strong cyclonic wind shear. Other studies (Lieder and Heinemann, 1999) have suggested that the fronts associated with mesocyclones can give areas of strong wind shear. With the third low there was a steep positive wind speed gradient ($c.$ 5 m s^{-1} (100 km)$^{-1}$) extending out from the centre of the low, where the wind speed was only 5 m s^{-1}. All three mesocyclones had wind speeds in excess of gale force with the strongest wind speed observed, as indicated above, being 22 m s^{-1}. All three mesocyclones examined in this study were well-developed vortices and much more vigorous than most Antarctic mesocyclones.

Lieder and Heinemann (1999) found that the most vigorous mesocyclone they investigated had surface wind speeds of greater than 20 m s^{-1}, with the peak wind speeds being to the west of the frontal cloud bands, although strong winds were also apparent along the front. While this low had gale force winds, they were well below the storm force winds found associated with some of the most vigorous polar lows in the Arctic. With the smaller air–sea temperature difference found in the Antarctic compared to the Arctic and the infrequent occurrence of deep convection, it seems unlikely that many Antarctic mesocyclones have the very strong winds of their northern counterparts.

The peak surface wind speed associated with a mesocyclone has been used as one of the criteria by which a mesocyclone is classified as a more intense *polar low*, such a system having winds of above gale force (15 m s^{-1}) for part of its existence. The most vigorous mesocyclones are obviously the most interesting to study and have received attention in a number of studies. Two intensively investigated polar lows occurred over Halley Station and have been examined by Turner *et al.* (1993b) and Lieder and Heinemann (1999), but from the studies carried out so far it would appear that these vigorous mesocyclones were exceptional events.

The wind scatterometers flown on a number of polar orbiting satellites, including Seasat (Allan and Guymer, 1984) and the ERS series (Offiler, 1994) provide valuable data for the investigation of the surface wind fields of mesocyclones. Scatterometers are radar instruments that measure the backscatter from the unfrozen ocean from a number of directions, and with suitable processing, these values can be converted into wind speeds and directions. However, there are a number of problems with the use of scatterometer data. In particular, the many mesocyclones that develop over the sea ice cannot be studied, the instruments cannot measure wind speeds greater than 25 m s^{-1} and the instruments only have a narrow 500 km swath. Marshall and Turner (1997b), in their study of mesocyclones around the Antarctic observed by the ERS-1 scatterometer, found that only 9% of the vortices apparent in satellite imagery were 'seen' by the scatterometer because of a combination of the development of the lows over the ice and the narrow swath. In addition there are problems in correctly determining some of the wind directions, the so-called 'ambiguity' problem. This makes the conduct of some case studies with scatterometer data difficult. Nevertheless, the scatterometer data when used with care can contribute to mesocyclone studies and have provided useful information on the surface circulation of these lows. For example, Marshall and Turner (1997) found, in a study of one year's mesocyclone activity around the Antarctic Peninsula, that 25% of the systems did not have a surface circulation.

The scatterometer wind vectors also allow the calculation of the vorticity field associated with mesocyclones and Marshall and Turner (1997) found that systems they examined had relative vorticity values between -100 and -400×10^{-6}s^{-1}, an order of magnitude lower than those calculated for an Arctic mesocyclone.

Mesocyclones have often been investigated as isolated phenomena, but they can affect the wind field on a regional basis. For example, it has been found that when mesocyclones pass to the east of Ross Island they can contribute to the development of barrier winds along the Transantarctic Mountains and give gale force winds at McMurdo Station (O'Connor *et al.*, 1994). The frequency of barrier winds may also be linked to mesocyclone activity in the Byrd Glacier and Terra Nova Bay areas (Carrasco and Bromwich, 1994), although further research is needed to confirm this.

Data from individual satellite instruments can provide information on the wind field associated with mesocyclones, but it is also possible to examine these systems with multi-sensor data sets. For example, both radar altimeter and passive microwave radiometer surface wind speeds have been used together and an example of collocated Geosat and SSM/I winds across a mesocyclone is shown in Figure 3.87.

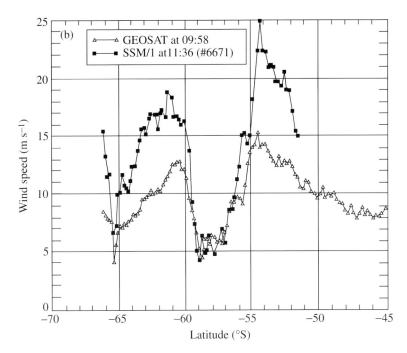

Figure 3.87. Collocated Geosat and SSM/I winds along the Geosat track over a mesocyclone on 4 October 1988. (From Claud *et al.*, 1996, *The Global Atmosphere and Ocean System*, Taylor & Francis Ltd (http://www.tandf.co.uk/journals).)

The Geosat pass at 0956 GMT 4 October 1988 crossed the centre of a mature mesocyclone and gave a very valuable cross-section of wind speed. In the latitude range 57–59° S the wind speed was close to 5–6 m s^{-1}, although on the other side of the system there were much stronger winds with a maximum speed of 13 m s^{-1} on the southern side and 16 m s^{-1} on the northern. The wind speed gradient was steepest on the southern side of the low with an increase of 10 m s^{-1} in 1° latitude compared with 10 m s^{-1} in 3° latitude on the north side. It can be seen from Figure 3.87 that the SSM/I wind speeds were greater than those produced from the Geosat data. This discrepancy can be partly explained by the fact that both sets of data were collected during the intensification phase of the mesocyclone, with the Geosat passes preceding those of the SSM/I. The steepest gradients in wind speed were found to the north or northeast of the low centres. Such multi-sensor investigations are very valuable in increasing the viewing frequency of the lows and also help in the error assessment.

In his study of eastern Weddell Sea mesocyclones using aircraft observations, Heinemann (1996b) collected wind data at the 100 m level across one low on 8 February 1990. These wind vectors are shown superimposed on the AVHRR imagery in Figure 3.88a, with cross-sections of raw aircraft

280   3 Observational studies

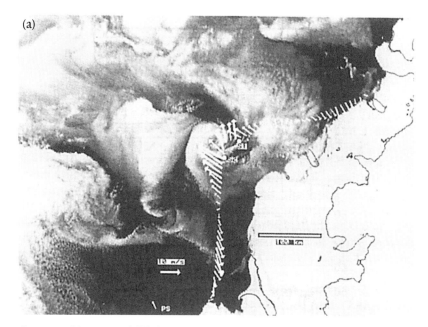

Figure 3.88. (a) AVHRR visible imagery of a mesocyclone at 1840 UTC 8 February 1990 with superimposed wind vectors at the 100 m level from an aircraft overflight. (b) Unfiltered aircraft measurements of wind speed, wind direction, temperature and relative humidity at 100 m through a mesocyclone. (From Heinemann, 1996b, *The Global Atmosphere and Ocean System*, Taylor & Francis Ltd (http://www.tandf.co.uk/journals).)

observations in Figure 3.88b. These data indicate increasing wind speed towards the centre of the mesocyclone, but no closed vortex at the surface. However, two cyclonic shear zones were present (indicated as S1 and S2). S1 was encountered at the beginning of the dry slot east of the low centre and was associated with a veering of the wind direction by about 60° and a decrease in relative humidity. The second shear zone (S2), was located at one of the curved cumulus bands south of the low centre and had the characteristics of a cold front, with the wind speed increasing from 8 to 15 m s^{-1} in association with a marked temperature decrease of 3 K (30 km)$^{-1}$. The wind direction also changed by about 30° at this location and the wind shear had values of up to 10^{-2} s^{-1} over about 1 km.

Marshall and Turner (1997b) examined four mesocyclone case studies in detail using surface wind vectors produced from ERS-1 scatterometer measurements and one example will be presented here. This case is illustrated in Figure 6.4 (p. 509) via AVHRR imagery, the scatterometer wind vectors and the field of derived relative vorticity. This mesocyclone is not necessarily typical of systems occurring around the Antarctic but serves to illustrate the value of

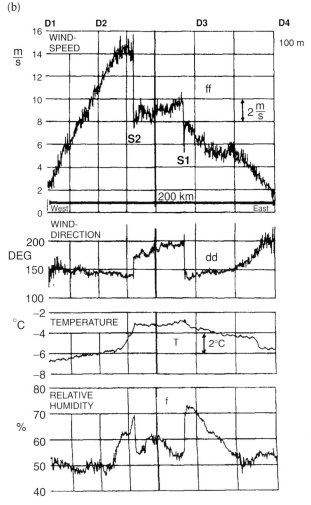

Figure 3.88 (cont.).

the scatterometer winds. It can be seen in Figure 6.4a that the cloud signature consisted of a comma of medium-level cloud with some convective elements present in the tail of the system. The nearest pass of scatterometer data collected some 2 hours after the imagery is shown in Figure 6.4b. The areas of missing data are a result of problems with the dealiasing algorithm, largely as a result of the loss of data from one antenna. Nevertheless, the wind field of the mesocyclone was well represented in the data and the wind vectors show clearly the cyclonic curvature around the equatorward side of the low with a maximum velocity of 15.8 m s^{-1}. There was no indication within the scatterometer data that the low had a closed circulation at this stage, although the light winds south of the centre may have resulted in directional errors, a problem

282   3 Observational studies

that is more likely to occur with scatterometer data of low wind speed. The field of relative vorticity (Marshall and Turner, 1997b, fig. 5c) indicates a maximum value of $-400 \times 10^{-6}\,\text{s}^{-1}$ in the trough extending to the north-northeast of the low centre. The form of the surface trough associated with this low was very similar to that expected by theoretical considerations when a moderate cyclonic vorticity anomaly is superimposed on a uniform flow (Marshall and Turner, 1997a).

The scatterometer data examined so far do not indicate sharp turning of the winds across the frontal cloud band associated with the mesocyclones. This may be a result of the problems in processing the scatterometer data, but more likely comes about because in most cases the fronts are rather weak.

Sea state

The sea state associated with Southern Hemisphere mesocyclones has received little attention to date, although Claud et al. (1996) have considered this question using satellite data. They used Geosat altimeter data to derive significant wave height (SWH) and minimum swell height for a low occurring on 4 October 1988 (Figure 3.89). This figure shows that a gradient in SWH was

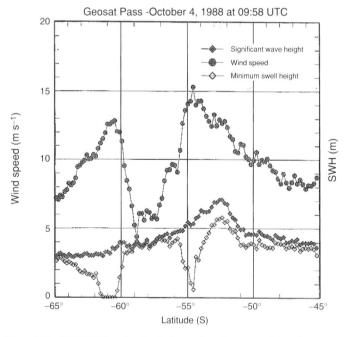

Figure 3.89. GEOSAT low-level wind speed and sea state (significant wave height and maximum swell height) at 0956 GMT 4 October 1988. (From Claud *et al.*, 1996, *The Global Atmosphere and Ocean System*, Taylor & Francis Ltd (http://www.tandf.co.uk/journals).)

present on the northern side of the low with values increasing from 4.5 to 7 m over the area, corresponding to the region of strong wind gradient. Over the remainder of the Geosat track the sea state was swell dominated with wave heights varying between 3 and 5 m. Passive microwave data have also shown, not surprisingly, that the highest waves are close to the region with the highest wind speed (Carleton *et al.*, 1995).

Other studies using SSM/I data have resolved swell levels of up to 10 m associated with mesocyclones (Carleton *et al.*, 1995), although the swell was generated by systems over the previous days and at some distance from the low itself.

3.2.4 Precipitation associated with mesocyclones

An important question regarding mesocyclones is the amount of precipitation that they have associated with them. When a low passes over a research station the synoptic observations can be used to determine the precipitation although, because the winds can often be strong, care needs to be taken to differentiate between falling precipitation and blowing snow. In the case of the vigorous mesocyclone that crossed Halley Station on the eastern side of the Weddell Sea, the low gave over 24 h of snowfall, most of which was associated with a comma-shaped cloud band (Turner *et al.*, 1993b). The snowfall was generally reported at continuous slight or moderate intensity. In the case investigated by Bromwich (1987) a pair of mesocyclones over the southwestern Ross Sea gave poor weather conditions at McMurdo Station with blowing snow resulting in horizontal visibility of less than 2 km and light snow falling.

Despite a well-defined cloud signature being apparent in the infra-red imagery, the *liquid* precipitation associated with the lows as determined from SSM/I data appears to be very light (Claud *et al.*, 1996). In this study a small cell with values of less than 0.8 on their precipitation index scale, suggests slight rain. As with the integrated water vapour, the precipitation algorithm used in this work was not tuned for polar use and may give precipitation rates that are too low. Partial filling of the SSM/I fields of view may also have led to underestimation of the precipitation rate. However, with the three cases examined by Carleton *et al.* (1995) it was found using SSM/I data that they all had light rain associated with them, although because of problems with availability of satellite data it was not possible to estimate the amount of solid precipitation associated with these cases.

Rockey and Braaten (1995) used satellite imagery and *in situ* precipitation observations to estimate the contribution of mesocyclones to the total precipitation on the Ross Ice Shelf. They found that over 38% of the precipitation at McMurdo came from mesoscale systems, with an average of 1.65 cm of snowfall per low. They found that snowfall accumulation of at least 2.5 mm was associated with 26.9% of the mesocyclones, trace amounts with 46.2% of systems and ice crystal events with 26.9%. These results suggest that mesocyclones are associated with the second largest percentage of snowfall received at McMurdo. The amount of snowfall from mesocyclones at the station is only slightly smaller than that from synoptic-scale Pacific cyclones, indicating the importance of these features to the regional climatology. It should also be noted that the climatological maximum precipitation at McMurdo Station occurs during March (Bromwich, 1988), which is also when the peak in mesocyclone activity is found at the end of the Austral summer (Carrasco and Bromwich, 1996). Since other studies suggest that the surface pressure perturbations and wind speeds associated with mesocyclones in this area are quite weak, studies need to be undertaken to determine why the snowfall they give is so great.

Mesocyclones over the eastern Weddell Sea do not generally give prolonged or heavy snowfall since most are fairly weak disturbances. However, one winter season low that affected von Neumayer Station gave moderate/heavy continuous snow for almost 24 h from active convective cloud bands.

3.2.5 Concluding remarks

Although there have been far fewer observational studies of mesoscale vortices in the Southern Hemisphere than in the north we can draw some preliminary conclusions about the systems that exist and their structure and intensity. The studies based on satellite imagery show the very large number of often small vortices that exist over the Southern Ocean, but the scatterometer data indicate that many of these do not have a surface circulation and are therefore very minor features. The early investigations of mesocyclones using passive microwave data also suggest that the precipitation associated with most mesoscale lows is very limited, with slight snow or rain being detected. However, some more vigorous lows do exist and these can give gale force winds and moderate or heavy precipitation, and are therefore very important to those concerned with forecasting for the Antarctic. Nevertheless, the extremely vigorous polar lows, with winds of 40–50 kt and deep convective cloud, such as occur in parts of the Northern Hemisphere, have not been observed in the Antarctic coast zone, although lows similar to this type of system may occur at the more northerly latitudes of 40–50° S, when cold outbreaks reach these latitudes.

3.2.6 Future observational studies needed

Our knowledge of mesocyclones over the Antarctic and the Southern Ocean still lags behind that of the Arctic and more observational studies are required to redress this problem. Satellite studies offer one of the most important means of making advances, since the data can provide high spatial and temporal coverage and, if multi-sensor investigations are carried out, then most of the important meteorological parameters can be determined. However, it is essential that the algorithms used to determine the geophysical parameters are tuned for the polar regions since the nature of the microphysical properties of the cloud and precipitation are quite different from the lower latitude areas. It will also be very important in such work to consider assimilating the various observations into a high resolution model since the greatest insight can be obtained if observational studies are tied closely to modelling investigations. As the operational models gain higher horizontal resolution so more mesocyclones and polar lows will be represented in the routine analysis, but limited area models with even higher resolution will still have their role.

It is also very important that more aircraft flights should be made through mesocyclones so that high density *in situ* data can be collected. These campaigns should include dropsondes being deployed to gain information on the vertical temperature, humidity and wind field structure of the lows, along with information on the cloud and precipitation within the systems.

A. VAN DELDEN, E. A. RASMUSSEN,
J. TURNER AND B. RØSTING

4

Theoretical investigations

4.1 Introduction

During the 1970s, research into the theoretical understanding of high latitude mesocyclones was focused on the basic mechanisms of development of the more intense systems, known as polar lows. The aim was to explain the striking differences between polar lows and other extratropical cyclones, namely the small size and rapid growth rates of polar lows, and their favoured formation within cold air masses over the oceans in winter. It will become apparent by the end of this chapter that these fundamental questions have not been completely answered. However, considerable progress has been made, and new areas of research have been opened up regarding the life cycle of polar lows, and their interaction with the broad-scale atmospheric flow.

The construction of mathematical and theoretical models of mesocyclones is not simple, because there are many types of vortices occurring in the high latitude areas. They vary widely in horizontal and vertical extent, in intensity and in structure. A mesocyclone may be a powerful system, extending through the depth of the troposphere, with intense deep convection and hurricane force winds, or a weak swirl in the boundary-layer cloud, clearly visible on satellite imagery but with little significant weather at the Earth's surface. The environment in which the vortex forms may differ widely being, for example, a low-level frontal zone, or a flaccid low-pressure region at the centre of a decaying synoptic cyclone. The lows sometimes resemble more familiar weather systems, such as baroclinic waves or tropical cyclones, but between the more identifiable systems lies a variety of transitional types (see Table 3.1).

The resemblance of mesocyclones to other, more studied, weather systems provides a useful starting point for an analysis of their structure and mechanisms of formation/development, and indeed many of the advances in the theoretical understanding of polar lows have occurred in the context of work done on other maritime cyclones. These include rapidly intensifying synoptic cyclones, where the importance of latent heat release has been demonstrated, frontal wave cyclones, synoptic systems that are initiated on low-level frontal zones, and tropical cyclones.

This chapter will draw on work from all of the above fields, in addition to studies directed explicitly towards polar lows. The first two sections that follow will discuss individually two of the principle fluid dynamical instabilities that are believed to contribute to mesocyclone formation, namely baroclinic and barotropic instability. We then examine the insight that can be gained into polar low formation and structure using the powerful tool of potential vorticity. We then consider the important role that thermal instability plays in some of the more vigorous systems. Finally, we attempt to draw together the preceding material into a current picture of our understanding of polar low development and structure and try to identify the most pressing questions for future research.

4.2 Baroclinic instability

Baroclinic instability is a type of dynamical instability associated with a baroclinic region of the atmosphere, i.e. an area where the density depends on both the temperature and pressure, or, expressed slightly differently, temperature varies along the pressure surfaces. Baroclinic instability is associated with the vertical shear of the mean flow, which is related to the horizontal temperature gradient by the thermal wind equation. Instabilities in a baroclinic region grow by converting potential energy associated with the mean horizontal temperature gradient into kinetic energy through ascending warm air and descending cold air. In the following, only a short overview will be given presenting features pertinent for the arguments elsewhere in the book. For a detailed discussion of baroclinic instability theory the reader is referred to textbooks, such as Holton (1992) or Bluestein (1993).

The mechanism through which a baroclinic wave amplifies within a region of strong north–south temperature gradient is illustrated by Figure 4.1. Assuming that a weak wave-like perturbation is initiated by some process in an otherwise uniform zonal flow, the meridional motions associated with the perturbation will distort the originally straight east–west oriented isotherms causing a wave in the temperature field to form. This wave will be displaced one quarter of a wavelength to the west of the wave in the pressure (geopotential) field. In the absence of other influences, horizontal temperature advection associated with the geostrophic wind field will further distort the isotherms

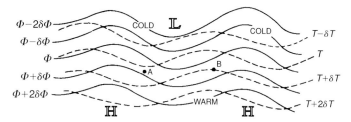

Figure 4.1. The distribution of geopotential height (solid lines) and temperature (broken lines) on a constant pressure surface in a developing baroclinic wave in the Northern Hemisphere. The pressure surface is located near the level where the speed of the wave is the same as the speed of the mean zonal flow (from Wallace and Hobbs, 1977).

from their original east–west orientation, causing the temperature wave to further amplify. In order for the wave to grow, the kinetic energy must increase. The mechanism by which this happens is a thermally direct circulation in which cold air at A (see Figure 4.1) sinks and warm air at B rises thus lowering the centre of gravity of the fluid, i.e. converting potential energy into kinetic energy.

By the 1950s, it was widely accepted that baroclinic instability acting within the belt of the mid-latitude westerlies was responsible for the development of the majority of extratropical, synoptic-scale cyclones. From theoretical as well as observational studies it was possible to infer the various stages of development of a baroclinic cyclone and the schematic flow in a typical system is shown in Figure. 4.2.

In this sequence of diagrams the developing low (Figure 4.2a) is seen as a perturbation on the baroclinic zone. During the rapid development phase there is a cooperative interaction between the flows at upper levels and near the surface, with strong, low-level cold air advection west of the low and weaker warm advection to the east. This particular pattern of thermal advection derives from the fact that the 500 hPa trough is to the west of the surface low with the mean geostrophic wind in the 1000–500 hPa layer being across the 1000–500 hPa thickness lines towards larger thickness values west of the surface low and towards smaller thickness east of the surface low. As the system continues to develop (Figure 4.2b) the distortion of the upper-level flow leads to a sharper 500 hPa trough and growing upper-level divergence with pressure falls at the surface leading to further development of the low (self development).

Figure 4.2 suggests that the low-level cyclone and the upper-level system (trough) develop simultaneously as a continuous process as part of the baroclinic development. In many cases, however, cyclone development starts in such a way, that a baroclinic wave that is well defined at upper or middle levels

4.2 Baroclinic instability 289

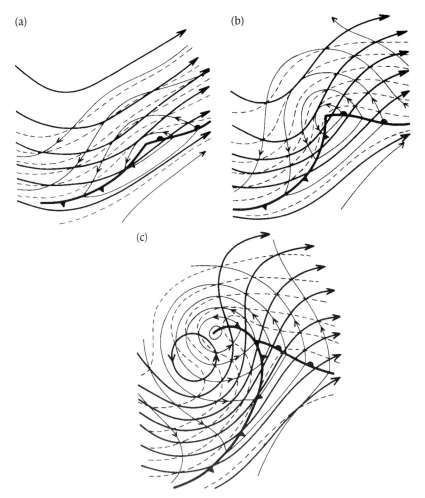

Figure 4.2. Schematic 500 hPa contours (heavy solid lines), 1000 hPa contours (thin lines), and 1000–500 hPa thickness (dashed lines), illustrating the 'self-development' process during the growth of an extratropical cyclone (from Palmén and Newton, 1969).

(an upper-level, cold short-wave trough, Figure 4.3a), but weak in the lower atmosphere, moves over a pre-existing low-level frontal zone. When the region of upper-level vorticity advection associated with the short-wave trough approaches the low-level front with its associated strong horizontal temperature gradients, low-level thermal advection will become increasingly important leading to hastening of low-level cyclogenesis in a manner similar to that illustrated on Figure 4.2 (Figure 4.3b and c).

The role of upper-level troughs for the development of baroclinic waves was elaborated by Pettersen and Smebye (1971). They distinguished between

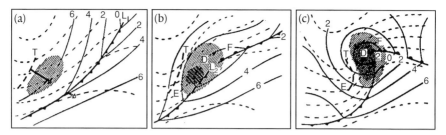

Figure 4.3. Stages in a Type B extratropical cyclone development. As the upper-level cold trough advances and the area of vorticity advection aloft (hatched area) spreads over the low-level frontal zone, the imbalance created results in convergence at low levels. When the thermal field has become distorted through cyclonic circulation the system can develop further through the self-development process illustrated in Figure 4.2 (from Palmén and Newton, 1969).

two types of development leading to the formation of extratropical cyclones. One (Type A) was characterized by an initial development under a more or less straight upper-level current and an initially large, low-level amount of baroclinic instability that decreases as the wave occludes. The other type of development (Type B) commences, according to Pettersen and Smebye, 'when a pre-existing upper trough, with strong vorticity advection on its forward side, spreads over a low-level area of warm advection in which fronts may or may not be present'. For the 'pure form' of Type B development the amount of baroclinicity initially is relatively small but increases as the storm intensifies. In the present text we use the term 'Type B development' simply to designate a baroclinic development initiated by the arrival of an upper-level disturbance, generally an upper-level trough (upper-level potential vorticity anomaly), over a low-level baroclinic zone or front.

A schematic picture of cyclogenesis associated with the arrival of an upper-level positive PV (potential vorticity) perturbation over a low-level baroclinic region illustrating the same process as Figure 4.3 but from a PV perspective is illustrated on Figure 4.4 (the PV approach is discussed in more detail in Section 4.4).

Although the role of baroclinic instability on the development of synoptic-scale depressions was well understood by the 1970s, it was unclear as to which process or processes were responsible for the development of polar lows and other mesoscale lows found in the polar regions and the degree to which baroclinic instability played a part. The role of baroclinic instability in the development of polar lows has been examined via different types of theoretical/numerical investigations. First, simple linear models, in both dry and moist forms, have been applied based on data from case studies. These have used normal mode techniques to examine the growth rates of

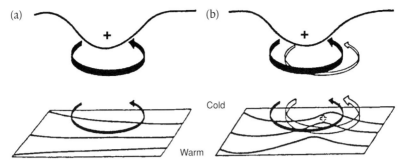

Figure 4.4. A schematic picture of cyclogenesis associated with the arrival of an upper-level positive PV anomaly over a lower-level baroclinic region. (a) The circulation induced by the upper-level vorticity anomaly is shown by solid arrows. The thin lines show potential temperature contours at the lower boundary. The advection of potential temperature by the induced low-level circulation leads to the formation of a warm anomaly slightly east of the upper-level PV anomaly. This in turn will induce a cyclonic circulation as shown by the open arrows in (b). The induced upper-level circulation will reinforce the original upper-level anomaly and can lead to amplification of the disturbance (after Hoskins *et al.*, 1985).

disturbances of different wavelengths. The 'problem' that polar lows have a much shorter wavelength than that which follows from the standard form of baroclinic instability theory has been explained by the fact that that the cold air masses involved are quite shallow, being confined to the lowest 1000–2000 m (e.g. Mansfield, 1974; Wiin-Nielsen, 1989). Another type of approach is the initial value method in which the general perturbations that trigger the storms have a more complex structure (type B developments). Also, full primitive equation mesoscale models have been used to simulate selected cases with experiments being undertaken with different parameterization schemes. Such work has shed light on the role of baroclinic instability and other instability mechanisms, such as Conditional Instability of the Second Kind (CISK).

Considering baroclinic instability in polar regions, it is possible to distinguish between low-level instability mainly confined to shallow boundary layers, typically, but not always, along the ice edges, and 'deep instability' involving deeper baroclinic layers. Furthermore, it is useful to distinguish between 'ordinary' baroclinic instability, where the direction of the surface wind, the thermal wind and the direction of wave propagation are all the same, and the so-called 'reverse shear instability' where the thermal wind is opposite to the direction of the surface wind and to the direction of the wave propagation. In this section we will consider these different types of investigation carried out over the last 30 years and relate the results to the observational studies presented in Chapter 3.

292   4 Theoretical investigations

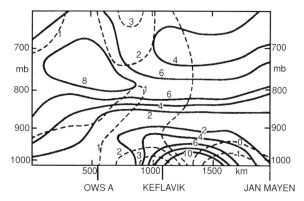

Figure 4.5. A cross-section perpendicular to the direction of travel of a polar low on 0000 GMT 7 December 1967. Isopleths of the vertical gradient of potential temperature $(\partial \theta/\partial z)$ in °C km^{-1} (solid lines) and horizontal gradient of θ in °C (100 km)$^{-1}$ (broken lines): positive θ increases to the left (from Mansfield, 1974).

The first attempt to apply baroclinic theory to polar lows using a linear model was by Mansfield (1974). This study was motivated by the observational work of Harrold and Browning (1969), who documented the structure and evolution of a polar low that crossed the British Isles using Doppler radar and conventional radar and synoptic observations. From their analysis, they concluded that the system was basically a baroclinic cyclone. As would be expected for a baroclinic cyclone, the polar low was, during its passage over the British Isles, predominantly associated with slantwise precipitation along a narrow tongue, rather than convective precipitation (see Section 3.1.2).

Mansfield (1974) constructed a typical cross-section of the region of formation of the Harrold and Browning polar lows, based on station and ocean weather ship radiosonde ascents. This cross-section, showing the horizontal and vertical gradients of potential temperature, is reproduced in Figure 4.5.

Mansfield noted the existence of a shallow layer of reduced static stability below a strong inversion at 850 hPa, and suggested that an instability would likely be confined to this layer. Both the low vertical stability and the shallow depth could significantly modify a baroclinic instability. To quantify these effects, Mansfield applied the Eady model (Eady, 1949; see also Gill, 1982) for the growth of perturbations in a plane-parallel flow. The basic state was assumed to have a constant static stability, constant vertical wind shear, and to be bounded above by a rigid lid. The parameter values that were chosen on the basis of the cross-section in Figure 4.5. were Brunt–Väisälä frequency $N^2 = 1.6 \times 10^{-4}$ s^{-2}, the horizontal gradient of potential temperature, $\partial \theta/\partial y = 1.4$ K (100 km)$^{-1}$ and layer depth $H = 1.6$ km. The solution of the Eady model is an exponentially growing wave. Using the above values and a Coriolis parameter of $f = 1.22 \times 10^{-4}$ s^{-1}, Mansfield calculated various parameters for the growing wave. He found that the fastest growing normal mode would have a wavelength of $l = 645 \pm 10$ km, a phase speed of $c = 6 \pm 1$ m s^{-1}, and a growth rate with e-folding time of $t = 28 \pm 1$ h. These values agree remarkably well with the estimates from observations of $l = 650 \pm 100$ km, $c = 6 \pm 4$ m s^{-1}, and $t = 24 \pm 6$ h. It is particularly notable that the calculation predicted the small wavelength and rapid growth rate of the system. The wavelength of the fastest growing Eady mode is proportional to the Rossby radius of deformation NH/f, which shows that the small scale predicted for the polar low was due mainly to its shallow depth. The growth rate, $\sigma = t^{-1}$, is proportional to the isentropic slope $N(g/\theta_0)(\partial\theta/\partial y)$ and thus has a relatively large value owing to the large horizontal temperature gradient. Low values of static stability could also have contributed to rapid growth at a short wavelength, but the value of N was unexceptional.

In retrospect, the agreement between model and observations may perhaps have been fortuitous. As Mansfield shows (Figure 4.6), factors not included in this analysis, such as friction and surface heat fluxes would tend to damp the growing wave. On the other hand, it will be seen later that latent heat release will tend to increase the growth rate. It should be noted that Rasmussen (1979) reproduced basically the same growth rates and horizontal scale of the disturbance studied by Harrold and Browning, and Mansfield using a CISK-type model.

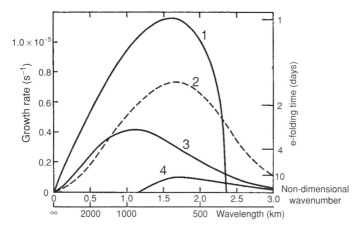

Figure 4.6. The effect of the inclusion of surface friction and heating in Eady's model. Growth rate versus wavelength for: 1, surface wind speed (U_0) zero (no heating or friction); 2, heating only $U_0 = 15$ m s^{-1}; 3, friction only $U_0 = 15$ m s^{-1}; 4, friction and heating $U_0 = 10$ m s^{-1} (from Mansfield, 1974).

In his analysis, Mansfield assumed that a stable layer acting as a lid confined the disturbance to the layer below 1.6 km. The data presented by Harrold and Browning, on the other hand, show that the low during its passage over the British Isles was a fairly deep system with cloud reaching above 5 km.

Baroclinic developments confined to very shallow layers may be found along the ice edges in the Southern as well as the Northern Hemisphere. During winter shallow baroclinic zones tend to form along the ice edges bordering the ice- or snow-covered polar regions. Over the years, developments along these low-level, shallow baroclinic zones have been considered essential for the formation of polar lows, cf. for example the classification scheme of Businger and Reed (1989b) in which one of the three main types of 'elementary polar low developments' is the so-called 'Arctic-front type' associated with ice boundaries.

A type A mesoscale cyclone formation over the northern part of the Fram Strait along a shallow frontal zone situated over the marginal ice zone along the northeast coast of Greenland, was studied by Rasmussen *et al.* (1997). The mesoscale cyclone in this case formed in a region covered by sea ice, but within an area of a very strong horizontal temperature gradient (Figure 4.7a). During the formation of the mesoscale cyclone a sharp low-level cold front formed. As the front passed the two observation camps in the region (Camp A and Camp O indicated on Figure 4.7a) the temperature dropped from around $-5\,°C$ to $-30\,°C$ and the wind increased from a few metres per second to nearly $10\,\mathrm{m\,s^{-1}}$.

The air over the ice was extremely stable and convection could entirely be ruled out as contributing to the development.

The vertical structure of the atmosphere within the region where the disturbance developed is illustrated by the radiosonde ascent shown on Figure 4.7b. The ascent shows a very shallow cold air mass which was only around 200 m deep and neutrally stratified, capped by a somewhat deeper frontal zone extending up to around 500 m. While the development of the weak low was accompanied by the formation of a sharp low-level cold front, the advance of warm air preceding the passage of the cold front had the character of a 'warm surge' and only a weak warm front could be identified.

The shallow character of the air masses involved in such systems indicates that only modest development in terms of strength of the resulting circulation can be expected. After its formation south of Camp A early on 11 April the low moved north and disappeared over the ice during the evening (Figure 4.7c).

During the time of development, the upper-air flow was mainly anticyclonic and no discernible upper-level system triggered or influenced the development of the near-surface system, which in this case must be classified as a baroclinic

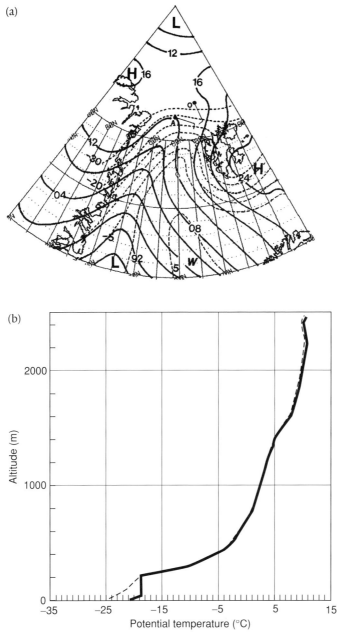

Figure 4.7. (a) Surface pressure and temperatures at 0000 GMT 11 April 1989. Pressure in 4 hPa intervals and isoterms in 5 °C intervals. Dots marked with the letters A and O show positions of observation camps. (b) Radiosonde ascent from Camp O showing potential temperature (solid) and dew point potential temperature (dashed) as a function of height at 2254 GMT 11 April. (c) Surface pressure and temperatures at 1800 GMT 11 April. Pressure and isoterms as in (a) (from Rasmussen *et al.*, 1997).

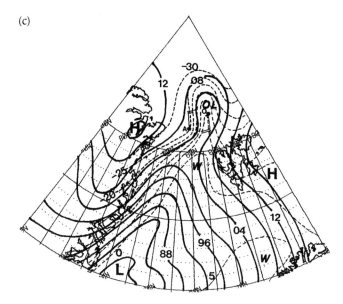

Figure 4.7 (cont.).

type A system being confined to the lowest few hundred metres above the surface.

No satellite images showing this disturbance were available. Figure 4.8, however, shows a wave train of two mesoscale cyclones within the same region, albeit over the sea. The satellite image shows that the meso-vortices forming the wave train in this case were characterized by low-level cloud. Another example of a baroclinic wave formed in the pack ice region along the northeast Greenland coast is shown on Figure 1.22.

Other minor mesoscale cyclones with spiral cloud structures are regularly observed on satellite imagery over the sea between northeast Greenland and Svalbard. Some of these vortices form as the result of barotropic instability along minor shear lines (see Section 4.3). They are rather insignificant weather features accompanied only by low wind speed.

Tsuboki and Wakahama (1992), in a study of mesoscale cyclones with a diameter of 200–700 km developing off the west coast of Hokkaido, Japan, found that baroclinicity in the lower troposphere was important for the formation of these cyclones. Based on an analysis of satellite images, the mesoscale cyclones could be classified into two types according to their horizontal scale: Type I (200–300 km in diameter) and Type II (500–700 km). From a linear instability analysis with a basic flow based on observed wind profiles, two unstable modes were found: Mode I, with a wavelength of 200–300 km and another, Mode II, of 500–700 km. Comparisons between theoretical and observational results

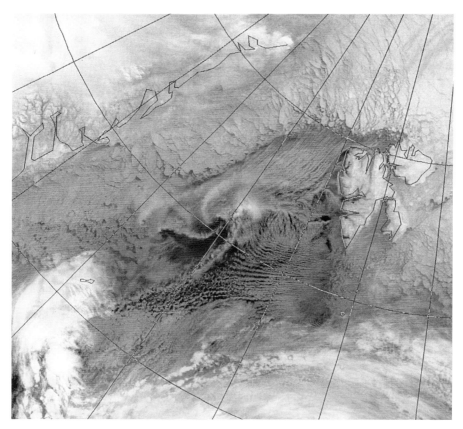

Figure 4.8. An infra-red satellite image for 1137 GMT 19 December 1989 showing two mesoscale cyclones that formed within the shallow baroclinic zone at the edge of the pack ice along the northeast coast of Greenland. (Image courtesy of the NERC Satellite Receiving Station, University of Dundee.)

indicated that the modes I and II could account for the characteristic properties of Types I and II systems respectively. The energetics showed that both modes I and II were maintained by the increase of eddy available potential energy and its conversion into eddy kinetic energy. The Type I systems were shallow disturbances confined below 850 hPa, while Type II systems were rather deep, extending to $c.$ 500 hPa. The lifetimes of Type I and Type II systems were 0.4–1.6 and 1.0–3.0 days, respectively.

On Figure 4.9 is shown the growth rate diagram of the unstable waves, the abscissa representing the magnitude of the non-dimensional wavenumber vector and the ordinate its direction. The figure has two maxima: one corresponding to a growth rate of $c.$ 2.4 day^{-1} located at wavenumber 8, and another corresponding to $c.$ 2 day^{-1}, at wavenumber 2.7. The corresponding dimensional wavelength of the former wave is $c.$ 240 km (Mode I), and that of the latter

298 4 Theoretical investigations

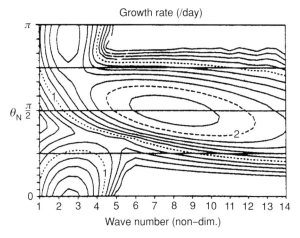

Figure 4.9. Growth rate diagram of unstable waves. The abscissa represents the magnitude of non-dimensional wavenumber vector and the ordinate represents its direction for representative parameters of the basic flow (from Tsuboki and Wakahama, 1992).

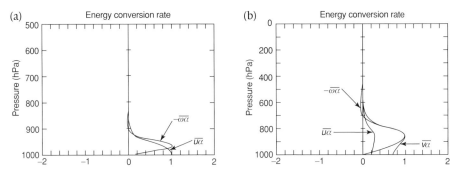

Figure 4.10. Vertical profiles of energy conversion rate of (a) Mode I, corresponding to shallow, short wavelength systems, and (b) Mode II, corresponding to a deeper system of longer wavelength (in arbitrary units). α, specific volume; ω, vertical velocity in isobaric coordinates; $\overline{u\alpha}$, zonal heat transport; $\overline{v\alpha}$, meridional heat transport (from Tsuboki and Wakahama, 1992).

c. 700 km (Mode II). Compared with the growth rates found by Mansfield (1974) and Duncan (1977) in their respective linear analyses of polar low growth, the growth rates are relatively large.

Figures 4.10a and b show the vertical profiles of energy conversion rate for Mode I and Mode II systems. The profiles show clearly that Mode I lows, corresponding to the short wavelength systems, are shallow, confined to a thin layer of a depth around 100 hPa, while the Mode II systems are about twice as deep.

Based on the data analysis of observations and the linear instability analysis, Tsuboki and Wakahama concluded that the mesoscale cyclones off the west coast of Hokkaido and Sakhalin were baroclinic disturbances. The systems studied by them developed over the Japan Sea within a northwesterly, light winter monsoon wind, and, judging from the information in the paper by Tsuboki and Wakahama, along a boundary layer front (BLF) rather similar to the BLFs frequently observed along the west coast of Svalbard. As argued in Section 4.3, the numerous minor vortices along such BLFs most probably owe their existence to barotropic instability. The results from the analysis discussed above do not necessarily contradict this statement since Tsuboki and Wakahama excluded small vortices of diameter less than 100 km from their study, i.e. those vortices most likely to be caused by barotropic instability.

The possible role of low static stability near the Earth's surface in the growth of polar lows was explored by Duncan (1977). He used a quasi-geostrophic model to look at normal mode solutions for unstable disturbances including both vertical and horizontal wind shears, with low static stability near the Earth's surface. Three cases, all of them developments at a fairly southerly latitude, were analysed. In each situation the developments took place within a baroclinic zone below an upper-level wind speed maximum. Good predictions were made for the wavelength of two of the observed disturbances. In the third case, Duncan suggested that strong horizontal variations in the static stability, which cannot be represented in a quasi-geostrophic model, were responsible for the poor prediction of wavelength. In all three cases the perturbation amplitudes were very small at upper levels and largest close to the surface, implying that the growth of the perturbations by baroclinic processes should be expected primarily at low levels.

Duncan concluded that polar air depressions could be considered as shallow baroclinic waves and that the conversion of available potential energy to eddy kinetic energy occurs in the lowest 200–300 hPa of the atmosphere when the low-level static stability is small. No significant growth could be ascribed to barotropic processes. The rate at which the energy conversion occurred, and therefore the growth rate, depended on the thermal wind and the static stability. In general, the largest vertical wind shear at low levels on the synoptic scale exists below the jet stream so that this is a likely region for polar low developments.

Duncan, in his 1977 paper discussed above, noted that a necessary condition for polar air depressions to develop 'is that a vertical wind shear exists such that the thermal wind and the mean flow at low levels are almost parallel'. The concept of the development of polar lows in a *reverse shear flow* (where the thermal wind and the mean flow are still parallel but of opposite direction) was first put forward by Duncan (1978). While most large, baroclinic waves develop when the surface wind, the thermal wind and the progression of the systems

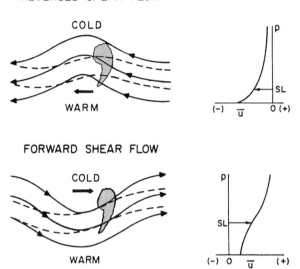

Figure 4.11. Comparison of the structure of disturbances in reverse-shear flow (above) and (normal) foreward-shear flow (below). Solid lines show streamlines, broken lines isotherms at the steering level; heavy arrows show phase propagation vector and steering level (SL) wind; stippling indicates the extent and position of an associated comma cloud (from Businger and Reed, 1989b, in Twitchell *et al.*, 1989. © A. Deepak Publishing).

are in the same direction, Duncan considered disturbances where the surface wind and thermal wind were in opposite directions, and the magnitude of the horizontal wind decrease with height. The configurations of reverse shear and forward-shear systems are illustrated in Figure 4.11.

In this situation, relative to the motion of the system, there is warm air to the left of the path and colder air to the right. The effects of horizontal advection will be to move warm air behind the trough so that kinetic energy will be gained at the expense of available potential energy if ascending motion predominates behind the trough with descending motion in the cold air ahead. Duncan used the model described above to investigate the growth of disturbances in a reverse shear flow using atmospheric conditions from a polar low development from 10 December 1976. It was found that the most unstable wave in the model had a wavelength of 900 km and a phase speed of 10.8 m s^{-1} resembling the observed polar low. Based on this, Duncan suggested '...that the observed disturbance was baroclinic in nature and that its structure was probably similar to that of the 900 km wave in the numerical model.'

Haugen (1986) used a three-dimensional primitive equation model to simulate a reverse shear polar low development within a channel over the Norwegian

4.2 Baroclinic instability 301

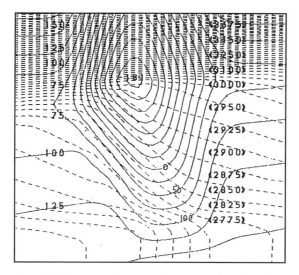

Figure 4.12. East–west cross-section across the Norwegian Sea, with Greenland to the left and the Norwegian coast to the right. Solid lines show wind components perpendicular to the cross-section and broken lines the potential temperature at the start of simulation. Labels on contours show the basic current in units of 0.1 m s^{-1} (positive wind velocities towards south) and the potential temperature in units of 0.1 K (from Haugen, 1986). (Figure used courtesy of the Polar Lows Project, Norwegian Meteorological Institute.)

Sea from 15 to 17 December 1982. A large-scale low with its centre over the northern part of Scandinavia advected polar air southwards over the Norwegian Sea, the wind direction in the lower and middle troposphere being opposite to the thermal wind direction. An east–west cross-section across the basic flow showing the wind component perpendicular to the cross-section and the potential temperature, is shown on Figure 4.12.

Greenland is situated to the left in the figure and the Norwegian coast to the right. The capped boundary layer with its constant potential temperature in the vertical can be maintained and increase in thickness when the polar air mass moves southwards. The east–west baroclinicity, concentrated in the middle of the channel with the coldest air to the west (close to Greenland) caused the low-level northerly basic flow to decrease with height, creating a weak southerly jet stream at tropopause height. The troposphere was initially stable on the cold side (except within the shallow, cold boundary layer) and in the middle of the region, but close to being conditionally unstable on the warm side.

The development of a fairly large polar low (diameter *c.* 1000 km) from the initial basic conditions shown on Figure 4.12 is illustrated in Figure 4.13, showing geopotential height and potential temperature after 24 and 36 h of simulation.

302   4 Theoretical investigations

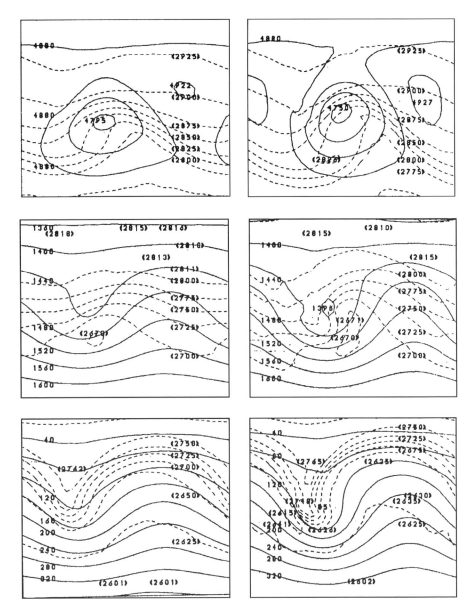

Figure 4.13. Horizontal contours of geopotential height (solid lines) and potential temperature (broken lines) after, respectively, 24 h of simulation (left column), and 36 h (right column) illustrating a reverse shear development corresponding to the basic flow shown on Figure 4.12. Pressure levels, from top, 500, 800 and 900 hPa. Labels on contours show heights (in m) and potential temperatures (in units of 0.1 K) with intervals 40 m and 2.5 K, respectively. The Greenland side is at the bottom of the figures (from Haugen, 1986). (Figure used courtesy of the Polar Lows Project, Norwegian Meteorological Institute.)

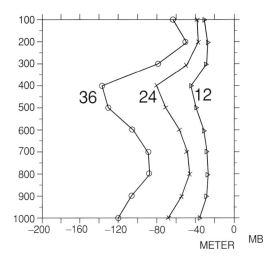

Figure 4.14. The vertical variation of the maximum amplitude of the low seen on Figure 4.13 as a function of time (12, 24, 36 h). The values describe the maximum height difference of a pressure surface relative to that of the initial (after Haugen, 1986). (Figure used courtesy of the Polar Lows Project, Norwegian Meteorological Institute.)

Greenland is at the bottom of the figures and the low moved southwards in the direction of the basic low-level flow. The development *seems* small at the surface, because of the strong pressure gradient of the basic flow. The basic flow across the channel was smallest at the 500 hPa level where a closed circulation developed. As a measure for the growth rate of the low, Haugen used the difference in height of a pressure surface from the initial stage. The vertical variation as a function of time is shown on Figure 4.14, with greatest amplification at the surface and at the tropopause level, in accordance with Eady (1949) in his analysis of the development of a baroclinic wave.

During the simulation of the reverse shear development the static stability changed because of differential horizontal advection with heating near the surface, as well as due to vertical stretching. As a result of this, the low levels, in particular, became conditionally unstable (Figure 4.15). Haugen interpreted this as a possibility of enhanced growth of such systems through release of latent heat from convection. This result agrees well with the observation of reverse shear systems (see Section 3.1.4) in the way that widespread deep convection often accompanies these developments.

A further study of reverse shear instability, by Reed and Duncan (1987), examined the development of four polar lows that had formed in January 1993 over the Greenland Sea in a shallow baroclinic zone at a time when the flow was from the northeast at the surface and the thermal wind was from the opposite

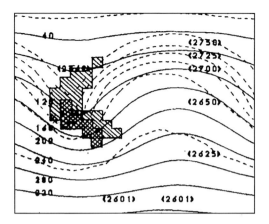
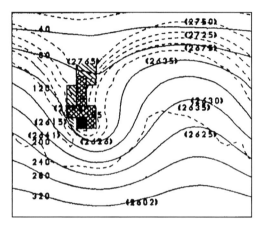

Figure 4.15. Horizontal contours of geopotential height (solid lines) and potential temperature (broken lines) at the 950 hPa surface after 24 h (upper) and 36 h (lower) of simulation (units as Figure 4.13). Areas with conditional instability are hatched. Single-hatched indicates instability over 900–850 hPa; cross-hatched, 900–800 hPa; solid, 900–700 hPa (from Haugen, 1986). (Figure used courtesy of the Polar Lows Project, Norwegian Meteorological Institute.)

direction. At the 500 hPa level, only a weak flow, generally from the north, was observed. Reed and Duncan applied their quasi-geostrophic model (Duncan, 1977) and obtained solutions for the fastest growing mode for wavelengths in the range 300–1000 km. The growth rates of these modes are shown in Figure 4.16 and indicate that wavelengths around 500 km are the most unstable. These model disturbances were confined almost entirely to the fastest moving lower layers adjacent to the surface as illustrated on Figure 4.17. Reed and Duncan suggested that the propagation of the disturbances in the opposite direction to the shear was explained by the fact that short baroclinic waves moved approximately with the mean wind in the layer in which they are embedded.

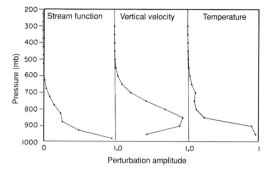

Figure 4.16. Growth rate of pertubations within a reverse shear flow as a function of wavelength for an experiment with no horizontal shear and vertical wind profile observed at the storm track (from Reed and Duncan, 1987).

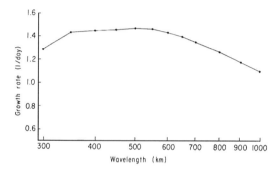

Figure 4.17. Perturbation amplitudes for the stream function, vertical velocity and temperature, for the 500 km wave of Figure 4.16. The magnitudes are in normalized arbitrary units (from Reed and Duncan, 1987).

Moore and Peltier (1989) extended the study of baroclinic wavetrains by Reed and Duncan arguing that these wavetrains should be considered as a new cyclone-scale mode of baroclinic instability discovered by them (Moore and Peltier, 1987). This new mode, however, is filtered out by both the quasi-geostrophic and geostrophic momentum approximations to the primitive equations. Specifically, Moore and Peltier in their 1989 study argued that the quasi-geostrophic approximation used by Reed and Duncan was invalid because of two factors: first, the static stability in the environment in which the polar lows grew varied strongly in the horizontal, and second, the background Richardson number field was of the order unity.

Using the primitive equations, Moore and Peltier considered the stability of a two-dimensional baroclinic zone to three-dimensional, small amplitude perturbations and demonstrated that the environment in which the polar low wavetrain was observed to develop was unstable to the 'cyclone scale branch of baroclinic instability' found by them. The predicted doubling time and wavelength of the most unstable wave were in good agreement with the observations made by Reed and Duncan (1987).

The short wave-length of the disturbances found in theoretical studies were in good agreement with the observed systems, but as noted by Reed and Duncan the predicted propagation speed was too fast (that is, the steering level in the reverse shear flow was too low). They suggested that cumulus convection might have served to deepen the system in the vertical, thus slowing its motion. They also suggested that baroclinic instability alone was not capable of explaining the rapid development and that another mechanism, possibly latent heat release from deep convection, also played an important role.

To date there has been no work on reverse shear flow in relation to mesoscale lows in the high latitude areas of the Southern Hemisphere. However, there is no reason to assume that such vortices are confined only to the Arctic.

Further work on baroclinic instability, using the Eady model but extended to two layers, has been carried out by Blumen (1979). In this study the static stability was horizontally uniform but different in each layer and the wind shear was uniform throughout both layers. An analysis of the unstable growth rates showed that the instability was associated with the delta function distribution of potential vorticity at one boundary and at the interface between the two layers. This work showed that the short- and long-wave baroclinic instabilities depend on the relative layer depths, along with the jump in static stability between the two layers. He found that the model gave a maximum of baroclinic instability of much shorter wavelengths when the lower layer had a stratification close to an adiabatic lapse rate. This work was important in confirming the earlier results of Mansfield and Duncan in showing that small-scale instabilities can grow as a result of abrupt changes in static stability and/or wind shear.

The Blumen model was successfully used by Rasmussen and Aakjær (1992) to explain a baroclinic polar low development over the North Sea and Denmark on 28–29 March 1985 (see Section 1.6.6). A radiosonde ascent close to the region of development showed that the conditions as set up by Blumen for this type of development were clearly fulfilled. The growth rate curve computed from the data observed is shown on Figure 4.18. The left branch of the curve with a 'most preferred wave length' around 4000 km corresponds to the 'normal' one-layer Eady solution, whereas the right branch, with a maximum at a much shorter wavelength, is a feature caused by the extra degree of freedom in the vertical. Waves within this interval may, provided they form within a cold air mass, as in the case studied by Rasmussen and Aakjær, be interpreted as short-wavelength, baroclinic polar lows.

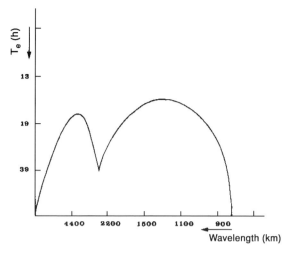

Figure 4.18. e-folding time (decreasing upwards) as a function of wavelength for unstable waves obtained from Blumen's model (Blumen, 1979) for a baroclinic polar low development in the North Sea (from Rasmussen and Aakjær, 1992).

In the studies discussed above a normal mode approach has been used to find the growth rate and the wavelength of the most rapidly growing polar lows within a baroclinic unstable environment. An alternative method of instability analysis is the initial value approach motivated by the recognition that in general the perturbations from which storms develop cannot be described as single normal mode disturbances, but may have a more complex structure. As discussed above strong dependence of cyclogenesis on initial conditions occurs when an short-wave, upper-level trough or a corresponding upper-level potential vorticity anomaly is advected into a region with a pre-existing meridional temperature gradient at the surface.

From the very start of the intensive research on polar lows in the 1970s and 1980s it was realized that virtually all high latitude polar low developments were triggered by upper-level disturbances in the form of short-wave cold troughs (e.g. Rasmussen, 1985b; Businger, 1985). Following Pettersen's ideas, it was assumed that the short-wave, upper-level troughs interacted with the shallow baroclinic zones along the ice edges resulting in the formation of a low-level circulation. Such a low-level circulation, generated through baroclinic instability, subsequently might act as a trigger for a 'convective' polar low (two-stage development).

While this type of baroclinic development triggered by short-wave, upper-level troughs undoubtedly occurs leading to the formation of polar lows (an example is presented below) then on the other hand its importance probably has been exaggerated. Forcing from an upper-level trough will generate a response whose vertical scale is proportional to the horizontal scale of the upper-level system. As shown by Holton (1992) for quasi-geostrophic systems, forcing at a given altitude will generate a response whose vertical scale is given by λ^{-1}, where λ is defined by $\lambda = (k^2 + l^2)\sigma f_0^2$, k and l being the wavenumbers in the x and y direction respectively, f_0 the Coriolis parameter and σ the vertical stability. Upper-level vorticity advection associated with disturbances of large horizontal scale and/or a small vertical stability in the environment will thus produce geopotential tendencies that extend down to the surface with little loss of amplitude, while for disturbances of small horizontal scale and large static stability the response is confined close to the levels of forcing. From a PV perspective, this is expressed in the way that the Rossby penetration depth, which describes to what extent an upper-level PV anomaly may penetrate towards lower levels, is given by $H = fL/N$, L being the horizontal scale of the system and N the Brunt–Väisälä frequency. It is doubtful, therefore, whether the small-scale, upper-level troughs can trigger low-level baroclinic developments, unless the Arctic air mass has become de-stabilized due to adiabatic heat transfer from the surface (see Sections 4.4.2 and 4.4.4 for more details).

Another factor that may limit the influence of the above discussed 'self development process' is the fact that the ice edge-generated baroclinic zones are very shallow. Some scientists have questioned the importance of baroclinic instability for polar low developments involving shallow air masses and their corresponding low-level baroclinic zones generated along the edge of the polar ice. Økland (1989), considering the baroclinicity over the ocean associated with cold air outbreaks from snow/ice-covered regions, pointed out that the baroclinicity 'is caused by the downstream increase in temperature and depth of the convective layer, and it is highly questionable if this baroclinicity can support baroclinic wave development'.

Albright et al. (1995), in a study of a polar low development over the Hudson Bay (see Section 5.1 for a discussion of this case), likewise questioned the role of these shallow, ice edge-generated baroclinic zones as a source of low-level baroclinic instability. The low studied by them formed over a relatively small ice-free region in the eastern part of the otherwise ice-covered bay. Albright et al. concluded that this case provided an example in which baroclinicity appeared to have played a minor role in the polar low development. Instead they pointed to latent heating released in deep convection as the overwhelming cause of the intensification.

In some cases, however, when the horizontal scale of the upper-level disturbance is sufficiently large and the low-level baroclinic zone is deep enough, baroclinic developments through mutual interaction between an upper-level trough and a low-level baroclinic zone may take place as illustrated schematically on Figures 4.3 and 4.4. Nordeng (1990) in a diagnostic study of two polar low developments over the Norwegian Sea found that both of the relatively weak developments were baroclinic and noted that 'the main triggering mechanism for both cases was approaching upper-level troughs which interacted with low-level baroclinic zones created by 'fixed surface forcing', i.e. sea surface temperature anomalies and ice-edges'. In both cases the main baroclinic zone, i.e. the polar front, or a branch of this zone was situated close to the place of development near Svalbard. As such, the two developments studied by Nordeng were representative of only a rather small group of polar lows, i.e. lows that develop close to the main baroclonic zone (for more details see Section 5.1.2, 'Nordeng's cases').

Occasionally, strong vortices qualifying as polar lows form along the semi-permanent baroclinic zone over and off the pack ice adjacent to the northeast Greenland coast. An example of this was presented by Rasmussen and Cederskov (1994). In this case a strong polar low, which dominated the weather in the region for more than two days, formed near the ice edge. The development was dominated by baroclinic forcing involving a marked low-level baroclinic zone as well as strong upper-level forcing, i.e. a type-B development. Convection, on the other hand, played a very minor role throughout the lifetime of this low, which achieved wind speeds around 25 m s^{-1}.

The development was triggered on 2 March 1989 as an upper-level, synoptic-scale cold trough approached the region around Scoresbysund (70.4° N, 21.4° W) from the west after crossing the ice cap. As the region with differential vorticity advection ahead of the upper-level trough spread out over the low-level baroclinic zone along the northeast Greenland coast a strong polar low developed within 12 h. The structure of the low-level baroclinic zone along the northeast Greenland coast just prior to the polar low development is illustrated through the 2 m temperature field shown in Figure 4.19a. The small wave (indicated by an arrow) and the associated increased gradient in the isotherm field near 75° N, 12° W northeast of Scoresbysund marks the centre of a cyclonic disturbance being formed within the zone of pronounced temperature gradient. The precise location of the development in this case was probably determined by the distribution of the sea ice in the region, the cyclonic disturbance initially forming in a region with only low ice concentration. In the following hours the cyclonic disturbance developed rapidly into a (baroclinic) polar low with well-defined low-level frontal zones (Figure 4.19b). A nearby radiosonde ascent (Figure 4.19c) from Danmarkshavn (76.8° N, 18.7° W, position indicated on Figure 4.19a) at 1200 GMT 2 March 1989, only about 200 km away from the location of the incipient low, illustrates the very stable conditions within which this low developed.

(a)

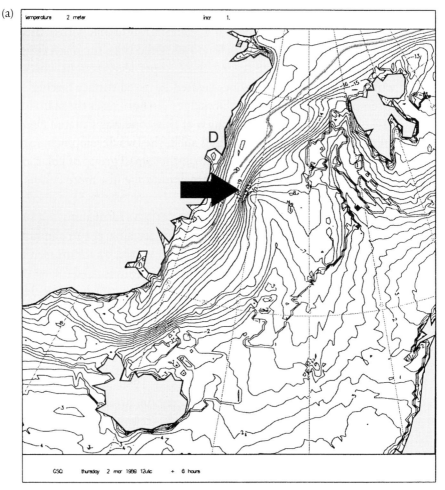

Figure 4.19. Meteorological fields and satellite imagery for the 2–4 March 1989 polar low case. (a) 2 m temperature (intervals 1 °C) at 1800 GMT 2 March. A zone of pronounced temperature gradient along the pack ice adjacent to the northeast Greenland coast shows the position of a strong, low-level baroclinic zone. A region of increased temperature gradient near 75° N, 12° W indicates the position of the centre of a cyclonic disturbance which rapidly developed into a polar low. D indicates the location of Danmarkshavn. (b) Surface pressure (thin solid lines, 2 hPa interval) and 850 hPa temperatures (dotted lines, 1 °C interval) at 0000 GMT 3 March showing a polar low northeast of Scoresbysund on the low-level baroclinic zone along the northeast Greenland coast. (c) Radiosonde ascent from Danmarkshavn (76.8° N, 18.7° W) (position indicated on (a)) at 1200 GMT 2 March. (d) 2 m temperature (intervals 1 °C) at 1800 GMT 3 March. The southern part of the zone of pronounced temperature gradient situated along the coast on 2 March (a) had now, following the development of the polar low, been displaced east. (e) Surface (10 m) wind field at 1800 GMT 3 March. The winds are plotted in the conventional way, a long barb signifying 10 kt, a short barb 5 kt. Thin solid lines show isotacks at 1 m s^{-1} intervals. (f) Surface pressure (thin solid lines, 2 hPa intervals) and 850 hPa temperature (dotted lines, 1 °C intervals) at 0000 GMT

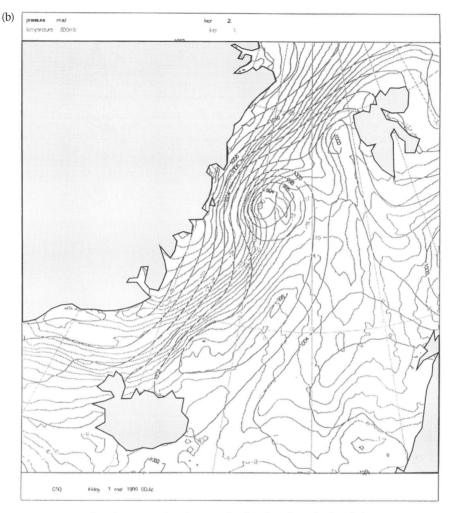

Caption for Figure 4.19 (*cont.*). 4 March, showing the polar low in its mature stage between northeast Greenland and Svalbard. (g) An infra-red satellite image for 1316 GMT 3 March showing the mature polar low between northeast Greenland and Svalbard. (Image courtesy of the NERC Satellite Receiving Station, University of Dundee. Charts and the radiosonde ascent courtesy of the Danish Meteorological Institute.)

Following the formation of the low, a strong outbreak of Arctic air, lead by a sharp Arctic front, affected the region north of Iceland, while advection of warm air took place further east and north. On the afternoon of 3 March the surface temperature field (Figure 4.19d) showed the characteristic 'T-bone structure' of some large, synoptic-scale cyclones as described by Shapiro and Keyser (1990). Very pronounced horizontal wind shear was present over the leading edge of the frontal zones (Figure 4.19e). The model-derived wind field at this time showed wind velocities exceeding 28 m s^{-1}. Late on the evening

312   4 Theoretical investigations

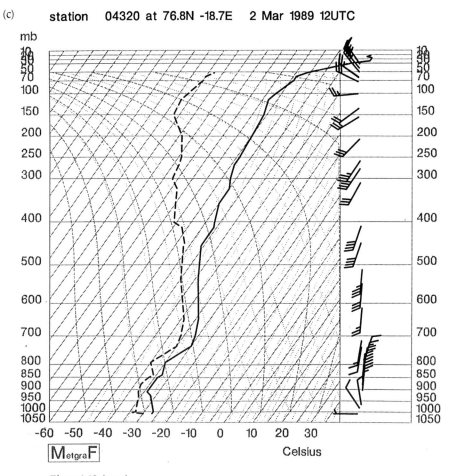

Figure 4.19 (cont.).

of 3 March the surface pressure within the centre of the low reached a minimum around 980 hPa. The surface pressure and 850 hPa temperature field from 0000 GMT 4 March, illustrating the mature stage of the low, is shown in Figure 4.19f.

This baroclinic development was further documented by a number of satellite images that confirmed the results from the numerical model. The polar low, as seen from a satellite during its mature stage around noon on 3 March, is shown in Figure 4.19g. The centre of the low, seen as a cloud spiral composed of a large number of low-level cloud streets converging into a common centre, was situated close to the ice edge. No deep convection can be seen within the region of the low in contrast to most polar low developments further east.

The relative importance of baroclinic instability versus latent heat release associated with deep convection was investigated by Sardie and Warner

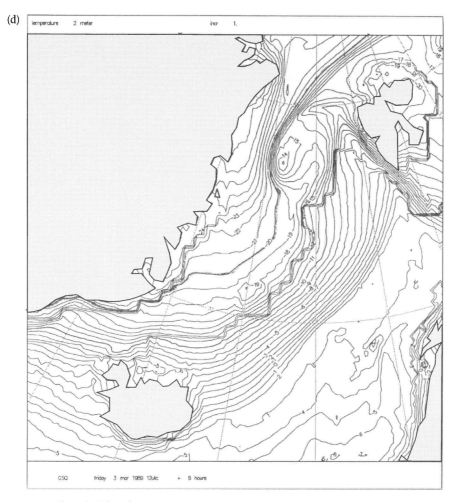

Figure 4.19 (cont.).

(1983) who used a three-layer, two-dimensional, quasi-geostrophic model that incorporated both these processes. The model included parameterizations of stable precipitation associated with moist baroclinic processes and convective precipitation associated with CISK. Seven case studies were used in the investigation, the model being run six times for each case. The model runs incorporated pure dry baroclinicity, pure CISK, moist baroclinicity and a combination of these processes. The study showed that the moisture in the boundary layer, the vertical distribution of convective latent heating and the mode of heat release (moist baroclinicity or CISK) were important parameters in determining the form of polar lows in their early stages. For the moist baroclinic modes, greater release of latent heat increased the maximum growth rate and decreased the wavelength of the system.

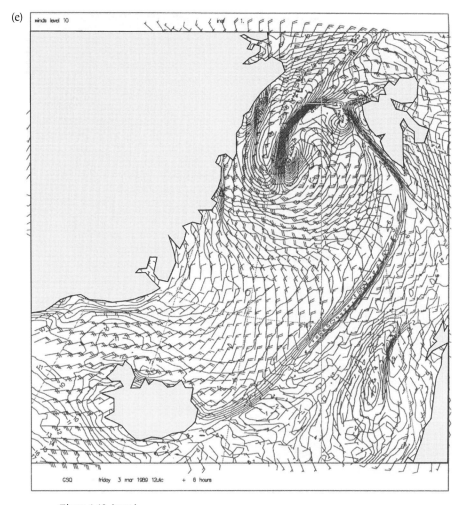

Figure 4.19 *(cont.)*.

Overall the model results of Sardie and Warner (1983) showed that neither CISK nor dry baroclinicity on their own provided the necessary forcing to allow instabilities to grow to the observed wavelengths at the observed rates. They found that baroclinicity was important in the formation of polar lows in both the Pacific and the Atlantic, noting that the average latitude of the maximum of baroclinicity was 30–40° N in the central and western Pacific, 30–50° N in the western Atlantic and 20–40° N over Europe. Therefore polar lows in the Pacific forming around 40° N benefit from the presence of strong, deep baroclinicity, while polar lows over the Atlantic, developing around 60° N near Iceland are too far north to take advantage of any strong baroclinicity. Concerning the latter developments they suggested that CISK must operate in conjunction

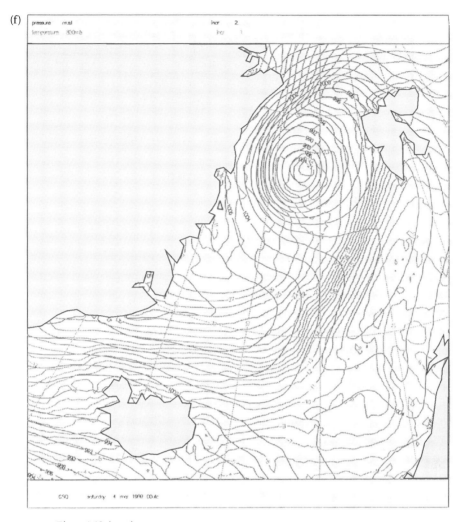

Figure 4.19 (cont.).

with shallow baroclinicity from residual circulations or occluded depressions. In this way moist baroclinic processes, not restricted to low levels, on their own may explain the genesis of polar lows observed in the Pacific, such as the systems with well-developed comma-shaped clouds observed within the strong baroclinic zone (Reed, 1979). These conditions provide a source of available potential energy (APE) throughout the whole troposphere to allow the perturbation to grow. On the other hand, moist baroclinicity as well as CISK were felt to be necessary in the development of Atlantic systems. The low-level wind shear, providing a low-level source of APE, complements an upper-level source due to CISK.

316   4 Theoretical investigations

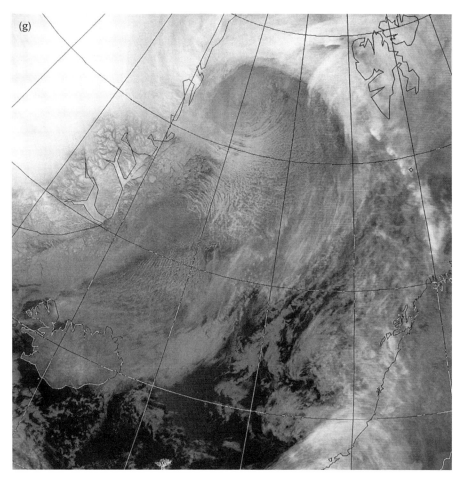

Figure 4.19 (cont.).

Another investigation into the nature of moist quasi-geostrophic instability was carried out by Mak (1982). In this study condensational heating was parameterized within a simple analytical model allowing the instability properties to be examined. It was found that as the heating intensity parameter was increased, the most unstable wave's growth rate increased significantly, its wavelength decreased significantly and its phase speed increased. The study also suggested that there is an upper limit for the growth rate, of about five times the dry model value, and a lower bound for the wavelength of the most unstable wave of about one-third the dry model value. This study also supported the concept that the baroclinic forcing in a disturbance could organize the condensational heating on a scale comparable to the wave itself.

High resolution, mesoscale models have now been applied to a number of polar low cases and used to examined the role of various instability mechanisms and synoptic situations. Grøndås *et al.* (1987a) used such a model to investigate developments within a reverse shear flow as discussed above. They found that the model was successful in reproducing the development of polar lows in such a flow with the output of the model being very similar to the observational data.

Building on their earlier study with a simple linear model, Sardie and Warner (1985) employed a full mesoscale model to examine two polar low developments in the Pacific and Atlantic Oceans. The Atlantic polar low examined developed on an intense but shallow baroclinic zone in the Denmark Strait region. The Pacific low also developed in a very strong baroclinic zone, but in this case the zone was deep. They found that the dry baroclinic effects accounted for much of the development in the Pacific case but that CISK was needed to correctly model the full development. It was also found that the development in the model was very sensitive to the shape of the vertical heating profile. Their simulation of the Atlantic case suggested that the baroclinicity was sufficient to allow a realistic initial development to take place while the polar low was close to the baroclinic zone. In this case baroclinicity was the dominant mechanism for development during the first 24 h of the simulation. However, it was found that both convective and non-convective latent heating and surface fluxes of sensible and latent heat were needed for simulation of the observed development after the polar low moved away from the baroclinic zone, which occurred during the 24–48 h period of the simulation. So moist baroclinicity and CISK were both found to be important to the observed development of both polar lows.

4.3 Barotropic instability

'Barotropic instability is a wave instability associated with the horizontal shear in a jet-like current. Barotropic instabilities grow by extracting kinetic energy from the mean flow field' (Holton, 1992). A number of researchers have, over the years, considered the possible role of barotropic instability in the formation of polar lows. Duncan (1977), in his study of three polar lows, found no significant growth due to barotropic processes. Reed (1979), Mullen (1979) and Sardie and Warner (1985) all concluded that, although barotropic instability may be present in a number of polar low developments, it nevertheless represents a minor contribution or no contribution at all to these developments. These results, however, were all based on considerations of the structure of the upper-level (polar) jet stream, and, according to Reed (1979), 'it seems doubtful that a jet stream can ever be sharp enough to account for the very small-scale systems that develop over the oceans in winter...'.

Rasmussen (1983), on the other hand, in a discussion of polar low developments along BLFs and their associated shear zones west of Svalbard pointed out that polar lows may form as 'shear vortices', i.e. through 'low-level barotropic instability', along these lines. As discussed in Section 3.1, such shear lines are frequently observed to 'roll up' forming numerous vortices on different scales from the very small, with a horizontal scale of a few kilometres, up to much larger vortices, occasionally on the scale of a polar low (see Figure 3.10).

Bond and Shapiro (1991) considered barotropic instability as a possible mechanism for the formation of polar lows over the Gulf of Alaska, but could not conclude decisively whether this effect was important.

Nagata (1993) in a study of meso-β-scale vortices along the Japan Sea Polar Air mass Convergence Zone (JPCZ) noted that barotropic shear instability may be expected to work dominantly for the development of relatively small (meso-β-scale) vortices when a large amount of vorticity is concentrated into a narrow shear zone within a few tens of kilometres. Using a high resolution (6 km horizontal grid) model, Nagata simulated the formation of meso-β-scale vortices along the convergence zone. The simulated vortices appeared as waves on belts of concentrated positive vorticity of a width of a few tens of kilometres along the JPCZ (Figure 4.20a). They became increasingly sharp, and eventually the troughs of the vorticity belt formed mesoscale lows with pressure deficits of 2–4 hPa. The disturbances were characterized by spiral cloud bands around a 'dry eye' with a warm core structure. The vortices had a core of large positive vorticity of around 80 km diameter.

To confirm that the barotropic process dominated in the energetics of the developing disturbances, Nagata calculated the barotropic and baroclinic energy conversion rates within a 'strip region' around vortex V1 (see Figure 4.20). The basic zonal flow within the strip, as well as the eddy momentum fluxes, are shown on Figure 4.21a.

In a mean flow U with zonal and meridional perturbations u' and v' the variation of eddy kinetic energy (K_E) (Eqn. (4.1)):

$$K_E = \overline{1/2(u'^2 + v'^2)} \tag{4.1}$$

across the strip region on four pressure levels is shown on Figure 4.21b, while the barotropic and baroclinic energy conversions calculated as:

$$C(K_Z \rightarrow K_E) = -\overline{u'v'}dU/dy \tag{4.2a}$$

$$\text{and} \quad C(A_E \rightarrow K_E) = \overline{-\omega'\alpha'} \tag{4.2b}$$

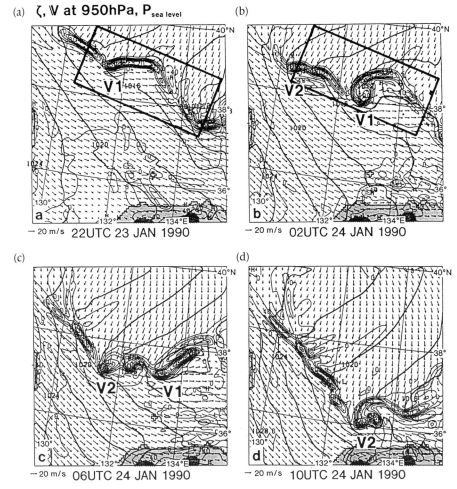

Figure 4.20. Evolution of simulated mesoscale vortices V1 and V2 along a shear line, seen in the 950 hPa relative vorticity field (thin lines; units 10^{-5} s^{-1}, contour intervals 20×10^{-5} s^{-1}; broken lines denote negative values). Wind vectors at every three grid points on the same level and sea level pressure (hPa) with 2 hPa contour intervals (thick lines) are also shown. Shading shows low land areas: (a) $t = 22$ h, (b) $t = 26$ h, (c) $t = 30$ h, (d) $t = 34$ h (after Nagata, 1993).

are shown on Figure 4.21c. The figures show clearly that the barotropic energy conversion dominates in most of the shear zone where the eddy kinetic energy is concentrated, while the baroclinic energy conversion gives a minor contribution. From the results of the energy conversion analysis, together with an agreement in spatial scale and growth rate between theory and the numerical simulation, Nagata concluded that the meso-β-scale vortices developed mainly due to barotropic shear instability.

320   4 Theoretical investigations

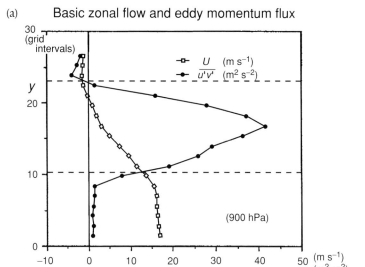

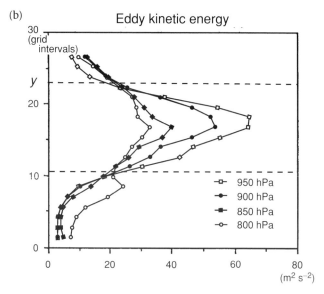

Figure 4.21. Basic zonal flow, eddy momentum flux, eddy kinetic energy and energy conversion rates for the strip region shown on Figure 4.20a. (a) Basic zonal flow (m s^{-1}) and eddy momentum flux (m^2 s^{-2}) at 900 hPa. (b) Eddy kinetic energy (m^2 s^{-2}) distribution along y on four pressure levels. (c) Barotropic ($K_Z \to K_E$) and baroclinic ($A_E \to A_E$) energy conversion rates ($\times 10^{-3}$ m^2 s^{-3}) along y at 900 hPa (from Nagata, 1993).

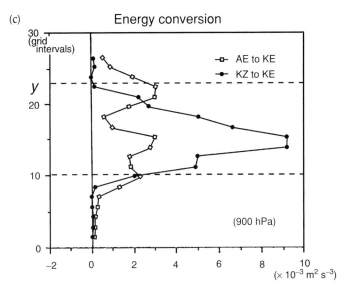

Figure 4.21 (cont.).

Detailed studies, such as the one by Nagata, have not yet been carried out for the Nordic Seas region. Satellite images, however, on numerous occasions have documented the presence of minor vortices along narrow shear lines within polar air outbreaks prior to polar low developments (see Section 3.1.4, 'Example: 25–27 January 1982'). Theory as well as observations therefore indicate that barotropic vortices forming along a shear line within a convectively unstable environment may 'focus' deep convection within a limited region affected by the vortex and in this way trigger a convectively driven polar low. On the other hand, there is no evidence that barotropic instability alone can lead to the formation of a polar low.

4.4 Potential vorticity thinking

4.4.1 Introduction

The PV (potential vorticity) concept has over the last decade received increased attention in theoretical work as well as in forecasting. PV is conserved in adiabatic, frictionless flow, and makes this dynamical parameter a very useful tool in identifying upper-air forcing that may trigger cyclogenesis. The effects from latent heating are probably more clearly understood by the PV approach, than by more traditional quasi-geostrophic methods.

For consistency, some important parts of the theory are repeated here.

Quasi-geostrophic PV (q_p) is given by the expression (see e.g. Holton 1992, pp. 164–5):

$$q_p = \nabla^2 \psi + f + f_0^2 \frac{\partial}{\partial p}\left(\frac{1}{\sigma}\frac{\partial \psi}{\partial p}\right) \tag{4.3}$$

Where ψ is the streamfunction, the constant f_0 is the planetary vorticity or Coriolis parameter at a particular standard latitude, f the Coriolis parameter at the latitude we consider, σ is a static stability parameter, which is given as $\sigma \equiv -\frac{\alpha}{\theta}\frac{\partial \theta}{\partial p}$, α is specific volume, θ potential temperature and p pressure. q_p is thus expressed by, from left to right, the relative vorticity, planetary vorticity and stretching vorticity. If diabatic effects and friction are neglected, quasi-geostrophic PV is conserved following the geostrophic flow:

$$\frac{dq_p}{dt} \equiv \left(\frac{\partial}{\partial t} + \vec{v}_g \cdot \nabla_p\right) q_p = 0 \tag{4.4}$$

The unit of q_p is the same as for vorticity, s^{-1}.

The last term on the right hand side (r.h.s.) of Eqn. (4.3), the stretching vorticity, can be written as:

$$f_0^2 \frac{\partial}{\partial p}\left(\frac{1}{\sigma}\frac{\partial \psi}{\partial p}\right) = \frac{f_0}{S_p}\frac{T}{\theta}\left(-\frac{\partial \theta}{\partial p}\right) \tag{4.5}$$

In this relation $S_p \equiv -\frac{T}{\theta}\frac{\partial \theta}{\partial p}$ is another expression for the static stability parameter. S_p varies slowly with height in the troposphere.

An air column that moves adiabatically, confined between two selected isentropic surfaces, is stretched vertically as it moves into a region where the isentropic surfaces have wider separation. With downward motion in the lower portion of the column and upward motion in the upper part, the upper part must cool and the lower part warm adiabatically (see e.g. Holton, 1992, fig. 4.7). In such a region the expression (4.5) above becomes smaller (increasingly negative), and relative vorticity, given by the first term on the right hand side of Eqn. (4.3), has to increase in order to conserve q_p assuming that the planetary vorticity changes are small. In the case of shrinking, a decrease of relative vorticity takes place.

Quasi-geostrophic PV is used for describing large-scale (synoptic-scale) flow and it is necessary to obtain an expression for PV which describes more general flow systems, including those with large curvature and large Rossby numbers (Ro $\approx$ 1).

For this purpose the Ertel potential vorticity (EPV $\equiv q$; in the following referred to as PV) is used and is given by

$$q = \frac{1}{\rho}(\vec{\zeta}_a \cdot \nabla\theta) \qquad (4.6)$$

The changes of q due to diabatic heating and friction are expressed by the following important relation (Hoskins *et al.*, 1985):

$$\frac{dq}{dt} = \frac{1}{\rho}(\vec{\zeta}_a \cdot \nabla\dot{\theta}) + \frac{1}{\rho}(\nabla \times \vec{F} \cdot \nabla\theta) \qquad (4.7)$$

where ρ is air density, $\vec{\zeta}_a$ is the total vorticity, $\dot{\theta}$ denotes diabatic heating, and $\vec{F}$ is the friction force. Relation (4.7) shows that if the gradient of diabatic heating has a component along the total vorticity vector, the first term on the right hand side of (4.7) contributes to increased PV below the level of the diabatic heating maximum and a decrease of PV above. The spatial location of the diabatically-induced PV anomalies tends to be oriented along the direction of the absolute vorticity vector.

Thus an upper, negative PV anomaly tends to develop due to diabatic heating. As will be discussed in the next paragraph, an anticyclonic circulation is associated with a negative PV anomaly and this circulation may counteract and delay an advancing upper positive PV anomaly, contributing to a prolonged deepening phase of the cyclone (Stoelinga, 1996).

The second term on the right hand side of (4.7) is the friction term. It generally contributes to a decrease of PV, but in regions where the low-level wind has a component directed opposite to the thermal wind (warm fronts) it will contribute to an increase of low-level PV (Stoelinga, 1996). It is seen that for adiabatic, frictionless flow, q is conserved, as for the quasi-geostrophic case.

The expression for EPV above can be written in a somewhat simpler (and more familiar) way. It may be obtained by introducing isentropic coordinates in relation (4.6) for q. We thus have the relation for PV, referred to as *isentropic potential vorticity*:

$$q = (\zeta_\theta + f)\left(-g\frac{\partial\theta}{\partial p}\right) \qquad (4.8a)$$

ζ_θ is the relative vorticity on an isentropic surface. q is usually expressed in PV units defined by

$$\{q\} = 10^{-6}\,\mathrm{m^2\,s^{-1}\,K\,kg^{-1}} \equiv 1\,\mathrm{PVU} \qquad (4.8b)$$

(see Hoskins *et al.*, 1985).

The conservation of q has important implications. As shown in e.g. Holton (1992), conservation of PV, $\frac{d}{dt}q = 0$, for adiabatic, frictionless conditions describes how flow over a large-scale mountain barrier produces a stationary Rossby wave, partly explaining important features of the observed mean tropospheric wave pattern.

Another important consequence of the conservation of q is seen in the exchange of air between the lower stratosphere and the troposphere. At the upper jet core, air from the lower stratosphere may easily be advected into the troposphere. Assuming adiabatic conditions, a requirement frequently met at the tropopause, this advection takes place along isentropic surfaces. If an air column confined between two isentropic surfaces is advected from the lower stratosphere, into the troposphere, the lower static stability (decrease of $-g\frac{\partial \theta}{\partial p}$ in relation (4.8a) above) in the troposphere gives a compensating increase of relative vorticity (ζ_θ). Thus air descending from the stratosphere into the troposphere tends to acquire a cyclonic rotation. q is nearly conserved in the stratosphere, and frequently during the initial descent into the troposphere (radiative cooling is important, but works rather slowly). However, after onset of cyclogenesis, diabatic and frictional effects become essential, as described in relation (4.7).

4.4.2 The invertibility principle

The invertibility principle provides the streamfunction or geopotential height field from which wind and temperature fields are obtained once a PV anomaly and suitable balance and boundary conditions are provided. This may in a simplified way be illustrated by considering the relation for quasi-geostrophic PV (relation (4.3)) which is rewritten

$$\nabla^2 \psi + f_0^2 \frac{\partial}{\partial p}\left(\frac{1}{\sigma}\frac{\partial \psi}{\partial p}\right) = q_p - f \tag{4.9}$$

f may be considered as a reference PV field and $q_p - f$ a PV anomaly. Relation (4.9) provides the streamfunction associated with the PV anomaly, i.e. the departure from the average streamfunction.

If we assume, for simplicity, that σ is constant, and we further assume that the ψ field is expressed by a single component $\psi = \Psi(t)\sin kx \sin ly \sin mp$ (real fields may be expressed as a sequence of such terms; a Fourier sequence), we obtain by inserting in (4.9):

$$-K^2 \psi \approx q_p - f \tag{4.10}$$

where $K^2 \equiv k^2 + l^2 + \frac{f_0^2}{\sigma}m^2$. Static stability ($\sigma$) is assumed positive, a necessary condition when using the quasi-geostrophic condition.

The assumption of constant σ is less reasonable since generally σ is changing in the troposphere. However, the result of this discussion would qualitatively be the same if σ were allowed to vary.

If q_p in an area is larger than the reference PV, f, the right hand side of (4.10) expresses a positive PV anomaly that yields a negative ψ or negative geopotential ϕ. Since these values are the departures from average, there is a lower geopotential associated with a positive PV anomaly than in the surroundings. Likewise we find a higher geopotential associated with a negative PV anomaly. Generally, inversion is carried out by solving Eqn (4.9) by standard numerical methods (relaxation). Usually there are several PV anomalies and inversion is done for each one separately. Since (4.3) and (4.9) are linear, adding the associated geopotential fields yields the total flow-field. This is an advantage of quasi-geostrophic PV. When the more general Eqn. (4.6) is used, the appropriate balance condition is Charney's nonlinear balance condition (Charney, 1955):

$$\nabla^2 \varphi = \nabla \cdot (f \nabla \psi) + \left(\frac{\partial^2 \psi}{\partial x^2} \frac{\partial^2 \psi}{\partial y^2} - \left(\frac{\partial^2 \psi}{\partial x \partial y} \right)^2 \right) \qquad (4.11)$$

By omitting the non-linear term and letting f be constant, we get geostrophy as balance condition, as for the case above (relation 4.9). For a circular vortex relation (4.11) reduces to the gradient wind relation. Relations (4.6) and (4.11) and suitable boundary conditions provide the equations for PV inversion in the more general case. Unfortunately, the nonlinearity of (4.6) introduces an ambiguity in the solution. The flow fields obtained from inverting a number of PV anomalies do not generally add to give the total flow. However, mathematical methods have been developed that cope with this problem, see Davis and Emanuel (1991), and Davis (1992).

To illustrate the magnitude of PV and its distribution and variability in the troposphere and the stratosphere we consider a typical north–south cross-section across the northern part of continental Europe and Scandinavia, aligned along the 10° E meridian (Figure 4.22). The cross-section represents a 24 h simulation by the Norwegian HIRLAM. The tropopause is represented by the PV = 2 contour.

If we follow the 330 K isentrope, we identify a negative PV anomaly in the region 57° N to 52° N and consequently a high tropopause (at 200–250 hPa) is present. There is a high pressure area south of 57° N and this is associated with the negative upper PV anomaly. The positive PV anomaly between 57° N and 64° N is identified by the lowered tropopause, which makes a 'dip' in this region. As an example, the 310 K isentrope is intersecting the PV = 2 contour (tropopause) at 58° N and at 64° N, defining a positive PV anomaly on this

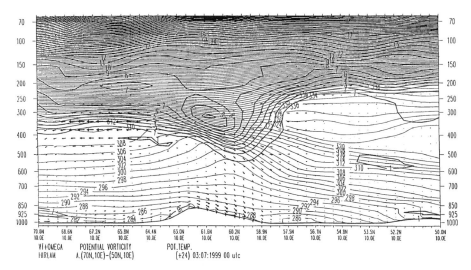

Figure 4.22. North–south cross-section along 10° E showing typical PV distribution in the northern part of continental Europe and Scandinavia (in PV units, thick lines) and potential temperature (in degrees K. thin lines). A negative (anticyclonic) PV anomaly is seen between 52° N and 57° N and a positive (cyclonic) PV anomaly between 58° N and 64° N. Arrows represent tangential winds and omega (vertical wind component in *p*-coordinates).

isentropic surface. The positive PV anomaly is associated with an upper tropospheric trough. The isentropes are seen to bow upwards below the positive PV anomaly and they are more widely separated as well, meaning that there is colder and less stable air in the region. Thus an upper trough is colder and contains less stable air than in the surroundings. Below the negative PV anomaly the isentropes are bowing downwards and they are generally squeezed more together (though less so above 400–500 hPa). The key point is that a deep tropospheric anticyclone is warmer and the air is more stable in the mid- and lower troposphere than in the surroundings.

These patterns are nearly identical with the idealized cross-sections presented by Thorpe (1985) and also discussed by Hoskins *et al.* (1985, their fig. 15). The results are obtained from a PV inversion considering a circular vortex and gradient wind used as balance condition. Fig. 15 in Hoskins *et al.* (1985) also presents the associated wind field, which is strongly cyclonic in the positive PV anomaly case, and strongly anticyclonic for a negative PV anomaly.

The degree of vertical penetration of the wind field associated with a PV anomaly is expressed by the following characteristic dimensions of a system:

$$H = \frac{fL}{N} \tag{4.12}$$

(see Section 4.5.7) where H is the Rossby penetration depth, which indicates to what extent the associated wind field of an upper PV anomaly is able to penetrate towards lower levels of the troposphere. The Rossby penetration depth depends on the horizontal dimension of the system, L, the planetary vorticity (Coriolis parameter) f, and the static stability expressed by the static stability parameter N, which is referred to as the Brunt–Väisälä frequency given by $N^2 \equiv \frac{g}{\theta} \frac{d\theta}{dz}$, where θ is potential temperature.

A low static stability N contributes to a large Rossby penetration depth- and so does a horizontal large scale L of an upper disturbance (e.g. upper positive PV anomaly).

Due to air–sea interaction, deep convective boundary layers may develop during Arctic outbreaks. In such conditions, N may become very small and H may be fairly large, even for a small-scale upper disturbance L. Thus cyclogenesis tends to take place on a small scale.

Figure 4.23, from Thorpe (1985) and Hoskins *et al* (1985), gives the wind and temperature fields associated with a warm (a) and cold (b) surface temperature anomaly. The insets in the right lower corner of the figures illustrate the distribution of isentropes along the Earth's surface at the boundary of the temperature anomaly.

Warm and cold surface anomalies can be regarded as surface cyclonic and anticyclonic PV anomalies respectively. The wind field associated with the warm (cold) anomaly is seen to be strongly cyclonic (anticyclonic), weakening with vertical distance. The static stability above the surface warm (cold) anomaly is smaller (higher) than in the surroundings.

4.4.3 The omega equation

We have seen that a cyclonic vortex is associated with and remains stationary relative to an upper positive PV anomaly. The intensity is unchanged in the absence of diabatic effects and friction. Referring to the thought-experiment described by Hoskins *et al.* (1985), we may consider a low-level wind field, such as a reverse shear flow, which is superposed below a positive PV anomaly (like the one presented in Figure 4.22) and its associated cyclonic vortex. The advection term in the vorticity equation must give a large contribution to the vorticity budget. However, the cyclonic vortex must stay in place relative to the upper positive PV anomaly, according to the invertibility principle. From the vorticity Eqn. (37) presented in Hoskins *et al.* (1985), we see that the advection term must be exactly cancelled by the terms on the right hand side. These terms contain the vertical velocity ω and the vortex stretching term is the most important one. Thus ascent upstream of the

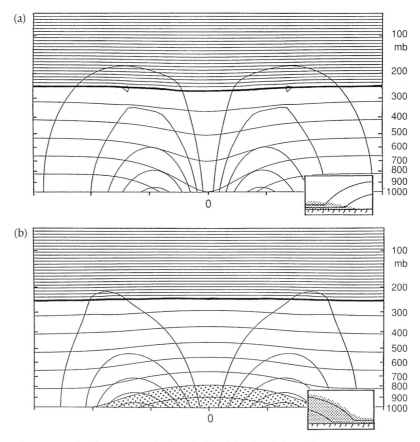

Figure 4.23. Circular symmetric flows induced by simple boundary temperature anomalies. Thick lines represent the tropopause and the two sets of thin lines, respectively, the isentropes for every 5 K and the transverse velocity for every 5 m s^{-1}. (a) A warm surface anomaly, interpreted as a cyclonic PV anomaly; (b) a cold surface anomaly, interpreted as an anticyclonic PV anomaly. The insets illustrate the distribution of isentropes along the Earth's surface at the boundary of the temperature anomaly (from Hoskins *et al.*, 1985).

upper PV anomaly and descent downstream must take place. By a change of coordinate system this thought-experiment may alternatively be described as a propagating upper positive PV anomaly. Ascent will take place ahead of the PV anomaly and descent at the rear. This is of course consistent with traditional quasi-geostrophic theory which gives ascent in response to upper positive vorticity advection. This process is described by the omega equation (Hoskins *et al.*, 1985):

$$\sigma \nabla^2 \omega + f^2 \frac{\partial^2 \omega}{\partial p^2} = f \frac{\partial}{\partial p}(\vec{v}_g \cdot \nabla q_p) \qquad (4.13a)$$

where σ is the static stability parameter, as given in relation (4.3), and ω the vertical velocity $\omega \equiv \frac{dp}{dt} \cong -g\rho\frac{dz}{dt} = -g\rho w$ and q_p is quasi-geostrophic PV, expressed by (4.3).

Or, in the more conventional form[1]:

$$\left(\nabla^2 + \frac{f_0^2}{\sigma}\frac{\partial^2}{\partial p^2}\right)\omega = \frac{f_0}{\sigma}\frac{\partial}{\partial p}\left[\mathbf{V}_g \cdot \nabla\left(\frac{1}{f_0}\nabla^2\phi + f\right)\right]$$
$$+ \frac{1}{\sigma}\nabla^2\left[\mathbf{V}_g \cdot \nabla\left(-\frac{\partial\phi}{\partial p}\right)\right] \quad (4.13b)$$

An advantage of (4.13a), compared with the traditional formulation (4.13b), is that there is a single term on the right hand side of the equation making qualitative interpretation easier since the cancellation problem is avoided. In the quasi-geostrophic omega equation written in the traditional form (4.13b), one sees that a cancellation between the forcing terms on the right hand side is possible, e.g. when cold air advection and upper level vorticity advection take place in the same region. If we assume that ω is a smoothly periodically varying function, we can write $\omega = W \sin kx \sin ly \sin mp$.

Inserting this expression in (4.13a) yields:

$$-K^2\omega \approx f\frac{\partial}{\partial p}(\vec{v}_g \cdot \nabla q_p) \quad (4.14)$$

where $K^2 \equiv \sigma(k^2 + l^2) + f^2 m^2$.

We now see from relation (4.14) that in, for example, regions where there is positive advection of potential vorticity increasing with height, there is ascent, while descent occurs in regions where negative advection of PV is increasing with height. In addition to the effects of PV advection, there is temperature advection in the boundary layer which also affects the vertical velocity field (Hoskins et al., 1985). Thus possible cancellation effects must be considered when the total vertical velocity field is assessed.

4.4.4 Some applications of PV

In high latitudes conditions can frequently be described as a combination of those seen in Figures 4.22 (an upper-level positive PV anomaly) and 4.23b (a low-level negative PV anomaly). Snow-covered land or vast expanses of ice are strongly cooled and a low level inversion develops. The wind field at low levels is anticyclonic, reflecting the higher pressure observed over snow-covered land or extensive ice fields in winter.

---

[1] Note that for a qualitative estimation of the vertical velocity the left hand side of Eqn. (4.13b) can be set proportional to $-\omega$. See Holton, 1992, sect. 6.4.

Upper troughs sitting above or moving across these areas may thus be visualized by a combination of important features seen in Figures 4.22 and 4.23b: an upper positive PV anomaly, a low-level surface temperature anomaly (corresponding to a negative PV anomaly) and their associated temperature and wind field.

The cyclonic wind field associated with a positive PV anomaly, such as the PV anomaly seen to the left in Figure 4.22 is counteracted by the anticyclonic wind field, mainly at lower levels. Further, the high static stability in the lower troposphere prevents the wind field associated with the upper PV anomaly from reaching the ground. From Eqn. (4.12) we see directly that the large N (static stability) contributes to a small Rossby penetration depth. Thus an upper trough is present, but at low levels the anticyclonic conditions prevail and no cyclonic activity takes place.

As described earlier, conditions in the Arctic, particularly in the regions stretching from Greenland across Iceland to Svalbard and the Barents Sea, are dominated by large contrasts between cold snow-covered land and sea ice on one side and open sea with comparatively high sea surface temperature (SST) on the other. This is illustrated in Figure 2.13, which gives the locations of ice edge and SST in December 1982. Thus strong low-level thermal contrasts are present.

As an upper positive PV anomaly moves from over ice fields or snow-covered land out across expanses of open, relatively warm water, the tropospheric static stability may become very low, resulting in a large Rossby penetration depth. This allows efficient communications between the upper disturbance (e.g. PV anomaly) and low-level features, such as shallow fronts, forced by strong SST gradients near the ice edge, areas of convergence and pre-existing low-level troughs. With such conditions, cyclogenesis may take place.

If the static stability is very low, the characteristic horizontal dimension, L, of the system may be fairly small and still produce a large Rossby depth, H. Consequently, cyclogenesis tends to take place on a smaller scale. The planetary vorticity is large at high latitudes and this could have a positive effect on rapid cyclogenesis as well, contributing to a large H. Polar lows appear to develop in regions where the synoptic-scale flow is cyclonic, and with average relative vorticity comparable with the planetary vorticity or even larger. This gives a modified Rossby depth (Section 4.5.7):

$$H^2 = \frac{f(f + \zeta_{av})L^2}{N^2} \tag{4.15}$$

where ζ_{av} denotes the average relative vorticity associated with the synoptic-scale flow. In some numerical studies this quantity has been found to be more than $3f$. Assuming a relative vorticity of $2f$, we obtain a Rossby penetration depth nearly twice as large as the one expressed by relation (4.12).

As an example of using PV in explaining a polar low development, we will now describe a case study from Sunde *et al.* (1994). Figure 4.24a shows the synoptic situation at 1200 GMT 4 February 1992 while Figure 4.24b presents the 500 hPa contour and 1000–500 hPa thickness 12 h prognosis for 0000 GMT 5 February. Frontal systems associated with the polar front are seen over the southwestern and eastern parts of the surface charts. The secondary, but fairly strong, frontal zone in the Norwegian Sea and around Bear Island is mainly

Figure 4.24. (a) Synoptic analysis for 1200 GMT 4 February 1992. (b) 12 h prognosis for 0000 GMT 5 February showing the 500 hPa contours (solid lines) and 1000–500 hPa thickness (broken lines). D indicates the position of the cross-section referred to in the text. (c) 6 h prognosis for 1800 GMT 4 February for cross-section D showing potential vorticity (thick lines, PV units) and potential temperature (thin lines, K). (d) 24 h prognosis for 1200 GMT 5 February for cross-section D showing potential vorticity and potential temperature. (e) 15 h prognosis for 0300 GMT 5 February showing wind and potential vorticity on the 285 K isentropic level. B indicates location of Bear Island. (f) 24 h prognosis for 1200 GMT 5 February for cross-section D with arrows indicating the velocity and thin solid lines the potential temperature. The other cross-section lines A, B, C are described in Sunde *et al.*, 1994, from which these figures are taken.

332　4 Theoretical investigations

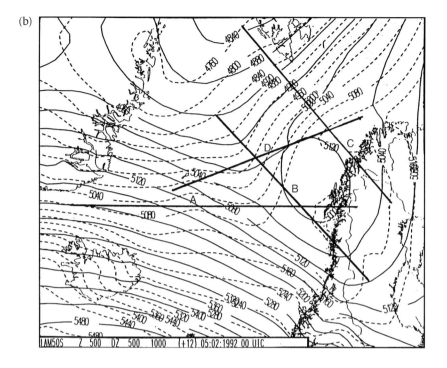

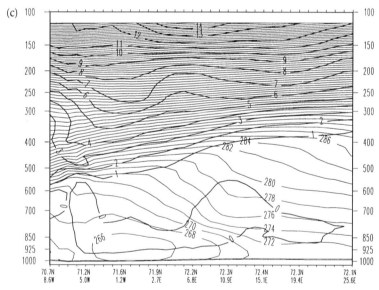

Figure 4.24 (cont.).

4.4 Potential vorticity thinking   333

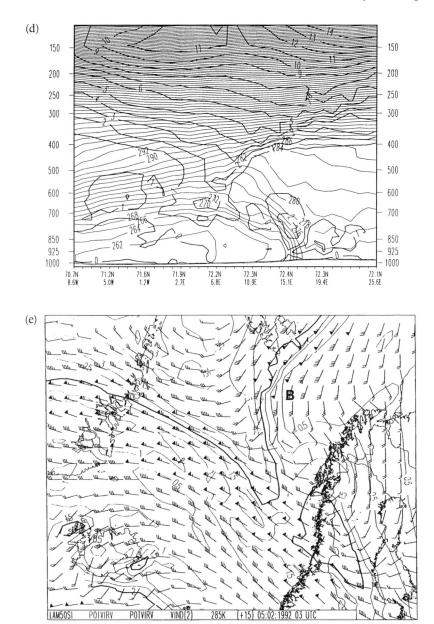

Figure 4.24 (cont.).

confined to the region below 700 hPa, as shown in Figures 4.24c and d. The cross-sections, with location indicated by the bar D in Figure 4.24b, show potential vorticity (in PV units) and potential temperature contours (in degrees K). Figure 4.24e (15 h prognosis) presents wind and potential vorticity on the 285 K isentropic level at 0300 GMT 5 February. The PVU = 2 contour is

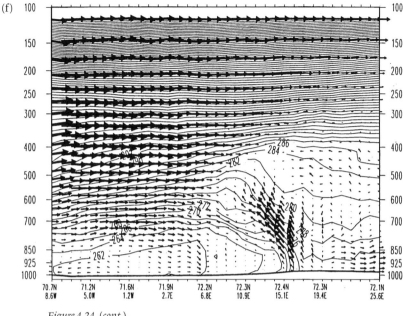

Figure 4.24 (cont.).

emphasized. Figure 4.24c shows the positive PV anomaly over the western part of the Norwegian Sea, the rising isentropes indicate colder air and their wider vertical separation means that the static stability was smaller than further east.

There was pronounced upper PV advection approaching the Bear Island region, as seen in Figure 4.24e (Bear Island is indicated by B). Referring to the omega equation (Eqn. 4.13), ascent must take place in the region of upper positive PV advection (PV advection at lower levels is zero or very small), which is readily verified in Figure 4.24f. This is a 24 h prognosis for 1200 GMT 5 February showing a cross-section along line D in Figure 4.24b. Tangential winds, vertical velocity and potential temperature are shown. The cross-section illustrates the classical picture of strong ascent of warm air in the frontal zone. There are also indications of descent in the cold air and upper part of the front, at 700 hPa. In Figure 4.24d, the positive PV anomaly had become more pronounced, and the descending air in the upper part of the front was advecting PV along isentropes, a process referred to as tropopause folding. Thus upper PV was brought further down towards the low-level frontal zone and incipient polar low at Bear Island.

The strong ascent seen in Figure 4.24f, resulting from upper positive PV advection, created release of latent heat and thus increase of PV below the diabatic heating maximum. The vorticity increased in the lower part of the frontal zone, exceeding 4×10^{-4} s^{-1} (not shown). The low-level PV anomaly, exceeding 2 PV units is seen in Figure 4.24d, at 800–850 hPa levels. This PV anomaly was due to latent heating and was not of stratospheric origin. Since a cyclonic wind field is associated with a positive PV anomaly, the low-level PV anomaly must have been an important contribution to the polar low cyclogenesis. Figures 4.25a and b present satellite images at 1225 GMT 4 February and 0359 GMT 5 February, respectively. They show development of an organized cloud system, with embedded convective clouds at Bear Island as the polar low was developing. The polar low was forced by upper, positive PV advection and, as is usual in the case of many polar lows, with a strong contribution from convection.

4.5 The role of thermal instability in polar low formation and maintenance

4.5.1 Introduction

Since the early investigations of polar lows, it has been recognized that convection plays a significant role in the development of these systems. Examples of this are numerous. Dannevig (1954) linked the formation of 'instability lows' to an organized release of thermal instability and Businger and Reed (1989b) in their definition of a polar low specifically pointed out that the main cloud masses of these systems were 'largely of convective origin'. Also, numerous satellite images of polar lows, of which a large number are shown in Chapter 3, document the occurence of convective clouds in most significant polar low developments.

The precise way in which convection affects the development of a polar low has been disputed over the years. Prior to Dannevig's hyphothesis, some meteorologists believed that a polar low could be considered as one huge cumulonimbus cloud. On the other hand, Harrold and Browning (1969), and a number of other British authors, more or less discarded the role of convection for polar low developments. Rasmussen (1977, 1979) and Økland (1977) revived the idea of the importance of deep convection for polar low developments applying the CISK[2] theory put forward simultaneously by Charney and Eliassen (1964) and Ooyama (1964) to explain the growth of hurricane depressions. According to this theory, a hurricane may intensify through a cooperative interaction between convection and a large-scale, balanced system (for a detailed discussion see the following sections). According to the studies of Rasmussen and Økland, this process might work even over polar/Arctic seas, albeit the

---

[2] For a discussion of the basic ideas regarding CISK and WISHE, see Section 4.6.1.

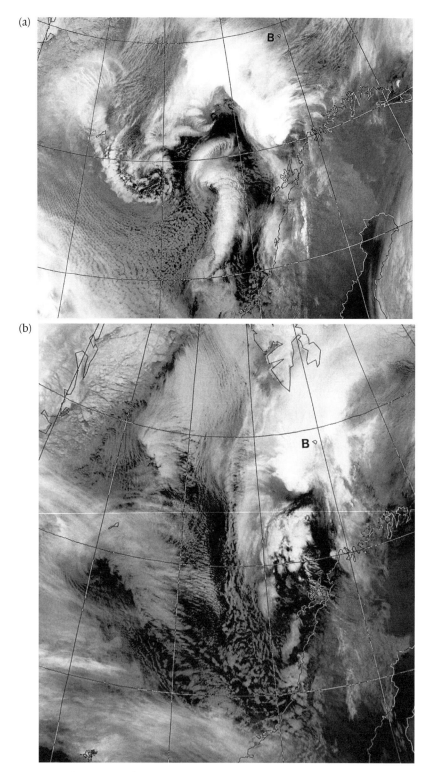

Figure 4.25. Infra-red satellite images for (a) 1224 GMT 4 February 1992 and (b) 0359 GMT 5 February 1992. B indicates location of Bear Island. (Image courtesy of the NERC Satellite Receiving Station, University of Dundee.)

sea surface temperatures in these region are far below the threshold value for the formation of tropical cyclones of around 26 °C.

Following the early studies by Rasmussen and Økland, it was gradually accepted that CISK might be the main dynamic mechanism for a significant group of polar lows, assuming that a reservoir of convective available potential energy (CAPE) was available. However, from the mid-1980s the idea of CISK and its role in the development of tropical cyclones (and polar lows as well) was challenged in a number of papers (Emanuel, 1986a; Emanuel and Rotunno, 1989). These authors argued that the tropical atmosphere was nearly neutral to deep moist convection, and that the reservoir of available potential energy assumed by Charney, Eliasen and Ooyama in their original concept of CISK, apparently did not exist in the tropical atmosphere. According to Emanuel (1986a), tropical cyclones can intensify and be maintained through air–sea interaction instability (ASII, later denoted WISHE (wind induced surface heat exchange); Emanuel, 1986a) *without* ambient conditional instablity (CAPE) providing a starting disturbance of sufficient amplitude exists. In their 1989 paper, Emanuel and Rotunno extended this point of view to polar lows, claiming that 'it seems likely that the convection observed in the environment of polar lows similarly serves to maintain a nearly moist adiabatic lapse rate, with no substantial stored convective available potentiel energy'.

While results from Betts (1982) indicated that the tropical atmosphere was very nearly neutral to deep convection when viewed in a proper thermodynamical framework, the situation may be different in polar/Arctic regions where low-level heating, due to strong air–sea interaction, occurring simultaneously with upper-air cold advection may lead to the formation of significant amounts of CAPE. Rasmussen (1979), in a study of a polar low that formed near Iceland and subsequently moved south over a warmer sea surface, argued that air particles ascending pseudo-adiabatically would achieve a temperature excess relative to their environment of up to 6 °C, corresponding to significant CAPE. In the same paper, another strong polar low development close to the Norwegian coast on 12 October 1971 was studied (the surface chart showing the polar low on 13 October is shown as Figure 3.1). This development took place close to the coast and relatively far south, near 65° N, over an exceptionally warm sea, with sea surface temperatures around 10 °C, as a major upper-level cold trough with 500 hPa temperatures as low as −44 °C approached the coastal region. A radiosonde ascent from the weather ship *Polar Front* (66° N, 2° E) from 0000 GMT 13 October and upstream, but close to the place of formation, is shown on Figure 4.26. Coastal measurements representative of offshore conditions close to the coast where the development took place, showed surface temperatures/dew point temperatures of 8 °C and 0 °C respectively. Air parcels

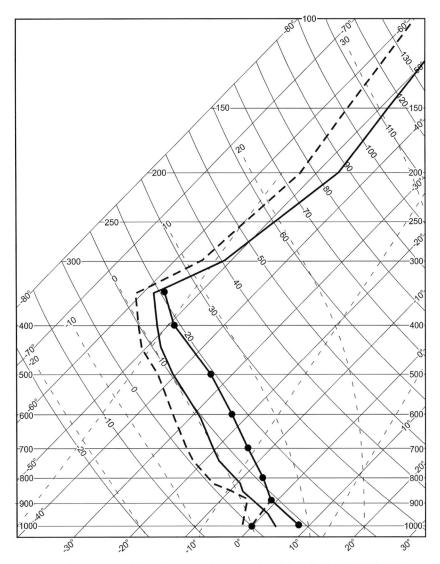

Figure 4.26. Radiosonde ascent from weather ship *Polar Front* (66° N, 2° E) at 0000 GMT 13 October 1971 showing temperatures (solid line) and dewpoint temperatures (broken lines). The solid line with dots shows the temperatures of an air parcel ascending pseudo-adiabatically from the surface with temperature and dewpoint at the start of the ascent as observed in the onshore flow at coastal stations.

ascending from the surface with these values would experience a significant temperature excess relative to the environment, corresponding to a CAPE value of around 1100 J kg^{-1} (a CAPE value of 1100 J kg^{-1} corresponds to a maximum vertical velocity of a pseudo-adiabatically ascending air parcel of around

4.5 The role of thermal instability in polar low formation

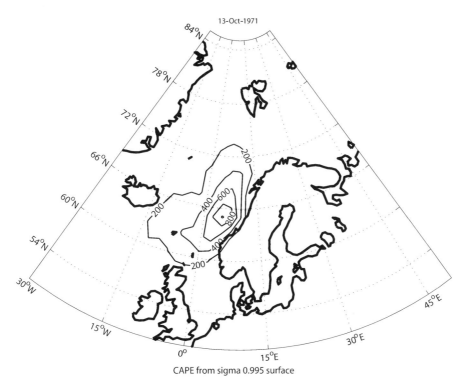

Figure 4.27. CAPE field for 0000 GMT 13 October 1971 (units J kg^{-1}, contour intervals 200 J kg^{-1}, maximum value 1000 J kg^{-1}).

47 m s^{-1}). Assuming that the neutral (dry adiabatic) low-level layer was well mixed (as actually shown by the radiosonde ascent) *all* the parcels within this layer would be approximately equally positively buoyant and the sounding as such 'unambiguously conditional unstable' as defined by Emanuel *et al.* (1994).

The extent of the region associated with high values of CAPE is illustrated by Figure 4.27 showing the CAPE field at 0000 GMT 13 October 1971 (J. Rytter, personal communication).

The CAPE values shown on Figure 4.27 were computed from the NCEP/NCAR re-analysis data set. The CAPE distribution was computed for a number of polar low developments assuming that the particles ascended undiluted from a level near the surface (sigma level 0.995), initially dry adiabatically, and later, after condensation, moist adiabatically carrying their condensation products along.

The CAPE values associated with the 13 October 1971 development are probably the highest documented for a polar low so far and the development as such should not be considered as typical. Wilhelmsen (1986a), on the other hand, considered 38 cases of gale-producing polar lows and found a conditionally unstable lapse rate between the surface and the 500 hPa level for *all* cases, which indicate that CAPE may be significant for other developments.

In addition to the October 1971 development, the CAPE fields for several other significant polar lows were studied. According to Rytter, the CAPE fields were highly sensitive to the initial conditions, the values being strongly dependent on the particular level from which the particles started their ascent. With this uncertainty in mind, the data indicated that a number of significant polar low developments seem to have been associated with moderate amounts of CAPE, a moderate amount here being defined as $c.400$–$600 \, \text{J kg}^{-1}$ corresponding to a maximum vertical velocity within the convective clouds of around 30–35 m s^{-1}. Occasionally, however, larger CAPE values were found.

The study indicated that the polar lows did not necessarily form within *pre-existing* reservoirs of CAPE but that the reservoirs and the polar lows developed more or less simultaneously as the lows moved over a warm sea surface.

The polar low near Bear Island on 12–14 December 1982 has been discussed in a number of papers (see Section 3.1.5), the development of the system being described as a two-step process; an initial phase followed by a convective stage during which convection was assumed to be important either in connection with CISK or with WISHE.

At the time when the impressive cloud spiral seen on Figure 3.29 had formed in the morning of 13 December only small CAPE values could be detected in the region. Later in the day at 1800 GMT when the cloud spiral had moved south over a warmer sea surface, and just prior to the second phase of development of the polar low, an intensification of the CAPE was indicated by the NCEP/NCAR re-analysis data, resulting in the formation of a rather strong maximum, north to northwest of North Cape (Figure 4.28). As the polar low intensified, some distance east of the centre of the cloud spiral but close to the region of maximum CAPE, the magnitude of the CAPE further increased to around 800 J kg^{-1}. Judging from the data presented by Rytter (EGS Polar Lows Working Group Meeting, Paris, October 2001), the large increase in CAPE was primarily due to the formation of a rather deep, low-level, neutrally (dry adiabatic) stratified layer which formed due to strong sensible heat fluxes from the warm sea surface.

Some preliminary, general conclusions can be drawn from the examples presented above and other cases investigated by Rytter. First, the combination of differential temperature advection and low-level heating may lead to the generation of, and to significant variations in the amount of, CAPE in polar/Arctic regions.

4.5 The role of thermal instability in polar low formation

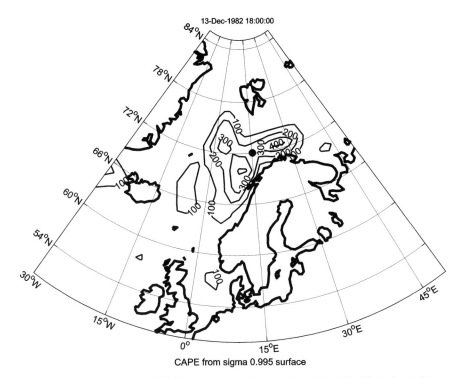

Figure 4.28. CAPE field for 1800 GMT 13 December 1982. The black dot indicates the position of the parent polar low at 1800 GMT. An intense vortex (a new polar low) developed a few hours later east of the position of the parent circulation, close to the centre of maximum CAPE.

Second, the CAPE fields over the Norwegian and Barents Seas seem highly structured, showing well-defined maxima and minima, and CAPE is *not* consumed as quickly as it is generated by large-scale processes. The maxima are well correlated with the positions of the polar lows showing good continuity in space and time.

The amount of CAPE is highly variable during the development of a polar low, being typically rather small during the initial stage, but growing as the lows mature. The reasons for this are not clear, but several factors may be significant, including that most polar lows form initially at high latitudes with relatively low sea surface temperatures. As they move south the sea surface temperature and the surface fluxes will increase significantly modifying mainly the layer adjacent to the surface, leaving the layers above relatively unaffected. A modification of this type will generally lead to an increase of CAPE.

Apart from the question of the mere existence of CAPE, the extent to which it contributes to polar low developments is still an open question as will be apparent from the discussion in the following sections.

4.5.2 Adjustment to heating and cyclone intensification

The supply of water vapour to the atmosphere from the ocean and the heating of the atmosphere when water vapour condenses or freezes in clouds are crucial requirements for the growth and maintenance of polar lows. Sections 4.5.2 to 4.5.13 aim to provide physical insight into the relation between heating and vortex intensification. Employing a variety of model problems, we investigate in detail how heating affects the pressure and potential vorticity distributions in the atmosphere and how this, in turn, affects the air motion. The frequent reference to tropical cyclones, although seemingly inappropriate in a book devoted to polar lows, is motivated by two facts: (some) polar lows and tropical cyclones bear a strong resemblance to each other and most theories discussed in this section were first constructed to explain tropical cyclone growth, but were later adopted by various authors to also explain a significant part of the intensification of polar lows.

Sawyer (1947) summarized the accepted (pre-1947) view of the mechanism for the growth of cyclonic storms in the tropics as follows (Figure 4.29):

> Most meteorologists who have studied the mechanics of the tropical cyclone agree that its formation is connected with the vertical instability in the atmosphere, and that the energy of the cyclone is derived from the latent heat of condensation released during the ascent of warm moist air from the surface. It is explained that the formation of a warm column of air by convection causes the isobaric surfaces at the top of the column to be raised, and an outflow of air at this level results from the rise in pressure. The outflow causes a reduction in total weight of the column

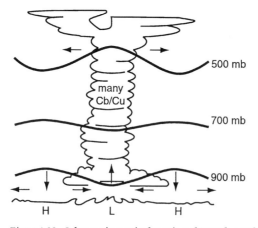

Figure 4.29. Schematic vertical section through a polar low. The arrows indicate the flow direction. The thick solid line indicates the position of a specific pressure level (vertical scale in this case is exaggerated) (based on fig. 8 of Rasmussen, 1979).

and produces a fall of pressure; the surface winds then converge towards the newly-formed 'low', and produce a cyclonic circulation as a result of the Coriolis effect.

This represents a short summary of the so-called 'thermal theory of cyclones', which had been in existence since the nineteenth century as a theory for the growth of cyclones (Austin, 1951). In more modern terms, Sawyer is in fact referring to the process of hydrostatic adjustment (Bannon, 1995, 1996; van Delden, 2000), and in his final sentence he is referring to the process of geostrophic adjustment (Gill, 1982). Both these processes are more complex than is implied by Sawyer's summary. They involve sound waves, gravity-inertia waves and, usually, convection.

In the 1950s researchers who worked on the thermal theory were frustrated by their inability to make a clear conceptual distinction between buoyant cumulus convection connected to the vertical instability in the atmosphere and the process of adjustment to balance (Bergeron, 1954). The theory of convection resulting from hydrostatic instability, which is based on the Boussinesq approximation, cannot account for the pressure decrease at the Earth's surface.

After preliminary work by Kleinschmidt (1951) and Eliassen (1952), the paper by Charney and Eliassen (1964) marked the beginning of a new approach to the problem. Charney and Eliassen made the following three important simplifying assumptions.

1. The diameter of a tropical cyclone is small compared with its distance from the Equator with the result that the Coriolis parameter can be regarded as independent of the geographical latitude.
2. The ideal tropical cyclone is rotationally symmetrical.
3. Above the friction layer near the ground the hydrostatic and gradient wind balance is achieved to a great degree.

The first assumption is very reasonable and should not lead to much controversy, but the other two assumptions are far from trivial. Although the theory behind the justification of the second assumption is interesting (see e.g. Melander *et al.*, 1987) and is, in fact, still the subject of intensive research, here we will be concerned chiefly with the theoretical justification of the third assumption and its important consequences. Charney, in his conversation with George W. Platzman (see Lindzen *et al.*, 1990, pp. 69–70), states:

> … I became interested in the mechanism of generation of hurricanes and this led to the notion of CISK (Conditional Instability of the Second Kind) … I worked on the problem and one of the things that led me to the formulation of CISK was the idea that the forces in a hurricane must

be in essential balance, that you were dealing with a balanced flow, not an inertial gravity oscillation ... [Earlier authors] had dealt with hurricane motions as a sort of gigantic convection cell. I knew that couldn't be correct, but I was still puzzled by the existence of conditional instability. But it was really Ooyama who pointed out that, despite the fact that the individual cumulus cells were conditionally unstable, that the hurricane as a whole was stable.

Charney and Eliassen (1964), Ogura (1964), Kuo (1965) and Ooyama (1969) constructed models of a tropical cyclone in which the motion was assumed to be in thermal wind balance (i.e. hydrostatic balance and gradient wind balance) at all times. It was hypothesized that the growth of a tropical cyclone is essentially the result of a process of continuous adjustment to thermal wind balance in the presence of processes (such as heating) disturbing this state of balance.

Incorporating (parameterizing) the heating associated with latent heat release in clouds and sensible heat fluxes at the Earth's surface in the balanced cyclone model poses great theoretical problems. Many authors have hypothesized that widespread and prolonged convective precipitation only exists in connection with 'large-scale' lifting of air (the term 'large-scale' is placed within quotation marks because it is not well defined; see the caption of Figure 4.30). One important cause of 'large-scale' lifting in a cyclonic vortex is frictional convergence in the boundary layer, the intensity of which can be shown to be proportional to the vorticity of the balanced flow just above the frictional boundary layer (Holton, 1992). This offers the possibility to theoretically link the heating to the intensity (i.e. the vorticity) of the vortex, yielding a feedback loop between heating and the balanced flow (the vortex). Figure 4.30 visualizes this feedback loop. A closed loop must include the segments shown by the heavy curves. The upper half is referred to as 'control' and the lower half is referred to as 'feedback'. Charney and Eliassen (1964) showed that this loop

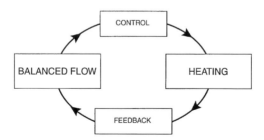

Figure 4.30. Schematic figure showing the interaction between balanced motion and heating (inspired by fig. 1.1 of Arakawa, 2000). Arakawa used the term 'large-scale processes' instead of 'balanced flow' and 'cumulus convection' instead of 'heating'.

4.5 The role of thermal instability in polar low formation

can exhibit unlimited growth, and used this fact to explain the intensification of the balanced flow. The instability resulting from this positive feedback was called 'Conditional Instability of the Second Kind'. CISK was also proposed as an explanation of the growth of polar lows (Rasmussen, 1977, 1979; Økland, 1977) and later employed to explain the growth of a certain type of Mediterranean cyclone occurring over the warm waters in the autumn and early winter (Rasmussen and Zick, 1987).

In the formulation of CISK due to Charney and Eliassen (1964), the 'control' (Figure 4.30) is frictional convergence. In more specific words, the intensity of the heating is controlled by (i.e. is proportional to) the vertical motion at the top of the turbulent boundary layer. Ooyama (1969) assumed that the 'constant' of proportionality was a function of the availability of moisture in the boundary layer. Obviously the amount of moisture in the boundary layer depends on the intensity of flux of moisture from the ocean. If this flux is large enough, the reservoir of CAPE can be replenished and continuously released or 'activated' due to frictionally induced upward motion.

Emanuel (1986a) has called into question this particular form of the 'control', which is referred to by Mapes (1997) as a form of 'activation control'. Emanuel argued against the existence of a reservoir of CAPE, pointing out that CAPE is destroyed by convection as quickly as it is created by surface fluxes. This implies that the lapse rate in the convecting part of the atmosphere is constrained to follow the moist adiabatic lapse rate. Simultaneously, the surface fluxes are enhanced by the high wind speeds within a polar low or a tropical cyclone. Therefore, as the vortex intensifies, so does the transfer of heat from the ocean to the atmosphere. The line in a thermodynamic diagram, representing the thermodynamic state of the atmosphere in a tropical cyclone or polar low, would then appear to shift slowly towards the right (increasing temperature) and slightly upward (decreasing pressure). This is visualized in Figure 4.31.

According to Emanuel, this 'convective adjustment' constitutes the 'control' in Figure 4.30. Mapes (1997) referred to this type of control as 'equilibrium control'. Emanuel's theory is now known as 'Wind Induced Sensible Heat Exchange' or 'WISHE' (earlier named 'air–sea interaction instability' (ASII)). Emanuel and Rotunno (1989) investigated how far the theory of WISHE (ASII) applies to polar lows.

Both theories (CISK and WISHE) hypothesize that the vortex evolves through a succession of balanced states. With this hypothesis the 'balanced' dynamics of the cyclone is separated from the much more complicated 'unbalanced' dynamics of cumulus convection. The numerical simulations due to Ooyama (1969) and Sundquist (1970) demonstrated that the structure and growth of a tropical cyclone can be explained by considering explicitly only the

346   4 Theoretical investigations

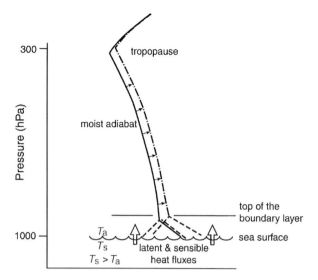

Figure 4.31. Skew-T log p thermodynamic diagram showing schematically two consecutive hypothetical thermodynamic states within a polar low or tropical cyclone, according to the equilibrium control theory. The basic assumption is that the lapse rate above cloud base must always adjust to the saturated adiabat, which is a function of temperature and pressure. Due to heating the temperature profile is shifted slowly towards the right.

balanced part of the motion. The reasons for this are quite subtle and merit a discussion, which will be provided in the following sections.

The question concerning us first is, how does heating affect the balanced flow (the so-called *feedback* in Figure 4.30)? In order to gain a better understanding of this feedback, we must turn to the basic theory of adjustment to balance (hydrostatic balance, geostrophic balance and thermal wind balance).

4.5.3 Hydrostatic adjustment

The first sentences of the introduction to a nearly 50-year-old paper by Scorer (1952) serve to clearly state the problem at hand: 'It is not obvious what will be the consequences of heating the lower layers of the atmosphere over a large region. What will happen to the air over the heated layers? At what levels will pressure gradients be generated and what air motion will ensue? Authors are not unanimous in their answers.'

It is the purpose of this section to shed some light on the answers to these questions. Our starting point is the equation of continuity for an ideal gas, which can be written as follows (Van Delden, 1992, or Durran, 1999):

$$\frac{d\Pi}{dt} = -\frac{R\Pi}{c_v}\vec{\nabla}\cdot\vec{v} + \frac{RJ}{c_v\theta} \qquad (4.16)$$

where t is time, $\vec{v}$ is the air velocity, J is the heating per unit mass, per unit time, θ is the potential temperature (defined as $\theta = T(p_{ref}/p)^\kappa$, where T is the temperature, p is the pressure, p_{ref} is a constant reference pressure, R is the specific gas constant of air, $\kappa = R/c_p$ with c_p the specific heat at constant pressure), c_v is the specific heat at constant volume, $d/dt = (\partial/\partial t + \vec{v}\cdot\vec{\nabla})$ and Π is the Exner function, defined as

$$\Pi \equiv c_p \left(\frac{p}{p_{ref}}\right)^\kappa \qquad (4.17)$$

Eqn. (4.16) demonstrates that the pressure inside an air parcel changes due to divergence (first term on the r.h.s.) and due to heating (second term on the r.h.s.).

When an air mass is heated at constant volume, potential energy is introduced in the form of available elastic potential energy (Bannon, 1995). A pressure gradient is set up at the edge of the air mass. The force resulting from this pressure gradient represents the germinal force for the excitation of a wave of expansion, i.e. a thermally forced sound wave. As a result of this force the air mass expands (i.e. $\vec{\nabla}\cdot\vec{v} > 0$) and, in turn, compresses the immediate environment. In this manner the positive pressure perturbation propagates outwards with the phase speed of sound waves (about 300 m s^{-1}). Due to the decrease in its density, the air mass acquires positive buoyancy, which ultimately tends to come in balance (if the environment is stably stratified) with a negative *perturbation* pressure gradient force such that

$$\frac{\partial p'}{\partial z} = -\rho g \qquad (4.18)$$

where p' represents the pressure perturbation relative to the hydrostatically balanced state previous to the heating and g is the acceleration due to gravity. These are the essential ingredients of the process of hydrostatic adjustment.

In order to obtain some insight into the process of hydrostatic adjustment to heating we will review some important results of theoretical investigations into the prototype problem of hydrostatic adjustment due to Bannon (1995, 1996) (Figure 4.32). This problem consists of the response of a stably stratified atmosphere to a vertically confined but horizontally uniform instantaneous heating. The problem was originally advanced by Lamb (1908) (see also Lamb, 1932, sections 309–311) and has therefore been termed 'Lamb's problem' by Bannon. However, Lamb did not actually solve the problem of adjustment to hydrostatic balance, but only investigated the properties of the waves excited by a point source of hydrostatic imbalance.

348   4 Theoretical investigations

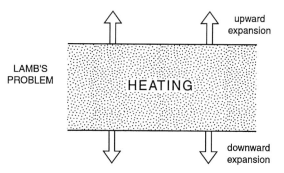

Figure 4.32. Lamb's problem. Horizontally homogeneous heating is applied to a layer of air, as a consequence of which it expands vertically.

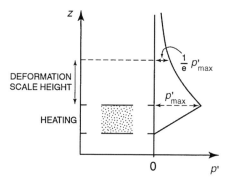

Figure 4.33. Schematic figure showing the pressure perturbation, induced by heating a vertically confined layer of air, as a function of height after adjustment to hydrostatic balance.

A first interesting result, due to Bannon (1996), states that if the heating is horizontally uniform, layers of air *below* the heated layer are not displaced in the final equilibrium state, relative to the initial state. The layers above the heated layer are lifted *uniformly* upward (or downward if there is cooling) with no change in their state variables. At a fixed point above the heated layer, the density increases and the potential temperature decreases adiabatically to produce a continuous pressure field in hydrostatic balance. A heat source in the lower troposphere affects the pressure distribution throughout the entire atmosphere aloft. Of course, the question is how strong this effect is at upper levels.

Inspired by Gill's (1982) linear technique of tackling the geostrophic adjustment problem, Bannon (1995) derived an equation for the pressure as a function of height in the final hydrostatic equilibrium state in an isothermal atmosphere. He found that the pressure increase, relative to the basic state pressure prior to the heating, is greatest at the top of the heated layer and decreases exponentially above the heated layer (see Figure 4.33).

4.5 The role of thermal instability in polar low formation

The e-folding distance associated with this exponential decrease is proportional to the density scale height in an isothermal atmosphere,

$$H_s = \frac{RT}{g} \tag{4.19}$$

This characteristic height scale, which in the troposphere has a value of about 8 km, is referred to as the *radius of deformation for hydrostatic adjustment*, or perhaps better as the *deformation scale height* in an *isothermal* atmosphere. The effect on pressure of heating in the lower layers of the troposphere will hardly be noticed in the pressure at a height equal to several times the vertical deformation scale height. It appears that the vertical deformation scale in an atmosphere with a realistic thermal structure does not differ very much from the vertical deformation scale in an isothermal atmosphere (Van Delden, 2000). This implies that the effect on pressure of heating in the lower layers of the troposphere will be noticed principally in the troposphere.

The time scale for the purely vertical adjustment (i.e. neglecting horizontal inhomogeneities) is proportional to the time required for the wave of expansion to travel a distance of the order of several times (say 10 times) the deformation scale height. With a typical phase speed of about 300 m s^{-1} and a deformation scale height of about 8 km, the adjustment time is nearly 5 minutes. In other words, within 5 min the pressure perturbation induced by the heating and the vertical readjustment process, has spread out over a characteristic vertical distance equal to 10 times the deformation scale height, and the heated layer has adjusted closely back to hydrostatic balance (Eqn. 4.18).

Lamb's problem gives considerable insight into the process of hydrostatic adjustment to heating, but is a very idealized problem to study. It is worthwhile to imagine what would happen in the more realistic situation of horizontally inhomogeneous heating (Figure 4.34).

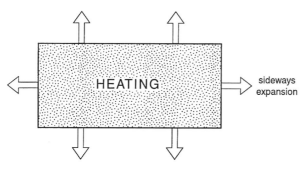

Figure 4.34. Extension of Lamb's problem to the case of horizontally inhomogeneous heating.

If heating is limited in its horizontal extent, for instance, to a circular area with radius, r, it would not only excite a vertically propagating horizontal wave-front at the upper edge of the heated area, but also a spherical wave-front at the outer edge of the heated area propagating horizontally and vertically. Due to its horizontal propagation, this spherical wave-front would affect the one-dimensional (vertical) hydrostatic adjustment process in the centre of the heated area after a time of the order of r/c with c the phase speed of sound waves. If $r \approx 500$ km, this time would be nearly 30 min. If the heating takes place over a sufficiently large region, the adjustment in the centre of a large area experiencing heating, would appear to be a two-step process: first there is a 'vertical' adjustment, involving a horizontal wave-front (i.e. the process described by Lamb's problem), after which there is a 'horizontal' adjustment, involving the spherical wave-front.

This process is illustrated in Figure 4.35a, which shows the pressure perturbation in the vertical plane, calculated with a two-dimensional (x–z) linear numerical model (see Tijm and Van Delden, 1999), 90 s and 270 s after instantaneously heating the lowest 2 km of the atmosphere over a horizontal distance of 100 km.

A basic temperature profile was prescribed as follows:

$$T_0(z) = T_s - \beta z, \tag{4.20}$$

with $T_s = 300$ K, $\beta = (g/c_p) - 0.005$ K m^{-1} for $z < 12$ km, $\beta = 0$ for 12 km $\leq z \leq 50$ km and $\beta = 0.005$ K m^{-1} for $z > 50$ km. The heat source was prescribed such that the lowest 2 km becomes neutrally stratified. The initial potential temperature perturbation, $\theta_i(z)$, is given by:

$$\theta_i(z) = \Theta \left(1 - \frac{z}{H}\right) \text{ for } z \leq H \text{ and } \theta_i(z) = 0 \text{ for } z > H, \tag{4.21}$$

where $\Theta = 10$ K and $H = 2$ km.

Three wave-fronts are formed (see Figure 4.35): a horizontal wave-front moving upwards and two circular wave-fronts (in x–z space) emanating from the edges of the heated layer. These two wave-fronts meet in the centre of the heated layer after about 160 s. The interference of these waves of compression is accompanied by a pressure decrease in the lower half of the heated layer. This is illustrated in Figure 4.35 and also in Figure 4.36, which shows the Exner function perturbation as a function of height in the centre of the domain after 90 s, 180 s and 270 s, respectively.

The profile after 90 s is identical to the profile in Lamb's problem in which there is only a vertically propagating wave. Note that the pressure at the Earth's surface ($z = 0$) does not decrease. However, after 160 s the effect of horizontal

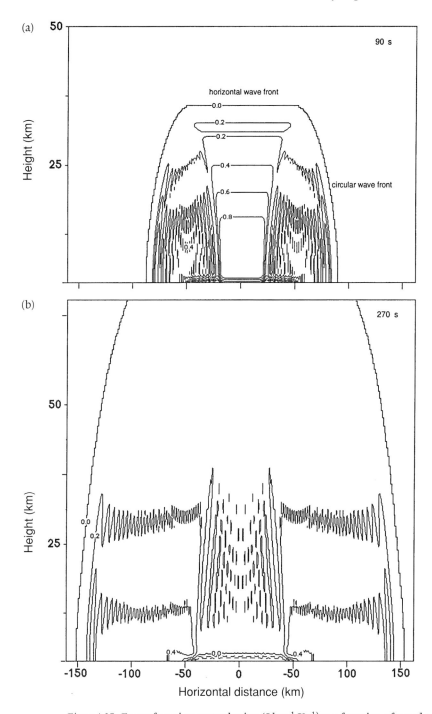

Figure 4.35. Exner function perturbation (J kg^{-1} K^{-1}) as a function of x and z, (a) after 90 s and (b) after 270 s, after the instantaneous heating of the lowest 2 km over a horizontal distance of 100 km in the middle of the domain.

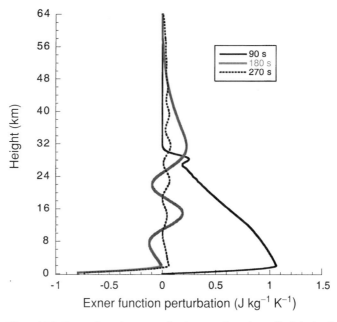

Figure 4.36. Exner function perturbation as a function of height in the middle of the domain ($x = 0$) at three points in time.

expansion of the heated air is felt in the centre of the heated area, leading to a decrease of the total mass above this point and a pressure decrease at the Earth's surface ($z = 0$). The decrease in the Exner function at the Earth's surface after 270 s of 0.8 J kg^{-1} K^{-1} is equivalent to a pressure decrease of about 2.7 hPa (if we assume that the pressure at the Earth's surface is 1000 hPa initially).

In reality, heating does not take place instantaneously. If the heat is added over a finite period, a weaker wave front travels a certain distance from the source before the heating is completed. The wavefront is less well defined than in the case of instantaneous heating. But, qualitatively, the result is the same. A pressure decrease is observed in the lower half of the heated boundary layer while a pressure increase is observed aloft and over the area where no heating has taken place (Tijm and Van Delden, 1999). The vertical pressure gradient force is approximately in (hydrostatic) balance with the buoyancy force, while the horizontal pressure gradient drives a circulation, which, if we apply the model to the core of a cyclone, can be identified with the radial circulation. It is important to note now, that such a circulation arises in an absolutely statically stable atmosphere. That is, CAPE is not required. As soon as the heating stops the non-rotating atmosphere will eventually return to a state of rest. We will see in Section 4.5.4, 'Geostrophic adjustment and the invertibility principle' that this is not the case if the atmosphere is rotating.

In many theoretical models of atmospheric circulations, it is assumed that the atmosphere is in hydrostatic balance at all times. In such a model, hydrostatic adjustment is 'instantaneous'. By integrating the hydrostatic relation and using the ideal gas law, it can be shown that if a layer of fixed thickness Δz in an isothermal atmosphere is heated such that the temperature increases uniformly by ΔT and the pressure remains fixed at the bottom of this layer, the pressure at the top of this layer will increase by a factor $\exp\{\Delta z \Delta T / [H_s(T + \Delta T)]\}$. According to the hydrostatic approximation, the heating of a layer of air gives rise to an instantaneous pressure increase at all levels above the base of this layer of air. The pressure increase above the heated layer is the *same* at all heights, erroneously implying an infinite deformation scale height. This error is a consequence of not taking into account the lifting and consequent adiabatic temperature decrease of the atmosphere above the heated layer.

Assuming the temperature $T = 300$ K, $\Delta T = 5$ K and $\Delta z = 2$ km, we obtain a factor of 1.0037. If the atmospheric layer in question is located just above the Earth's surface, we obtain a pressure increase of about 3 hPa (assuming that the pressure at 2 km height is 800 hPa prior to the heating). If the consequent divergence of mass were to eliminate this upper-level pressure perturbation, which is not necessarily the case if the effects of rotation (i.e. 'Coriolis' effects) are taken into account, the pressure decrease at the Earth's surface would attain a maximum possible amplitude of about 3 hPa. Of course, convergence of mass near the Earth's surface would tend to partly eliminate this negative pressure perturbation. Eventually, a balance between the inertial forces associated with rotation and the pressure gradient force (i.e. geostrophic balance or gradient wind balance) will be reached. We are, in fact, interested in determining the exact distributions of pressure, temperature and wind associated with this balanced state. The question now is, do we need to know all the details of the sound waves and other slower moving waves emanating from the heated region in order to determine the balanced state? In the next section we will show that (and expain why) this, fortunately, is not the case.

4.5.4 Geostrophic adjustment and the invertibility principle

In this section we illustrate some typical characteristics of adjustment to geostrophic balance using a simple one-layer model of a rotating, density stratified fluid. We intend to demonstrate that the adjustment process is dominated by a materially conserved quantity called potential vorticity. We will show that we need not know the details of the waves and oscillations, excited as a consequence of an imbalance of forces, in order to determine the ultimate state of geostrophic balance to a relatively high degree of accuracy.

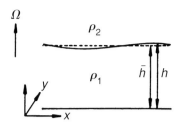

Figure 4.37. Geometry of the shallow slab-symmetric rotating layer of fluid.

We will not adopt the model used in the previous section to make our point, because this would be computationally prohibitive. Instead, we adopt the most simple model of a rotating stratified fluid, which illustrates some essential characteristics of adjustment to geostrophic balance (Figure 4.37). This model consists of a layer of incompressible fluid (depth h and constant density ρ_1) below an infinitely deep motionless layer with density $\rho_2 < \rho_1$. The fluid is rotating with a constant angular velocity equal to $\Omega = f/2$. Assuming hydrostatic balance, the equations of motion and continuity for the lower layer are (Gill, 1982)

$$\frac{dv}{dt} = -\partial u \tag{4.22a}$$

$$\frac{du}{dt} = -g'\frac{\partial h}{\partial x} + \partial v \tag{4.22b}$$

$$\frac{dh}{dt} = -h\frac{\partial u}{\partial x} \tag{4.22c}$$

In these equations $u(x, t)$ and $v(x, t)$ are the x- and y-components of the velocity, respectively, $h(x, t)$ is the height of the lower layer, $d/dt = \partial/\partial t + \partial u/\partial x$, and $g' = g(\rho_1 - \rho_2)/\rho_1$ is the reduced gravity. We have neglected derivatives with respect to y.

Suppose we extract a specified volume of mass from the lower layer. To incorporate this effect into the model we set $h = \bar{h} + h'$ at $t = 0$, where $\bar{h}$ is a constant reference height and where

$$h' = h'_{\max}\exp\left\{-\left(\frac{x - x_0}{a}\right)^2\right\} \tag{4.23}$$

This represents a bell-shaped perturbation in the height of the free surface centred at $x = x_0$ with a horizontal scale represented by the parameter a and a maximum amplitude equal to $h'_{\max}$. Due to this perturbation, horizontal pressure gradients are created in the lower fluid layer leading to convergence of mass

4.5 The role of thermal instability in polar low formation 355

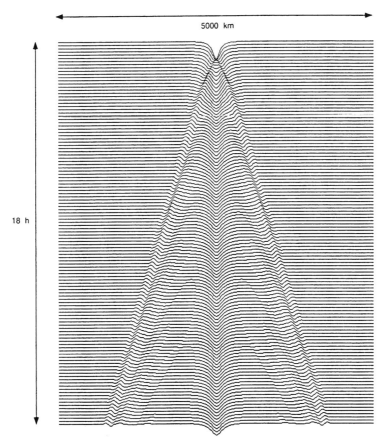

Figure 4.38. Height of the free surface as a function of time and horizontal distance. The initial perturbation in the free surface has a horizontal scale a of 60 km and a maximum amplitude of 50 m at $t = 0$. The height, $\bar{h} = 1000$ m, $f = 0.0005$ s^{-1} and $g' = 1$ m s^{-2}. The Rossby radius is 63.2 km. The waves observed are gravity-inertia waves.

towards $x = x_0$. A gravity-inertia wave is the result. This wave propagates towards $x = \pm\infty$ (sound waves are not possible here because of the condition of incompressibility). At the same time adjustment to geostrophic balance, given by

$$g'\frac{\partial h}{\partial x} = fv; \quad u = 0 \tag{4.24}$$

takes place in the region where the perturbation was inserted. Figure 4.38 visualizes the gravity-inertia waves and the adjustment to geostrophic balance in the centre of the domain.

Although the exact functional relation between h (or v) and x in the geostrophically balanced state can, in principle, be derived from Eqn. (4.22) by numerical integration, it can also be determined directly from Eqn. (4.24) with the definition of potential vorticity,

$$\zeta_{\text{pot}} \equiv \frac{\zeta + f}{h} \qquad (4.25)$$

(the relative vorticity here is $\zeta = \partial v/\partial x$). If we differentiate Eqn. (4.24) with respect to x (assuming g' and f are constant) and substitute the result into Eqn. (4.25), and then again differentiate the resulting equation with respect x and again use Eqn. (4.24), we obtain the following equation for v in the balanced state:

$$\frac{d^2 v}{dx^2} - \frac{f \zeta_{\text{pot}}}{g'} v = h \frac{d\zeta_{\text{pot}}}{dx} \qquad (4.26)$$

Eqn. (4.26), which is of the elliptic type if $\zeta_{\text{pot}} > 0$, is an expression of the so-called *invertibility principle*. This principle states that the velocity distribution in the balanced state can be determined exactly given the potential vorticity distribution and boundary conditions. In the example discussed here (see Figure 4.38), the potential vorticity in the final balanced state is, of course, not known. However, we do know that ζ_{pot} is materially conserved. This can easily be deduced from Eqn. (4.22). If we neglect horizontal advection of ζ_{pot}, the potential vorticity distribution at $t = 0$ is identical to the potential vorticity distribution at any later time. We can then solve Eqn. (4.26) numerically by successive over-relaxation assuming that $h(t \to \infty)$ on the r.h.s. of Eqn. (4.26) is equal to $\bar{h}$, which is reasonable if $h' << \bar{h}$ initially. This yields the solution shown by the thick broken line in Figure 4.39, which, within a certain distance from the place of insertion of the perturbation, apparently is nearly identical to the solution of the time-dependent Eqn. (4.22) after 96 h of integration (the thin solid line in Figure 4.39).

This remarkable fact implies that the potential vorticity, which is inserted initially, indeed practically stays in place. In other words, the potential vorticity perturbation does not propagate away from the source region with the waves. Since the potential vorticity determines the balanced state, this balanced state must therefore be nearly identical in both cases, in spite of the presence of large amplitude waves in one case.

The solution of the homogeneous part of Eqn. (4.26) is of the form

$$v \approx \exp\left(\pm \frac{x}{\lambda}\right) \qquad (4.27)$$

with

$$\lambda \equiv \sqrt{\frac{g'}{f \zeta_{\text{pot}}}} \qquad (4.28)$$

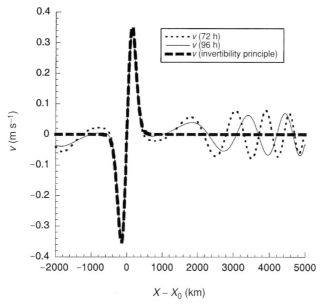

Figure 4.39. Velocity v as a function of horizontal distance according to the invertibility principle (Eqn. 4.26) (thick broken line) compared to the time-dependent solution for $t = 72$ h (dotted line) and $t = 96$ h (thin solid line). The initial perturbation in the free surface has a horizontal scale, a, of 180 km and a maximum amplitude of 50 m at $t = 0$. Height, $\bar{h} = 1000$ m, $f = 0.0005$ s^{-1} and $g' = 1$ m s^{-2}. The Rossby radius is 63.2 km.

the so-called *Rossby radius of deformation* for geostrophic adjustment. The solution Eqn. (4.27) in effect states that horizontal variations in potential vorticity force or 'induce' a velocity field with a characteristic horizontal scale equal to λ. If $\zeta \ll f$ the Rossby radius can be expressed in the more familiar form as

$$\lambda \equiv \frac{\sqrt{g'\bar{h}}}{f} \qquad (4.29)$$

i.e. as the ratio of the phase speed of surface gravity waves and the Coriolis frequency.

The Rossby radius of deformation can be viewed as the analogue of the deformation height for hydrostatic adjustment (see the previous section). The Rossby radius of deformation is the *e*-folding distance characterizing the horizontal scale of the pressure perturbation in the centre of the domain in the final geostrophic equilibrium (Figure 4.38).

The robustness of the invertibility principle is illustrated by Figure 4.40. The solid curve is the result of many 72 h integrations of the shallow water equations (Eqn. 4.22), with varying values of the Coriolis parameter (f) or

4 Theoretical investigations

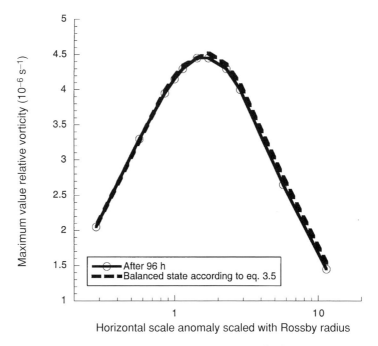

Figure 4.40. The maximum value of ζ (in units of 10^{-6} s^{-1}) as a function of the horizontal scale of the perturbation a, divided by the Rossby radius at $t = 96$ h, according to Eqns. (4.22a, b, c) with Eqn. (4.23) as initial condition and with $h'_{max} = 50$ m and $a = 180$ km (solid line), and according to the invertibility principle (Eqn. 4.26) using the initial potential vorticity distribution (broken line) $\bar{h} = 1000$ m, and $g' = 1$ m s^{-2}).

equivalently with varying values of the Rossby radius. The maximum value of the relative vorticity after 72 h of integration is shown as a function of the Rossby radius. Also shown is the maximum value of the relative vorticity according to the invertibility principle (Eqn. 4.26), assuming the potential vorticity distribution at $t = 0$ is fixed in time (the dashed curve). The fact that the two curves in Figure 4.40 practically coincide demonstrates that the motion associated with the waves generated by the initial imbalance (i.e. the 'unbalanced' part of the motion) is of negligible importance to the 'balanced' part of the motion, which is represented by the invertibility principle. This is also found in more realistic (i.e. more complex) models of the atmosphere (Davis *et al.*, 1996). We will return to this important conclusion in a later section.

The fluid adjusts to the new distribution of potential vorticity by converting potential vorticity associated with the perturbation in h into potential vorticity associated with relative vorticity, ζ. Apparently (Figure 4.40) this conversion attains an optimum when the scale, a, of the initial perturbation is slightly greater than the Rossby radius. If a is significantly greater than the Rossby radius, the effect of the *rotational stiffness* (or inertial stability) is relatively strong, so that very little conversion of height-related to vorticity-related potential vorticity is possible. If a is significantly smaller than the Rossby radius, the perturbation spreads out over a distance of the order of the Rossby radius, which is also less effective for the above-mentioned conversion.

The processes just discussed must be considered as an analogue of what happens in the real atmosphere when it is heated locally. The mass extracted is an analogue for the heating, because both effects disturb the pressure distribution as well as the potential vorticity distribution, as we will see in the next section. In the real atmosphere the balanced state need not be in geostrophic balance. For instance, in an axisymmetric cyclone it is gradient wind balance (Section 4.5.6), or in a more complicated flow situation it is described by the nonlinear balance equation (Holton, 1992, p. 387). Nevertheless, in all these cases it is possible to formulate an invertibility principle, which relates potential vorticity to the balanced flow.

4.5.5 How heating changes the distribution of potential vorticity

Since potential vorticity determines the wind field in the balanced state, changes in the potential vorticity distribution will determine changes in the wind field. It is therefore of considerable interest to examine how heating changes the distribution of potential vorticity in a cyclone. Following Kleinschmidt (1951) and many others we assume that the cyclone is axisymmetric, so that we need only examine the structure of the vortex in the radius–height plane. If we further assume hydrostatic balance, it is possible and convenient to use potential temperature as vertical coordinate instead of height. Hydrostatic balance with potential temperature (θ) as vertical coordinate is expressed as (Anthes, 1971)

$$\frac{\partial \Psi}{\partial \theta} = \Pi \tag{4.30}$$

where Π is the Exner function, defined in Eqn. (4.17), and Ψ is the isentropic streamfunction, defined according to $\Psi = c_p T + gz$, with z the height of the isentropic surface.

The time-dependent equations of motion and the continuity equation in cylindrical/isentropic coordinates (r, θ), where r is the radius, are (Anthes, 1971)

$$\frac{dru}{dt} = -rfv \tag{4.31a}$$

$$\frac{dv}{dt} = u\left(f + \frac{u}{r}\right) - \frac{\partial \Psi}{\partial r} \tag{4.31b}$$

$$\frac{d\sigma}{dt} = -\sigma \left(\frac{1}{r} \frac{\partial rv}{\partial r} + \frac{\partial}{\partial \theta} \frac{J}{\Pi} \right) \qquad (4.31c)$$

Again, we assume that f is constant. Note the mathematical similarity between Eqn. (4.31) and Eqn. (4.22). In Eqn. (4.31), v is the radial velocity, u is the azimuthal velocity (positive if the flow is cyclonic; negative if the flow is anticyclonic), σ is the inverse of the static stability (defined below), and $d/dt = \partial/\partial t + v \partial/\partial r + (J/\Pi)\partial/\partial \theta$. The radial derivative is taken at constant θ. The 'vertical velocity' in this coordinate system is equal to (J/Π). The inverse of the static stability is defined as

$$\sigma = -\frac{1}{g} \frac{\partial p}{\partial \theta} \qquad (4.32)$$

From Eqn. (4.31a) we find that

$$\frac{dM_a}{dt} = 0 \qquad (4.33)$$

where $M_a \equiv ru + fr^2/2$ is the angular momentum per unit mass. Eqn. (4.33) implies that M_a is materially conserved, even in the presence of heating.

It is easily deduced from Eqn. (4.31a) and Eqn. (4.31c) that

$$\frac{dZ_\theta}{dt} = Z_\theta \frac{\partial}{\partial \theta} \left(\frac{J}{\Pi} \right) - \frac{1}{\sigma} \frac{\partial u}{\partial \theta} \frac{\partial}{\partial r} \left(\frac{J}{\Pi} \right) \qquad (4.34)$$

where Z_θ is the isentropic potential vorticity, defined as

$$Z_\theta \equiv \frac{\zeta_\theta + f}{\sigma} \qquad (4.35)$$

where $\zeta_\theta = \partial u/\partial r + u/r$ is the relative vorticity on an isentropic surface. Eqn. (4.34) tells us *that isentropic potential vorticity is materially conserved if there are no heat sources* (remember: we have neglected friction). The first term on the r.h.s. of Eqn. (4.34) can be interpreted as a 'stretching' or shrinking effect (a vertically-confined source of heat affects the distance between isentropic surfaces). The second term on the r.h.s. of Eqn. (4.34), can be interpreted as a 'tilting' effect. It is non-zero if the vortex is baroclinic and if the heat source is horizontally confined. Isentropic potential vorticity appears partly as absolute vorticity and partly as static stability. The proportions of this partitioning are adjusted through radial and vertical motions.

It is interesting to note that the absolute vorticity can be expressed in terms of M_a as

$$\zeta_{\theta_{abs}} \equiv \zeta_\theta + f = \frac{1}{r} \frac{\partial M_a}{\partial r} \qquad (4.36)$$

The further interpretation of the terms on the r.h.s. of Eqn. (4.34) is made easier if we first discuss the structure of a balanced cyclone in terms of potential vorticity (Z_θ). But, before we proceed with this, it is perhaps interesting to mention an additional constraint on the potential vorticity budget.

Using the fact that div curl of any vector field is zero and remembering that we have axisymmetry, Eqn. (4.34) can be rewritten as

$$\frac{dZ_\theta}{dt} = \frac{1}{\sigma} \vec{\omega}_a \cdot \vec{\nabla} \left(\frac{J}{\Pi} \right) = \frac{1}{\sigma} \vec{\nabla} \cdot \left(\vec{\omega}_a \left(\frac{J}{\Pi} \right) \right), \tag{4.37}$$

where $\vec{\omega}_a$ is the absolute vorticity vector. If we integrate this equation over a material volume V, with a surface S, and use Gauss's divergence theorem, we find

$$\int_V \sigma \frac{dZ_\theta}{dt} dV = \int_S \left(\frac{J}{\Pi} \vec{\omega}_a \right) \cdot d\vec{S} \tag{4.38}$$

The r.h.s of this equation is zero if there are no heat sources on the boundary S. In other words, for every positive material tendency of isentropic potential vorticity there is, within a layer bounded by two isentropic surfaces, and with no heating at the lateral edges, a 'compensating' negative material tendency of isentropic potential vorticity (weighted by σ and integrated over a material volume, $2\pi r \, dr \, d\theta$). The interpretation of Eqn. (4.38) is discussed at length by Haynes and McIntyre (1987, 1990) and by Danielsen (1990).

4.5.6 The potential vorticity structure of a circularly symmetric cyclone

Let us assume a vortex which is in hydrostatic balance Eqn. (4.30) and in gradient wind balance:

$$u \left(f + \frac{u}{r} \right) = \frac{\partial \Psi}{\partial r} \tag{4.39}$$

The equations for hydrostatic balance (Eqn. 4.30) and for gradient wind balance (Eqn. 4.39) together yield the equation for thermal wind balance:

$$\frac{\partial \Pi}{\partial r} = \frac{c_p}{\theta} \frac{\partial T}{\partial r} = f_{\text{loc}} \frac{\partial u}{\partial \theta} \tag{4.40}$$

where $f_{\text{loc}} = f + (2u/r)$. Thus, we see that in a warm core vortex (with $[\partial T/\partial r]_\theta < 0$), u must decrease with increasing potential temperature or height (i.e. the thermal wind is anticyclonic), while in a cold core vortex (with $[\partial T/\partial r]_\theta > 0$), u must increase with increasing potential temperature or height.

Let us assume that the vortex has an azimuthal velocity profile given by

$$u(r, \theta) = \frac{2B(\theta)\hat{u}r}{\hat{r}\left[1 + \left(\frac{r}{\hat{r}}\right)^2\right]} \qquad (4.41)$$

where $\hat{r}$ is the radius of maximum wind, $\hat{u}$ is the maximum azimuthal wind and $B(\theta)$ is the so-called baroclinicity of the vortex. In a cyclone, the value of $B(\theta)$ varies between 0 and $+1$. If $B(\theta)$ decreases with height, the balanced vortex has a warm core.

We now specify the following boundary conditions. The pressure and the potential temperature at the Earth's surface ($z = 0$) at $r = 0$ are specified by $p = p_s = 1000$ hPa and $\theta = \theta_s = 275$ K. The potential temperature lapse rate, $\partial\theta/\partial p$, at $r = 0$ is then fixed such that the temperature in the centre of the cyclone decreases from 275 K at the Earth's surface to 225 K at $z = 10\,200$ m ($\theta = 335$ K) and subsequently decreases more slowly to 210 K at $z = 18\,095$ m ($\theta = 445$ K). Therefore, the potential temperature lapse rate in the lower 10000 m at $r = 0$ is c. 5 K km^{-1}.

With the balance conditions (Eqns. 4.30 and 4.39) we can now obtain the distribution of M_a and Z_θ as a function of θ and r, belonging to a particular choice of the parameters $\hat{r}$, $\hat{u}$ and $B(\theta)$. The surface pressure, $p_s(r > 0)$, which is required in order to diagnose σ at the lowest isentropic level ($\theta = 280$ K), is obtained from gradient wind balance using (Eqn. 4.41) and the boundary condition that the thermal wind is equal to zero in the layer between the Earth's surface and the lowest isentropic level ($\theta = 280$ K). We therefore neglect the effects of friction and assume that the Earth's surface is an isentropic level. The lowest isentropic level above the ground ($\theta = 280$ K) does not reach the ground at any point within the model domain. We thus avoid the difficulties associated with translating potential temperature anomalies at the Earth's surface to potential vorticity anomalies (see e.g. Hoskins *et al.*, 1985; Bleck, 1990).

A prototype cyclone in the atmosphere has a cold core in the troposphere. The associated thermal wind is cyclonic in the troposphere, while it is anticyclonic in the stratosphere. If we choose $\hat{u} = 20$ m s^{-1} and $\hat{r} = 240$ km and specify $B(\theta)$ as displayed by the thick solid line in Figure 4.41, we obtain a distribution of Z_θ shown in Figure 4.42a. In terms of the wind velocity, the intensity of a cold core cyclone is a maximum at the top of the troposphere. However, as far as the potential vorticity is concerned, this cyclone is most intense in the stratosphere, where it is in fact a 'warm' core cyclone.

A polar low or tropical cyclone typically has a warm core in the lower troposphere implying an anticyclonic thermal wind. Its horizontal scale is

4.5 The role of thermal instability in polar low formation   363

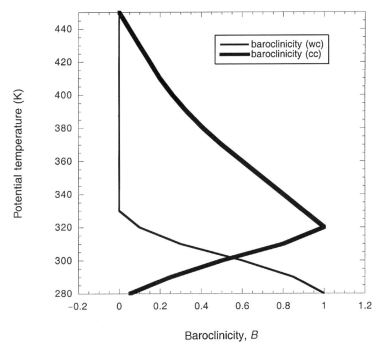

Figure 4.41. Baroclinicity parameter B of the prescribed vortex as a function of potential temperature in the case of a warm core (wc) and in the case of a cold core (cc).

relatively small. A representative value of the radius of maximum azimuthal wind is $\hat{r} = 100$ km. If we choose $\hat{u} = 10$ m s^{-1} and specify $B(\theta)$ according to the thin curve in Figure 4.41, we obtain from the balance conditions (Eqns. 4.30 and 4.39) the distribution of Z_θ as shown in Figure 4.42b. We see that the polar low is a relatively shallow phenomenon. The maximum potential vorticity is found in the lower troposphere. We will see in Section 4.5.8, that this property of warm core cyclones is advantageous for their growth due to heating.

4.5.7 The invertibility principle applied to an axisymmetric cyclone

We have seen that a particular wind distribution is associated with a particular potential vorticity distribution if the atmosphere is in some kind of dynamic balance. Of course, as was explained in the section 'Geostrophic adjustment and the invertibility priciple', the inverse problem is also of relevance: a particular potential vorticity distribution is associated with a specific wind distribution. In this section we explain some dynamical consequences of this so-called invertibility principle. Following Hoskins *et al.*

364   4 Theoretical investigations

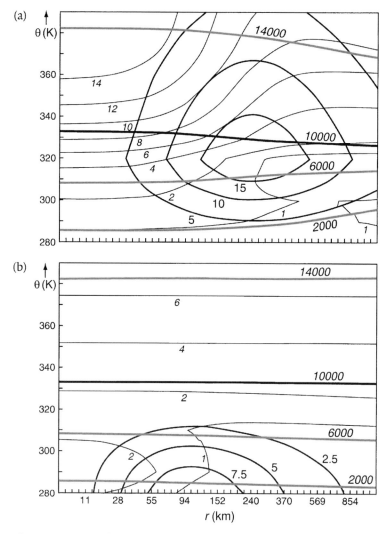

Figure 4.42. Isentropic potential vorticity (labelled in small italic numbers in PV units; 1 PVU is 10^{-6} K m^2 kg^{-1} s^{-1}), height (labelled in large italic numbers in m) and azimuthal velocity (labelled in m s^{-1}) as a function of potential temperature and radius, for (a) a balanced cold core cyclone and (b) a balanced warm core cyclone. The azimuthal velocity is specified by Eqn. 4.41. $B(\theta)$ for both cases is plotted in Figure 4.41. Parameters and static stability at $r = 0$ are specified in the text.

(1985), let us differentiate Eqn. (4.35) with respect to r. Using Eqn. (4.32), this yields

$$\frac{\partial}{\partial r}\left\{\frac{1}{r}\frac{\partial(ru)}{\partial r}\right\} + \frac{Z_\theta}{g}\frac{\partial^2 p}{\partial r \partial \theta} = \sigma \frac{\partial Z_\theta}{\partial r} \qquad (4.42)$$

Now, using the definition of the Exner function (Eqn. 4.17) and the equation of state ($p = \rho RT$), we can express thermal wind balance (Eqn. 4.40) as

$$\frac{\partial p}{\partial r} = f_{\text{loc}} \rho \theta \frac{\partial u}{\partial \theta} \qquad (4.43)$$

(f_{loc} was defined below Eqn. (4.40)). Using Eqn. (4.43) to eliminate p from Eqn. (4.42) yields

$$\frac{\partial}{\partial r}\left\{\frac{1}{r}\frac{\partial (ru)}{\partial r}\right\} + \frac{Z_\theta}{g}\frac{\partial}{\partial \theta}\left(\rho\theta f_{\text{loc}} \frac{\partial u}{\partial \theta}\right) = \sigma \frac{\partial Z_\theta}{\partial r} \qquad (4.44)$$

This equation, which is another version of the 'invertibility principle' discussed earlier in the context of the one layer model, describes the flow pattern, $u(r, \theta)$, which is associated with a specific pattern of the isentropic potential vorticity in a balanced axisymmetric cyclone. If $f_{\text{loc}} Z_\theta > 0$, Eqn. (4.44) is elliptic. Given the distribution of Z_θ and suitable boundary conditions, the coupled system of Eqns. (4.43) and (4.44) can be solved for the wind $u(r, \theta)$, for instance by relaxation methods (see e.g. Wirth, 2001, p. 29). The term on the r.h.s. of Eqn. (4.44) can be interpreted as 'forcing'. In other words, a radial gradient in Z_θ 'forces' or 'induces' an azimuthal flow.

In order to investigate the implications of the invertibility principle in a little more detail, let us neglect effects of curvature in Eqn. (4.44) by taking the limit $r \to \infty$ and further assume that $\rho\theta f_{\text{loc}} \simeq \rho\theta f$ is a constant. The equation then transforms into the following equation

$$\frac{\partial^2 u}{\partial r^2} + A\frac{\partial^2 u}{\partial \theta^2} = B \qquad (4.45)$$

with

$$A \equiv \frac{\rho\theta f(f + \zeta_\theta)}{g\sigma}; \quad B \equiv \sigma \frac{\partial Z_\theta}{\partial r} \qquad (4.46)$$

Assuming that A is constant (which makes Eqn. (4.45) linear) and that u depends on r according to

$$u = U(\theta) \sin \frac{2\pi r}{L} \qquad (4.47)$$

where L is the typical horizontal scale of the response to the forcing, B, we get

$$A\frac{\partial^2 U}{\partial \theta^2} - \frac{4\pi^2}{L^2} U = B \qquad (4.48)$$

The solution of the homogeneous part of the above equation (i.e. with the forcing, $B = 0$) is

$$U = C_1 \exp\left(\frac{2\pi\theta}{L\sqrt{A}}\right) + C_2 \exp\left(\frac{-2\pi\theta}{L\sqrt{A}}\right) \qquad (4.49)$$

where C_1 and C_2 are constants determined by the boundary conditions and the nature of the inhomogeneous term (i.e. the forcing, B). If $A > 0$ (which is the case if $Z_\theta > 0$) this solution describes an exponentially decaying function of θ with a maximum or minimum value in the region where B (i.e. the forcing) is non-zero. If B is negative at a certain height, $U > 0$ (cyclonic flow), whereas if B is positive at a certain height, $U < 0$ (anticyclonic flow). The idea is that a potential vorticity anomaly *induces* (or forces) a wind field, $u(r, \theta)$, with a vertical dependence given approximately by Eqn. (4.49) (Figure 4.43). Stated differently, local changes of the potential vorticity (due to for example heating) must be accompanied by local changes in the wind (in order to preserve thermal wind balance). Thorpe (1997) coined the term *attribution* as a slightly weaker form of 'cause-and-effect' to characterize the relation between potential vorticity and the induced wind field. In other words, a particular change in the wind field can be *attributed* to a particular change in the potential vorticity field. The associated vertical scale, $\Delta\theta$, of the response in u is easily distilled from Eqn. (4.49):

$$\Delta\theta \equiv L\sqrt{A} \qquad (4.50)$$

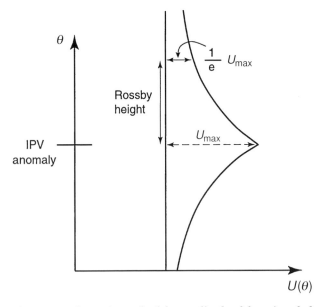

Figure 4.43. Schematic graph of the amplitude of the azimuthal wind as a function of potential temperature, induced by a PV anomaly.

$\Delta\theta$ is referred to by Hoskins *et al.* (1985) as the *Rossby height*. It measures the vertical penetration (in isentropic coordinates) of the flow structure above and below the location of the potential vorticity anomaly, induced by thermal wind adjustment to the anomaly. The potential vorticity anomaly, in turn, may be induced by heating. The concept of Rossby height is complementary to that of the deformation scale height for hydrostatic adjustment (see Eqn. 4.19) or that of the Rossby radius of deformation for geostrophic adjustment (see Eqn. 4.28). Stated shortly, we may say that a potential vorticity perturbation induced by heating 'induces' a perturbation in the wind field with a characteristic vertical scale equal to the Rossby height. The Rossby height for the potential vorticity anomaly shown in Figure 4.42b (with $L \approx 200$ km), is of the order of 10 K in the troposphere and the lower stratosphere.

We can transform the expression for the Rossby height to physical space, using hydrostatic balance, written as $\Delta p = -\rho g \Delta z$, and Eqn. (4.32), written as $\sigma = -\Delta p/(g \Delta \theta)$. This yields

$$\Delta z \equiv \frac{\sqrt{f(f+\zeta_\theta)}L}{N} \qquad (4.51)$$

where N is the buoyancy frequency, defined as

$$N \equiv \sqrt{\frac{g}{\theta}\frac{\Delta\theta}{\Delta z}} \qquad (4.52)$$

Expression (4.51) for the Rossby height (Δz) demonstrates that the cyclonic wind field 'induced' by potential vorticity anomalies caused by heating has the greatest vertical penetration if the static stability in the environment is low and/or if the absolute vorticity of the environment is high and/or if the horizontal scale of the perturbation is large.

Another interpretation of (4.51) is that, since in general $N >> \sqrt{f(f+\zeta_\theta)}$, balanced circulation systems typically have a large aspect ratio, $L/\Delta z$.

4.5.8 Axisymmetric adjustment to a circular heat source

We now know that there is a specific relation between the isentropic potential vorticity distribution and the wind distribution in a balanced axisymmetric cyclone (Eqn. 4.44). We also know that heating changes the isentropic potential vorticity of an air parcel in a specific way, depending on its potential vorticity (Z_θ) and the local baroclinicity ($\partial u/\partial\theta$) (Eqn. 4.34).

These two facts imply that heating will affect the wind. Let us investigate this effect more in detail by assuming that the atmosphere is in rest initially

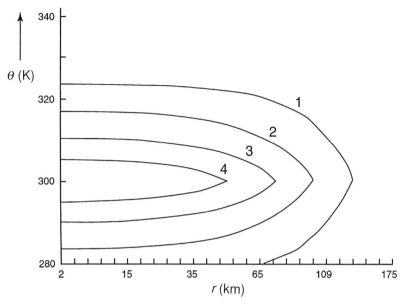

Figure 4.44. Prescribed heating as a function of potential temperature and radius (labelled in units of 10^{-4} K s^{-1}) ($r^* = 100$ km).

and specifying a circularly symmetric heat source, whose intensity depends on r and θ as

$$\frac{d\theta}{dt} \equiv \dot{\theta} = \frac{J}{\Pi} = Q_0(\theta) \exp\left\{-\left(\frac{r}{r^*}\right)^2\right\} \tag{4.53}$$

with $Q_0(\theta)$ specifying the vertical distribution of the heating. Note that the heating according to Eqn. (4.53) has a maximum value at $r = 0$ and that the amplitude of the heating falls off exponentially with increasing r. The associated e-folding distance is $r = r^*$. The net latent heat released to the air at any height depends on the difference between condensation and evaporation and between any freezing and melting at that height. Since evaporation and melting in a precipitating cloud occurs principally below the cloud and condensation and freezing principally higher up within the clouds, the net heating will possess a maximum value at mid-levels in the troposphere. $Q_0(\theta)$ is specified accordingly, yielding a schematic heating distribution as shown in Figure 4.44.

It is important to note that the Earth's surface is assumed to be an isentropic surface, i.e. $Q_0(\theta = \theta_s) = 0$ at the Earth's surface. The maximum value of J/Π is 0.0005 K s^{-1} which corresponds to about 1.8 K h^{-1}.

Note also that the vertical scale of the heating is significantly larger than the

4.5 The role of thermal instability in polar low formation

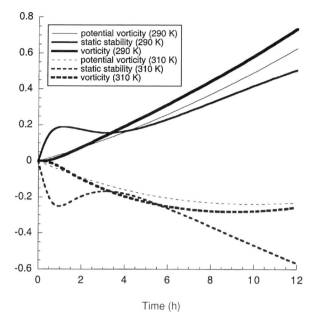

Figure 4.45. Perturbations (deviations from the initial state) as a function of time of potential vorticity in PV units (1 PVU = 10^{-6} K m^2 kg^{-1} s^{-1}), static stability in units of 10^{-4} K Pa^{-1} and vorticity in units of 10^{-4} s^{-1} at two points, i.e. at $r = 19$ km, $\theta = 290$ K (solid lines) and at $r = 19$ km, $\theta = 310$ K (broken lines). The heating is specified as in Figure 4.48.

Rossby height associated with an imposed anomaly with a horizontal scale of 200 km (i.e if we specify $r^* = 100$ km), implying that the balanced response to the heating will have approximately the same vertical scale as the vertical scale of the heating itself.

The closed set of Eqns. (4.30) and (4.31) is approximated by finite differences on a grid within a domain defined by $280 \leq \theta \leq 450$ K and $0 \leq r < 6215$ km. The pressure at the Earth's surface is diagnosed from $\sigma(r, \theta)$, which is a prognostic variable, assuming that $\theta_s(r)$ is constant and that the pressure at the highest isentropic level is constant.

Figure 4.45 shows the evolution in time at $r = 19$ km of the perturbations in, respectively, isentropic potential vorticity, static stability and relative vorticity at, respectively, $(r, \theta) = (19\text{ km}, 290\text{ K})$ (below the level of maximum heating) and $(r, \theta) = (19\text{ km}, 310\text{ K})$ (above the level of maximum heating). At both points we see an initial oscillation in the static stability associated with the excitation of a Lamb wave as well as internal gravity-inertia waves. Both the potential vorticity and the vorticity increase below the level of maximum heating while they decrease above the level of maximum heating.

These results are as expected, but are not as obvious as they may seem at first sight. A close look at the budget of potential vorticity makes this clear. The budget of isentropic potential vorticity at a fixed point in r-θ space is made up of the following terms (Eqn. 4.34):

$$\text{term 1} = Z_\theta \left(\frac{\partial \dot{\theta}}{\partial \theta}\right); \text{term 2} = -\frac{1}{\sigma}\frac{\partial u}{\partial \theta}\left(\frac{\partial \dot{\theta}}{\partial r}\right);$$

$$\text{term 3} = -\dot{\theta}\left(\frac{\partial Z_\theta}{\partial \theta}\right); \text{term 4} = -v\left(\frac{\partial Z_\theta}{\partial r}\right) \quad (4.54)$$

The sum of terms 1, 2 and 3 is called the diabatic PV forcing (Edouard *et al.*, 1997). Term 4 represents forcing due to advection of potential vorticity along isentropes and is called adiabatic PV forcing. Term 3 represents vertical advection of potential vorticity in a reference frame with potential temperature as a vertical coordinate. In this coordinate system a heat source appears as a vertical velocity which advects potential vorticity and angular momentum. Actually (i.e. in physical space), there is no vertical advection (i.e. vertical transport of air parcels) due to heating: rather, isentropes descend.

Figure 4.46 shows the evolution of the total budget of potential vorticity as well as that of the individual terms listed in Eqn. (4.54) at, respectively, $(r, \theta) = (19 \text{ km}, 290 \text{ K})$ (below the level of maximum heating) and $(r, \theta) = (19 \text{ km}, 310 \text{ K})$ (above the level of maximum heating). As expected, term (1) is dominant at both isentropic levels. In fact, since this term is proportional to Z_θ, and because positive and negative perturbations in Z_θ are formed at, respectively, lower and upper levels, it becomes more effective. The effects of terms 2 and 4 are negligible, while term 3 is negative at both levels. This is due to the basic positive vertical gradient of Z_θ. However, because a low-level positive perturbation in Z_θ, centred below 310 K, is created by the heating, term 3 changes sign at this level after about 4 h of steady heating. In the lower troposphere term 3 remains negative because low values of Z_θ are advected diabatically towards higher values of θ. We will see in the next section that term 3 may dominate the local budget of potential vorticity in a cyclone of realistic intensity.

The negligible amplitude of the only adiabatic PV forcing term (term 4) implies that the potential vorticity perturbation, induced by the heating, hardly propagates away from the source region. The robustness of this remarkable implication can be tested by repeating the numerical experiment with the heating intensity increased 12-fold, but turned off after 1 h. In that case the heating induces an approximately identical potential vorticity

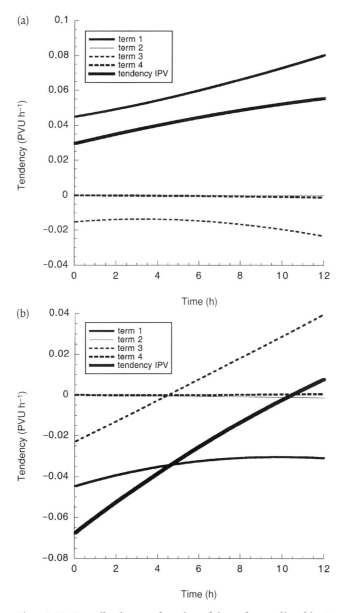

Figure 4.46. Contribution as a function of time of terms listed in Eqn. (4.54) to the total local tendency of the isentropic potential vorticity (also shown) at (a) $r = 19$ km, $\theta = 290$ K, and (b) $r = 19$ km, $\theta = 310$ K.

perturbation in 12 h, but also excites waves of significantly larger amplitude (see Figure 4.47). Despite this, the induced potential vorticity perturbation after 12 h is approximately identical (Figure 4.48). The induced vorticity after 12 hours is also practically identical in both cases, despite

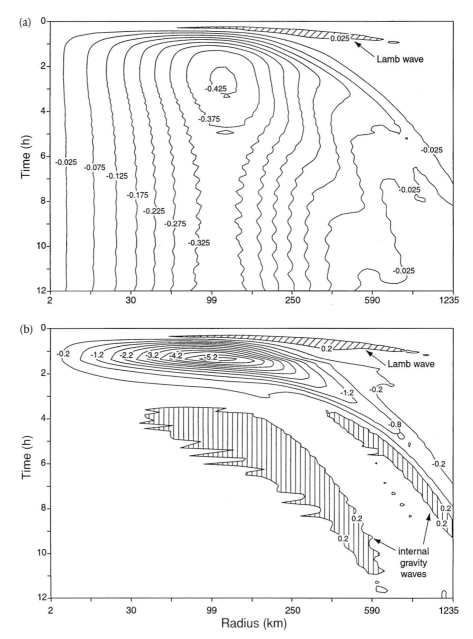

Figure 4.47. Radial velocity at $\theta = 290$ K (c. 2500 m) as a function of time and radial distance in, respectively, case (a) (gentle, slow forcing) and case (b) (fast, intense forcing). In both cases the initial condition (rest) is identical and the final states are practically identical (a weak vortex). Labels indicate values in m s^{-1}. Hatched areas indicate areas where the radial velocity, $v > 0.0025$ m s^{-1} (a) or $v > 0.2$ m s^{-1} (b).

4.5 The role of thermal instability in polar low formation 373

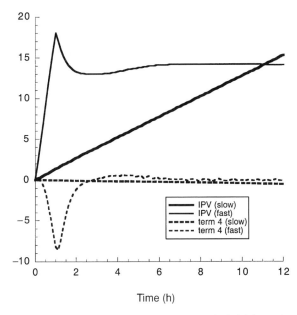

Figure 4.48. Perturbation (deviation from the initial state) as a function of time of isentropic potential vorticity in units of 0.01 PVU (solid lines), and the contribution of term 4 to the total budget of potential vorticity in units of 0.01 PVU h^{-1} (broken lines), at $r = 98$ km, $\theta = 290$ K for, respectively, the experiments with intense forcing (indicated by 'fast') and gentle forcing (indicated by 'slow').

significant departures from gradient wind balance in the case of intense forcing (Figure 4.49).

Figure 4.49 reveals the presence of regular oscillations around gradient wind balance with decreasing frequency, implying that waves with progressively lower frequencies are left in the centre of the cyclone, while the higher frequency waves propagate outwards. In this way the balanced state will eventually be reached.

From these numerical experiments we conclude that the waves and oscillations are of little or negligible importance to the balanced dynamics (represented by the potential vorticity)! This agrees with the results obtained earlier with the shallow water equations (Figure 4.40). Presumably this is the basic theoretical foundation of the success of Ooyama's (1969) balanced model in simulating the life-cycle of a tropical cyclone. In other words, a balanced model, which filters out all motions associated with waves and hydrostatic instability, is sufficient to describe the dynamics of the growth of a cyclone by heating, even quantitatively. Of course, problems and controversy arises (Emanuel *et al.*, 1994; Stevens *et al.*, 1997; Smith, 1997) when one tries to link the heating to the motion, because, for this knowledge of the unbalanced (convective) motion is

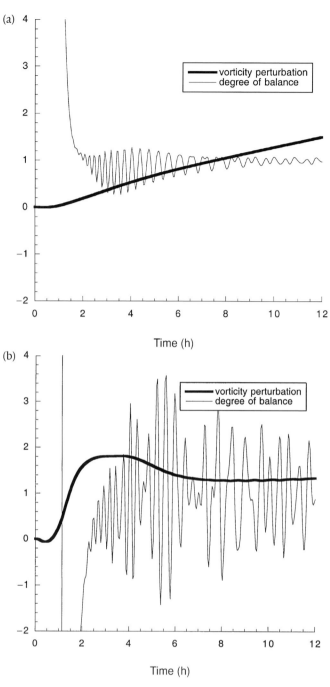

Figure 4.49. Perturbation (deviation from the initial state) as a function of time of the relative vorticity in units of 10^{-5} s^{-1} (thick solid lines) and the degree of balance (thin solid lines), defined as the ratio $\partial \psi/\partial r/[u(f + u/r)]$ (see Eqn. 4.39), at $r = 98$ km and $\theta = 290$ K for, respectively, the experiments with gentle forcing (a) and intense forcing (b). The vortex is in perfect gradient wind balance if the degree of balance is equal to 1. Note that, despite large departures from balance in case (b), the relative vorticity at $t = 12$ h is approximately equal in both cases.

presumably required. A further discussion of this problem will be given in the section 'Processes controlling the heating in a cyclone'.

4.5.9 The intensification of cyclones by heating

On the basis of what we learned from the experiments described in the previous section, we now expect that a warm core cyclone (Figure 4.42b) will react differently to heating than a cold cyclone (Figure 4.42a). In this section we investigate this further by taking the two examples shown in Figure 4.42 and imposing the axisymmetric heating distribution shown in Figure 4.44.

Figure 4.50 shows the isentropic potential vorticity and azimuthal flow as a function of r and θ after 12 h of steady heating for the two cases; one case (case a) with the cold core vortex of Figure 4.42a as initial condition and the other case (case b) with the warm core vortex of Figure 4.42b as initial condition.

It should be stressed that the spatial distribution and intensity of the heating is identical in both experiments. It should also be stressed that these results are meant to illustrate and provide insight into the basic physical mechanism of cyclone growth by heating. The actual heating distribution in a cyclone, such as a polar low, will be discussed further in Section 4.5.11, 'Processes controlling the heating in a cyclone'.

The most remarkable result (Figure 4.50) is that the potential vorticity in the warm core cyclone increases significantly, while the potential vorticity in the cold core cyclone decreases. Associated with this, the maximum azimuthal wind velocity in the cold core cyclone decreases from 20 m s^{-1} to 18.7 m s^{-1}. In the warm core cyclone, on the other hand, the maximum azimuthal wind velocity increases from 10 m s^{-1} to 13.4 m s^{-1}.

The increase in potential vorticity in the centre of the warm core cyclone in this reference frame is again principally due to the effect of terms 1 and 3. This is shown in Figure 4.51.

Term 2 hardly contributes to the local potential vorticity budget. The absolute value of this term initially does not exceed 0.05 PVU per hour. The absolute value of term 4 is zero initially and in both cases never exceeds 0.05 PVU per hour during the 12-hour integration. In this context, it must be noted that the second term in Eqn. (4.54) could become important in a balanced warm core cyclone (with $\partial u/\partial \theta < 0$) if the cyclone possesses an eye with most of the (latent) heating taking place near the radius of maximum wind, as is frequently observed.

In a cold core cyclone, heating is seen to create an intense negative potential vorticity anomaly centred at $r = 0$ and $\theta = 320$ K. Comparatively weak positive potential vorticity tendencies are found outside the core of the vortex, implying that the cold core cyclone will expand due to heating.

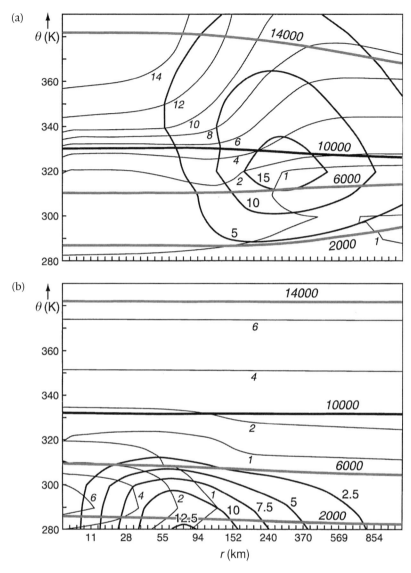

Figure 4.50. Isentropic potential vorticity (labelled in small italic numbers in PV units; 1 PVU is 10^{-6} K m^2 kg^{-1} s^{-1}), height (labelled in large italic numbers in m) and azimuthal velocity (labelled in m s^{-1}) as a function of potential temperature and radius, after 12 h constant heating according to Figure 4.44, for a case (a) with the cold core vortex of Figure 4.42 as initial condition, and for a case (b) with the warm core vortex of Figure 4.42 as initial condition.

4.5 The role of thermal instability in polar low formation

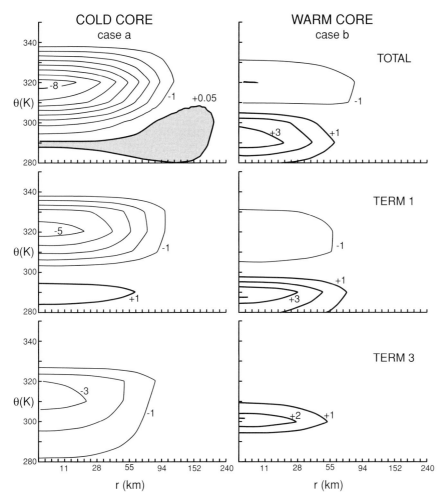

Figure 4.51. Contribution as a function of potential temperature and radius of terms listed in Eqn. (4.54) to the local tendency of the isentropic potential vorticity (Eqn. 4.34) at $t = 0$. The top diagrams show the total tendency for, respectively, the cold core cyclone (case a) and the warm core cyclone (case b). The middle diagrams show the contribution of term 1, while the diagrams below show the contribution of term 3. Labels are in units of 0.05 PVU h^{-1}. Thick contours represent positive values; thin contours represent negative values.

Thus, tropospheric heating in the centre of a warm core cyclone produces a relatively intense positive anomaly of the potential vorticity at low levels and a relatively weak anomaly (positive and negative) at upper levels (a similar result was obtained by Schubert and Alworth, 1987), while tropospheric heating in the centre of a cold core cyclone is likely to produce a negative potential vorticity anomaly at all heights (a similar result was obtained by Wirth, 1995).

Therefore, assuming that the invertibility principle is valid at all times, we may conclude that, in general, the intensity of a warm core cyclone will increase due to heating while the intensity of a cold core cyclone will decrease due to heating. Similar conclusions were reached by Emanuel and Rotunno (1989) and van Delden (1989).

An additional understanding of the mechanism by which the vortex intensifies (i.e. the vorticity increases) can be gained by observing the changes in the angular momentum distribution, since changes in this can easily be translated to changes in the absolute vorticity distribution using Eqn. (4.36), and since angular momentum is materially conserved (even in the presence of heating: Eqn. 4.33). Therefore, these changes are governed only by advection (horizontal and vertical in $r - \theta$ space). Thus, assuming vorticity (or the radial gradient of angular momentum) is a measure of cyclone intensity, following the movement of isopleths of angular momentum and noting where isopleths of angular momentum converge or diverge radially provides insight into the process of cyclone intensity changes.

Figure 4.52 gives an impression of the movement of isopleths of angular momentum in the two examples discussed in this section. The radial displacements (25 km at the most) of air parcels during a 12 h period of constant heating are small compared to the total radial scale of the cyclone. It is obvious that these small displacements will have very little effect on the potential vorticity anomaly induced by the heating.

In the warm core cyclone the static stability-related potential vorticity perturbation caused by the diabatic PV forcing is partially converted into vorticity related potential vorticity by inward radial advection of high values of the angular momentum and consequent concentration of isopleths of angular momentum.

In the cold core cyclone the situation is very different. The negative perturbation in potential vorticity is accommodated by upward and outward radial advection of angular momentum throughout the whole cyclone, except at low levels near the radius of maximum wind. Note that even above the heat source (well into the stratosphere) there is a displacement of isopleths of angular momentum. Clearly, the cold core cyclone expands as a result of heating in the centre. This is the consequence of the theorem expressed in Eqn. (4.38), which states that a negative (potential) vorticity anomaly induced by internal heating must be 'compensated' in an integral sense, by a positive (potential) vorticity anomaly elsewhere (remember: we did not impose heating at the boundaries). The opposite is of course also true, implying that a warm core cyclone will contract as a consequence of internal heating.

4.5 The role of thermal instability in polar low formation 379

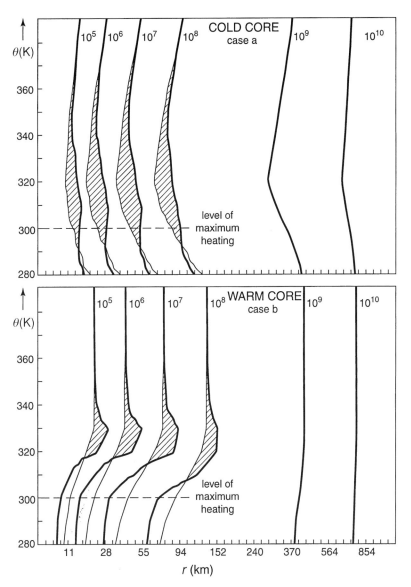

Figure 4.52. Angular momentum per unit mass (labelled in units of m² s⁻¹) as a function of potential temperature and radius in case a (above) and in case b (below), initially (thin lines) and after 12 h (thick lines).

In reality, cyclones do not have such a simple potential vorticity structure as illustrated in Figure 4.42. For example, there may be a warm core structure embedded in a larger-scale cold core structure. Indeed, many polar lows develop near or at the centre of a deep, cold core, cut-off cyclone aloft (Businger, 1985; Rasmussen, 1985a). Although cold air aloft is of course favourable for the development of deep convection and attendant intense latent heating, it is clear now that this will not automatically lead to the growth of a polar low embedded in the cold core cyclone. Of course, if heating continues for a sufficiently long time, a shallow warm core system will eventually develop, embedded within the cold core system, and the chances of producing a positive potential vorticity anomaly due to heating will increase. For example, if we add the azimuthal wind field shown in Figure 4.42a to the azimuthal wind field shown in Figure 4.42b, we are effectively embedding a small scale warm core cyclone in a larger-scale cold core cyclone. The resulting distribution of isentropic potential vorticity is shown in Figure 4.53a.

This potential vorticity distribution does not differ greatly from the distribution shown in Figure 4.42a, except that the $Z_\theta = 2$ PVU isopleth is depressed. However, the effect of a heat source as specified by Figure 4.44 is very different. Term 1 is now much more effective as a source of low-level potential vorticity, while term 3 is less effective as a sink of potential vorticity (Figures 4.53b and 4.54).

Emanuel and Rotunno (1989) performed similar numerical experiments and showed that the growth by heating of a warm core vortex embedded (at low levels) in a tropospheric cold core vortex is possible if the warm core vortex already has a sufficiently large amplitude. Figure 4.55 shows the central surface pressure as a function of time for the three 12 h integrations discussed in this section.

4.5.10 Physical interpretation of Ooyama's balanced model

So how far is the physics outlined in the previous sections represented in the theory of CISK (or cooperative intensification theory) put forward by Ooyama (1969)? We will investigate the answer to this question in this section employing Ooyama's (1969) tropical cyclone model.

Ooyama's tropical cyclone model consists of three axisymmetric superposed incompressible layers (see Figure 4.56). The density of the boundary layer is equal to the density of the layer just above, while the density of the upper layer is a factor ε smaller (i.e. $\varepsilon < 1$). The thickness of the lower layer, i.e. the boundary layer, is assumed constant. Frictional convergence in this layer will give rise to a mass flux into the layer above. The thickness of the upper two layers may vary due to convergence or divergence, or mass fluxes between the layers. Due to the incompressibility, compression waves are not possible. Gravity-inertia waves are not permitted either, because the flow is assumed to be in thermal wind balance permanently. Disturbances to thermal wind balance are accommodated instantaneously and continuously by means of a radial flow. This radial flow is responsible for cyclone intensity changes.

4.5 The role of thermal instability in polar low formation

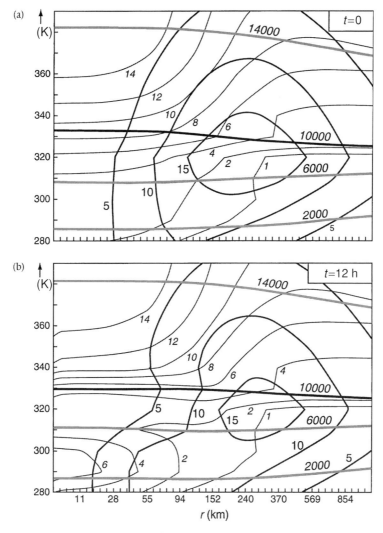

Figure 4.53. Isentropic potential vorticity (labelled in small italic numbers in PV units; 1 PVU = 10^{-6} K m^2 kg^{-1} s^{-1}), height (labelled in large italic numbers in m) and azimuthal velocity (labelled in m s^{-1}) at (a) $t = 0$ and (b) $t = 12$ h as a function of potential temperature and radius, after 12 h of constant heating according to Figure 4.44, for a small-scale warm core vortex embedded in the centre of a large-scale cold core vortex.

The model contains no explicit thermodynamics. How then are the effects of temperature changes due to heating incorporated? The answer to this question nicely illustrates the concept of *parameterization*. Heating in Ooyama's model is represented as a *mass flux* from layer 1 to layer 2, denoted by "Q" in Figure 4.56. Due to this, the thickness of the upper layer grows at the cost of the thickness of

382   4 Theoretical investigations

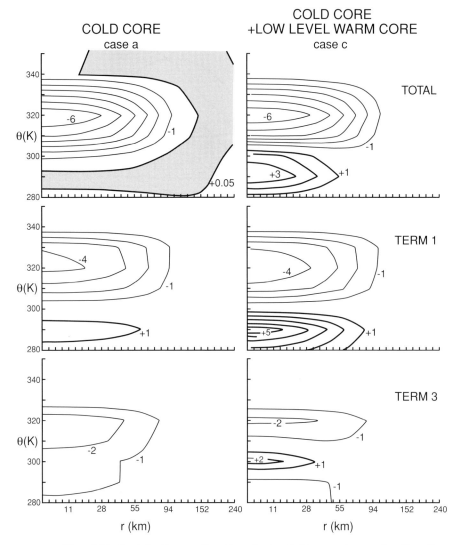

Figure 4.54. Contribution as a function of potential temperature and radius of terms listed in Eqn. (4.54) to the local tendency of the isentropic potential vorticity (Eqn. 4.34) at $t = 3$ h. The top diagrams show the total tendency for, respectively, the cold core cyclone (case a) and the warm core cyclone embedded in the cold core cyclone (case c). The middle diagrams show the contribution of term 1; the lower diagrams show the contribution of term 3. Labels are in units of 0.05 PVU h^{-1}. Thick contours represent positive values; thin contours represent negative values.

4.5 The role of thermal instability in polar low formation   383

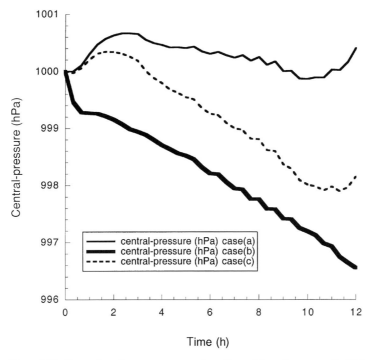

Figure 4.55. Central surface pressure as a function of time for cases a (initial cold core), b (initial warm core) and c (initial warm core embedded in a cold core vortex).

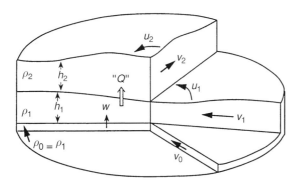

Figure 4.56. Structure of Ooyama's (1969) model (see text for further explanation).

the lower layer. Thus, the pressure in the upper layer increases while the pressure at the Earth's surface remains unchanged. Therefore, the mass flux, "Q", crudely mimics the effect of heating and hydrostatic adjustment in the vertical on the pressure distribution in the atmosphere (Lamb's problem, discussed earlier). This relatively simple parameterization of heating also crudely mimics the effect of diabatic potential vorticity forcing (the effect of term 1 in Eqn. 4.54) in a warm core cyclone. The potential vorticity for each layer is defined according to Eqn. (4.25). Therefore a mass flux from the lower layer to the upper layer represents a source of potential vorticity in the lower layer and a simultaneous sink of potential vorticity in the upper layer.

Ooyama (1969) assumed that "Q" was proportional to the upward velocity, w, at the top of the boundary layer by

$$\text{``}Q\text{''} = \eta w^+, \quad w^+ = w \text{ i} > 0_{w^+} = \text{i} \leq 0 \tag{4.55}$$

where η is a function of the latent and sensible heat fluxes from the sea, which, in turn, were assumed to be proportional to the wind speed in the boundary layer. The vertical velocity at the top of the boundary layer was the result of frictional convergence, which increased as the intensity of the azimuthal balanced flow in the free atmosphere increased. Therefore, the balanced vortex controlled the heating or forcing. Hence, we have a feedback loop involving continuous, perfect adjustment to thermal wind balance of a vortex to self-induced disturbances to this state of balance. Ooyama (1969) demonstrated that this feedback loop could become unstable if $\eta > 1$, i.e. if the diabatic mass flux functions as a source of potential vorticity in the lower layer of the (model) free atmosphere.

Ooyama (1969) and many others thereafter interpreted the condition, $\eta > 1$ as a criterion for the existence of conditional instability or positive CAPE. This led to the paradigm: 'CAPE is required for CISK'. Aside from the question whether the condition, $\eta > 1$, can be interpreted in this way in a hydrostatic model consisting of two incompressible fluid layers, it is clear, in the light of the theory discussed in earlier sections, that this interpretation is not needed as an explanation of the intensification. The physical interpretation that does justice to the physical content of Ooyama's model is that the vortex intensifies because the diabatic mass flux induces a positive potential vorticity perturbation at low-levels. This intensification is demanded by the invertibility principle. Relatively few authors have brought potential vorticity into the discussion. One of the early exceptions was Schubert *et al.* (1980), who concluded (p. 1482) that 'the effects of clouds on the potential vorticity field (i.e. the apparent potential vorticity source) is the important ingredient in understanding the feedback of clouds on the large scale fields.'

During the past 10 years more authors have recognized this fact. For example, Montgomery and Enagonio (1998) have taken an approach whereby the vorticity budget of an ensemble of cumulus clouds is parameterized in a quasi-geostrophic model by potential vorticity anomalies having a horizontal scale of approximately 200 km. Both in Ooyama's model and in the approach taken by Montgomery and Enagonio, vertically-confined, mid-tropospheric heating is seen to create a positive potential vorticity anomaly at low-levels and a negative potential vorticity anomaly aloft. According to the analysis presented in earlier sections, the effect of heating on the potential vorticity is more subtle.

4.5.11 Processes controlling the heating in a cyclone

Can the balanced flow indeed 'take control' of the heating to produce a positive feedback loop between the intensity of the balanced flow and the heating? In other words, how does the upper part of the feedback loop shown in Figure 4.30 work? This question has occupied many researchers since the pioneering papers by Charney and Eliassen (1964) and Ooyama (1969). Basically, two mechanisms have been suggested. As mentioned in Section 4.5.2 and explained in the previous section, Charney and Eliassen (1964) and Ooyama (1969) suggested that frictional convergence in the boundary layer could force moisture (transferred from the sea over a large surrounding area) in and up the core of the vortex leading to condensation of water vapour and, higher up, freezing of cloud water. Both condensation and freezing are accompanied by latent heat release. Thus, they emphasized latent heat release as the principal heat source responsible for the cyclone intensity changes. Ooyama (1969) tried to improve the physical reasoning given by Charney and Eliassen and extended CISK theory to the finite amplitude, non-linear regime. Shapiro and Willoughby (1982), Schubert and Hack (1982) and van Delden (1989) showed that the feedback loop, suggested by Ooyama, was most effective in an intense baroclinic (warm core) vortex. Nevertheless, Emanuel (1986a) called into question this mode of thinking. The essence of Emanuel's criticism is summarized in the following quote from the abstract of Emanuel *et al.* (1994):

> The dominant thinking about the interaction between large-scale atmospheric circulations and moist convection holds that convection acts as a heat source for the large-scale circulations, while the latter supply water vapour to the convection. We show that this idea has led to fundamental misconceptions about this interaction, and offer an alternative paradigm, based on the idea that convection is nearly in statistical equilibrium with its environment. According to the alternative paradigm, the vertical temperature profile itself, rather than the heating, is controlled by the convection, which ties the temperature directly to the sub-cloud-layer entropy. The understanding of large-scale circulations in convecting atmospheres can therefore be regarded as a problem of understanding the distribution in space and time of sub-cloud-layer entropy. We show that the sub-cloud-layer entropy is controlled by the sea surface temperature, the surface wind and the large scale vertical velocity in the convecting layer.

In other words, the 'alternative paradigm' states that convection acts to constrain the temperature profile towards the moist adiabat (see Figure 4.31). This process is called 'convective adjustment'.

Convective adjustment can be incorporated into the calculations described in Sections 4.5.8 and 4.5.9 if we modify Eqn. (4.31c) as follows:

$$\frac{d\sigma}{dt} = -\sigma \left(\frac{1}{r} \frac{\partial r v}{\partial r} \right) - \lambda (\sigma - \sigma_{\text{ref}}) \tag{4.56}$$

where λ is a relaxation constant determining the time scale of adjustment of the static stability $[-\partial \theta / \partial p] = 1/g\sigma$ of a model layer towards a reference value of the static stability, $[-\partial \theta / \partial p]_{\text{ref}} = 1/g\sigma_{\text{ref}}$.

The reference thermodynamic state, represented by the parameter, σ_{ref}, depends on the cloud top and is different for deep (precipitating) and shallow (non-precipitating) convection. If the atmospheric layer is conditionally unstable and saturated this reference state could be represented by the moist adiabat, i.e. $\sigma_{\text{ref}} = \sigma_s$. However, in many model studies a slightly different reference state is adopted in order to correct for effects of cloud water on buoyancy (Betts, 1986). Thus, the problem of convective adjustment consists of determining the value of λ, the value of σ_{ref} as well as finding a criterion to determine whether the adjustment scheme is activated (when it is not activated we may set $\lambda = 0$). This requires an equation for conservation of water vapour in order to determine whether the air at a particular point is supersaturated.

Eqns. (4.56) and (4.31c) imply that

$$\frac{\partial \dot{\theta}}{\partial \theta} = \frac{\lambda}{\sigma} (\sigma - \sigma_{\text{ref}}) \tag{4.57}$$

This equation can be integrated from the Earth's surface upward with the heating prescribed at the Earth's surface, yielding the heating as a function of r and θ. The heating parameterization problem in this theory, therefore, is linked directly *to the problem of parameterizing the surface heat flux*. In this context, there are two effects that may provide a 'control' as implied in Figure 4.30. First, the intensity of the heat and moisture transfer from the sea surface to the atmosphere is proportional to the wind speed just above the sea surface. Hence, as the wind speed increases, so does the heat transfer, which in turn induces a further intensification of the wind speed and so on. Second, the decrease of the surface pressure in the core of the cyclone is accompanied by a decrease in the air temperature (owing to adiabatic expansion). This leads to a temperature gradient between the sea surface and the air just above. Since the heat transfer is also proportional to the temperature difference between the sea surface and the air just above, this effect will further enhance the heat transfer.

If true, the idea of 'equilibrium control' is very attractive as a mode of thinking about the interaction of cumulus convection with the balanced motion, because it ties the heating to a simple, physically plausible and observationally well documented constraint. Another clear advantage is related to the fact that the convective heating is not related in any way to the divergent part of the flow, which makes it independent of exact knowledge of the 'unbalanced' part of the motion.

It must be remarked, however, that in cases of cold air outbreaks over the oceans in polar regions, the stratification in the lower half of the atmosphere can sometimes be in a state which is very far from the thermodynamic reference equilibrium state. Nevertheless, in principle it is possible to handle this problem by using Eqn. (4.56) and specifying an appropriate value of the relaxation constant, λ.

4.5.12 Travelling upper-level disturbances and cyclogenesis

There is little doubt about the existence in general of strong conditional instability and associated large values of CAPE in regions where, and at times when, polar lows form (Section 4.5.1). Thus, some polar lows may form simply due to heating associated with a sudden release of latent heat associated with CAPE in a limited area of horizontal dimensions comparable to the Rossby radius of deformation. We have seen that if the heat associated with CAPE is released quickly in an area with a low-level positive potential vorticity anomaly, a cyclone may form within a few hours (Figure 4.49). A process that may serve as a 'spatial focusing mechanism' for the release of latent heat is an upper-level potential vorticity anomaly travelling into a region of reduced static stability. Montgomery and Farrell (1992) have investigated this process with application to polar low formation.

By virtue of the invertibility principle, the wind field (or vorticity field) below and above a moving potential vorticity anomaly must undergo changes such that relative vorticity increases in advance of the approaching anomaly, while relative vorticity decreases at the trailing edge of the anomaly. This implies ascending motion in advance of the travelling upper-level anomaly and descending motion at the trailling edge.

This effect can be easily demonstrated with the two-layer model shown in Figure 4.57. This model was adopted by Phillips (1951) and Bretherton (1966) to investigate baroclinic instability. The equations of motion and mass continuity for each layer are similar to Eqns. (4.22a–c), i.e.

$$\frac{du_i}{dt} = fv_i - \frac{\partial \phi_i}{\partial x} \tag{4.58a}$$

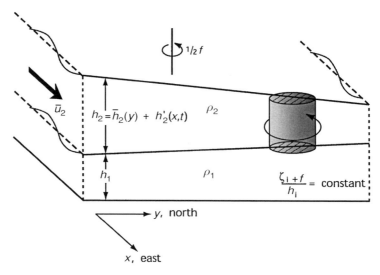

Figure 4.57. Schematic diagram of the two-layer model (see the text for further explanation). The basic potential vorticity gradient (south to north) is positive in the upper layer and negative in the lower layer.

$$\frac{dv_i}{dt} = f(\bar{u}_i - u_i) \qquad (4.58b)$$

$$\frac{dh_i}{dt} = -h_i \frac{\partial u_i}{\partial x} \qquad (4.58c)$$

with $i = 1$ referring to the lower layer and $i = 2$ referring to the upper layer (Figure 4.57). The geopotential, ϕ_i, defined as

$$\phi_1 = g(h_1 - \bar{h}_1) + \varepsilon g(h_2 - \bar{h}_2) \qquad (4.59a)$$

$$\phi_2 = g(h_1 - \bar{h}_1) + g(h_2 - \bar{h}_2) \qquad (4.59b)$$

(see Ooyama, 1969). We assume that there is a time-independent meridional potential vorticity gradient aloft, that induces a constant geostrophic velocity, $\bar{u}_i (i = 1, 2)$. In order to have geostrophic balance, this geostrophic velocity is associated with an imposed time-independent gradient in the thickness of the layers. Therefore, the time-mean thickness $\bar{h}_i$ depends only on the meridional coordinate, y. For simplicity, the perturbations are assumed independent of y. Given the potential vorticity distribution, the induced geostrophic meridional flow can be computed by solving the following equation:

$$\frac{\partial^2 v_i}{\partial x^2} - A_i v_i = B_i \qquad (4.60)$$

with

$$A_i = \frac{f \zeta_{poti}}{g(1-\varepsilon)} \qquad (4.61a)$$

$$B_1 = h_1 \frac{\partial \zeta_{pot1}}{\partial x} - \frac{\varepsilon f v_2 \zeta_{pot1}}{g(1-\varepsilon)} \qquad (4.61b)$$

$$B_2 = h_2 \frac{\partial \zeta_{pot2}}{\partial x} - \frac{f v_1 \zeta_{pot2}}{g(1-\varepsilon)} \qquad (4.61c)$$

and

$$\zeta_{poti} = \frac{f + \zeta_i}{h_i} \qquad (4.61d)$$

This is the two-layer version of the invertibility principle, analogous to Eqn. (4.26). It is derived from the conditions of geostrophic balance in both layers and from the definition of potential vorticity (4.61d).

Let us assume the existence of a positive potential vorticity anomaly in the upper layer, as shown in Figure 4.58. The physical mechanisms that produce this anomaly are not addressed here. We only assert that polar lows are very frequently observed in the vicinity of such upper-level potential vorticity anomalies. In the lower layer the potential vorticity is constant. The solution of Eqn. 4.60 in terms of the relative vorticity in both layers for this potential vorticity distribution is also shown in Figure 4.58. We see that the potential vorticity anomaly in the upper layer induces a relative vorticity anomaly in the lower layer. The intensity of this relative vorticity anomaly increases with decreasing static stability, represented by the parameter, ε, in this model (Figure 4.59).

If we now set the potential vorticity anomaly in the upper layer into motion (by assuming that $\bar{u}_2 = 30$ m s^{-1}; $\bar{u}_1 = 0$), the potential vorticity anomaly will travel eastward (remember that potential vorticity is materially conserved), deforming slightly due to meridional advection of potential vorticity associated with the meridional gradient in the basic state thickness (compare the thick solid line in Figure 4.58 with the solid line in Figure 4.60). The induced vorticity in the lower layer will necessarily also travel eastward. The associated process of adjustment in the lower layer requires convergence in advance of the moving vorticity anomaly (where $\partial \zeta / \partial t > 0$) and divergence behind the moving vorticity anomaly (where $\partial \zeta / \partial t < 0$). This is illustrated in Figure 4.60. Hoskins *et al.* (1985, p. 907) have come up with the following instructive analogy of this process:

> One may think of an eastward-moving upper-air anomaly as acting on the underlying layers of the atmosphere somewhat like a broad very gentle 'vacuum cleaner', sucking air upwards towards its leading portion

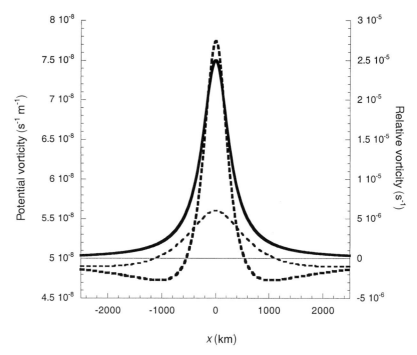

Figure 4.58. Potential vorticity in the upper layer (thick solid line) and in the lower layer (thin solid line) as a function of x, and the induced relative vorticity, according to the invertibility principle (Eqn. 4.60) in both layers (thick broken line: upper layer; thin broken line: lower layer). The potential vorticity anomaly is prescribed using the formula (4.23) (a bell-shaped perturbation) with $a = 300$ km. The static stability parameter, $\varepsilon = 0.8$, $g = 9.81$ m s^{-2}, $f = 0.0001$ s^{-1} and $\bar{h}_i = 2000$ m (for $i = 1, 2$).

and pushing it downwards over the trailing portion. The vertical motion field arises in response to the need to maintain mass conservation and approximate balance. If a potential vorticity anomaly were to arrive overhead without any adjustment taking place underneath it, then the wind, temperature and pressure fields would be out of balance to an improbable extent.

The induced upward motion in advance of the approaching anomaly is very likely to generate precipitating clouds, thereby generating potential vorticity in the lower layers of the atmosphere. On the basis of this idea, Montgomery and Farrell (1992) proposed a conceptual model of polar low development. In the first stage of development, called 'induced self-development', a mobile upper trough initiates a rapid low-level spin-up. This spin-up is especially strong if the trough moves into an area of reduced static stability (see Figure 4.59). A secondary development follows, called 'diabatic destabilization', which is

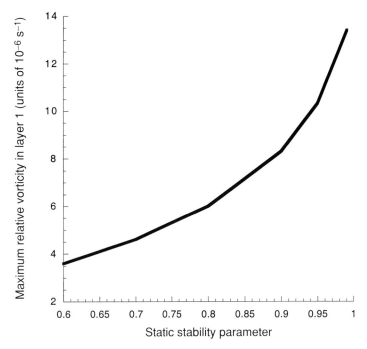

Figure 4.59. Maximum value of the relative vorticity induced in the lower layer by a fixed anomaly in the upper layer (see Figure 4.58) as a function of the static stability parameter, ε. The intensity of the induced vorticity increases with decreasing stability (increasing ε) as was also predicted by Eqn. 4.51.

associated with the production of low-level potential vorticity by heating within clouds forced by the induced upward motion *ahead* of the oncoming upper-level potential vorticity anomaly.

4.5.13 Summary and concluding remarks regarding the role of heating

Although a real polar low is not exactly a balanced flow structure, it is contended that only the 'balanced dynamics' is of importance to understand the intensification and structure of a polar low. Because of this, we can adopt the 'potential vorticity viewpoint' to disentangle from the primitive equations the most important physical principles governing the dynamics of cyclone intensification due to heating. The principle that heating disturbs the state of balance by altering the potential vorticity distribution, the principle of invertibility, which defines the balanced state, and the principle of material conservation of potential vorticity in the absence of friction and heating, offer physical insight into the reasons why a cyclone, such as a polar low, can grow as a result of heating.

392   4 Theoretical investigations

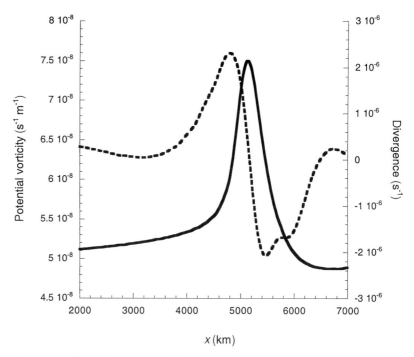

Figure 4.60. The numerical solution of Eqns. 4.58 with $\bar{u}_1 = 0$ and $\bar{u}_2 = 30$ m s^{-1} at $t = 48$ h, in terms of potential vorticity in the upper layer (solid line) and divergence $(\partial u_1/\partial x)$ in the lower layer (broken line). The initial condition is a potential vorticity anomaly in the upper layer at $x = 0$ (as shown by the thick solid line in Figure 4.58), conforming to the invertibility principle (Eqn. 4.60). The amplitude of the potential vorticity anomaly remains constant, but the anomaly is deformed slightly at both the leading and trailing edge by meridional advection of basic state potential vorticity.

Because of material conservation of potential vorticity in adiabatic circumstances, the 'balanced dynamics' retains high quantitative accuracy even in the presence of large-amplitude unbalanced motions, such as those associated with gravity waves or convection. This is related to the fact that the 'unbalanced motion' is practically incapable of changing the potential vorticity distribution, i.e. potential vorticity is not carried away or radiated by gravity inertia waves as is the case with energy. We have seen this very clearly in Sections 4.5.4 and 4.5.8.

The intensity and the sign of the perturbation of the potential vorticity depends in particular on the vertical stratification of potential vorticity at the location of the heat source. Heating in the mid-troposphere appears to be most effective as a local source of positive potential vorticity if the cyclone has a warm core in the lower troposphere, because in that case the diabatic advection of potential vorticity contributes to creating a positive tropospheric potential vorticity perturbation. Therefore, the chances of further intensification of a cyclone due to heating increase as the cyclone acquires a warm core.

Many polar lows are observed in the vicinity of, or directly under, a larger-scale cold core low. The low static stability under such a low stimulates convection and latent heating, but the numerical experiments described in Section 4.5.8 (see also Emanuel and Rotunno, 1989) clearly indicate that the growth of a polar low-like cyclone requires a low-level positive potential vorticity anomaly (i.e. a low-level warm anomaly) that can interact constructively with this heating. A synoptic structure, often observed in association with the early stages of polar low formation, which may play the role of this positive potential vorticity anomaly, is a low-level shear line.

It should be mentioned that deviations from axisymmetry and the consequent excitation of vortex Rossby waves could give rise to changes in the symmetric potential vorticity distribution and, thus, play a role in the evolution of the vortex. This subject is currently under investigation (Möller and Montgomery, 2000; Shapiro, 2000).

The coupling between the heating and the balanced flow has been a controversial topic for many years (Stevens *et al.*, 1997; Smith, 2000). Basically, there are two theoretical views on this problem. The first is based on the assumption that the heating is proportional to the moisture flux convergence, principally in the boundary layer. The second view is based on the assumption that the temperature in the atmosphere, in the presence of moist convection, adjusts to a reference temperature profile with a vertical gradient close to that associated with the moist adiabat. The second view has considerable advantages in a balanced model because knowledge of divergent motion is not required to determine the heating intensity. Agreement on this topic has yet to be reached.

4.6 Further theoretical considerations

In order to summarize and clarify some of the subjects and mechanisms referred to in the preceding sections some additional comments are presented below.

4.6.1 CISK and WISHE

Observations have shown that the development of most polar lows is closely associated with deep convection. Reflecting this fact, two mechanisms, CISK and WISHE, have been suggested to explain the development of a 'large-scale' balanced system, such as a polar low in the presence of deep convection. The persistence of the two modes of thought which have existed in

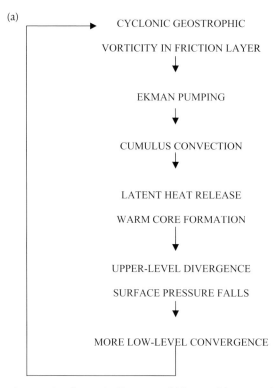

Figure 4.61. Schematic diagrams of (a) CISK (b) WISHE (ASII) (adapted from Bluestein, 1993).

parallel for about 20 years is due at least in part to the difficulty in distinguishing between them with currently available observations (Craig and Gray, 1996).

The two mechanisms are illustrated by the schematic diagrams shown on Figure 4.61 based on Bluestein (1993). Considering Figure 4.61a illustrating the CISK mechanism, it should be kept in mind that a necessary prerequisite for CISK is the presence of CAPE as discussed in Section 4.5.1, so that air parcels, after being lifted to their level of free convection will continue their ascent. In this case the Ekman pumping may trigger and/or enhance cumulus convection. The latent heat release associated with the convection will, through the formation of a warm core, induce upper-level divergence (outflow), falling pressure at the surface and more low-level convergence. Convergence acting on the existing low-level cyclonic vorticity produces more vorticity, which induces more Ekman pumping, and so on.

In CISK the cumulus heating (Q) is controlled by the balanced flow and is often taken to be proportional to the low-level convergence, i.e. $Q = \eta w^+$ where

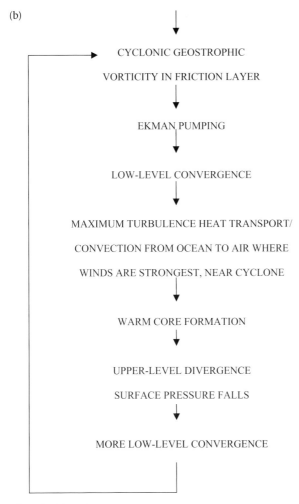

Figure 4.61 (cont.).

w^+ is the vertical velocity at the top of the boundary layer. Ooyama (1969) assumed that η was a function of the conditional instability of the atmosphere, while Charney and Eliassen (1964) assumed it to be proportional to the boundary layer specific humidity. In linear CISK theories, η is assumed to be constant, which is equivalent to assuming that surface fluxes act rapidly to replenish any depletion of CAPE or boundary layer moisture by the convection. In nonlinear CISK models, surface fluxes of heat and moisture may contribute to a time evolution of the coefficient η. The possibility of a variable η is especially relevant for polar lows which during their lifetime may move from regions with a relatively cold sea surface to regions characterized by much higher sea surface temperatures and fluxes.

In the ASII/WISHE mechanism (Figure 4.61b) the low-level inflow towards the centre of an incipient tropical cyclone or polar low results in increased surface fluxes from the sea surface in the regions of high wind speed near the centres of the cyclones. Sensible and latent heat is transported upwards by turbulent heat transport and convection. The WISHE theory assumes that the atmosphere is basically convectively neutral, i.e. the lapse rate is being constrained to follow a moist adiabatic lapse rate (see Section 4.5.2 and Figure 4.31) which means that there will be no (or little) CAPE. Also, in this case, a warm core will be formed. It should be noted that although the energy source of the cyclone according to the WISHE theory is through surface fluxes, a small amount of CAPE is not inconsistent with this theory (Emanuel *et al.*, 1994).

Concerning the initial formation of a vortex (polar low) from an 'infinitesimal disturbance', results from simple, linear CISK-driven models (i.e. Rasmussen, 1979; Pedersen and Rasmussen, 1985, Bratseth, 1985) showed growth rate curves with preferred wavelengths and a relatively short e-folding times corresponding to those for low-level baroclinic instability found by Mansfield (see Figure 4.6). These results were interpreted in such a way that small, 'natural' disturbances within a conditionally unstable air mass might grow due to CISK. Disturbances with a wavelength equal to or near the most unstable wavelength would amplify more quickly than disturbances of other wavelengths, and therefore, as long as nonlinear interactions could be neglected, would tend to dominate (Rasmussen 1979).

However, theoretical results have indicated that neither CISK nor WISHE can explain *the initial* development from an infinitesimal disturbance to a polar low. Van Delden (1989) found that a relatively weak cyclone is rather insensitive to diabatic heating, and underlined that '*CISK must be interpreted as a finite-amplitude instability* accounting for the rapid or explosive intensification of a balanced cyclone by diabatic heating'.

Concerning WISHE (ASII), Emanuel and Rotunno (1989) argued, that '*disturbances of substantial amplitude are apparently necessary to initiate intensification by air-sea interaction*'. To the extent that this is a valid finding, it points to the necessity of some presumably non-axisymmetrical dynamical process that operates in the early stages of cyclogenesis. They mention baroclinic instability as an obvious candidate and point out, that the development of polar lows might be thought of as a *two-stage process* for which also other disturbances, such as topographically generated cyclones, might act as starting disturbances.

The theoretical results mentioned above have been supported by earlier observational results. While satellite images have shown that most polar lows are associated with deep convection, then on the other hand many of these systems initially form in a highly (low-level) baroclinic environment near the ice edges. The coincidence of baroclinicity and convection led Rasmussen (1985a), among others, to propose that convection and baroclinic instability co-operate to produce the polar lows. The presence of an upper-level disturbance in the form of a short-wave trough or a vortex, prior to a number of significant polar low developments, led to the formulation of the widely accepted scenario that high latitude polar lows typically were initiated through a baroclinic process involving an upper-level short-wave trough and a low-level, ice edge-generated, baroclinic zone. As discussed elsewhere in this volume (e.g. Section 4.2), this mechanism may not be so dominant as once believed. The role of the low temperatures associated with the upper-level, cold short-wave troughs may be a crucial factor (see the following discussion of a model proposed by Økland (1987, 1989).

Økland (1987) pointed out that 'although the winter Arctic air mass originally is very stable in its lower layers, the same is not necessarily the case at greater heights'. It should be noted, though, that according to Økland, areas characterized by small vertical stability may be limited to the regions around small-scale, upper-level cold troughs, or emerge during reverse shear baroclinic developments (see Section 4.2 and Figure 4.15). According to Økland, the horizontal scale of circulations (polar lows) will be determined mostly by the size of the region of enhanced convection, i.e. indirectly of the size of the upper-level cold trough, and not through a scale selection for maximum growth rates as envisaged by some linear CISK theories.

Økland (1989) elaborated these ideas arguing that deep convection within Arctic air masses characterized by a low-level inversion is reached only in those cases where a layer of weak stability is present on top of the low-level inversion. When the air moves over the warm ocean, the low-level inversion will gradually be eroded away by a convective boundary layer which increases in thickness and temperature downstream. If this convective boundary layer reaches a height above which the atmosphere is conditionally unstable the increase in depth of the convective layer will be 'explosive', resulting, according to Økland, in a much greater effiency of the CISK process.

Concerning the question of the source of the low mid-tropospheric stability sometimes found in Arctic air masses, Økland, as noted above, pointed towards the role of the small-scale cold, upper-level troughs. Such troughs, as shown by several case studies (e.g. Rasmussen, 1985a; Businger, 1985), are able to trigger polar low developments. Concerning the nature of the triggering process, Økland (1987, 1989), was probably the first to stress the role of the low static stability due to the presence of cold air aloft. Økland (1989) summarized his ideas of the development of a polar low due to latent heat release in deep

398   4 Theoretical investigations

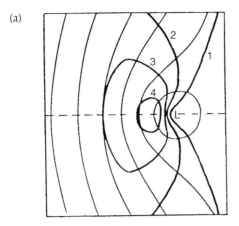

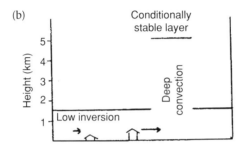

Figure 4.62. Sketch of a polar low caused by convection. (a) Shows isobars (thin lines) and isotachs (thick lines). The numbers on isotachs are relative. (b) A vertical cross-section through the low centre (from Økland, 1989, in Twitchell *et al.*, 1989. © A. Deepak Publishing.)

convection in the schematic diagram shown on Figure 4.62. The polar low is embedded in a northerly flow or located near the shear zone between a strong northerly flow to the west and weak winds towards the east (Figure 4.62a). The asymmetry results in an area of strong wind to the west of the low. In this wind maximum the heat flux from the sea is large. In the centre of the developing vortex the low-level stable layer is missing and if the air mass at greater heights has small static stability the convection will become deep. This will, according to Økland, happen over a fairly restricted area where the mid-level stability has a minimum due to the presence of an upper-level trough (Figure 4.62b). On the figure is shown how, outside the area with deep convection, the low-level inversion usually persists in a large part of the peripheral sections of the low. However, also in this region, the heat and moisture fluxes from the sea surface are comparatively large because of the strong winds associated with the low. The Ekman layer flux towards low pressure brings this air into the central part of the low where the heat and moisture feed the deep convection.

Emanuel and Rotunno (1989) argued that polar lows in general and the 13–14 December 1982 Bear Island case in particular were driven by the WISHE mechanism. However, significant CAPE was indicated by the NCEP/NCAR data (see Section 4.5.1) during the second intense phase of the Bear Island, December 1982 polar low development, pointing towards a contribution from CISK. The very rapidity, only a few hours, of development of this hurricane-like polar low seems to favour a CISK type of mechanism utilizing a source of accumulated CAPE rather than a comparatively slow WISHE mechanism. In the numerical experiments simulating this development using the WISHE mechanism, Emanuel and Rotunno found that the development of the maximum azimuthal velocity around the polar low required a time of the order of two days, whereas the observed time of development was of the order of a few hours.

Craig and Gray (1996) pointed out that the mere presence of CAPE does not provide a conclusive test of CISK and that a small amount of CAPE is not inconsistent with WISHE. To test the relative importance of CISK and WISHE they noted that the sensitivity of the intensification rate of tropical cyclones and polar lows to surface properties, such as surface friction and moisture supply would be different for the two mechanisms. Experiments with an axisymmetric model and explicit convection (to avoid the possibility of prejudicing the results through the choice of a particular parameterization scheme) showed that the intensification rates of a simulated polar low had a strong dependence on the heat and moisture transfer coefficients, while remaining largely insensitive to the frictional drag coefficient. Their results imply that the rate-limiting process for cyclone intensification is the rate of heat and moisture fluxes while frictional convergence is of secondary importance. Based on this they concluded 'that the intensification of the numerically simulated tropical cyclones and polar lows is due to WISHE', and that 'It is anticipated that a similar conclusion would apply in nature'. While it is very likely that WISHE will be the dominant mechanism for some polar low developments it is more doubtful whether this statement is valid in a more general way as advocated by Craig and Gray. They simulated an axisymmetric polar low characterized by a well-developed eye and with only small values of CAPE. In nature the lows will be highly asymmetric, often without an eye-region, and form in regions with relatively large values of CAPE. As seen from satellite images such as Figure 3.31, the intense Bear Island polar low, which formed late on 13 December 1982, had no eye but was characterized by a small cluster of convective clouds within the central region of the low. Økland (1987) argued that polar lows forced by heating due to surface fluxes proportional to surface wind speeds (WISHE) show a tendency to displace the active part of the low from the centre to more peripheral areas forming an 'eye', whereas heating proportional to the relative vorticity (CISK) tended to form an intense vortex in the central part of the low.

It is also doubtful whether a general conclusion concerning the relative importance of the two mechanisms can be drawn alone based on results on the use of an axisymmetric model. For example, in nature most polar lows are asymmetrical including the presence of shallow low-level fronts. The Arctic air masses in which polar lows typically develop are often quite dry at low levels, which inhibits widespread deep convection. It has been documented from satellite images that deep convection during polar low developments tends to be organized along shallow Arctic fronts along which air parcels are forced to ascend to their level of free convection. The effect of this lifting, which is likely to be important, is limited to polar/Arctic regions and is not included in the traditional formulation of CISK.

4.6.2 The Montgomery–Farrell model

Montgomery and Farrell (1992) in their work utilized observations such as presented by Reed (1979), Rasmussen (1987) and Businger (1987) showing that polar lows are generally initiated by an upper-level disturbance in the form of a mobile, short-wave cold trough. Based on a study utilizing a three-dimensional nonlinear geostrophic momentum model that incorporated moist processes and strong baroclinic dynamics, they proposed, as briefly considered in Section 4.5.12, a conceptual model in which polar low development occurs in two stages. The first stage, the induced self-development, comprises an interaction between an upper-level potential vorticity anomaly, i.e. an upper-level trough, and a low-level potential vorticity anomaly in a nearly moist neutral atmosphere. The enhanced omega response (vertical velocity) in the nearly neutral atmosphere initiates a rapid low-level spin-up and generation of low- and middle-level potential vorticity anomalies that augment the baroclinic interaction between the low-level system and the system aloft. The subsequent secondary development, called diabatic destabilization or diabatic intensification, is associated primarily with the diabatic production of low-level potential vorticity in regions of ascending air and cloud formation (see Section 4.5.12).

Montgomery and Farrell ascribe the fact that polar lows often maintain their intensity, or slowly intensify until they reach land to diabatic destabilization. In exceptional cases instances of polar cyclogenesis with negligible upper-level forcing, diabatic destabilization may also explain the gradual intensification of small-scale vortices in regions of sustained neutrality and surface baroclinicity.

While cooperative intensification processes associated with either CISK or WISHE may be operative in the latter phase of development of polar lows, neither, according to Montgomery and Farrell, appear essential for describing the basic formation process of polar lows. CISK or WISHE may, however, according to them, play the role of a cyclone 'afterburner'.

In conclusion, Montgomery and Farrell hypothesized that the formative mechanisms for polar lows are the same as those for mid-latitude cyclones, with moist processes playing a more central role in the polar low case.

Montgomery and Farrell assumed as a basis for their work that polar lows form in regions of 'strong ambient baroclinicity'. The cases they refer to – comma clouds, polar lows along occluded cyclones and reverse shear systems – may, corresponding to their basic premises, all be characterized as forming within a baroclinic environment. It is doubtful, however, whether the theory put forward by Montgomery and Farrell can be applied to the 'real' or 'true' polar lows which develop at high latitudes away from the main baroclinic zone. In spite of this shortcoming, the ideas put forward by Montgomery and Farrell have been an important contribution to our understanding of polar low dynamics.

4.6.3 The Fantini model

Emanuel (1994) discussed a non-modal mechanism for a polar low WISHE-type development based on work by Fantini (1990). According to this theory, which in many ways is reminiscent of the theory of Montgomery and Farrell discussed above, a surface cyclone is initiated when an upper-level PV anomaly approaches a region of low static stability and high sea–air thermodynamic disequilibrium (Figure 4.63a). The low-level cyclone then amplifies rapidly by the WISHE mechanism forming a highly concentrated, relatively shallow warm core surface cyclone, associated with an intense, small-scale low-level PV maximum. The subsequent overtaking of this low-level PV anomaly by the upper-level anomaly (Figure 4.63b) then results in rapid baroclinic deepening. Numerical experiments with this model (Fantini, 1990) of the development of large-scale maritime storms in a baroclinic, saturated environment resulted in 'explosive' growths with deepening rates near or above the conventional definition of explosive cyclones. The rapid deepening stage was (Emanuel, 1994) 'almost purely baroclinic and should be thought of as a rapid baroclinic growth greatly enhanced by the presence of a low-level PV anomaly created early in the evolution by the WISHE mechanism triggered by an approaching upper-level PV anomaly'.

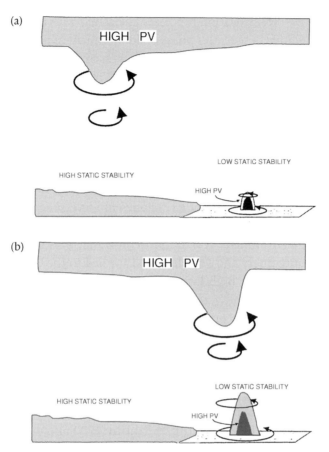

Figure 4.63. An upper-level PV anomaly approaches a region of large sea–air thermodynamic disequilibrium and low static stability and initiates a surface cyclone (a). The WISHE mechanism forms a concentrated low-level PV maximum, similar to that of a hurricane but more shallow and not as intense; this subsequently interacts with the approaching upper-level PV anomaly (b) to produce baroclinic cyclogenesis (from Emanuel, 1994).

4.6.4 The Craig and Cho model

In Section 4.2, two studies of Sardie and Warner (1983, 1985) concerning the relative importance of baroclinic instability and CISK were discussed. Craig and Cho (1989) explored the role of cumulus heating and CISK in a baroclinic environment using a simple linear model that incorporated both baroclinic and convective processes. The model applied was a semi-geostrophic Eady model of baroclinic instability with a continous vertical structure, including a simple parameterization of cumulus heating. Within the context of their linear model they found that the interaction of baroclinic and convective processes could be understood as follows. If the heating rate was not too great a baroclinic wave would grow in the usual manner, the cumulus heating merely intensifying the existing circulation. The effects of convective heating in this case were essentially the same as a reduction in static stability, which resulted in a faster growing disturbance with a shorter wavelength. CISK occurred in the model when the circulation induced by the heating was sufficient to supply the same or a greater amount of heating. This is only possible when a heating parameter in the model, which is a measure of the efficiency of the heating in resupplying itself, is above a certain threshold. Since the convective and baroclinic processes cooperate at all stages, the onset of CISK is through a smooth transition and there is no precise point where one instability mechanism ceases to dominate and the other starts. Six polar air disturbances considered by Craig and Cho illustrated the full range of behaviour shown by the model. Two polar lows (including the 13–14 December 1982 Bear Island case discussed in several places above) tended to be CISK-dominated, two comma clouds tended to be primarily baroclinic, and the two remaining systems of a transitional nature. Craig and Cho concluded, in accordance with observations, that polar lows may change over time. For example, an initially baroclinic system could become CISK-dominated as surface fluxes of heat and moisture decrease the stability of the atmosphere and enhance cumulus convection.

4.7 Summary and concluding remarks

In the preceding sections of Chapter 4 we have discussed a number of physical mechanisms which, over the years, have been suggested as being responsible for the development of polar lows.

The areas in which polar lows have been observed to form range from highly baroclinic regions near the polar front and along the ice edges, to high latitude, nearly barotropic (or equivalent barotropic) environments. Because of this a variety of forcing mechanisms will be effective, and it is not surprising that polar lows appear in so many forms. The arguments from the late 1970s and early 1980s that polar lows were either baroclinic disturbances, *or* convective systems akin to tropical cyclones, were gradually replaced with the understanding that both mechanisms were important. This advance led to the idea of a 'polar low spectrum' with purely baroclinic systems at one end, and purely convectively driven systems at the other (see Section 3.1.4, The 'Polar Low Spectrum'). In between the two 'pure types', there was room for a variety of hybrid systems for which both mechanisms, i.e. baroclinic instability and a 'convective mechanism' such as CISK or ASII/WISHE, could play a smaller or greater role depending on the precise circumstances under which a particular development took place. Experience has shown that the hybrid types are by far the most common.

Compared to the two principal fluid dynamic instabilities mentioned above, baroclinic instability and thermal (convective) instability, the third main mechanism of barotropic instability, is thought to play a minor role in connection with polar lows. However, increasing evidence indicates, that while this mechanism in isolation may seem only of minor importance, and as such only important for the development of rather insignificant vortices, in combination with other factors it may play an important role as a trigger for subsequently stronger developments as discussed in Section 4.3. In addition, the two main mechanisms of baroclinic and thermal instability may appear in different forms. Concerning baroclinic instability, these forms include low-level baroclinic instability, deep baroclinic instability, reverse-shear baroclinic instability and processes involving a cooperation between upper-level baroclinic systems and low-level baroclinic zones (type B developments).

While it has been generally accepted that convection plays an important role in the development of a number of polar lows, the form through which this influence takes place has been a subject of much discussion. During the 1980s the point of view that CISK was the main dynamic mechanism for a significant group of polar low developments was challenged by a number of authors (i.e. Emanuel and Rotunno, 1989). Arguing that the polar atmosphere, like the tropical one, was nearly neutral to deep, moist convection, they rejected the CISK theory and proposed the WISHE mechanism (Section 4.5.1). In WISHE, sensible and latent heat fluxes from the sea surface maintain the atmosphere in a convectively neutral state with correspondingly negligible amounts of CAPE. There is evidence though that in the polar regions, the combined action of surface fluxes of heat and moisture and upper-air cold advection occasionally may lead to the generation of significant amounts of CAPE. The presence of CAPE plus the fact that some polar lows develop very rapidly indicates that CISK may contribute to these developments. Based on the material presented in this book a preliminary conclusion on the ongoing discussion of the relative importance of CISK versus WISHE for the developments of tropical cyclones and polar lows must be, that for polar lows both mechanisms may, according to the specific circumstances, contribute to the development of the lows. Both theories, CISK and WISHE hypothesize that the large-scale vortex, i.e. the polar low, develops through a succession of balanced states. The question how the heating associated with either the CISK or WISHE affects this balanced flow was dealt with in detail in Sections 4.5.5 to 4.5.10. Clearly future research is required on the role of convection in the development of polar lows.

S. GRØNÅS, E. RAUSTEIN AND G. HEINEMANN

5

Numerical simulation

5.1 The Arctic

5.1.1 Introduction

Numerical models of the atmosphere, which predict future conditions from an analysis of an initial state, have proved to be a tool of growing importance in the study and forecasting of polar lows. To some degree, the life cycles of some polar lows are now simulated operationally by numerical weather prediction (NWP) centres. Some special cases of polar lows have been simulated more extensively in an *a posteriori*, non-operational mode with models suited for this purpose. Such simulations have provided a new form of data for the study of the formation and evolution of these vortices.

The history of the development of NWP, along with the growth of computing capacity in super-computers, is well known. For a long time the resolution of the numerical models was too coarse to describe polar lows. A breakthrough for the simulation of polar lows came in a polar low project organized by the Norwegian Meteorological Institute (DNMI) in the first half of the 1980s (Lystad, 1986; Rasmussen and Lystad, 1987). As part of this project, a mesoscale NWP system was established (Grønås *et al.*, 1987b; Grønås and Hellevik, 1982; Nordeng, 1986), which gave the first realistic numerical simulations of polar lows (Grønås *et al.*, 1987a; Nordeng, 1987). As operational NWP systems were further developed, reliable guidance for the prediction of polar lows was eventually advanced at several meteorological centres.

Special mesoscale observations of polar lows were made over a few hours from NOAA research aircraft over the Arctic seas (Shapiro *et al.*, 1987). High resolution simulations were made for these situations (e.g. Grønås *et al.*, 1987a; Aarhus and Raustein, 1998). In these papers the predictability and dynamics of polar lows was studied using simulations. Similar studies have also been made for situations when polar lows made landfall in areas with a relatively dense operational network of observations (Grønås *et al.*, 1987a; Nordeng and Rasmussen, 1992; Albright *et al.*, 1995; Nielsen, 1997). When the simulations were in reasonable agreement with the available observations, the predicted life cycle has been considered to represent probable estimates of the true evolution. In this way studies of successful simulations have given valuable insight into the development mechanisms of polar lows.

This chapter presents a review of the studies based on simulations of real polar lows with numerical models. The small scale of the vortices, the relatively large influence of internal forcing, such as latent heat release, the rapid development, the large range of intensities and the lack of observations make such simulations a challenging task. Results from some well-known Arctic case studies are reviewed in Section 5.1.2. We concentrate on the formation, structure and evolution of the lows, rather than on details of the numerical modelling. Most of the cases are from the Norwegian and Barents Seas, but cases from the Labrador Sea, the Bering Strait and the Gulf of Alaska are also presented.

In this section we will restrict our analysis to the type of polar low that has been defined as an Arctic front polar low (Businger and Reed, 1989b; Blier, 1996; Nielsen, 1997) or an Arctic outbreak polar low (Grønås and Kvamstø, 1995; Mailhot *et al.*, 1996). For this type, Arctic air masses, connected to the Arctic inversion over the sea ice, play an important part in the development. When Norwegian meteorologists mention polar lows, they nearly always mean this kind of polar low.

When the results of such simulations are synthesized, it should be borne in mind that simulations of some cases, which have shown poor agreement with observations, have probably not been reported in the literature. An example is the so-called Bear Island (December 1982) polar low investigated by Rasmussen (1985a) and discussed elsewhere in this book (see Section 3.1.5). Several workers, including one of the authors, have tried to simulate this case, but have failed to predict the strong deepening that was observed.

5.1.2 Simulation of selected Arctic polar lows

Regions where polar lows frequently form in the Northern Hemisphere are the Nordic Seas (Barents Sea, Norwegian Sea, Greenland Sea), the Bering Sea/Gulf of Alaska, the Labrador Sea/Hudson Bay and in the Pacific and over the Japan Sea. In the following we will give a review of some successful simulations of polar low developments in some of these regions.

Cases where the development started in the Barents Sea, and where the evolution progressed as a disturbance into the Norwegian Sea, have been simulated by, amongst others, Nordeng (1990), Nordeng and Rasmussen (1992), Grønås and Kvamstø (1995) and Nielsen (1997).

The first numerical simulation of a polar low in the Greenland/Norwegian Sea was made by Sardie and Warner (1985). Grønås *et al.* (1987a) and B. Aarhus and E. Raustein (unpublished data) simulated another polar low development in this area, the so-called ACE (Arctic Cyclone Expedition) case from 1984, when Shapiro *et al.* (1987) made measurements with a research aircraft. Qualitatively, the results from these simulations compare very well. Results from Aarhus and Raustein's simulations will be reviewed in a later subsection.

Relatively few simulations of polar low developments have been made in the Bering Sea/Gulf of Alaska region. The first successful one known to the authors is that of Sardie and Warner (1985). Blier (1996) simulated a development that started in the Bering Sea, and evolved further into a polar low in the Gulf of Alaska. Bresch *et al.* (1997) simulated a polar low development in the Bering Sea in 1977. The two last-mentioned simulations will be reviewed in the following.

For the Labrador Sea/Hudson Bay region, only three simulations of polar low development are found in the literature, although it is known that polar lows form here quite frequently. A polar low development that took place in Hudson Bay in 1988 was successfully simulated by Albright *et al.* (1995). The other case was a polar low that developed in the Labrador Sea, and was simulated by Mailhot *et al.* (1996). Another Labrador Sea polar low development successfully simulated was briefly discussed in Rasmussen *et al.* (1996). The first two of these simulations will also be reviewed later in this section.

Nordeng's cases from 25–26 January 1987 and 26 January 1988

Nordeng (1990) studied the life cycles of two polar lows in the Norwegian Sea. He combined the analyses based on observations with (a) operational forecast results for the January 1987 case, and (b) numerical simulation results for the January 1988 case. The model used was the same as that used by Nordeng and Rasmussen (1992).

For the 1987 case, a polar low developed in the Norwegian Sea between 1800 GMT 25 January and 0000 GMT 26 January. In order to study the role of baroclinic versus diabatic processes, Nordeng chose to show the vorticity advection with the thermal wind (the forcing term in the adiabatic and frictionless ω equation) at 850 hPa. Ascending motion is found in regions where the thermal wind blows from high to low vorticity. In regions of strong vertical velocity, vortex stretching and intensification is expected. From Figure 5.1 it can be seen that locally there was strong vorticity advection with the thermal wind, while 6 h later it was very weak (not shown). From this, Nordeng concluded that since the ω-forcing was largest during the initial phase of the development, the finite amplitude initial value theory of Farrell (Farrell, 1985) was important.

The polar low that formed on 26 January 1988 resembled a common baroclinic development (Figure 5.2). However, the forecast results indicated that

408   5 Numerical simulation

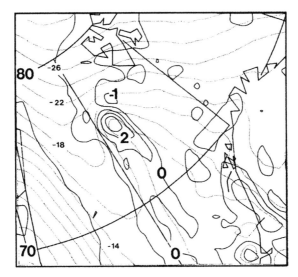

Figure 5.1. 850 hPa relative vorticity at contour intervals of 10^{-4} s^{-1} (solid lines) and mean temperature in the layer between 700 and 925 hPa at contour intervals of 2 K (dotted lines). A 6 h forecast from 1800 GMT 25 January 1987 (from Nordeng, 1990).

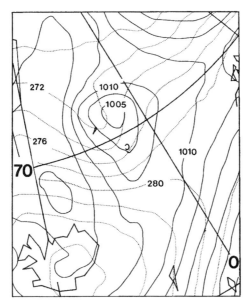

Figure 5.2. A 6 h forecast from 0000 GMT 26 January 1988 of mean sea level pressure (contour interval 2.5 hPa, solid lines) and 700 hPa geopotential height (contour interval 2 dam, broken lines). (From Nordeng, 1990, fig. 8a).

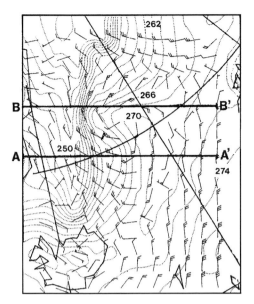

Figure 5.3. Potential temperature in the lowest model layer (*c.* 40 m above surface) (contour intervals of 2 K, dotted lines) and winds in lowest model layer (every second gridpoint plotted). A 6 h forecast from 0000 GMT 25 January 1988 (from Nordeng, 1990).

the polar low started to deepen as a result of a combination of events. First, the polar low formed close to the ice edge, where the low-level baroclinicity was strong (Figure 5.3). As with the 1987 case, the intensification may be explained by traditional quasi-geostrophic theory. Both polar lows intensifed as upper-level vorticity maxima approached low-level baroclinic zones, as in the Type B cyclones of Petterssen *et al.* (1962).

As for the maintenance and mature phases of the polar lows, Nordeng found that for the 1987 case the adiabatic ω-forcing term decreased with time while ω remained strong, indicating that the diabatic heating gradually took over as the main forcing term. Nordeng's results suggest that air–sea interaction instability (Emanuel, 1986b; Emanuel and Rotunno, 1989) might have occurred for this case. Backward trajectory calculations for air particles entering the inner regions of the polar low, show that the particles' equivalent potential temperatures had increased from 260 K to 280 K over 24 h. This corresponds to a considerable amount of heat transferred from the warmer sea surface to the air, confirming the assumption of air–sea interaction instability. This kind of instability requires a finite amplitude precursor. If the theory is applicable to this polar low, baroclinic instability release from the approaching upper-level wave seems to have played the role of a precursor. The polar low rapidly dissipated after making landfall, supporting the air–sea interaction instability theory, or at least showing that the warm sea was a necessary condition for the maintenance of the low.

The 1988 polar low showed no signs of further growth after the initial baroclinic development phase. Thus, air–sea interaction instability did not seem to be present for this case, which is in agreement with application of the above-mentioned theory to this polar low (air parcels ending up in the area of strongest updraught (those from the north) mainly passed over ice-covered sea and picked up very little energy from the underlying surface).

Nordeng and Rasmussen's case of 26 February 1987

During the afternoon of 26 February 1987 a synoptic-scale low and a corresponding occluded frontal system formed over the Norwegian Sea, centred at around 74° N, 30° E. Within this low pressure area, a polar low started to develop during the following 24 h. At an early stage of the development, radiosonde ascents from nearby Bear Island showed that the surface layer in the region where the vortex formed was capped by a pronounced inversion which inhibited deep convection from developing. In the region west and southwest of the centre, over the sea, strong cold-air advection took place. The air in the region east of the cyclone, characterized by shallow convection, had been modified by strong surface fluxes. Whether the shallow baroclinic field between the fresh cold air to the north and west of the low, and the modified warmer air mainly to the east helped trigger the development of the polar low, will be discussed later.

Theoretical investigations by, for example, van Delden (1989) and Emanuel and Rotunno (1989) have shown that the environment of polar lows is normally stable to small-amplitude perturbations, and that in order for polar lows to grow, disturbances of substantial amplitude appear to be necessary to initiate intensification through either CISK (Conditional Instability of the Second Kind) or air–sea interaction (see Section 4.5).

The satellite image in Figure 5.4 from 0418 GMT 27 February 1987 shows

- a striking similarity to a tropical hurricane;
- regions of deep convection in the spiral arms;
- a well-defined eye of a rather large diameter.

One important difference, however, between this polar low and a tropical hurricane is the contrasting air masses in the boundary layer, which are very obvious from the structure of the cloud fields as seen in the image.

Immediately after making landfall, satellite images showed that the polar low lost its hurricane-like structure and well-defined eye. The barogram traces from some stations close to where the landfall took place (see Nordeng and

Figure 5.4. An infra-red satellite image taken by NOAA 9 at 0418 GMT 27 February 1987 (Image courtesy of the NERC Satellite Receiving Station, University of Dundee.)

Rasmussen, 1992, their fig. 6) show that the pressure disturbance associated with the polar low was of the order of 5 hPa and that the polar low had a horizontal scale of a few hundred kilometres.

To study the development of this polar low in detail, Nordeng and Rasmussen (1992) used (with small modifications) the then operational mesoscale numerical model NORLAM of the Norwegian Meteorological Institute. For this simulation the horizontal resolution was 25 km and 18 levels were used in the vertical.

The initial start time of the simulation was 1200 GMT 26 February 1987. Initial mean sea level pressure and 500 hPa geopotential height fields are shown for the integration domain in Figure 5.5. In Figure 5.6 is shown the simulated evolution of the low system over a period from 1200 GMT 26 February (corresponding to 6 h of integration) to 0300 GMT 27 February (15 h of integration). We see that a mesoscale low (a polar low) developed on the northwestern part of the parent low, and moved anticlockwise while it deepened. After 18 h of integration, at 0600 GMT 27 February, the polar low made landfall after which time it quickly disappears.

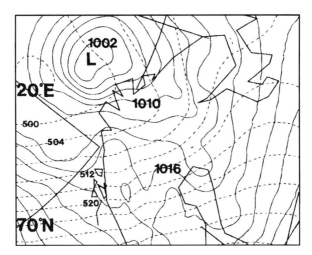

Figure 5.5. Mean sea level pressure at contour intervals of 2 hPa (solid lines) and geopotential height of 500 hPa at contour intervals of 40 m at the start of the integration (1200 GMT 26 February 1987) (from Nordeng and Rasmussen, 1992).

The model description of the polar low, compared with observations, clouds from satellite images versus model precipitation, and model versus observed winds, confirms that by and large we can use the model diagnostics to try to understand the physics leading to the development of the polar low.

The polar low looked very symmetric in the satellite images, with a striking resemblance to a tropical hurricane.

Due to the distribution of the surface fluxes of sensible and latent heat (see Nordeng and Rasmussen, 1992, their fig. 10), a direct secondary flow was generated approximately perpendicular to the isobars in the zone where the gradient in the surface fluxes was a maximum. This flow was frontogenetic, and is shown clearly in the cross-sections in Figure 5.7.

The polar low formation will now be looked at from a PV perspective. Figure 5.8 shows the PV at the $\theta = 278$ K surface at the initial time of the simulation. This surface is situated between 500 and 550 hPa. We see that there was a large-scale PV anomaly to the west of the synoptic-scale low, but in addition there was a superimposed small-scale anomaly where the polar low started to develop. In general, a PV anomaly will set up a secondary circulation in the vertical plane, where the ascending motion is found and where there is positive PV advection. From Figure 5.8 we notice that the small-scale anomaly may have been responsible for the deep circulation seen in Figure 5.7a, while the large-scale anomaly did not contribute to the vertical motion in the area of interest. From Figure 5.7 we see that there was a low-level, small-scale warm anomaly, which according to Hoskins *et al.* (1985), is equivalent to a cyclonic

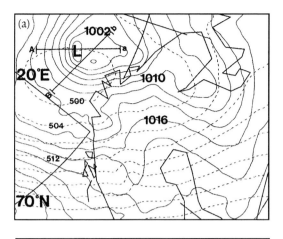

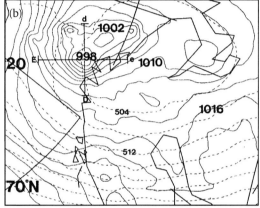

Figure 5.6. Mean sea level pressure at contour intervals of 2 hPa (solid lines) and geopotential height of 500 hPa (dash-dot lines) at contour intervals of 40 m after (a) 6 h and (b) 15 h of integration (from Nordeng and Rasmussen, 1992).

PV anomaly concentrated at the surface. In this case, the upper- and low-level anomalies reinforced each other and gave rise to a significant low-level PV anomaly. Provided the upper-level anomaly comes close enough to the low-level anomaly the two may interact forming a deep circulation, such as seen in Figure 5.7a.

The small static stability in the column with strong vertical velocity in Figure 5.7a favours the penetration of an upper-level PV anomaly to low levels. Since small static stability is connected to small values of PV, one obtains the conceptual model that PV anomalies are required at upper and low levels to give vertical circulation, but this circulation becomes strong only if it is established in a region of low PV between the anomalies. Observations (Emanuel, 1988) and numerical simulations show that the vertical circulation tends to be aligned along absolute momentum (M) surfaces, which was also the case here.

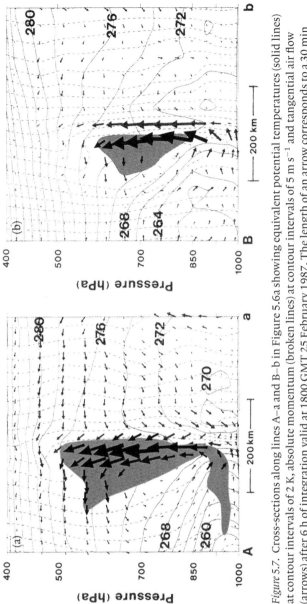

Figure 5.7. Cross-sections along lines A–a and B–b in Figure 5.6a showing equivalent potential temperatures (solid lines) at contour intervals of 2 K, absolute momentum (broken lines) at contour intervals of 5 m s^{-1} and tangential air flow (arrows) after 6 h of integration valid at 1800 GMT 25 February 1987. The length of an arrow corresponds to a 30 min trajectory. The shaded regions have relative humidity above 90% (From Nordeng and Rasmussen, 1992).

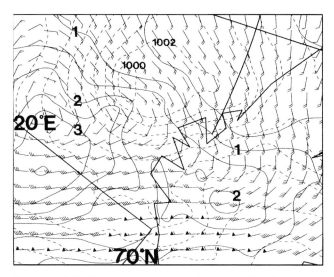

Figure 5.8. Solid lines are potential vorticity at the 278 K isentropic surface from 1200 GMT 26 Feb 1987, at contour intervals of 0.5 potential vorticity (PV) units (1 PVU = 10^{-6} m^2 s^{-1} K kg^{-1}). Also shown are horizontal winds at the 278 K surface valid at the same time, a flag is 25 m s^{-1}, a full barb 5 m s^{-1} and a half barb is 2.5 m s^{-1}. Broken lines are mean sea level pressure at contour intervals of 2 hPa valid at 1800 GMT 26 February (i.e. 6 h later) (from Nordeng and Rasmussen, 1992).

PV is conserved for adiabatic, non-viscous motion. From Chapter 4, release of latent heat is a sink of PV above the diabatic heating maximum and a source below, contributing to a low static stability and hence strong vertical circulation above the heating. If the upper-level PV is close to zero, the M and θ surfaces will be nearly parallel; which will promote the outflow aloft and hence facilitate the surface pressure fall.

Montgomery and Farrell (1992) found that in a semi-geostrophic model with latent heat release, in addition to the 'classical' strong baroclinic development where an upper-level anomaly couples with a low-level anomaly so that they mutually intensify, a class of growing disturbances from PV generation at low levels due to latent heat released in ascent regions, was possible. The developments of the second class were slow compared with the classical type, but they are not dependent on the presence of large amplitude perturbations. They suggest that some polar low developments consist of an initial baroclinic growth phase followed by a prolonged slow intensification due to diabatic effects.

To understand the spin-up of the polar low through destruction of PV in the outflow regions, the spatial distribution of latent heating must be explained. From Figure 5.9 one notices that air parcels that followed the selected

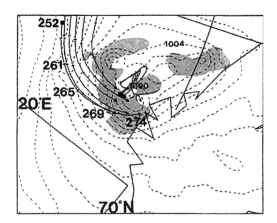

Figure 5.9. Surface air trajectories ending up in the vicinity of the low at 0300 GMT 27 February 1987. The numbers along the trajectory are equivalent potential temperature after 0, 6, 9, 12 and 15 h of integration. Areas with heavy precipitation (>1 mm $(3\text{ h})^{-1}$) are stippled. Mean sea level pressure after 15 h of integration valid at 0300 GMT 27 February 1987 is indicated by broken lines at contour intervals of 2 hPa (from Nordeng and Rasmussen, 1992).

trajectory experienced considerable heating from surface fluxes of latent and sensible heat. In order for deep convection to develop in the central region of a polar low, air parcels must acquire large amounts of heat and moisture from the surface as they spin towards the centre. Soundings along the trajectory (see Nordeng and Rasmussen, 1992, their fig. 14) showed that the air was extremely stable when it left the ice, but quickly destabilised due to sensible heat fluxes from the 'warm' ocean so that when it entered the updraught region it was marginally stable. The low stability favoured strong vertical motion. A capping inversion around 850 hPa seen at the time when the air left the ice was thought to have been important in localizing the spatial distribution of the heating. It limited the amount of air to be heated, so that the temperature could increase considerably, and furthermore, convection was not taking place over the whole region, but was limited to the areas in the vicinity of the polar low. Økland (1989) advocates that such a phenomenon is the main factor determining the scale and location of this polar low development.

Cross-sections (see Nordeng and Rasmussen, 1992, their fig. 15) through the centre of the low at 0300 GMT 27 February showed that the air was neutrally stable to slantwise convection in the region of strong vertical velocity, and the parcels closely followed the absolute momentum surfaces. Why the vertical motion was weak and even downward in some places with convective instability is not easily explained from the model fields. But in spite of this peculiarity, the model results showed the essential features of this polar low formation and development. The initial development was triggered by an upper-level PV anomaly in a region of strong low-level baroclinicity, while the driving mechanism for the further evolution was the transfer of sensible and latent heat from the sea surface with subsequent release of latent heat at low levels.

For more information on this case, see Section 3.1.5.

Grønås and Kvamstø's cases

Wilhelmsen (1985) and Businger and Reed (1989b) suggested that for the Norwegian and Barents Seas, a northerly flow and an Arctic inversion seemed to be *necessary* conditions for polar lows to develop (Businger and Reed, 1989b). Grønås and Kvamstø (1995) made simulations of four cases from late winter to early spring that satisfied those conditions. In two of the cases no polar low was observed, in neither simulations nor satellite images, while in the other two cases polar lows formed in the simulations (and cloud features indicating their presence appeared in satellite images). In each of the cases a synoptic-scale cyclone was situated over Scandinavia and steered a northerly surface flow over the Norwegian Sea. Cold air masses were being advected from the sea ice to the north, and a convective planetary boundary layer (PBL) had formed over the sea through strong fluxes of sensible and latent heat. The well-mixed PBL was separated from the upper cold, stable polar air masses by a strong Arctic inversion. The capping inversion descended to the sea surface at the leading edge of the cold air outbreak forming what is referred to as an Arctic front. According to Businger and Reed (1989b), these conditions should be favourable for what they called Arctic front polar lows. In their classification, this type was associated with an ice edge and characterized by shallow baroclinicity and strong surface fluxes. Since the disturbances that developed in Grønås and Kvamstø's (1995) cases did not seem to be directly associated with the leading edge of the Arctic inversion, they preferred to use the name Arctic outbreak polar lows instead.

The model used by Grønås and Kvamstø (1995) was the NORLAM model, mentioned earlier, with the addition of a hydrological cycle according to Sundqvist *et al.* (1989). The horizontal grid distance was 50 km, and in the vertical 18 sigma levels were used between 100 hPa and the surface. Analyses and lateral boundary conditions were taken from ECMWF analyses interpolated to pressure surfaces.

Grønås and Kvamstø (1995) focused on the vertical distance (H) between the tropopause (defined by them as the 2 PVU surface) and the top of the PBL. This distance will reflect the distance between the upper-level PV anomaly and the low-level positive temperature anomaly. For the two cases simulated when no polar lows developed the minimum of H was at least 2500 m. The main reason that the minimum of H was so large was that the horizontal distance between where the tropopause was at its lowest and where the PBL at its highest was too large. During the simulations, the synoptic conditions did not change to

418   5 Numerical simulation

give more favourable conditions for polar lows. Satellite images (Grønås and Kvamstø, 1995, their fig. 2a) confirmed that there were no signs of polar low development.

For the other two cases simulated, the synoptic conditions turned out to be favourable for the development of polar lows. Results for those cases are shown in Figure 5.10. The first case (Figure 5.10a, b) is the same as that simulated by Nordeng and Rasmussen (1992), discussed earlier in this section. For the second case (Figure 5.10c, d), we notice the small horizontal distance between the

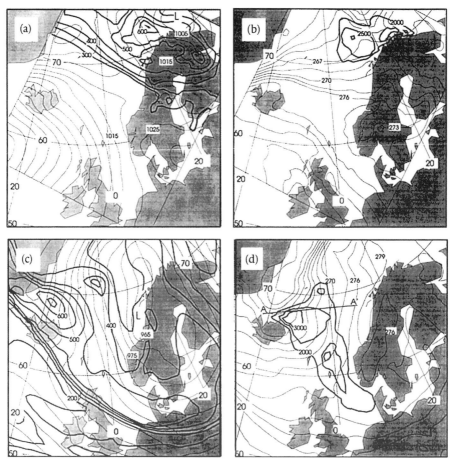

Figure 5.10. (a) Simulation (6 h) of MSLP (thin solid lines, contour step 5 hPa) and the pressure of the 2 PVU surface (thick solid lines, contour step 50 hPa), valid at 18 GMT 26 February 1987. (b) Simulation (6 h) of θ at 925 hPa (thin solid lines, contour step 3 K) and the height of the planetary boundary layer (thick solid lines contour step 500 m). The fields are valid at 1800 GMT 26 February 1987. (c) As (a) but valid at 1800 GMT 13 March 1992. (d) As (b) but valid at 1800 GMT 13 March (from Grønås and Kvamstø, 1995).

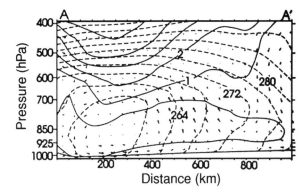

Figure 5.11. Cross-section along A–A′ in Figure 5.10d of θ_e (broken lines, contour interval 4 K), PV (1 PVU = 10^{-6} m^2 s^{-1} K kg^{-1}) and ageostrophic tangential flow (arrows) resulting from a 6 h simulation valid at 1800 GMT 13 March 1992. The length of an arrow corresponds to a 30 min trajectory (from Grønås and Kvamstø, 1995).

position of maximum PBL height and the ice edge off the coast of Greenland. According to studies of the convective PBL by Økland (1983), this height is far too large to be formed by heating from the ocean alone. Below the upper PV anomaly, static stability is generally low (Hoskins *et al.*, 1985) and the local maximum of the height of the convective layer is formed by the PV anomaly aloft. In this case, the PV anomaly moved towards the southeast, and the PBL was raised by the vertical motion caused by the so-called 'vacuum cleaner effect' of the approaching PV anomaly.

For the two last cases, in the region where polar lows developed, the minimum value of H was less than 1000 m, at the same time as the horizontal distance between where the tropopause was at its lowest and where the PBL top was at its highest was less than 200 km (see also Figure 5.11).

The earlier mentioned 'vacuum cleaner effect' is demonstrated in Figure 5.11, where cyclonic vorticity was created below the maximum updraught, and anticyclonic vorticity above. Up to 24 h into the simulation, the cyclogenesis seemed to follow a normal baroclinic development of a reversed shear flow, resulting in a mature warm core polar low. However, Grønås and Kvamstø (1995) demonstrated that, as is also demonstrated in all the simulations reviewed in this section, release of latent heat played a substantial role in obtaining the simulated strength of the polar low.

After 24 h of simulation the cold air had secluded the warm air and a well-developed polar low had formed, situated at about 61.5° N just west of the coast of Norway. The diameter of the low was about 350 km, the amplitude in surface pressure was 8 hPa and the temperature anomaly was nearly 7 K. The maximum surface wind generated by the polar low was 15 m s^{-1} (see Figure 5.12).

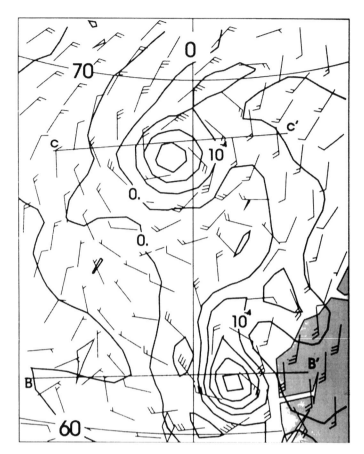

Figure 5.12. Simulation (24 h) of relative vorticity (solid lines, contour step 1×10^{-4} s^{-1}) and horizontal wind relative to the moving polar low (arrows) at 925 hPa. A full barb is 5 m s^{-1} and a half barb 2.5 m s^{-1}. Valid at 1200 GMT 14 March 1992 (from Grønås and Kvamstø, 1995).

The second disturbance seen on Figure 5.12 (centred at about 68° N, 1° W) does not appear in the analyses and satellite images. A further investigation of it is therefore not given. Whether it was real or not, it showed the model's ability to develop mesoscale lows when the large-scale conditions were favourable. There might be a delicate balance between different effects that determines whether mesoscale disturbances will develop into polar lows, even though everything seems to favour such development.

The simulations of Grønås and Kvamstø (1995) clearly showed their model's ability to develop mesoscale disturbances into polar lows in the way described when the conditions are favourable. Furthermore, the results of the simulations suggest a single parameter for prediction of the risk of polar low formation. This parameter is given by the height difference between the tropopause (they defined the tropopause as the 2 PVU surface) and the top of the convective PBL underneath. When this height difference is small, the tropopause is low and the static stability is small, conditions that both favour development of small-scale cyclones (the length scale of a cyclone is sometimes denoted $L \sim NH/f$, where N is the Brunt–Väisälä frequency indicating the static stability and H is the height of the tropopause).

Woetmann Nielsen's case of 13–16 October 1993

An unusually long-lasting polar low occurred in the Barents Sea/Norwegian Sea over the period 13–16 October 1993. A successful simulation of the case has been performed by Nielsen (1997).

Woetmann Nielsen used an experimental version of the operational forecasting model (Hirlam) from the Danish Meteorological Institute, details of which model are given in Woetmann Nielsen (1997).

The satellite image in Figure 5.13 gives a picture of the synoptic conditions at a relatively early stage of the development of the polar low. The

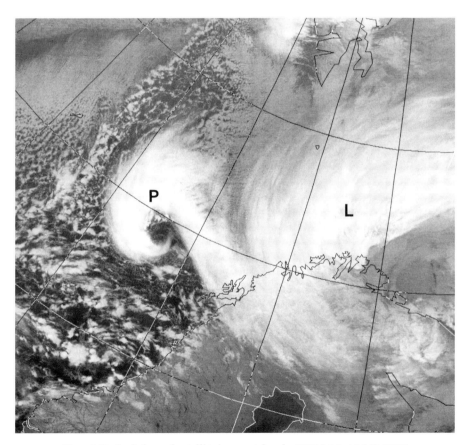

Figure 5.13. An infra-red satellite image, taken by NOAA 11 at 1349 GMT 14 October 1993. P indicates the polar low. (Image courtesy of the NERC Satellite Receiving Station, University of Dundee.)

comma-shaped cloud system (P in Figure 5.13) at about 70°N, 5°E belongs to the depression which developed into the polar low, while the larger-scale cloud vortex L further east belonged to the synoptic-scale cyclone over Northern Scandinavia. From the image we also see that the evolution of P occurred in a convectively-active air mass on the poleward side of the polar front.

Woetmann Nielsen's (1997) study was based on satellite cloud images, conventional observations and simulation products. Fortunately, P moved into a region around the Shetland Islands with abundant surface observations, enabling a comparison of the near-surface structure of the simulated low with that of the observed polar low.

Woetmann Nielsen (1997) started his simulations at 0000 GMT 14 October 1993, at a time when some development had already taken place. His simulations are in excellent agreement with subjective analyses, e.g. Figures 5.14a and b show how well the 48 h simulation corresponded with the surface analysis at 0000 GMT 16 October.

Figures 5.15a–d together with Figure 5.14a give an impression of the development of the disturbance into the polar low, while it moved equatorward, approximately parallel to the Norwegian coast at an average speed of 10 m s^{-1}. In the same period, the synoptic-scale cyclone moved slowly northeastward, and the two systems separated.

Figures 5.15a–c show the disturbance P in its deepening phase, during which time the surface trough amplified and a closed circulation formed. The mature stage of the polar low is shown in Figure 5.15d, while Figure 5.14a shows the polar low in its filling stage.

A remarkable change took place in the cloud structure from the early to the late phase of the polar low, as demonstrated by Figures 5.13 and 5.16. The relatively sharp cloud boundaries toward the cold air edge of P, shown in Figure 5.13, are typical of a baroclinic development (Carlson, 1991). In Figure 5.16 the polar low has lost its sharp cloud boundaries. The cloud system then consisted of individual cumulonimbus clouds, organized in an irregular spiral structure, typical of a mature or decaying polar low (Rasmussen *et al.*, 1992). At the time of Figure 5.15d, the simulated mesoscale low fulfilled the criteria for being classified as a polar low: it formed a closed mesoscale surface low (radius $\simeq$ 175 km) with near-surface winds of at least gale force (Figures 5.14a and 5.15d), and it developed in a cold air mass with 500 hPa temperatures below $-40\,°$C on the cold side of, and isolated from the polar front.

In order to discuss the evolution of the simulated polar low, Woetmann Nielsen (1997) used the generalised quasi-geostrophic tendency and ω equations, although the low had such a small scale that the quasi-geostrophic approximation becomes inappropriate.

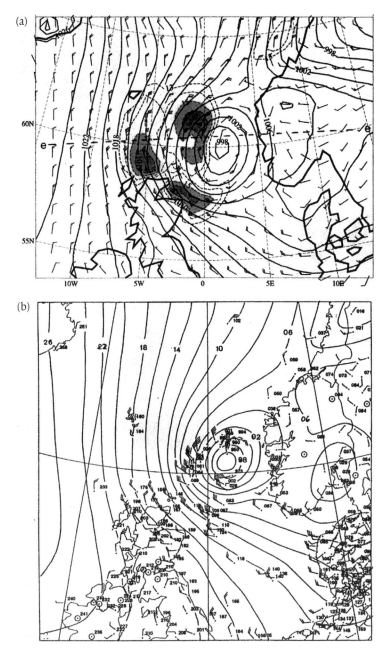

Figure 5.14. (a) 48 h forecast of mean sea level pressure and 10 m wind velocity (World Meteorological Organization (WMO) standard) valid at 0000 GMT October 16, 1993. Thin broken curves show wind speeds of 13, 15, 16 and 17 m s^{-1} are light, medium, and dark shaded, respectively. (b) Surface observations (WMO standard) together with a manual analysis of mean sea level pressure at 0000 GMT 16 October, 1993. Contour interval is 2 hPa (from Nielsen, 1997).

424　5 Numerical simulation

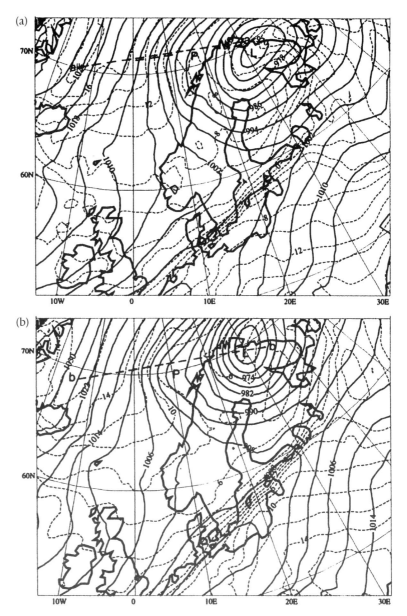

Figure 5.15. Forecasts of mean sea level pressure (solid curves) and 850 hPa temperature (broken curves) from the analysis time 0000 GMT 14 October 1993. (a) 6 h forecast valid at 0600 GMT 14 October, (b) 12 h forecast valid at 1200 GMT 14 October, (c) 24 h forecast valid at 0000 GMT 15 October, (d) 30 h forecast valid at 0600 GMT 15 October. Contour interval is 2 hPa for pressure and 2 °C for temperature (from Nielsen, 1997).

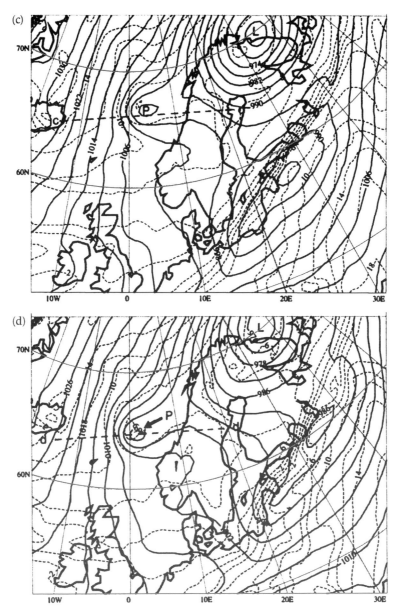

Figure 5.15 (cont.).

The disturbance P formed on the relatively warm and moist cyclonic shear side of a lower-tropospheric baroclinic zone (BB in Figure 5.17) over the Norwegian Sea. The cross-section in Figure 5.17 is taken along the dashed line a–a in Figure 5.18. The high PV area in the upper troposphere (denoted PV+) and the low-level area of large positive PV (BB) are shown in

426   5 Numerical simulation

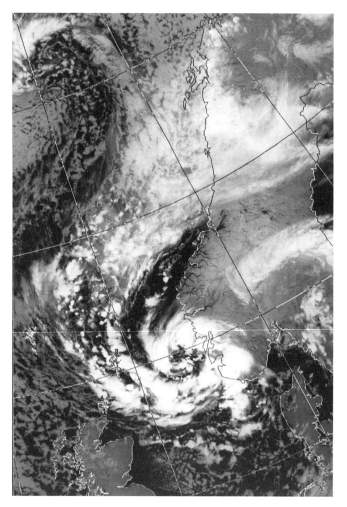

Figure 5.16. An infra-red satellite image taken by NOAA 11 at 0503 GMT 16 October 1993. (Image courtesy of the NERC Satellite Receiving Station, University of Dundee.)

Figure 5.17 to be separated by a layer of low static stability, where, in addition, middle-level warm air advection took place (see Nielsen's fig. 7c). The horizontal circulation induced by a PV anomaly will have the sign of the anomaly and tends to be centred on the anomaly and to have a penetration depth that increases with decreasing static stability (Hoskins *et al.*, 1985). From a PV perspective, then, it might be suggested that in the early phase of the evolution of P the upper-tropospheric positive PV anomaly interacted constructively with the low-level anomaly and thus contributed to the spin-up of the low-level disturbance.

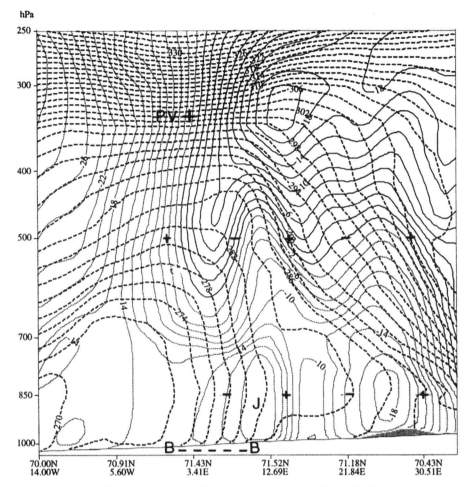

Figure 5.17. Vertical cross-section along the line a–a in Figure 5.18 approximately west–east at 71° N. Shown are 6 h forecasts valid at 0600 GMT 14 October, 1993, of equivalent potential temperature (broken curves) and wind speed normal to the cross-section (curves are solid and dotted for westerly and easterly winds, respectively). B–B marks a lower tropospheric baroclinic zone, while PV+ indicates an upper-level PV anomaly. Plus and minus symbols at 500 and 850 hPa denote positive and negative shear vorticity, and J marks the low-level northerly jet associated with B–B. Contour intervals are 2 K and 2 m s^{-1} (from Nielsen, 1997).

From a traditional baroclinic instability point of view, Nielsen (1997) found that there was a positive feedback between the adiabatic forcing terms (upper-level vorticity advection and lower-level temperature advection) in the ω equation.

The baroclinic instability just described occurred at a horizontal wavelength of only 350 km, which is considerably smaller than for an ordinary

428   5 Numerical simulation

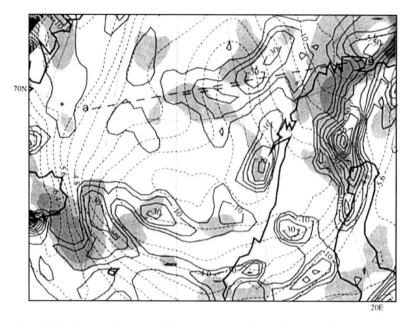

Figure 5.18. Six-hour forecast valid at 0600 GMT 14 October 1993 showing convergence at 850 (solid curves), divergence at 400 hPa (shading), and bulk dry static stability (broken curves) in the layer from 900 to 400 hPa. Contour interval for convergence is -10 U (1 U $= 10^{-6}$ s^{-1}), and for static stability it is 0.4×10^{-4} K Pa^{-1}. Light, medium and dark shading show 400 hPa divergence from 10–30, 30–60, and >60 U, respectively (from Nielsen, 1997).

extratropical cyclone evolution. However, the evolution into a polar low was in good agreement with the short-wave baroclinic theory of Blumen (1979), which lends support to the viewpoint that the initial evolution of P was by baroclinic instability.

Baroclinic instability was, however, not likely to be the only mechanism at work. A strong indication that other processes, after the initial phase, in fact dominated, is obtained from a parallel simulation with the latent heat of condensation set to zero. In that simulation, the disturbance P did not develop into a polar low (see Nielsen's fig. 11). Latent heat fluxes are obviously necessary for the polar low formation. The role of the forcing by heating was, in a way, opposite to that in a synoptic-scale extratropical cyclone. In the present case the forcing amplified the surface trough, while it was mainly the upper-level ridge that was amplified in a synoptic-scale cyclone. In both cases, the efficiency of the process increases with decreasing static stability. Furthermore, the simulated weak amplification of the upper-level ridge also indicated that after the early phase of the evolution the coupling between the upper-level vorticity advection and the low-level temperature advection was weak and thereby rapid growth by baroclinic instability was prevented.

CISK has been proposed as an important mechanism for polar low formation by E. Rasmussen (1989) and others. The environment of the disturbance P was nearly neutral to deep moist convection (Figure 5.17). Hence the amount of stored CAPE (Convective Available Potential Energy) was apparently small, which seems to exclude CISK as an important mechanism of evolution in this case.

Subsequent to Woetmann Nielsen's study, CAPE fields (for more detail, see Section 4.5) have been computed for this case using the NCEP/NCAR reanalysis data sets. These indicated that significant amounts of CAPE were present throughout these developments from the early stages over the northern part of the Norwegian Sea until the decline of the low in the North Sea. The discrepancy may be due to the convection scheme used in the model, which prevented the storage of significant amounts of CAPE. Provided that CAPE was present, CISK may have contributed to the development of the system together with the other factors mentioned by Woetman Nielsen.

Emanuel and Rotunno (1989) have suggested that the air–sea interaction mechanism may play an important role in at least some polar low formations. It is a necessary condition for air–sea interaction to operate properly that the conversion to kinetic energy is concentrated near the surface. Conditions for that have been formulated by Emanuel and Rotunno (1989) for the idealized case of an axisymmetric warm core vortex in a polar air mass with a stratification typical of the polar low environment. They concluded that disturbances of substantial amplitude apparently are necessary to initiate intensification by air–sea interaction. Furthermore, to prevent part of the generated kinetic energy from being used on spin-up of an upper-level vortex, which has no feedback on the surface fluxes, the radius at which maximum heating occurs must be small compared with the effective Rossby deformation radius. The disturbance P was highly non-axisymmetric, but for a qualitative discussion Woetmann Nielsen (1997) applied the results for an axisymmetric vortex referred to above.

The effective Rossby radius for the disturbance P was estimated to be about 150 km, which is near the scale of P, and accordingly, it is also near the radius at which maximum heating occurs. As mentioned above, these conditions seem to be unfavourable for air–sea interaction. It is, however, possible that air–sea interaction may work in conjunction with baroclinic instability and thereby intensify the low-level circulation to a level above that predetermined by baroclinic instability.

The following facts are in favour of air–sea interaction. First, the wind and the equivalent potential temperature patterns west of the centre of P (see Nielsen, 1997, his fig. 12) resembled those found in tropical hurricanes, in which air-sea interaction is the driving mechanism (Emanuel and Rotunno, 1989). Second, as in a hurricane, there is a close association between the low-level wind maximum and the vertically-coupled extremes of low-level convergence and upper-level divergence, showing the juxtaposition of the maximum heating and the low-level wind maximum. During the evolution of P, a band of maximum heating encircled its centre to form a spiral band in the mature and decaying phases of the polar low. In this process the distance from the centre to the heating maximum west of the polar low increased with time. The accompanying decrease in the low-level wind maximum, together with the spin-up of an upper-level vortex inside the heating maximum (see Nielsen, 1997, his fig. 12), indicated that air–sea interaction, due to the increase in the ratio of the maximum heating radius to the effective Rossby radius, no longer was able to sustain the high wind speeds.

Aarhus and Raustein's case of 27 February 1984

Research aircraft measurements within a polar low were taken during the Arctic Cyclone Expedition (ACE) in 1984. ACE was a collaborative effort organized and mainly conducted by NOAA. The polar low occurred on 27 February 1984 over the Norwegian Sea southeast of Jan Mayen. The results from ACE were presented by Shapiro *et al.* (1987). This polar low was not detected by conventional observations. To investigate whether it was possible to reconstruct the observed structure of the polar low by a numerical model, simulations were performed by Aarhus and Raustein (unpublished) at the University of Bergen, with the same model as used by Grønås and Kvamstø (1995), referred to earlier in this section. The only difference was that Aarhus and Raustein used horizontal grid lengths of 25 km and 12.5 km, with 30 σ layers in the vertical. The integration area for the 25 km simulation can be seen in fig. 1 of Aarhus and Raustein (1998). Simulations of the same case had been made earlier by Grønås *et al.* (1987a).

Initial conditions, as well as boundary conditions at subsequent 6 h intervals through the integration period, were defined through interpolation of the ECMWFs 1° in latitude and longitude global analyses to the model grid. The simulations were initialized at 1200 GMT 26 February.

The synoptic situation prior to the development of the polar low was characterized by a synoptic-scale low pressure system in the Greenland Sea producing an air stream from the west-southwest over Iceland. The low stayed almost stationary for a long period and was centred close to the ice edge east of Greenland. For the synoptic situation, see figs. 2 and 3 of Shapiro *et al.* (1987).

It appeared by 6 h into the simulation that the first signs of what later developed into the polar low, was a surface trough north of Iceland, which formed as a positive PV anomaly in the middle troposphere. This surged rapidly eastward and interacted positively with the considerable offshore low-level baroclinicity east of Greenland. After about 6 h of simulation, the synoptic-scale low started decaying and moved slowly eastwards, as the associated middle-level PV anomaly moved from the western to the eastern side of the low.

Six hours later, the mesoscale vortex that had evolved from the surface trough had reached the stage where it fulfilled most of the criteria for classification as a polar low. One criterion is that a polar low should be situated poleward of the main polar front. From the simulations, this front is not easily identified; however, satellite images (Figure 3.39) indicate that there was a polar front south of the cloud structures associated with the polar low (see figs. 27 and 28 of Shapiro *et al.*, 1987). From this, Aarhus and Raustein classified the mesoscale vortex simulated in the model as a polar low. The polar low developed further until about 27 h into the simulation. From then on the middle-level PV anomaly was positioned to the east of the polar low, and thus the conditions were no longer favourable for positive interaction.

The time of the ACE measurements corresponds approximately to 25 h into the simulation. It is convenient to compare the analyses based on these measurements with those after 24 h of simulation.

The simulated position of the polar low, of which a satellite image was shown on Figure 3.39, was about 200 km to the east of that observed by Shapiro *et al.* (1987). The simulation captured many of the essential features of the observed structure of the polar low as was seen through a comparison of the relative vorticity computed from the aircraft measurements with those computed from the simulations. The agreement was satisfactory, especially in the southwestern part of the polar low, although the simulated vorticity did not reach such high values as those computed from the measurements.

Both the observations and the simulations showed a warm air seclusion. As for temperature, the agreement was excellent. Also a reasonably good agreement between the vertical structure of equivalent potential temperature through the observed and simulated polar lows was noted.

The development from a weak trough in a low-level baroclinic environment northeast of Iceland was provided by the interaction between the surface trough and a middle-level trough (positive PV anomaly). Since the polar low seemed to dissipate after the middle-level trough had moved to its eastern side, Aarhus and Raustein concluded that in this case diabatic effects, such as air–sea interaction or CISK, did not play a decisive role in the later stages of the development. During the first 12 h, however, sensible heat fluxes of up to 900 W m^{-2} and latent heat fluxes up to 350 W m^{-2} from the sea surface were calculated, indicating that air–sea interaction played an important role, in that the heat flux contributed to easing the interaction between the upper and lower anomalies by lowering the static stability, and decreasing the horizontal scale of the developing disturbance.

Another interesting result found by Aarhus and Raustein was that as the middle-level trough moved eastward faster than the mesoscale low below, it caused the formation of new polar lows further downstream, called mesoscale downstream development by Ralph *et al.* (1994).

Blier's case from the Bering Sea/Gulf of Alaska over 10–12 January 1987

In the period 10–12 January 1987, a polar low formed in the southeastern Bering Sea, passed over the Aleutian Islands and developed further in the Gulf of Alaska. This development was simulated by Blier (1996). He employed the PSU-NCAR MM4 model, a hydrostatic and 3D high-resolution primitive equation model, using a grid spacing of 40 km. An overview of the model is given by Anthes and Warner (1978); an additional description of its formulation is found in Anthes *et al.* (1987). Initial conditions, as well as conditions at subsequent 12 h intervals through the integration period, were defined through an interpolation of the NMCs 2.5° latitude by 2.5° longitude global analyses to the model grid.

The synoptic conditions at the initial time of the simulation (0000 GMT 10 January 1987) were characterized by large-scale cyclonic flow and cold air passing from the sea ice and land over the Gulf of Alaska, with even colder air and a weak surface trough over the Bering Sea (Figure 5.19a). The 500 hPa analysis at the initial time (Figure 5.19b) shows a short-wave trough over the Bering Sea associated with the surface trough, the upper short-wave trough at 500 hPa being associated with temperatures below $-40\,°$C.

By hour 12 of the simulation the surface trough had sharpened and moved southeastward. At the same time, at 500 hPa, the initial short-wave trough over the Bering Sea had become more sharply defined as it moved to the southeast (see Blier, 1996, fig. 6).

To determine the relative importance of, and interaction between, the adiabatic dynamics and the surface heat fluxes and latent heat release, Blier (1996) conducted experiments in which the effects of these two processes were individually and in combination eliminated. He also performed two additional experiments, one without sea ice and one where the sea surface temperature

5.1 The Arctic 433

(a)
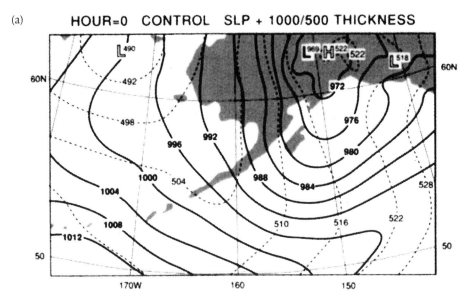

(b)
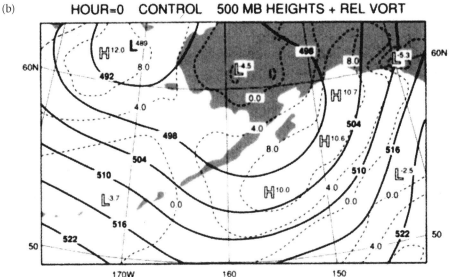

Figure 5.19. Control run initial model state (0000 GMT 10 January 1987). (a) Sea level pressure at 4 hPa intervals (solid lines) and 1000–500 hPa thickness (in dekametres) at 60 m intervals (broken lines), (b) 500 hPa heights at 60 m intervals (solid) and 500 hPa relative vorticity at 4×10^{-5} s^{-1} intervals (dashed). In both figures solid (outline) letters H and L correspond to maxima and minima, respectively, for fields contoured with solid (broken) lines (from Blier, 1996).

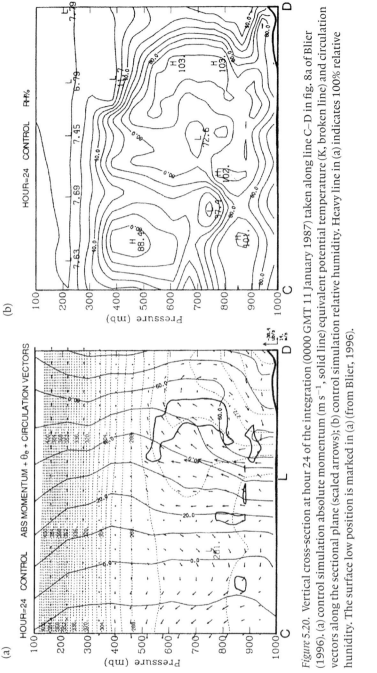

Figure 5.20. Vertical cross-section at hour 24 of the integration (0000 GMT 11 January 1987) taken along line C–D in fig. 8a of Blier (1996). (a) control simulation absolute momentum (m s^{-1}, solid line) equivalent potential temperature (K, broken line) and circulation vectors along the sectional plane (scaled arrows); (b) control simulation relative humidity. Heavy line in (a) indicates 100% relative humidity. The surface low position is marked in (a) (from Blier, 1996).

over all open ocean areas was raised by 4 K (see his table 2 for descriptions of the experiments). He was then able to demonstrate that the initial low-level troughing occurred as cold air was advected over the open water from off the Bering Sea ice and from the Alaskan land mass. Initial growth resulted from forcing from an upper-level, short-wave disturbance in an environment of low-static-stability tropospheric air. Rising motion was then organized along and to the south of the warm air boundary of the baroclinicity, and enhanced by the low static stability of the air. There was no indication of a relatively shallow convective boundary layer in this early growth period, in the region where the polar low formed.

By hour 24, the aforementioned small-scale low had deepened, and had moved to the Gulf of Alaska (see fig. 8a of Blier, 1996), in good accordance with subjective analyses. Its centre lay along the warm air boundary of a significant baroclinic zone (as seen in the 1000–500 hPa thickness field). There was little evidence of significant baroclinicity to the south of the low centre. Figure 5.20a indicates that baroclinicity was still confined to the lower troposphere north of the low centre. Along and near the warm edge of the baroclinic zone, the velocity vectors indicate the existence of a narrow and slightly sloping sheet of rapidly rising air. Figure 5.20b shows that the air was saturated or nearly so in the boundary layer, and from Figure 5.20a we see that there was a weak decrease in θ_e with height. It appeared, then, that moisture convergence occurred along the low-level baroclinic zone and was released through convection. Aloft, however, there appears to have been a larger-scale but weaker updraught area in a region that was closer to static neutrality, but within which the absolute momentum (M) surfaces inclined toward lower values of θ_e. Figure 5.20b shows the air to have been saturated within part of this region; as a result the presence of dry or moist symmetric instability (depending on whether or not the air was saturated) was indicated. The depicted circulation was consistent with the theoretical results of Emanuel (1985) and Thorpe and Emanuel (1985) regarding the effect of weak stability to slantwise convection on frontogenesis. The low had by hour 24 not yet reached the stage where it could be classified as a polar low, but it was still under development, so that by hour 30 it fulfilled the polar low criteria.

Blier's (1996) experiments indicated that after the previously described first stage of development, subsequent (and potentially much more substantial) amplification occurred as a result of an interaction between latent heat release (the required moisture was produced by surface latent heat fluxes) and the low-level baroclinicity. This mechanism is consistent with a CISK-like process (Holton, 1979), but also with air–sea interaction instability proposed by Emanuel and Rotunno (1989) to explain the growth of both hurricanes and polar lows. It does appear clear in Blier's case that concurrent surface fluxes of both sensible and latent heat are necessary for the development of the polar low. What should also be noted is that the polar low developed along a pre-existing low-level baroclinic zone along which the rising motion was organized in a fundamentally non-axisymmetric fashion.

Blier (1996) pointed out that there have been a few cases of polar air stream cyclogenesis that would appear to be difficult to explain within the framework he proposed. In these cyclogenetic events, a relatively intense, symmetric, small-scale disturbance forms in the absence of significant influence from either low-level baroclinicity or upper-level forcing. One example is the case studied by Albright *et al.* (1995), which is reviewed below.

Bresch et al.'s case of 7 March 1977

A polar low that developed over the western Bering Sea on 7 March 1977 and which tracked across St Paul Island, Alaska, was investigated by Bresch *et al.* (1997) using observations and fine-mesh model simulations. This polar low was not forecast, and it did not appear on weather charts until it struck the observing site on St Paul Island. A closer look at available satellite images might have given a hint at an earlier time that a polar low development was under way, and thus a better subjective forecast of it would have been possible.

Bresch *et al.* (1997) wished to investigate the development of the polar low by expanding the observational description and numerical simulation of Businger and Baik (1991) by including the period leading up to the polar low formation and by carrying out model experiments aimed at simulating the evolution and structure of the low and at elucidating the physical mechanisms responsible for its development.

A brief synoptic overview will be given, with reference to Figures 5.21 to 5.23. Prior to 0000 GMT 6 March 1977, an intense synoptic-scale low in the western Pacific moved northward to the tip of Kamchatka (for geographical names, see Figure 5.21), where it stalled and gradually filled. An offshoot of the low travelled eastward across the Bering Sea, leaving in its wake an elongated east–west trough that by 0000 GMT 6 March (Figure 5.22) stretched from the offshoot low, by then located in the Gulf of Alaska, to the parent low, still filling over Kamchatka. A front, analysed as an Arctic front, lay within the trough. Three disturbances were analysed along the front: a shallow wave labelled A, a low pressure system B, and a second wave C. This analysis was supported by satellite images (not shown). Immediately offshore from Kamchatka, north of C, a secondary cold front was analysed (Figure 5.22). At 1200 GMT 7 March the polar low was already well developed. The continued evolution is seen in Figures 5.23 and 5.24. A satellite image for 2054 GMT

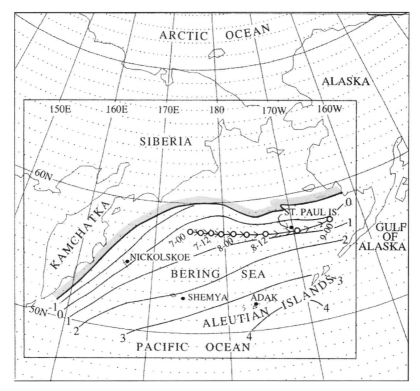

Figure 5.21. A map of the Bering Sea region including geographic names and sea surface temperature contours (°C). The location of the sea ice edge, obtained from visible and infra-red satellite imagery, is indicated by the shading. The track of the polar low is shown by the line with arrows and 6 h positions by the open circles, labelled at 12 h intervals. The inner rectangle represents the mesoscale model domain (from Bresch *et al.*, 1997).

8 March 1977 showed that the eye-like core was detached from the rest of the cloud mass and was centred just east of St. Paul Island (see Bresch *et al.*'s fig. 13).

The model used for the simulations was the non-hydrostatic version of the Pennsylvania State–NCAR mesoscale model (MM5) described by Dudhia (1993) and Grell *et al.* (1994). The model domain (Figure 5.21) was covered with a grid of 20 km horizontal spacing and 27 vertical levels. The NMC 2.5° gridded analyses were the basis for the initial fields.

Two sets of experiments were conducted (Table 5.1). The first set was initialized at 0000 GMT 6 March; the five simulations were designed to determine the model's ability to predict the development, beginning from a time well prior to the first appearance of the polar low, and to test the sensitivity of the development to model physics, ice edge location, and grid resolution. The second set

438   5 Numerical simulation

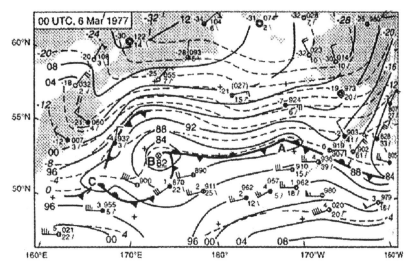

Figure 5.22. Surface analysis for 0000 GMT 6 March 1977. Isobars (hPa, solid lines) and isotherms (°C, broken lines). Station model and frontal symbols are conventional (from Bresch *et al.*, 1997).

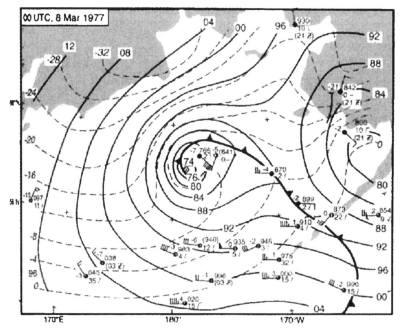

Figure 5.23. As in Figure 5.22 but for 0000 GMT 8 March 1977 (from Bresch *et al.*, 1997).

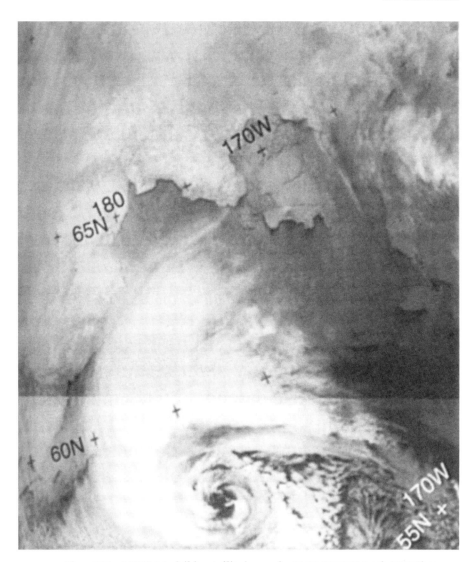

Figure 5.24. A NOAA 5 visible satellite image for 2143 GMT 7 March 1977 (from Bresch *et al.*, 1997).

of experiments was initialized at 0000 GMT 7 March during the polar low's incipient stage and run for 24 h; they were designed to test the sensitivity of the further polar low development to surface energy fluxes and latent heating.

The full-physics simulation FP6 was able to produce a realistic polar low. Figure 5.25a shows that after 12 h of simulation, the predicted leading surface low (corresponding to system B on Figure 5.22 and here labelled B′) had moved rapidly eastward and was *c.* 400 km east of the observed position. However, its overall structure and intensity were realistic and the surface trough/wind

440 5 Numerical simulation

Table 5.1. *Summary of numerical experiments. From Bresch et al. (1997)*

Name	Surface fluxes	Latent heat release	Duration (h)	Remarks
Set A: Simulations initialized at 0000 GMT 6 March 1977				
FP6	Yes	Yes	48	Full physics
NF	No	Yes	48	No flux
DF	No/yes	Yes	48	Delayed fluxes
AI	Yes	Yes	48	Adjusted ice edge
FG	Yes	Yes	45	Fine grid
Set B: Simulations initialized at 0000 GMT 7 March 1977				
FP7	Yes	Yes	24	Full physics
FDN	No	No	24	Fake dry/no flux
FDF	Yes	No	24	Fake dry/with flux
LH	No	Yes	24	Latent heating/no flux

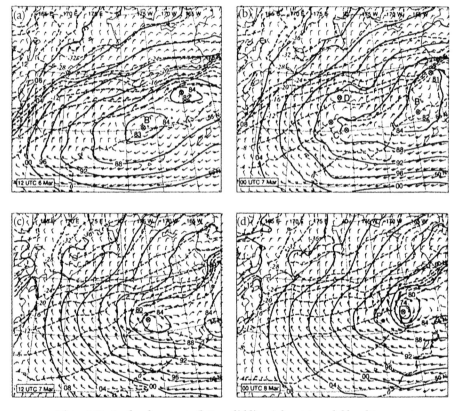

Figure 5.25. Sea level pressure (hPa, solid lines), lowest model level temperature (°C, broken lines), and lowest model level wind (full barb indicates 5 m s^{-1}) from model run FP6 at (a) 1200 GMT 6 March, (b) 0000 GMT 7 March, (c) 1200 GMT 7 March, (d) 0000 GMT 8 March 1977 (from Bresch *et al.*, 1997).

shear line was quite similar to the observed trough. By 0000 GMT 7 March (Figure 5.25b), the leading low had moved eastward to about 170° W, with the wind field continuing to suggest a double low structure, as was observed.

By hour 36 of the simulation (Figure 5.25c) a compact low with a central pressure of 980 hPa and winds over 25 m s^{-1} was located near 56° N, 176° W. This low developed from the southern circulation centre in Figure 5.25b, while in reality, it was the northern system D′ that evolved into the observed polar low. The development of the simulated low at a somewhat more southerly latitude, where the eastward progression of the surface cold front was more rapid than to the north, accounts for the excessive eastward movement of the low. By hour 48 of the simulation (Figure 5.25d), the simulated polar low had deepened to 974 hPa and winds on its west and southern sides reached values above 25 m s^{-1}.

Although the modelled polar low development was not an exact replica of the observed development, there is sufficient similarity between the two systems to justify treating the simulated polar low as a successful representation of the actual case. By comparing Figure 5.26 with Figure 5.27 we see that the

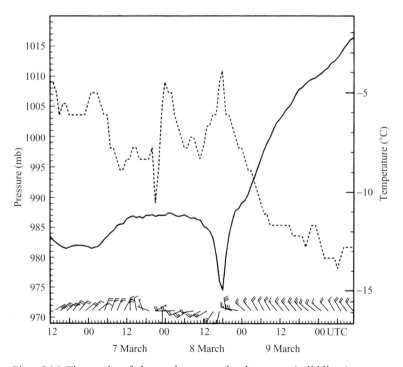

Figure 5.26. Time series of observed mean sea level pressure (solid lines), temperature (broken lines), and winds (conventional notation) at St Paul Island, Alaska (from Bresch *et al.*, 1997).

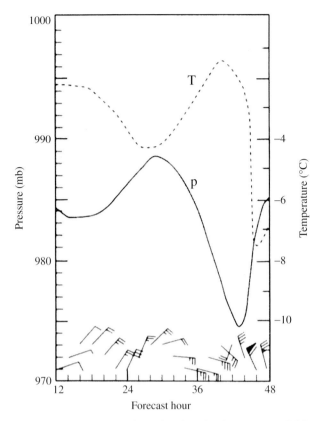

Figure 5.27. Time series of simulated sea level pressure (solid lines), temperature (broken lines) and winds (conventional notation) at point P in Figure 5.28d (from Bresch *et al.*, 1997).

simulated time series is quite similar to the observed, with the passage of the polar low signified by a sharp drop and even sharper rise in pressure, a temperature peak, and the onset of strong northwesterly winds.

From the analyses of the simulated surface fields, together with 850, 700 and 500 hPa fields (upper-level fields are not shown), it is clear that the polar low formed in a baroclinic environment. The baroclinic zone extended from the surface up to 700 hPa, and significant upper-level forcing, as judged by positive PV advection at 500 hPa, moved across the genesis region. Pronounced asymmetries were evident in the relative humidity and PV fields. This environment is quite unlike that of a hurricane, where thermal contrasts and upper-level forcing are weak, but is similar to that of other observed polar lows (e.g. Shapiro *et al.*, 1987; Nordeng, 1990; Grønås and Kvamstø, 1995; Douglas *et al.*, 1995).

Backward trajectory analysis showed that much of the air in the simulated polar low had its origin upstream of the low (figures not shown). Part of the current from upstream moved downstream of the region of formation and subsided into the boundary layer where it warmed and moistened. The now buoyant air was drawn into the developing polar low and ascended in the cloud shield to its east and north. Not all of the air emanated from upstream: some had an early history in the boundary layer well east of the polar low and also rose after being well modified over the open water. The anomalously high PV values in the interior of the polar low had two very different origins. In the vicinity of 850 hPa, the high values were produced by differential condensation heating in organized mesoscale convection; in the vicinity of 500 hPa, by advection of elevated values from far upstream and higher elevations.

Experiment NF (for an explanation, see Table 5.1) failed to produce even the semblance of a low in the cold air-stream behind the preceding synoptic-scale cyclone. In FP6, sensible heat fluxes exceeded 1000 W m^{-2} and latent heat fluxes were in excess of 250 W m^{-2} just south of the ice edge. The lack of development in NF demonstrates that surface fluxes were crucial for the development of the modelled polar low. Plots of 700 hPa vertical velocity and 500 hPa PV from the NF run (not shown) show evidence of upper-level forcing by positive PV advection over the polar low's genesis region. However, without the destabilization of the boundary layer by the surface fluxes, the upper-level forcing was unable to initiate development at the surface. The DF experiment yielded a low pressure system in nearly the same position as the polar low in FP6, but the low was weaker and more diffuse. It can be concluded that under the given initial conditions, surface fluxes are crucial to the development of a polar low, and that for the development to be intense it is necessary for the fluxes to be operative over a sustained period.

The results from the second set of experiments (see Table 5.1) may be summarized as (no figures shown): Experiment FP7 generated a polar low not very different from that in FP6. The warming of the low levels by surface sensible heat flow in experiment FDF and the warming of the low to mid-levels by latent heating in experiment LH contributed about equally to the development of the polar low, since polar lows of intermediate depth are developed in each of these experiments.

Structurally, the modelled polar low more closely resembled a small extratropical cyclone than an Arctic hurricane. In contrast to the circular symmetry of a hurricane, the polar low had pronounced asymmetric features in the wind, temperature, humidity and PV fields.

Viewing the experiments as a whole, Bresch *et al.* (1997) concluded that in order to form a polar low of the observed type and intensity, it is essential to have a substantial modification of the Arctic boundary layer, both before and during the development. Also essential is latent heating in clouds, at least some of which entails deep upright or slantwise mesoscale convection. The presence in the area beforehand of a baroclinic zone raises the question of whether baroclinic instability may also be an essential ingredient. Certainly, baroclinic energy conversion furthered the development in this case, as witnessed by the ascent of warm air and descent of cold air. But baroclinic growth was able to proceed only when the atmosphere possessed reduced stability at low levels.

Let us consider the polar low development within the PV paradigm of cyclogenesis. The required upper-level PV anomaly clearly existed. However, when the lower layers were highly stable, as in some of the experiments, the downward penetration of the upper-level anomaly was too weak to excite surface development. On the other hand, in the experiment with reduced stability in the lower atmosphere, the downward penetration was sufficient to induce a low-level thermal anomaly that through mutual interaction and phase-locking with the upper disturbance was able to form a significant low-level circulation and create a low-level PV anomaly by condensational heating in organized convection. The PV so created then added its contribution to the low-level circulation, further intensifying the low.

Bresch *et al.* (1997) rounded off their conclusion as follows:

> The fact that ideas commonly applied to mid-latitude cyclogenesis are also applicable to the development of polar lows suggests that the latter can be regarded in many instances as small, primarily Arctic counterparts of the larger mid-latitude systems, particularly the systems that form over warm waters offshore of cold continents in the winter half-year. Inasmuch as typical polar lows possess many physical characteristics in common with other extratropical cyclones, they are perhaps better viewed as part of a broad spectrum of ocean cyclones that differ in such respects as degree of baroclinicity, strength of upper-level forcing, tropospheric stability, amount of latent heat release, and the magnitude and arrangement of the surface energy fluxes, rather than as a truly distinct phenomenon. The polar low described here occupies the part of the spectrum in which the low-level baroclinicity and upper-level forcing are weak to moderate, a considerable depth of the troposphere is neutrally or only slightly stable to upright or slantwise ascent, and surface fluxes are of crucial importance in the immediate heating and moistening of the atmosphere and the promotion of latent heat release in deep convection.

Albright *et al.*'s case

During 8 and 9 December 1988 a rare occurrence of a polar low development over an ice-free region in eastern Hudson Bay took place. Roch *et al.*

(1991) described the results of numerical simulations of the low. They were able to demonstrate the importance of the open water to the formation of the low. More recently, Albright *et al.* (1995) made simulations of the same case, in order to expand the results of Roch *et al.* (1991). They used the same model as Blier (1996), described earlier in this section.

A synoptic background relevant to the development is shown in Figure 5.28, which depicts surface and 500 hPa charts at 0000 GMT 8 December prior to the polar low event (Figures 5.28a, and c) and at 1200 GMT 9 December (Figures 5.28b, and d) close to the time of maximum development. At the earlier hour, a broad current of Arctic air with upstream surface temperatures in the range $-25\,°C$ to $-30\,°C$ was flowing southeastward across Hudson Bay (Figure 5.28a). Aloft (Figure 5.28c) a series of short-wave troughs encircled a low centred over Baffin Island. Of special significance is the short-wave trough on the northwest shore of Hudson Bay. During the subsequent 36 h this trough amplified, forming a closed low over southern Hudson Bay (Figure 5.28d). As it did so, a small intense surface low appeared over open water adjacent to the east shore of the bay (Figure 5.28b).

The model experiments were run on a 20 km horizontal grid with 28 σ levels in the vertical, starting at 1200 GMT 8 December. The initial analyses were basically obtained from coarse NMC gridded data reanalysed to the mesoscale grid with the use of available surface and upper-air data. Water temperatures were also supplied by NMC analyses. Ice cover was according to the analyses of Environment Canada's Ice Centre. The first eight experiments were carried out on the 113×113 grid domain, while the final two experiments were conducted on the larger 113×155 grid domain, both centred in Hudson Bay. Time-dependent lateral boundary conditions were obtained from 12-hourly NMC analyses. A summary of the 10 experiments appears in Table 5.2.

The simulations were initialised at 1200 GMT 8 December; **initial** surface and 500 hPa charts are shown in Figures 5.29a, and c. Simulated 30 h surface and 500 hPa fields for the control simulation (HB1, full physics, observed ice cover) are shown in Figures 5.29b, and d. The 30 h simulated pressure field (Figure 5.29b) exhibited a general rise in pressure over ice-covered portions of Hudson Bay and the appearance of a small, intense cyclone over the ice-free region. The depth of the low is similar to that of the observed low 6 h earlier (Figure 5.28b), while its location is somewhat further south. The initial surface chart in the ice-covered experiment (HB2) featured the expected lower temperatures in eastern Hudson Bay resulting from the removal of the warm anomaly. Without this anomaly, and in the absence of subsequent surface energy fluxes of appreciable size, no surface low formed during the 30 h simulation (see fig. 5b of Albright *et al.*, 1995).

446   5 Numerical simulation

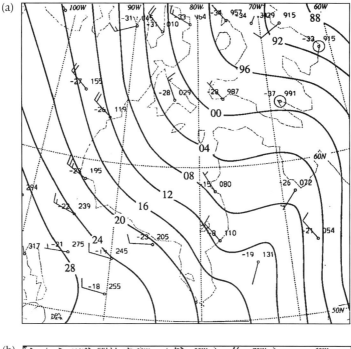

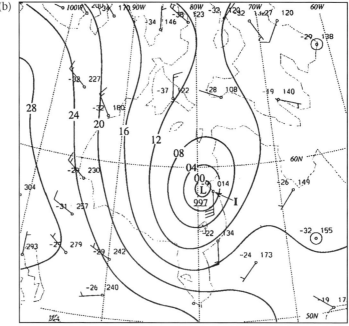

Figure 5.28. NMC surface and 500 hPa analyses for 0000 GMT 8 December and 1200 GMT 9 December 1988. (a) Surface and (c) 500 hPa for 0000 GMT 8 December; (b) and (d) same for 1200 GMT 9 December. The I in (b) denotes the Inukjuak station. The heavy broken line in (c) indicates the short-wave trough on the northwest shore of Hudson Bay (from Albright *et al*., 1995).

(c)

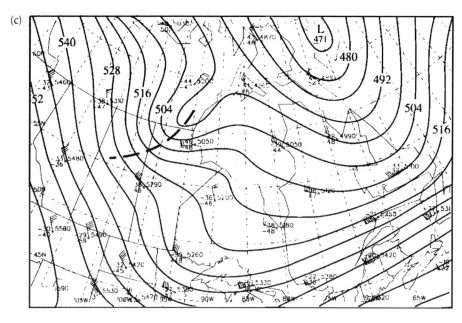

(d)

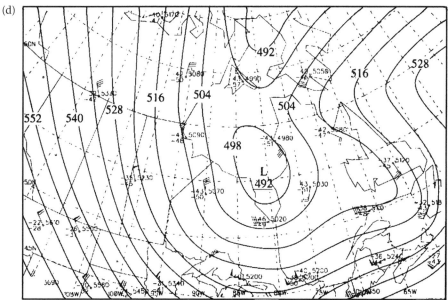

Figure 5.28 (cont.).

The initial 500 hPa chart (Figure 5.29c) shows that the upper-level, short-wave trough amplified in the 12 h period between 0000 and 1200 GMT 8 December (cf. Figure 5.28a). The amplification continued during the forecast period so that by 30 h (Figure 5.29d) a closed low had appeared over southern Hudson Bay, very similar to that observed at 1200 GMT 9 December

Table 5.2. *The experiments carried out by Albright et al. (1995)*

Name	30 h central pressure (hPa)	SST (°C)	Size of ice-free region (km^2)	Sensible flux	Latent heat flux	Latent heating	Brief description
HB1	997	0 °C	200000	Yes	Yes	Yes	Control, observed ice cover
HB2	–	–	0	Yes	Yes	Yes	Complete ice cover
HB3	997	0 °C	200000	Yes	Yes	Yes	No initial PBL modification
HB4	1003	0 °C	200000	Limited	Limited	Yes	Limited surface fluxes
HB5	1006	0 °C	200000	Yes	No	Yes	No latent heat flux
HB6	1006	0 °C	200000	Yes	Yes	No	No latent heating
HB7	981	8 °C	200000	Yes	Yes	Yes	Enhanced SST
HB8	996	8 °C	200000	Limited	Limited	Yes	Enhanced SST, limited fluxes
HB9	995	0 °C	700000	Yes	Yes	Yes	Enhanced ice-free region
HB10	1001	0 °C	700000	Yes	Yes	Yes	Enlarged sea, straight ice edge

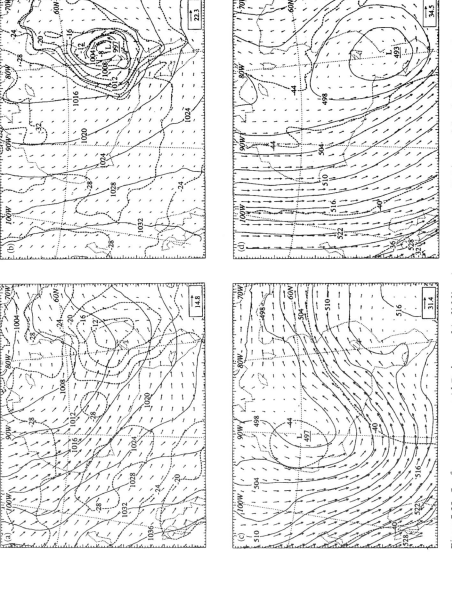

Figure 5.29. Surface pressure at 4 hPa interval (solid lines), temperature at 4 °C interval (broken lines) and vector wind in m s^{-1} at (a) initial hour, 1200 GMT 8 December, and (b) 30 h forecast time, 1800 GMT 9 December 1988, for control run HB1 and corresponding 500 hPa charts (c) and (d) of geopotential height at 3 dam interval, temperature at 4 °C interval, and winds (m s^{-1}). Wind velocity scales at lower right of each map (from Albright *et al.*, 1995).

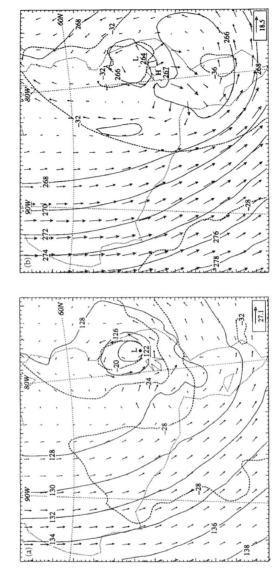

Figure 5.30. Upper-level charts for 1800 GMT 9 December 1988. Solid lines indicate geopotential height at 2 dam interval; dashed lines, temperature at 4 °C interval; vector winds in m s^{-1}, scale at lower right. (a) 850 hPa. L denotes low centre and dot indicates position of soundings in Figure 5.31. (b) 700 hPa (from Albright *et al.*, 1995).

(Figure 5.28d). The 500 hPa chart for the ice-covered case (see fig. 5c of Albright *et al.*, 1995) resembled closely that of the control run.

Intermediate-level 850 and 700 hPa charts (Figures 5.30a, and b) reveal that the upper and lower vortices were discrete entities. The surface vortex weakened upwards along the vertical so that by 700 hPa only a weak height minimum and temperature maximum remained.

In Figure 5.31 are shown the initial sounding, together with the soundings for the 30 h simulations from both HB1 and HB2 (the dot in Figure 5.30a indicates the location of the sounding). Comparison of the two soundings after 30 h illustrates well the heating of the lower atmosphere produced by the fluxes. Average sensible and latent heat fluxes in HB1 between 24 and 30 h in the simulation reached values of 1315 and 305 W m^{-2}, respectively, near the upwind ice edge (see fig. 9 of Albright *et al.*, 1995).

Examination of the accumulated precipitation during the integration period reveals that 100% of it stemmed from grid-resolvable precipitation rather than from subgrid-scale precipitation. Figure 5.31 provides an example (solid curve) of conditions at a grid point where resolvable-scale precipitation was occurring. The sounding is moist neutral up to 570 hPa, and the vertical motion was of the order of 1 m s^{-1}. These conditions are regarded as an indication of deep convection of mesoscale dimensions.

The traces of central pressure of the surface low are displayed in Figure 5.32 for experiments HB1 and HB3–10. For HB2, in which a low failed to develop, the curve depicts the change of surface pressure at the position of the low in experiment HB1. The curve is regarded as providing background pressures against which the deficit of pressure in the control experiment, and to a considerable degree in the other experiments, can be gauged.

Limiting the buildup of the fluxes with wind speed (HB4) resulted in a weaker system and a delay in the start of the deepening. It is clear that the flux feedback mechanism had a substantial effect on the intensification. Shutting off the surface latent heat flux (HB5) allowed only an exceedingly weak low to form. Evidently, the sensible heat flux alone was unable to produce a significant low. An almost identical weak development took place when the latent heating was withheld (HB6). Thus, to produce a vigorous polar low, surface sensible and latent heat fluxes must operate jointly to form deep precipitating clouds and thereby heat a substantial depth of the atmosphere.

A very intense polar low formed in the experiment with enhanced SST (HB7) for which hurricane force winds occurred in the central region of the polar low (see Albright *et al.*, 1995, fig. 12). However, the results presented here suggest that only under extreme conditions is it possible to create winds of hurricane force.

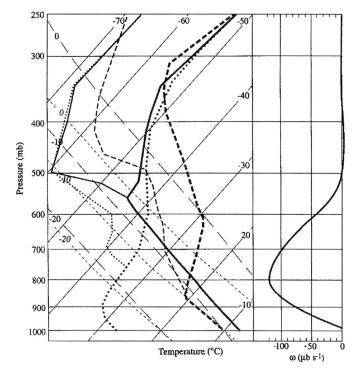

Figure 5.31. Soundings at the location of L in Figure 5.30. Initial temperature (heavy broken line) and dewpoint (medium broken line) at 1200 GMT 8 December for HB1. Final temperature (heavy solid) and dewpoint (medium solid) for 1800 GMT 9 December for HB1. Final temperature (large dots) and dewpoint (small dots) for HB2. Isotherms: thin slanting solid lines. Dry adiabats: pecked lines. Saturated adiabats: thin broken lines. Curve at right depicts omega profile in μb s^{-1} (from Albright *et al.*, 1995).

Experiment HB8 featured limited fluxes in the enhanced SST case and showed an even more dramatic effect than in the experiments (HB1 and HB4) with the true SSTs. The lowest pressure was only 994.5 hPa, and the maximum wind speed was well below hurricane force. These results lend support to the Emanuel and Rotunno (1989) ASII/WISHE theory of polar low intensification (see Section 4.5).

Displacement of the eastern boundary of the open water to a position further east (HB9) had only a slight impact on the polar low development. Experiment HB10 with a straight ice edge showed, by comparing with HB9 in Figure 5.32, that the polar low development was very sensitive to the shape of the ice edge.

From the foregoing results, it was concluded that the polar low was a moderately deep, vertical system, that owed its intensification to the fluxes of sensible

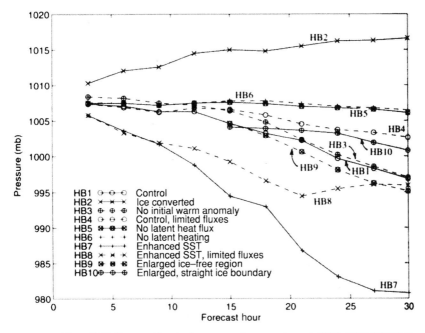

Figure 5.32. Traces of central pressure for experiments HB1, HB3-HB10. See text for further explanation (from Albright *et al.*, 1995).

and latent heat from the small region of open water in eastern Hudson Bay. Fluxes occurring during the development phase were sufficient to account for the observed intensity of the system.

In the absence of surface moisture fluxes, precipitation production was negligible so that only the effects of surface heating were felt. In view of the much larger sensible heat flux (3–4 times the latent heat flux) it is surprising that sensible heating had so little impact when acting by itself.

The central role of deep organized convection in the development of the type of polar low under study is clearly demonstrated. This is in accordance with theoretical work of, amongst others, Økland (1987). Fluxes of both sensible and latent heat are needed to produce deep convection, which in turn provides the condensational heating required for substantial intensification to occur.

The experiments demonstrating the dependence of flux intensity on wind speed lend strong support to the hypothesis of Emanuel and Rotunno (1989) that ASII/WISHE is a major factor in at least some types of polar low development.

Perhaps a surprising finding was the strong sensitivity of the polar low development to the shape of the upstream ice edge. The fact that many polar lows seem to have their source in the wedge-shaped region of open water to the west of Spitsbergen and extending southwards to Bear Island supports the hypothesis that the configuration of the ice edge can have an important influence on polar low formation.

Most studies on the formation of polar lows have implicated baroclinicity in their formation either through the occurrence of shallow Arctic fronts (e.g. Reed and Duncan, 1987; Shapiro *et al.*, 1987), Arctic outbreaks (e.g. Grønås and Kvamstø, 1995), or in connection with upper-level mobile troughs that give rise to upward motions and attendant large-scale destabilization and condensation heating.

Nordeng, Blier, and Mailhot *et al.* (see the cases elsewhere in this section) envisage polar low formation to entail two phases: a development phase in which baroclinic forcing initiates the spin-up, and a convective phase that maintains and strengthens the development.

The question arises as to whether baroclinic influences were important elements in the initiation of this polar low. First, with respect to boundary layer fronts, it seems questionable to characterize the elliptical zone of temperature contrast along the border of open water (Figure 5.28a) as a front that relates to the problem of baroclinic instability. Indeed, in the experiment (HB3) that was started without this feature, the zone quickly formed as a consequence of the intense sensible heat flux. It seems more logical to attribute the development to the fluxes, and the associated latent heat release in deep convection, than to a baroclinic instability of the shallow, flux-induced thermal gradient. Second, as regards forcing by an upper-level trough, such a feature did indeed exist (Figures 5.28c and 5.29c), and gave the appearance of constituting a potentially significant triggering mechanism. However, the same feature existed in the ice-covered experiment, in which a surface low failed to develop. Clearly, the upper-level baroclinic forcing was by itself insufficient to initiate a surface low.

In view of the evidence that baroclinic forcing by the upper-level trough had at most a small effect on the surface development, it must be asked what feature of the upper-level flow was conducive to the deepening, since it seems unlikely that the polar low and the cold low aloft appeared together by coincidence. Albright *et al.* suggested that the general cooling aloft accompanying the advance of the upper-level cold low (Figure 5.29d), was the key factor. This cooling, coupled with low-level warming and moistening by surface fluxes, resulted in a drastic destabilization of the layer between the surface and 550 hPa (Figure 5.31). From the point of view of the sensitivity experiments, the destabilization gave rise to substantial condensational heating in deep mesoscale convection, and in terms of PV diagnostics, it allowed a large, mid-level, positive PV anomaly (not shown) to induce cyclonic flow all the way to the surface.

In the inferred absence, or near absence, of baroclinic forcing as a triggering and organising mechanism, there is evidence that the configuration of the ice edge played an important role in the development.

Mailhot *et al.*'s case for the Labrador Sea

Mailhot *et al.* (1996) made a successful simulation of a polar low development that took place during 10–12 January 1989 in the Labrador Sea. The synoptic situation prior to the polar low development was characterized by a synoptic-scale low pressure system that moved from the coast of Labrador eastward to the east of Greenland from 9 to 10 January, and from then on remained relatively stationary east of Greenland (Figure 5.33a). A low-level trough developed over Hudson Bay on 10 January, and the associated 500 hPa short wave trough moved over the Labrador Sea on 11 January, with a dome of air with temperatures as low as $-47\,°C$ at 500 hPa (Figure 5.33b). This resulted in a weakening of the atmospheric static stability, favourable to the development of deep convection. At the time of the development of the polar low, the only observations available in the Labrador Sea were from a drifting buoy, so that the routine analyses contained no information of a polar low. The rapid evolution of a polar low was, however, well revealed from satellite images (Figure 5.34).

The model used for the simulations was a mesoscale (15 km) version of the Canadian regional finite-element (RFE) model. A summary of the model is given in Table 1 in Mailhot *et al.* (1996). The initial conditions for the 36 h simulations were obtained by interpolation from the archived operational objective analyses of the Canadian Meteorological Centre. Details of the objective analysis procedure can be found in Mitchell *et al.* (1990). The horizontal resolution of the data in the analysis was $2° \times 2°$ longitude/latitude, equating to about 200 km over the area of development.

The development of the polar low as simulated by the model is shown in Figure 5.35. By hour 24 (Figure 5.35a), a surface trough was present and intensifying over the Labrador Sea. The polar low appeared to develop along a convergence line extending eastwards, along $60°\,N$ from the ice margin east of the coast of Labrador. The convergence line formed as the northwesterly flow off Baffin Island and Davis Strait met the westerly flow to the east of Labrador. The development took place in an environment with conditions of reversed shear flow.

After 30 h of simulation, the surface trough had deepened to 990.5 hPa. The surface pressure values over the area were in agreement with the last buoy report of 989.5 hPa.

The deepening of the surface trough continued through the following 6 h (Figure 5.35c) and from about hour 24 it attained the strength necessary to be

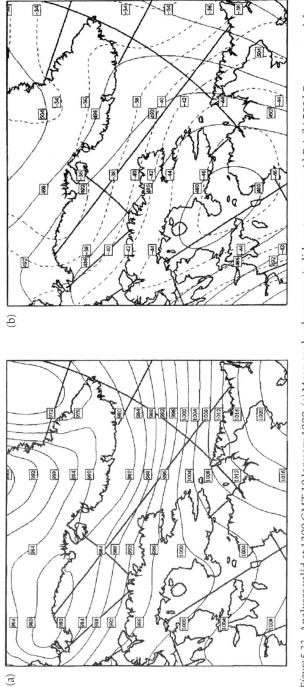

Figure 5.33. Analyses valid at 1200 GMT 10 January 1989. (a) Mean sea level pressure (solid contours every 4 hPa); (b) 500 hPa geopotential height (solid every 6 dam) and temperature (broken lines every 2 °C) (from Mailhot et al., 1996).

Figure 5.34. An infra-red satellite image for 1509 GMT 11 January 1989, showing a polar low over the Labrador Sea.

classified as a polar low. Except for the buoy, there were hardly any conventional observations in the area to confirm this development, but it was in good agreement with satellite data. Sensitivity experiments indicated that the polar low development was sensitive to SST, to the accurate location of the ice edge, and to the horizontal resolution.

From the simulation, the conclusion is that the polar low development was associated with an intense outbreak of cold Arctic air to the rear of the synoptic-scale cyclone that had become stationary east of Greenland. The polar low was associated with a short-wave trough aloft and a dome of very cold air favourable to the outbreak of deep convection.

As in the case simulated by Blier (1996) (also reviewed in this section), the development might be viewed as taking part in two stages, denoted as the early

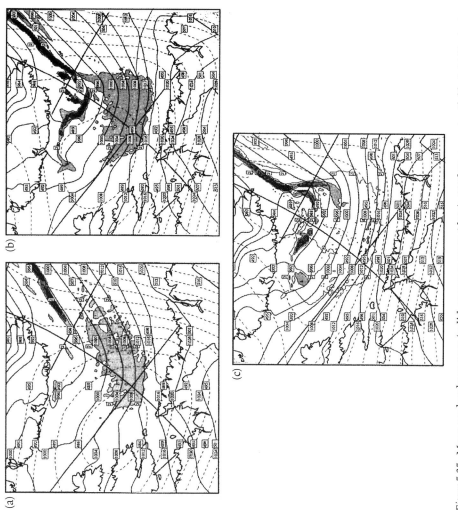

Figure 5.35. Mean sea level pressure (solid contours every 4 hPa), 500 hPa geopotential height (dashed contours 3 dam) and precipitation rate (shaded contours, every 2×10^{-7} m s^{-1}) from (a) a 24 h forecast valid at 1200 11 January 1989, (b) a 30 h forecast valid at 1800 GMT 11 January, (c) a 36 h forecast valid at 0000 GMT 12 January (from Mailhot *et al.*, 1996).

and the mature stages by Mailhot *et al.* (1996). During the early stage, baroclinic development took place in a basic state of reversed shear flow. The Arctic boundary layer was rapidly modified by strong sea surface fluxes from the relatively warm waters of the Labrador Sea. Simulated maximum values of the heat fluxes were 1400 W m^{-2} for sensible and nearly 600 W m^{-2} for latent heat fluxes, at the exit of Hudson Bay and east of Labrador. A stratocumulus layer then formed at the top of the rising convective boundary later. From a PV perspective, we might say that the interaction between an approaching upper-level PV anomaly and low-level baroclinicity associated with the Arctic front was favoured by a deep convective boundary layer; this could be the mechanism that spun-up the polar low, in agreement with the proposal of Grønås and Kvamstø (1995). The importance of a deep convective boundary layer and a low-level thermal anomaly in the triggering of the polar low was confirmed by a sensitivity experiment without surface heating from the ocean. The mature stage set in as the cold air dome (and hence the PV anomaly) approached the storm. Deep convection then broke out, and the polar low intensifed rapidly. This seems to confirm that some CISK-like process was present. The structure of the polar low was characterized by a warm core below 500 hPa due to the combined effects of warm air seclusion by cold air wrapping around the surface low, and diabatic heating due to convection. Sensitivity experiments without condensation processes produced a weak and shallow trough, and indicated that latent heat release from organized deep convection contributed to the major part of the rapid deepening in the mature stage. A sensitivity experiment without evaporation from the sea surface confirmed the major role of deep convection for the development of this polar low, and the importance of surface evaporation as the feeding mechanism for condensation processes.

5.1.3 Discussion

The polar low simulations presented above were quite successful and the studies give valuable insight into the formation and evolution of the vortices. For the synoptic conditions we stress that an outbreak of Arctic air mass from the sea ice (or continents) is a necessary condition for small-scale polar lows to form. In all the presented cases, upper-air forcing was significant for the development. We find that the normal synoptic preconditions are strong upper-air forcing, which gives low static stability in the upper air, a convective boundary layer at the surface formed by the heating of Arctic air masses over the ocean, and a shallow layer of Arctic air with strong stratification in between. For strong polar lows to develop we believe that the low-level convection must break through the layer of Arctic air locally to form an area with deeper convection.

5.2 The Antarctic

5.2.1 Simulations with high-resolution mesoscale models

Although, as described above, there have been a number of studies of Arctic polar lows using numerical models, only a few numerical investigations of polar mesocyclones exist for the Antarctic. In several studies, mesoscale limited area numerical models have been used in the form of idealized simulations of mesocyclone developments in the Antarctic, but only a few simulations of real mesocyclone cases can be found in the literature. In contrast to observational studies, numerical model studies are the only tool for investigating the importance of different physical processes for the development of polar mesocyclones.

The numerical simulations for the investigation of Antarctic mesocyclones can be classified into three categories:

Instability analyses

Instability analyses for Antarctic mesocyclones using a complex, non-linear mesoscale 3D grid point model based on the primitive equations were carried out by Münzenberg-St Denis and Schilling (1996).

Realistic simulations using high-resolution 3D mesoscale models

For the Antarctic, the NORLAM (Norwegian Limited Area model) has been used in several case studies of mesocyclones. These studies comprise summertime cases of near-coastal developments in the Weddell Sea area (Engels and Heinemann, 1996), summertime cases over the Ross and Amundsen Seas (Lieder and Heinemann, 1999), and wintertime cases near the sea ice edge and over the open water for both areas (Heinemann, 1998; Lieder and Heinemann, 1998, 1999).

Idealized simulations and process studies

Complex 3D mesoscale limited area numerical models have been used in several studies in the form of idealized simulations of mesocyclogenesis (Parish, 1992; Gallée and Schayes, 1994; Gallée, 1995; Heinemann, 1996a, 1997; Klein and Heinemann, 2001). The idea was to use a full physics high-resolution mesoscale model, but highly simplified initial and boundary conditions with respect to the atmospheric forcing (state of rest, or analytical fields). Process studies using complex 3D mesoscale models have the goal of investigating the impact of different physical processes and their parameterizations on the development of observed or simulated mesocyclones (e.g. Sardie and Warner, 1985; Nordeng *et al.*, 1989; Heinemann, 1998).

Description of the NORLAM model used at the Meteorological
Institute, University of Bonn (MIUB)

The model used at the MIUB for study of Antarctic mesocyclones is the NORLAM model, but without the Sundqvist extension (see Section 5.1.2). NORLAM is a hydrostatic primitive equation model and is generally run with 121×97 horizontal grid points and 30 vertical sigma levels for the Antarctic studies. Forecast runs can be carried out in a nested mode; a first run is made with a 50 km grid (LAM50) with the model domain centred over the area of interest, with initial fields and lateral boundaries provided by ECMWF analyses (with resolutions varying from $1.125° \times 1.125°$ to $2.5° \times 2.5°$). A model domain with 25 km horizontal grid distance (LAM25) can be nested in the integration area of LAM50. Topography data were taken from data sets with a resolution of $\frac{1}{6}° \times \frac{1}{6}°$ and $\frac{1}{12}° \times \frac{1}{12}°$, which resulted in relatively realistic coastal slopes near the Antarctic continent on the scale of the model grids. Different roughness lengths z_0 have been defined for the surface types of water, sea ice and continental ice (including ice shelf surfaces). For z_0 over water surfaces, a Charnock relation (Charnock, 1955) was used with a lower limit of 1.5×10^{-5} m, z_0 values for sea ice and continental ice were taken as 10^{-2} m and 10^{-4} m, respectively. This choice for z_0 is in accordance with experimental results for roughness lengths over the Antarctic ice surfaces (Heinemann, 1989; King and Anderson, 1994). The SST data from ECMWF analyses were used, but areas with unrealistically low SST were set to a lower limit of $-1\,°C$. The sea ice coverage was taken from daily maps produced by the Pelicon project (Heygster *et al.*, 1996), from SSM/I data processed at MIUB or from the weekly maps produced by the Navy-NOAA Joint Ice Center. Summaries of the model characteristics and references for the main parameterizations can be found in Heinemann (1997, 1998). Figure 5.36 shows the model topography and the sea ice front for the model domains of LAM50 and LAM25 used for the Weddell Sea studies for summertime conditions. Triangles indicate stations making radiosonde ascents and collecting surface observations.

Description of the MAR (Modèle Atmosphérique Régional) model
used at the Université Catholique de Louvain (UCL)

Idealized simulations of Antarctic mesocyclones have been performed by Gallée (1995, 1996) at the Université Catholique de Louvain (UCL). The model used for these studies was the limited area model MAR. MAR is a hydrostatic primitive equation model and was run with 160×160 horizontal grid points and 18 vertical sigma levels. The horizontal grid resolution was 10 km and covered the Terra Nova Bay area of the Antarctic. Figure 5.37 shows the model domain and the topography of the whole Antarctic (topography data

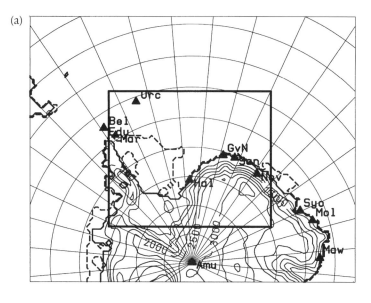

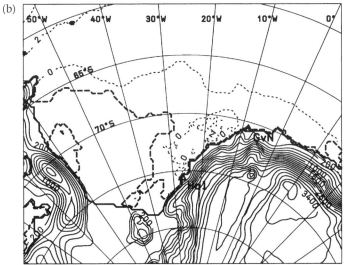

Figure 5.36. (a) Model domain of the NORLAM used for the Weddell Sea studies with 50 km resolution (LAM50) with the Antarctic topography (contour interval 500 m); the coastline (thick line) includes the ice shelf areas, the sea ice front (thick dashed line), and the model domain of LAM25 (box) are shown; triangles mark Antarctic research stations with radiosonde soundings (from Heinemann, 1997). (b) As (a), but for the model domain of LAM25 (topography with contour interval 200 m). Sea surface temperature (SST) (thin dashed isolines every 2 °C) is also shown; triangles mark the stations Halley (Hal) and Georg-von-Neumayer (GvN) (from Engels and Heinemann, 1996).

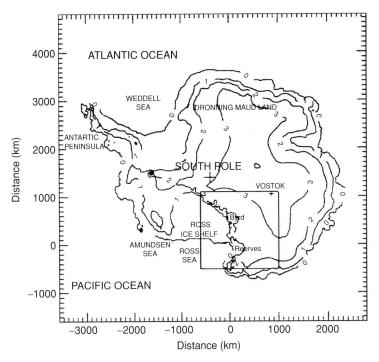

Figure 5.37. Model domain of the MAR Limited Area Model with 10 km resolution (inset box) with the Antarctic topography (contour interval 1000 m) (from Gallée, 1995).

from Drewry (1983). The model was either initialized by an atmosphere at rest or a constant geostrophic wind and an ideal thermodynamic sounding being horizontally homogeneous for the whole model domain. The SSTs were taken as the freezing point of sea water ($-2\,°C$). The sea ice coverage was prescribed as being ice-free and 95% for the simulations in Gallée (1995) and varying from 70% to 100% in Gallée (1996). A summary of the model characteristics and references for the main parameterizations can be found in Gallée and Schayes (1994), Gallée (1995) and Gallée (1996).

5.2.2 Idealized simulations: Weddell Sea area

Mesocyclone developments near the continent

The mesocyclone developments near the continent cannot be separated from the katabatic wind system developing over the continental ice slope. The interaction between mesocyclogenesis, katabatic wind and synoptic forcing has been studied at MIUB using the NORLAM model in a nested mode. As with forecast runs in a nested mode, a first run was made for a 50 km grid (LAM50) with the model domain centred over the Weddell Sea region

(Figure 5.36a). For the 25 km run (LAM25), forecasts of LAM50 were taken as initial fields (6 h forecast) and time-dependent boundary conditions during the simulation (see Figure 5.36b for the topography and domain of LAM25).

Heinemann (1997) used an atmosphere at rest with horizontally homogeneous fields of temperature and humidity as the initial state for his idealized simulations. Two different sets of vertical profiles of temperature and humidity were chosen representing a wintertime and a summertime situation, respectively. Two types of sea ice coverage were investigated for the wintertime situation: Weddell Sea without sea ice (WNI case) and Weddell Sea completely ice-covered (WI case). For a summertime case with realistic sea ice conditions (SRI case), the fields of surface parameters (sea ice conditions, SST) corresponded to the situation observed during February 1990 (see Figure 5.36). Apart from the different forcing by the daily cycle of the solar radiation, the main difference between these simulations was that strong convection occurred over the ocean for the WNI case, while almost no air–sea temperature gradient existed initially for the SRI case.

While the near-surface wind over the continent reached a quasi-stationary state (with a superposition of a daily course for the SRI case) for both cases after about 24 h, the development of mesoscale circulation structures could be found close to the coast for the SRI and WNI cases for longer simulation times. In the WNI case a couple of mesoscale circulations with horizontal scales of 100–150 km developed along the coast after two days of simulation. These developments took place at the baroclinic zone at the coast (horizontal temperature gradient about 8 K (100 km)$^{-1}$ at 975 hPa), where surface fluxes of sensible heat exceeded values of 120 W m^{-2} over a large area. The coastal convergence zone of the katabatic wind at low levels was associated with a belt of high cyclonic vorticity along the coast. The mechanism for the generation of low-level cyclonic vortices by the low-level convergence (associated with vortex stretching) and thermally-induced low-level circulations has similarities to the development mechanism of mesocyclones at the sea ice front investigated in the study of Heinemann (1996a) by discussing the vorticity budget terms. The main difference from developments close to the sea ice front, where the topography plays no role, is the fact that for the WNI case the structure of the low-level convergence associated with the katabatic wind and the temperature distribution were influenced by the topography.

For the SRI case, the first indications of the development of mesoscale cyclonic vortices could be found close to the coast after 72 h simulation time, where a baroclinic shear zone with a horizontal temperature gradient of up to 6 K (100 km)$^{-1}$ had been built up at 975 hPa. Surface fluxes of sensible and latent heat had values of up to 130 and 80 W m^{-2}, respectively. At a simulation time of 120 h, the mesoscale circulations had intensified and areas with a cyclonic vorticity exceeding -1.4×10^{-4} s^{-1} could be found at 975 hPa. These cyclonic mesoscale circulation systems could also be identified at the 900 hPa level, but vanished at levels above 850 hPa, thus being confined to the lowest 2 km. A significant influence of the sea ice distribution on the low-level temperature gradient and wind field was present and led to a quite different structure of the flow field over the ocean area compared with the WNI case.

The simulations described above were carried out without initial synoptic forcing (the atmosphere being at rest for the initial state) and highlighted the importance of the topographic and surface forcing for coastal mesocyclone developments, which occurred after at least two days of simulation time. Observational studies, however, have indicated a much faster intensification process and the importance of the synoptic forcing for mesocyclogenesis. Idealized simulations and process studies for different synoptic forcings are shown by Klein and Heinemann (2001). The initial synoptic situation was prescribed analytically assuming geostrophic balance for the initial fields, corresponding to that of the case of an observed near-coastal mesocyclone during summer. A large-scale cyclone was triggering off-ice flow over the eastern Weddell Sea (Figure 5.38); the real mesocyclone formed northeast of Halley. Two different

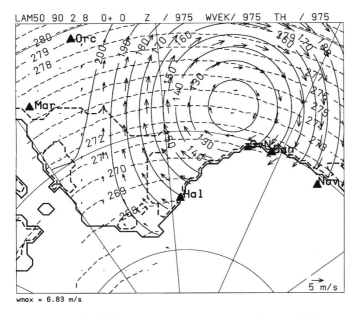

Figure 5.38. Initial fields for the CYCW experiment (see text) of the potential temperature (solid lines, isolines every 1 K), temperature (broken lines, isolines every 1 K), geopotential (every 10 gpm) and wind vectors at 975 hPa for a sub-area of the 25 km domain. A scaling vector is shown in the bottom right-hand corner. The sea ice front is marked by the thick broken line (from Klein, 1996).

466 5 Numerical simulation

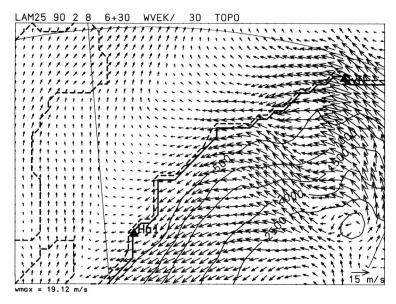

Figure 5.39. Topography (contour interval 500 m) and wind vectors at the lowest sigma level (30 m above the surface) for a sub-area of the LAM25 domain after 30 h simulation for the CYCW experiment (from Klein, 1996).

situations were investigated. The case CYCW corresponded to an initial temperature field with a meridional gradient and low-level temperatures being close to the SSTs (Figure 5.38); in the case CYCC the temperatures were 8 °C colder (but still typical for summertime situations in the Antarctic, since cold air outbreaks with air temperatures between −10° and −15 °C lie in a realistic temperature range). The results of these two simulations after a 30 h simulation of LAM25 were quite different and stressed the role of the boundary layer processes in the development of mesocyclones. Figures 5.39 and 5.40 show the wind vectors at about 30 m above the surface. In both runs the katabatic wind system was well developed and extended over the coast due to the synoptic forcing. While only a weak, large-scale trough developed along the coast in the CYCW run (Figure 5.39), a pronounced mesocyclone with wind speeds of more than 10 m s^{-1} developed in the CYCC run (Figure 5.40). The mesocyclone was confined to the lowest 2 km and had a warm core. The generation of this warm core was associated with a seclusion process, as is common with many observed mesocyclones in the Antarctic (see Section 3.2). Process studies show that the combined effect of the divergence of the turbulent fluxes of sensible and latent heat is essential for mesocyclone formation. But it can also be shown, that the generation of vorticity over the slope by the synoptically-enhanced katabatic flow is another important component for the initial formation of mesocyclones, which initially form close to the coast.

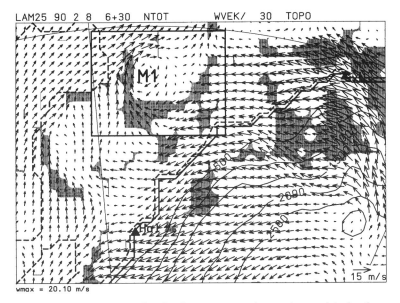

Figure 5.40. As Figure 5.39 but for the CYCC experiment. Areas with cloud coverage exceeding 95% are shaded; M1, centre of the mesocyclone (from Klein, 1996).

The governing mechanisms for the generation of the cyclonic vorticity maxima near the coast can be studied by means of the vorticity equation in pressure coordinates given in Eqn. (5.1):

$$\frac{\partial \zeta}{\partial t} = -\vec{v}_h \bullet \overline{\nabla}_p(\zeta + f) - \omega \frac{\partial \zeta}{\partial p} - (\zeta + f)\overline{\nabla} \bullet \vec{v}_h - \vec{k} \bullet (\overline{\nabla}_p \omega \times \frac{\partial \vec{v}_h}{\partial p}) + \text{DF}$$

LOCAL   HADV   VADV   DIV   TT

(5.1)

In this notation the local tendency of the relative vorticity (LOCAL) is given by the horizontal and vertical advection of relative vorticity (HADV and VADV), the divergence term (DIV), the tilting term (TT), and the local change of relative vorticity due to the vertical turbulent diffusion (DF). All values are computed as 10 min means at each grid point of the model. The main production term in the case of a synoptically-induced flow from the continent over the coast line is DIV, i.e. vortex stretching is the dominant process for the generation of pronounced cyclonic vorticity maxima near the coast.

The effects of the three forcing processes are summarized schematically in Figure 5.41. Without upper-level forcing, the situation corresponds to the SRI case of Heinemann (1997) and only the katabatic wind system and a belt of cyclonic vorticity at the coast is simulated during the first 30 h (the mesocyclones found by Heinemann (1997) developed after 72 h). An additional forcing by the CYCW cyclone led to increased vorticity production by vortex stretching, and

468   5 Numerical simulation

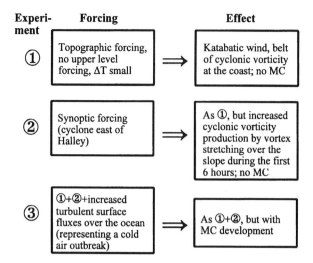

Figure 5.41. A diagram indicating the forcings within the experiments and the resultant effects.

the katabatic wind flow began to extend over the coastline, but again no mesocyclone developed. In contrast, a mesocyclone was simulated, if all forcing processes were combined.

The results of a series of sensitivity experiments led to the conclusion that the genesis of mesocyclones near the coast of the eastern Weddell Sea occurred in two phases: (1) barotropic generation of vorticity by synoptically-enforced vortex stretching near the continental slope, and (2) intensification by diabatic processes over the open water. The main difference to the simulations of Heinemann (1997) was that the mesocyclone development occurred with a more realistic spin-up rate of the circulation, i.e. within 24 h.

Mesocyclone developments at the sea ice edge

A large fraction of polar mesoscale cyclogeneses are observed near the sea ice front in both hemispheres during winter. In many cases, development of mesocyclones takes place at the low-level atmospheric baroclinic zone or boundary layer front (BLF) associated with the sea ice front, and shallow vortices with fronts and low-level jets are generated. The sea ice edge represents an area with strong gradients of surface temperature, roughness and albedo, and therefore also large differences in the energy budget of the boundary layer above the sea ice and the open water exist. A pronounced BLF (representing a low-level baroclinic environment where polar lows can develop) forms when the flow direction in the boundary layer is almost parallel to the sea ice front (Shapiro and Fedor, 1989; Shaw et al., 1991). Frontogenetic processes lead to the formation of a pronounced cloud band parallel to the sea ice front in many cases, which represents the signature of the BLF on satellite images. With the onset of substantial off-ice wind components connected with cold air outbreaks or cyclogenesis, the BLF can be advected over the open ocean. Polar lows observed near the sea ice front associated with a BLF as described by Fett (1989b) may be classified as the Arctic/Antarctic front type (shallow but strong baroclinicity in conjunction with intense surface energy fluxes) according to the classification of Businger and Reed (1989a).

The surface forcing on atmospheric mesocyclone developments in the transition zone close to the sea ice front was investigated by Heinemann (1996a) by means of idealized numerical experiments. Again the NORLAM model was used in a nested mode with the LAM50 domain centred over the Weddell Sea region. The initial state was an atmosphere at rest with horizontally homogeneous fields of temperature and humidity, and vertical profiles representative of the winter. The fields of surface parameters (sea ice conditions, SST) were taken for April 1989. While the katabatic wind system and the flow field close to the continent reached a quasi-stationary state after about 24 h of simulation, a considerable development took place close to the sea ice front. Due to the differential heating in this region, a pressure fall took place in a belt parallel to the sea ice edge, inducing a geostrophic wind along the sea ice front and an ageostrophic component directed towards the open water. The result was a belt of increased wind speed along the ice edge with a convergence zone over the open water after 24 h. The temperature gradient normal to the ice front reached values of about 5 K $(50 \text{ km})^{-1}$ at 975 hPa at that time, i.e. a distinct BLF had formed at low levels.

Mesoscale cyclonic vortices with scales of about 200 km were generated (Figure 5.42). Since the geostrophic wind component associated with the thermally-induced pressure gradient normal to the sea ice front followed the shape of the ice edge, the bay-like structures of the sea ice front were the preferred genesis areas for these cyclonic vortices. Above the most pronounced bay, an almost closed cyclonic circulation had already developed after 24 h, and after 48 h of simulation a further intensification of the circulation systems close to the surface had taken place. The highest wind speeds at 975 hPa, in the vicinity of the vortices were around 8–10 m s^{-1}.

The strongest increase in cyclonic vorticity associated with the vortices near the three bay-like structures of the sea ice front took place between 24 and 36 h. The vortices could be detected as pronounced maxima of cyclonic vorticity reaching values of up to -2.0×10^{-4} s^{-1}. Associated with the movement of the BLF over the ocean and the intensification of the wind speed after 48 h, values for the turbulent surface flux of sensible heat of *c.* 140 W m^{-2} were observed in the areas of cold air advection over the ocean. The structure of the field of

470 5 Numerical simulation

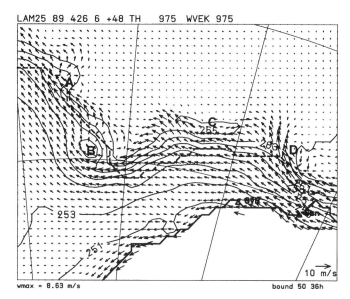

Figure 5.42. Potential temperature (solid lines, isolines every 2 °C) and wind vectors at 975 hPa for a sub-area of the 25 km domain after 48 h. A scaling vector is shown in the lower right corner. Small triangles indicate positions of GvN and Sanae (San), individual mesocyclones are marked by A to D (from Heinemann, 1996a).

turbulent surface flux of latent heat was similar, but values of only $c.$ 40 W m^{-2} were reached.

Since the mesocyclones were generated by surface inhomogeneities, it was expected that their vertical extent would be relatively shallow. Figure 5.43 displays wind vectors and potential temperature at 925 hPa after 48 h of simulation. As a visualization of the clouds at different levels, areas exceeding 90% relative humidity at 925 hPa (light shading) and between 800 and 850 hPa (darker shading) are displayed as well. Already, after 24 h of simulation the BLF was represented by a cloud band consisting of low clouds parallel to the ice edge at 24 h simulation time. This area coincided with the convergence zone close to the surface and a region with strong upward motion. After 48 h simulation time the cyclonic circulation of B (indicated in Figure 5.42) could clearly be seen (Figure 5.43). Low-level clouds were advected around the centre of B and at the convergence zone higher clouds formed. The intensification of the vorticity centres at higher levels was evident from the flow field at 925 hPa after 72 h (Figure 5.44). At that time, the cloud band of the BLF had broadened; the associated convergence line was also present at 925 hPa. No signal of mesoscale circulations connected with the sea ice front was found at the 700 hPa level even after 72 h of simulation, where only a weak large-scale cyclonic flow around the Antarctic continent could be observed. The vertical extent of the mesocyclones

5.2 The Antarctic 471

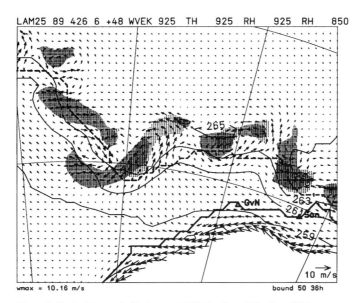

Figure 5.43. As Figure 5.42, but at 925 hPa after 48 h. In addition, areas of relative humidities exceeding 90% at 925 hPa (light shading, representing low clouds), at 850 and at 800 hPa (thicker shading, representing higher clouds) are shown (from Heinemann, 1996a).

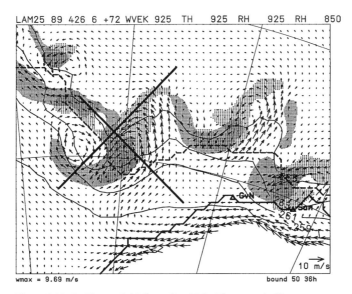

Figure 5.44. As Figure 5.43, but after 72 h. The cross indicates cross-sections referred to in Heinemann (1996a).

was confined to the lowest 2 km, which corresponded to that of several observational studies.

For the simulations described above, Heinemann (1996a) examined the formation of the vorticity centres by means of the vorticity budget terms computed in the numerical model. It could be shown that the wind field convergence at lower levels and its divergence at higher levels represented an important mechanism for the production of cyclonic and anticyclonic vorticity, respectively. Simulations without topography showed that the mesocyclone developments at the ice edge were only slightly coupled to the katabatic wind system.

Although being an idealized simulation study, the structures of the simulated mesocyclones resembled those of observed systems or realistic simulations in many respects. The simulated mesocyclones that developed near the sea ice edge were shallow vortices with fronts and jets. Observational evidence and the idealized numerical simulations show that the structure of the sea ice front can influence the position of local mesocyclogenesis, since the development of the most intense mesocyclones occur over the most pronounced bay in the sea ice front. This observed tendency for mesoscale cyclogenesis over ice-free areas or bays in the sea ice was also found by numerical studies of Roch *et al.* (1991), who performed sensitivity studies for a polar low development over the Hudson Bay. The idealized simulations starting from a state of rest cannot show the upper-level forcing that may occur in reality. Fett (1989b) concluded that mesocyclones at a BLF can develop into intense polar lows if they are supported by a cold 500 hPa trough. These conditions are met e.g. for the Antarctic case shown by Heinemann (1996c) and the Arctic polar low presented by Nordeng (1990). An example of the lack of upper-level forcing is the Arctic case near Jan Mayen discussed by Douglas *et al.* (1995), where the vortex remained shallow and did not develop into a very intense polar low (see also Section 3.1).

5.2.3 Idealized simulations: Terra Nova Bay

A formation of mesoscale circulations similar to those discussed above is shown by the idealized simulations of Gallée and Schayes (1994), Gallée (1995) and Gallée (1996) for the area of Terra Nova Bay (see Figure 5.37). The goal of the paper by Gallée and Schayes (1994) was to study katabatic winds over the Ross Sea region for a wintertime situation with full sea ice coverage, but also the initiation of a weak mesocyclone activity was found. The topographic gradients in this area are large compared to the Weddell Sea region. The simulations using MAR showed the development of a mesoscale cyclone with a weak pressure anomaly of 0.4 hPa in a convergence zone of the katabatic wind.

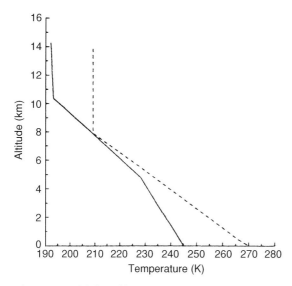

Figure 5.45. Initial profiles of potential temperature for the wintertime simulations (solid line) and summertime simulations (broken line) for the idealized MAR simulations (see text) (from Gallée, 1995).

Gallée (1995) used MAR for the same region, but for a summer situation without sea ice. The initial temperature profile is shown in Figure 5.45 as the dashed line. The potential temperature was −4°C at the surface and increased linearly with height (2.5 K km^{-1}) in the lowest 8 km. Thus the air temperatures near the surface were slightly below the SST. The initial air humidity was set to 40% for all levels in order to avoid the generation of high clouds. Starting from a state of rest, two different katabatic airstreams developed over the continental slope and formed a convergence zone over the Terra Nova Bay (Figure 5.46). After 48 h of simulation this confluence zone extended between the points M1 and M2 and was associated with a boundary layer front. M1 and M2 were also centres in the pressure anomaly field, since a rise in surface pressure took place in the areas where the katabatic wind propagated over the Ross Sea (Figure 5.47) and piled up cold air in the coastal zone. M1 undertook a considerable intensification during the next 24 h of simulation and developed into a mesoscale vortex of about 200 km diameter and a pressure anomaly of about 2 hPa formed after 72 h (Figure 5.48). The specific topographic structure around the Terra Nova Bay, which causes a cyclonic shear of two different katabatic airstreams, played an important role in the mesocyclogenesis. A comparable bay-like structure of the topography was not present in the area of the eastern Weddell Sea. Sensitivity experiments by Gallée (1995) showed that no significant mesocyclonic activity was simulated for the case of 95% sea ice coverage. In contrast, a synoptic forcing by a weak southerly geostrophic wind of

474   5 Numerical simulation

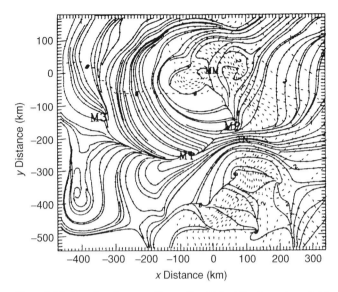

Figure 5.46. Simulated streamlines after 48 h at the lowest model level (10 m above the surface) for a sub-area of the MAR domain for the summertime case with an ice-free ocean. Topography (contour interval 200 m) is shown as broken lines (from Gallée, 1995).

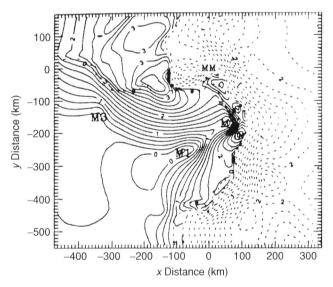

Figure 5.47. As Figure 5.46 but for the simulated surface pressure anomaly (difference to the initial pressure, shown only for sea level, contour interval 0.2 hPa) after 48 h (from Gallée, 1995).

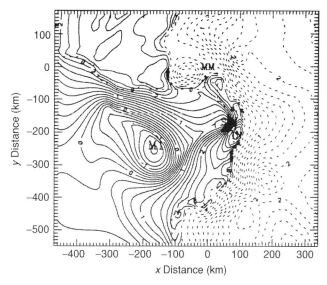

Figure 5.48. As Figure 5.47 but after 72 h (from Gallée, 1995).

5 m s^{-1} led to an intensification of the mesocyclones M1 and M2. Gallée (1995) attributed this effect to the process of vortex stretching of the air mass flowing from the Antarctic interior to the Ross Sea. The sensitivity studies of the diabatic forcing showed that the release of latent heat had little effect on the mesocyclone development (although snow precipitation was simulated), but that the increase of the temperature contrast by the sensible heat input from the ocean was important. Thus these mesocyclone developments at the BLF could be characterized as mainly dry baroclinic processes, which can be enforced by the large-scale flow field.

The wintertime studies of Gallée and Schayes (1994) were refined by Gallée (1996) by considering different lead fractions in the sea ice covering the Ross Sea, thereby investigating the effect of the strong heat input from small areas of open water during winter. The initial temperature profile is shown in Figure 5.45 as the solid line, which represents a mean sounding at Vostok for July (extrapolated to the surface). The potential temperature was −28 °C at the surface and was well below the SST (−2 °C). Starting from a state of rest, a strong katabatic airstream developed after 72 h of simulation at the northern side of the valley near Terra Nova Bay (marked TN) and propagated over the sea ice (Figure 5.49). A lead fraction of 20% was assumed for this simulation. The katabatic airstream at Terra Nova Bay was associated with a cold core (potential temperatures down to −37 °C) in the lowest 500 m. The strength of this cold core and the maximum wind speed increased with increasing lead fraction, reflecting the increase of the baroclinic forcing near the coast caused by the heat input of the leads. North and south of the katabatic stream, mesoscale

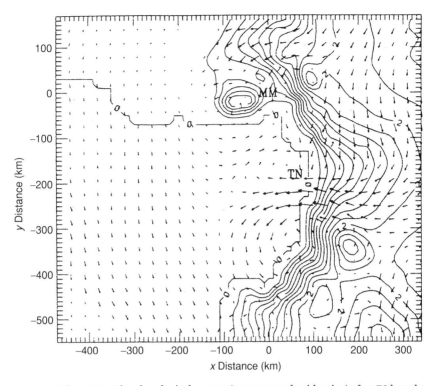

Figure 5.49. Simulated wind vectors (every second grid point) after 72 h at the lowest model level (10 m above the surface) for a sub-area of the MAR domain for the wintertime case with ice coverage of 20% for the ocean areas. Topography (contour interval 200 m) is shown as solid lines. TN and MM mark Terra Nova Bay and McMurdo Station, respectively (from Gallée, 1996).

circulations could be seen being anticyclonic and cyclonic, respectively. These circulations were associated with pressure anomalies of +0.4 and −1.6 hPa, the latter value being valid for the mesocyclone south of TN. Taking a lead fraction of 0%, this anomaly was only −0.8 hPa, while it was −2.2 hPa for 30% lead fraction. This well-marked mesocyclone had a horizontal scale of about 200 km and a depth of about 2.5 km. This is comparable to that of the summertime mesocyclone M1 simulated in Gallée (1995), which was found to develop over the Ross Sea, while the mesocyclone M2 (simulated at the same location as the wintertime mesocyclone) turned out to be much weaker. As for the summertime situation, the katabatic winds played an important role in the mesocyclogenesis.

5.2.4 Simulations of realistic cases: summertime cases

Halley vortices

The motivation for this modelling effort was the results of a satellite-based mesocyclone climatology for the Weddell Sea area by Heinemann (1990), which showed that during summertime, mesocyclones were frequently detected along the coast between the research stations Halley (Hal, 75.60° S, 26.78° W) and Georg-von-Neumayer (GvN, 70.67° S, 8.37° W). Compared to wintertime polar lows in Arctic regions, summertime coastal mesocyclones over the eastern part of the Weddell Sea are considerably weaker and shallower in general. But, for mesocyclones occurring in the coastal area between Halley and GvN, the steep coastal slope of the Antarctic orography plays an important role in the dynamical development, as was already demonstrated by the idealized simulations in Section 5.2.2. This type of topographically-forced mesocyclogenesis has also been observed in other regions of the Antarctic (Carrasco and Bromwich, 1993). Engels and Heinemann (1996) investigated the role of the katabatically- and synoptically-forced outflow of continental air in mesocyclogenesis over the eastern part of the Weddell Sea. Numerical simulations were carried out for two case studies of weak, orographically-influenced developments, as well as for the intense mesocyclone near Halley that occurred in January 1986. The first two cases had lifetimes of around 12 h and horizontal scales of about 200–300 km (Figure 5.50) and were observed by aircraft measurements during the Antarctic summer of 1989/90 (Heinemann, 1996b). The mesocyclone occurring on 8 February 1990 (Figure 5.50a) developed at a distance of about 400 km from Halley and GvN. The second mesocyclone developed near Halley station on 19 January 1990 (Figure 5.50b). Both mesocyclones had a shallow vertical extent of about 2 km. Detailed observational studies of these two cases are given in Heinemann (1996b). The third case of the summertime Halley vortex investigated by Engels and Heinemann (1996) is the Antarctic polar low near Halley with a horizontal scale of about 1000 km and a lifetime of more than three days in January 1986 as documented by Turner *et al.* (1993b). This type of mesoscale vortex was driven by baroclinic forcing due to upper-level cyclonic vorticity advection (CVA) without any indication of deep convection. For a description of such summertime Halley mesocylcones based on observations see also Section 3.2.

The numerical simulations show the important role of the synoptic-scale environment in the formation and the maintenance of these coastal mesocyclones. For the two cases with horizontal scales of 200–300 km and lifetimes of around 12 h, the katabatic flow associated with the Antarctic orography was found to be essential for the dynamical development and the three-dimensional structure. Two coastal regions of enhanced katabatic outflow were found between Halley and GvN. In one region the surface wind was channelled by a valley, whereas in a second region a stronger orographic gradient caused the increase of the gravity-driven katabatic outflow. Due to differences between the structures of the model orography (which was interpolated and smoothed

478 5 Numerical simulation

(a)
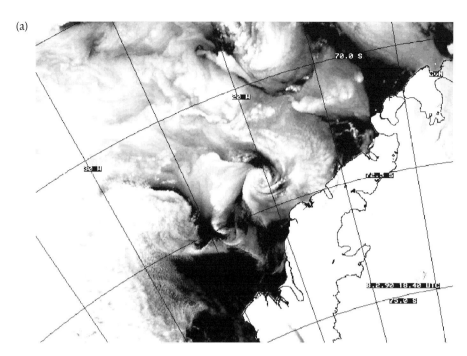

(b)
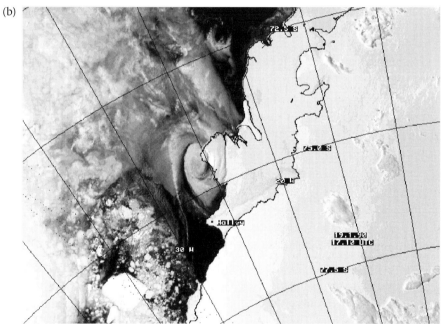

Figure 5.50. AVHRR visible wavelength images for (a) 1840 GMT 8 February, and (b) 1710 GMT 19 January 1990 (from Engels and Heinemann 1996, *The Global Atmosphere and Ocean System*, Taylor & Francis Ltd (http://www.tandf.co.uk/journals).)

to the grid resolution) and the real orography (which is not well known), a direct comparison of the low-level flight measurements of Heinemann (1996b) with model results may be difficult for these cases of coastal mesocyclones (in particular the flow structures related to the channelling of the flow by the valley). Nevertheless, the value of the numerical simulations lies in the ability of the model to yield insight into the formation mechanisms and the role of the synoptic-scale environment.

The numerical simulations show for the mesocyclone of 8 February 1990 that the outflow from the continent was increased by a transient synoptic-scale low, leading to the production of cyclonic vorticity mainly by the stretching mechanism. Figure 5.51 displays the 18 h forecast of 850 hPa geopotential height valid for 0000 GMT 9 February. During the first 18 h of integration, the large-scale low at 850 hPa moved eastward and led to a distinct change in the flow direction over the observed location of mesocyclogenesis. The 18 h forecast of the 850 hPa geopotential height shows the centre of the synoptic-scale cyclone northeast of GvN and a high pressure ridge over the central Weddell Sea. Supported by the synoptic-scale pressure gradient, an outflow from the continental ice slope between Halley and GvN towards the ice-free Weddell Sea was established.

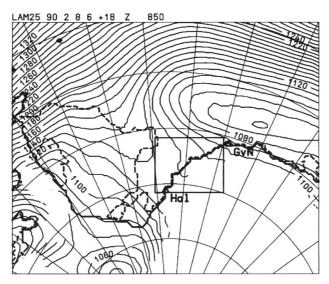

Figure 5.51. 850 hPa geopotential height (isolines every 10 gpm) after 18 h simulation of LAM25 (valid at 0000 GMT 9 February 1990). The rectangle indicates the subsection of the model domain shown in Figure 5.52 (from Engels and Heinemann 1996, *The Global Atmosphere and Ocean System*, Taylor & Francis Ltd (http://www.tandf.co.uk/journals).)

480　5 Numerical simulation

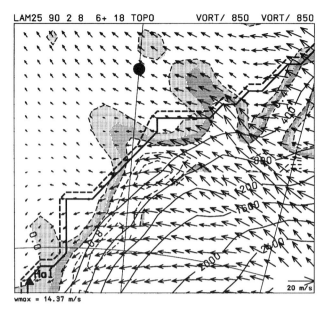

Figure 5.52. Topography (contour interval of 400 m), horizontal wind at the lowest level of the model (30 m above the surface) and relative vorticity at 850 hPa (dashed, contour interval 0.4×10^{-4} s^{-1}, only negative values, representing cyclonic vorticity) after 18 h simulation (valid at 0000 GMT 9 February 1990). The light and medium-shaded areas mark regions with a relative vorticity between 0 and -0.2×10^{-4} s^{-1}, and -0.2 and -0.4×10^{-4} s^{-1}, respectively. The filled circle marks the position of the observed mesocyclone at 1800 GMT 8 February 1990.

The synoptically-enhanced vorticity production over the slope and near the coast led to the formation of two cyclonic flow patterns near the coast between Halley and GvN, which were simulated as elevated maxima of cyclonic vorticity (CV) in a stably-stratified layer between 800 hPa and 950 hPa (Figure 5.52). Due to the synoptic-scale conditions (Figure 5.51), the low-level katabatic flow from the valley region northeast of Halley extended over the open water, and strongly influenced the low-level wind field beneath the most pronounced CV maximum lying close to the observed mesocyclone. The study of the vorticity budget confirms that mainly dynamical processes were responsible for the development of the two CV maxima. The vertical profiles of each term of Eqn. (5.1) shown by Engels and Heinemann (1996) demonstrate that the production of CV took place mainly below 800 hPa. Between 800 and 950 hPa, i.e. above the layer influenced by the low-level katabatic airflow, the local tendency of CV was mainly determined by the stretching mechanism during the phase of maximum mesocyclone development (and being diminished by HADV), while DF and TT were less important. These results agree qualitatively with the aircraft measurements where no cyclonic circulation could be detected close to the ground and at 700 hPa, but a distinct cyclonic cloud structure was present in the lowest 2 km. The lifetime of this simulated mesoscale cyclonic vorticity centre was less than 12 h due to the fact that the formation and maintenance were controlled by the synoptic-scale environment. The numerical simulations confirm the hypothesis of Heinemann (1996b), that at least some coastal mesocyclones may only be observed as elevated CV maxima. In contrast to their distinct cyclonic cloud structure, the low-level wind field can be highly influenced by orographic structures and no cyclonic rotation may be present at the surface. This result is of relevance for climatological investigations of mesocyclones, which are often based merely on the detection of cloud structures on satellite images.

The development of the mesocyclone on 19 January 1990 took place under synoptic conditions quite different from those on 8 February 1990. The main purpose of the model run for this case was to capture the developing phase of the vortex, which was observed during the first hours of 19 January 1990. Therefore, a 24 h simulation was started at 0600 GMT 18 January using LAM25. During the simulation time, the formation of a low-level trough could be noticed, extending parallel to the coast over the eastern Weddell Sea at 0600 GMT 19 January. Between 12 and 24 h integration time, i.e. during night-time conditions, the katabatic wind system developed over the continental slope. As in the simulation of 8 February, two regions of enhanced katabatic outflow could be found. Again, the low-level wind was channelled by the valley 500 km northeast of Halley. A second local maximum of enhanced winds was simulated in the coastal region with steep gradients of the orography southwest of the valley (Figure 5.53). In both areas a deflection of the outflow to the southwest occurred at the coast, caused by the low-level pressure trough parallel to the coast. The vertical turbulent fluxes over the ocean were quite small; in only a few areas did the sensible heat flux exceed 20 W m^{-2}.

The mesocyclone was considered to be represented by the southern tip of the coastal trough, associated with a pronounced mesoscale cyclonic vorticity maximum which was confined to levels below 700 hPa. In the area of the observed mesocyclogenesis, a local CV maximum of -1.0×10^{-4} s^{-1} could be detected at 850 hPa, but wind vectors did not show a closed mesoscale circulation cell (Figure 5.54). The largest amount of local CV tendency was provided by the horizontal advection of CV, but it was also supported by vortex stretching associated with downslope wind components. The fact that no closed circulation was resolved in the simulation of this case may be a result of the relatively coarse grid resolution of 25 km compared with the small horizontal scale of the observed mesocyclone of about 200 km, which was estimated to have a surface pressure anomaly of 1–2 hPa (Heinemann, 1996b).

482   5 Numerical simulation

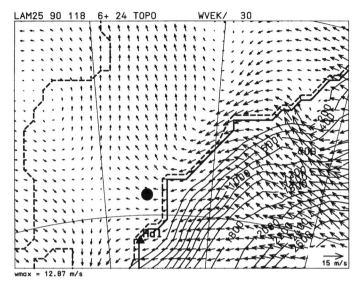

Figure 5.53. Topography (contour interval of 200 m) and horizontal wind at the lowest level of the model (30 m above the surface) for the 24 h simulation of LAM25 (valid at 0600 GMT 19 January 1990). The filled circle marks the position of the observed mesocyclone. A scaling vector is shown in the lower right corner (modified from Engels and Heinemann, 1996).

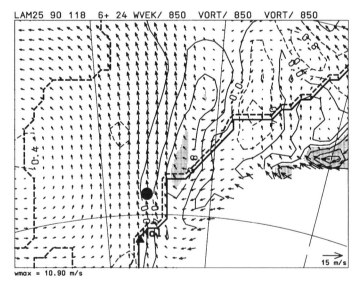

Figure 5.54. Relative vorticity (positive values broken lines, contour interval 0.4×10^{-4} s^{-1}, area with less than -0.4×10^{-4} s^{-1} shaded) and horizontal wind vectors at 850 hPa after 24 h simulation of LAM25 (valid at 0600 GMT 19 January 1990) (modified from Engels and Heinemann, 1996).

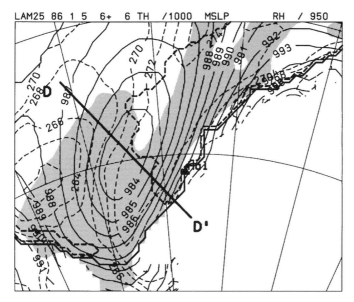

Figure 5.55. LAM25 forecast of the MSLP (solid lines, isolines every 1 hPa), potential temperature at 1000 hPa (broken lines, isolines every 2 K) and relative humidity of 950 hPa (area exceeding 90% shaded) valid at 1200 GMT 5 January 1986. The sea ice front is marked by a thick broken line, the triangle marks the station Halley (Hal) (modified from Engels and Heinemann, 1996).

The case occurring in January 1986 represents a quite different type of coastal mesocyclone with a lifetime of more than three days and a diameter of about 1000 km during its mature stage. The vortex had a pronounced baroclinic structure in the lowest 3 km and experienced a forcing by an upper-level trough. Figure 5.55 displays the fields of geopotential height and potential temperature at 1000 hPa for the subsection of the model domain covering the eastern Weddell Sea. It shows the 6 h forecast of LAM25 valid at 1200 GMT 5 January 1986. The centre of the mesocyclone was found close to the sea ice front about 300 km west of Halley. The isolines of the geopotential height at 1000 hPa reveal the strongest gradients along the coast. The wind vectors at this level indicate a pronounced confluence zone almost parallel to the coast. A belt of maximum winds was found over the area of Halley, and simulated winds at 10 m near Halley reached values of 16 m s^{-1} (compared with 12.5 m s^{-1} measured at Halley at that time). The mesocyclone had a well-defined baroclinic structure at 1000 hPa, with cold air over the central Weddell Sea on its western side and warmer air over the open water on its eastern side. Over this open water area, turbulent surface fluxes of latent and sensible heat remained relatively small (values around 20 W m^{-2}), only close to the coast-line did the fluxes reach magnitudes of about 60 W m^{-2}. The areas of possible cloudiness (shaded in Figure 5.55) in the southern part of the mesocyclone agree with the area of observed low-level clouds (see Turner *et al.*, 1993b, figs. 4 and 14). Even the pattern of high relative humidities at 500 hPa has similarities to the high clouds associated with the observed vortex. The centre of the surface low was found downstream of the upper-level trough at 500 hPa, indicating the baroclinic upper-level forcing for this mesocyclone. Interestingly, the structure of the temperature field at 500 hPa was quite different from that at 1000 hPa, since warm air advection was found over the central Weddell Sea, i.e. over a region with strong cold air advection at low levels. This reversal of the horizontal temperature gradient occurred even at 300 hPa.

The vertical structure of the temperature and wind field of the mesocyclone can be studied by a cross-section (Figure 5.56) from northwest to southeast through the surface centre of the mesocyclone, i.e. approximately perpendicular to the coast. The wind component normal to the cross-section clearly reveals the vertical structure of the belt of high wind speed close to the coast. The maximum with more than 24 m s^{-1} was found at 900 hPa close to the coastal slope in connection with an upward motion up to 6 cm s^{-1}. In accordance with the horizontal temperature gradient, the wind parallel to the coast revealed a strong vertical wind shear and at 500 hPa only very weak winds were present. This wind field structure with weak winds in the middle troposphere and a strong low-level jet along the continental slope resembled that of the 'barrier winds' observed on the eastern side of the Antarctic Peninsula (Schwerdtfeger,

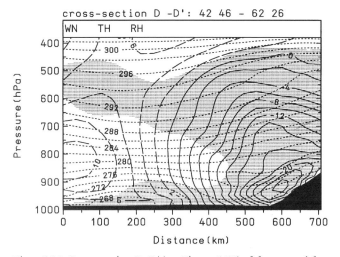

Figure 5.56. Cross-section D–D' (see Figure 5.55) of the potential temperature (thin broken isolines every 2 K) and the normal wind component (contour interval 2 m s^{-1}). Areas exceeding 90% relative humidity are shaded. The arrow marks the maximum of upward motion (*c*. 6 cm s^{-1}) (from Engels and Heinemann, 1996).

1984), in which a belt of cold air was formed by weak easterly winds over the central Weddell Sea, and strong southerly low-level winds parallel to the coast are generated due to the associated thermal wind. The cyclonic circulation northwest of the centre (near D in Figure 5.56) was found to be considerably weaker.

The cross-section displays a shallow layer of low-level cloud close to the centre, and shows a higher and thicker cloud layer near the coast reaching up to 450 hPa. The above mentioned reversal of the horizontal temperature gradient between lower and upper layers can also be detected in Figure 5.56. The low-level cold air advected over the ice-covered central Weddell Sea on the western side of the mesocyclone can be seen up to 850 hPa. Above that level, a weak warm air structure was present up to 300 hPa. This structure of a thick layer of relatively warm air overlying a shallow layer of cold air may explain in part the unusual pattern of the 500–1000 hPa thickness obtained by the retrievals from TOVS data as shown by Turner *et al.* (1993b), where a distinct warm anomaly was observed southwest and west of the mesocyclone on 3 and 4 January (see Section 3.2).

Although being a more classical type of baroclinic development, a topographic forcing for the initial development of this mesocyclone cannot be ruled out, since as described by Turner *et al.* (1993b) the mesocyclone formed inland of Halley and passed the station as a relatively weak vortex on 3 January with a scale of about 300 km.

A case over the Ross and Amundsen Sea

Lieder and Heinemann (1999) investigated a family of three mesocyclones developing during 11–12 January 1995 over the northern Bellingshausen and Amundsen Sea by means of AVHRR data, ERS data and SSM/I retrievals, and mesoscale numerical model data. The most pronounced mesocyclone was investigated in detail. It had a diameter of about 800 km, a lifetime of over 24 h and reached the intensity of a polar low. Figure 3.69 (p. 248) shows the AVHRR image for the mesocyclones at 1400 GMT 11 January 1995. The mesocyclone M3 at 112° W, 63° S had a pronounced cloud band and had moved eastward during the previous hours. The circulation associated with mesocyclone M2 (at 95° W, 59° S) seemed to be dissipating at that time. Low-level cloud near the centre and mid-level cloud south of the centre clearly indicated the circulation associated with M3. The appearance of M3 on the satellite image resembled a short baroclinic wave with shallow convection west and north of the centre. M1 (with its centre at 83° W, 60° S) was associated with a cloud band with a south/north orientation at that time. The centre of a large-scale cyclone had remained almost stationary at 90° W, 67° S during the

486   5 Numerical simulation

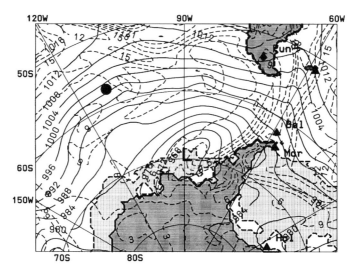

Figure 5.57. Simulated MSLP (full lines, isolines every 2 hPa) and integrated water vapour (IWV, broken lines, isolines every 3 kg m^{-2}) after 12 h simulation time for a subsection of the LAM50 domain (valid at 1200 GMT 11 January 1995). The centre of M3 at 1400 GMT 11 January obtained from satellite images is marked by the filled circle and L indicates the centre of a synoptic-scale cyclone (from Lieder and Heinemann, 1998).

past 12 h, but its front had moved further eastwards and began to pass over the Antarctic Peninsula. This was also reflected in the simulation of LAM50 valid at 1200 GMT 11 January (Figure 5.57), which shows the highest IWV gradients just west of the Antarctic Peninsula. The blocking effect of the topography and the initial formation of a lee cyclone east of the Antarctic Peninsula is also visible in the model simulations of the MSLP field. Along the 60° S latitude circle, two weak troughs were simulated at 110° W and 80° W, being approximately coincident with the positions of M3 and M1, respectively. The mesocyclones were also present as a distinct mesoscale signal in SSM/I retrievals of cloud liquid water (CLW), near-surface wind speed (WS) and integrated water vapour (IWV). ERS scatterometer data (ERS-SCAT) were available for this period, but ERS-derived wind vectors gave no insight into the structure of the mesocyclones, because the mesocyclones were not covered by the narrow ERS swaths. Nevertheless, these data were used to validate the numerical simulations.

The LAM50 simulations of the MSLP showed M3 and M1 as mesoscale troughs moving eastwards with the mean zonal flow. M3 was found to be best developed by the 18 h simulation valid at 1800 GMT 11 January. Figure 5.58 shows the relative vorticity at 950 hPa at that time for a subsection of the

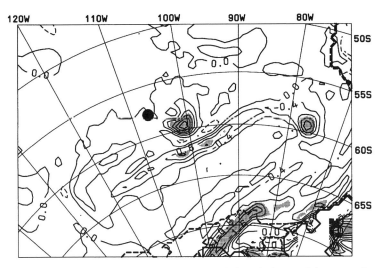

Figure 5.58. NORLAM forecasts (50 km grid) of the relative vorticity at 950 hPa (contour interval 0.4×10^{-4} s^{-1}, shaded for values lower than -1.0×10^{-4} s^{-1}) after 18 h simulation time (valid at 1800 GMT 11 January 1995). The centre of M3 at 1800 GMT 11 January obtained from satellite images is marked by the filled circle (from Lieder and Heinemann, 1999).

LAM50 domain. Several pronounced cyclonic vorticity maxima are present. The first one at 90° W, 70° S is associated with the synoptic-scale cyclone (values of -1.6×10^{-4} s^{-1}). The second one with values of -2.8×10^{-4} s^{-1} can be found east of the Antarctic Peninsula associated with the lee cyclone. M3 and M1 can be identified as the cyclonic vorticity maxima at 101° W, 59° S (-2.0×10^{-4} s^{-1}) and at 76° W, 59° S (-2.0×10^{-4} s^{-1}), respectively. M2 was not simulated, neither as a surface trough nor as a distinct mesoscale cyclonic vorticity maximum, but lay in a belt of moderate cyclonic vorticity south and southeast of M3.

A more detailed mesoscale analysis for M3 is given using LAM25 forecasts nested in the LAM50 domain. The LAM25 simulation after 12 h (valid at 1800 GMT 11 January) is shown in Figures 5.59 and 5.60. The wind vector field and potential temperature at 850 hPa (Figure 5.59) reflects the mesoscale circulation of M3 (superimposed on the strong zonal wind) with the circulation centre at about 100° W, 60° S. The temperature field reveals moderate baroclinicity and a wave-like structure. The result is an advection pattern resembling that of a short baroclinic wave. The intensification of the low-level circulation of M3 was triggered by a short-wave 500 hPa trough (Figure 5.60). The maximum upward vertical velocity at 700 hPa lay downstream of the 500 hPa trough axis and over the low-level circulation centre. A belt of pronounced lifting extended to

488   5 Numerical simulation

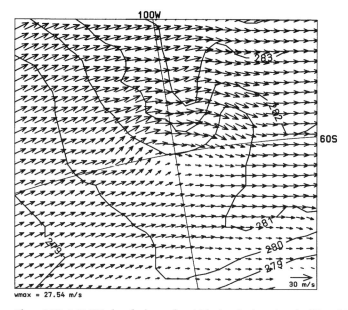

Figure 5.59. LAM25 simulations after 12 h simulation time (valid at 1800 GMT 11 January 1995) for the potential temperature (solid lines, isolines every 1 K) and wind vectors (every grid point) at 850 hPa. A scaling vector is shown in the lower right corner (from Lieder and Heinemann, 1999).

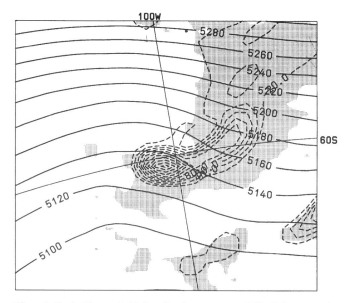

Figure 5.60. As Figure 5.59, but for the geopotential height at 500 hPa (solid lines, isolines every 20 gpm) and upward vertical velocity in pressure coordinates (broken lines, isolines every 20×10^{-2} Pa s^{-1}) at 700 hPa for the same area as in Figure 5.59a. Shaded areas indicate simulated cloud coverage at 500 hPa exceeding 90% (from Lieder and Heinemann, 1999).

the east and northeast with a maximum of 2.2 Pa s^{-1}. The areas with cloud coverage of more than 90% at 500 hPa revealed similarities to the mid-level clouds associated with M3 on the satellite image 4 h earlier (Figure 3.69). The same kind of triggering was also simulated for M1, but the upward vertical velocities were much smaller (0.8 hPa s^{-1}).

In conclusion, the numerical model results showed two of the mesocyclones as short-wave baroclinic developments triggered by an upper-level trough, while a less significant third mesocyclone was not simulated by the model. The development occurred far away from the sea ice front or topographic structures. The synoptic-scale environment was quite different from that associated with summertime mesocyclone developments over the Weddell Sea region (Heinemann, 1990), but had similarities with wintertime polar lows over the Weddell Sea (Heinemann, 1996a), as well as over the Ross and Bellingshausen Seas (Carleton and Fitch, 1993). The comparison between the numerical model simulations and the corresponding satellite-retrieved parameters IWV (SSM/I) and near-surface wind (SSM/I and ERS-SCAT) showed relative good agreement (Lieder and Heinemann, 1999). The differences between ERS and NORLAM wind vectors were 1.1 m s^{-1} and $-3.2°$ for the mean bias, and the standard deviation was 2.5 m s^{-1} and 25° for wind speed and direction, respectively. This finding gives confidence in the quality of the numerical simulations with NORLAM and the quality of the ECMWF analyses, in which the limited area model was nested.

5.2.5 Simulations of realistic cases: wintertime cases

A case over the Weddell Sea

From 30 July to 1 August 1986 several mesocyclones developed close to the sea ice front over the northern Weddell Sea, and were investigated by Heinemann (1998) by means of satellite data and simulations using NORLAM. This mesocyclone event occurred during the Winter Weddell Sea Program (WWSP) field phase 1986, when additional observational data were available, particularly observations from the German research vessel *Polarstern*, which was situated close to a mesocyclone on 31 July 1986 and these data allowed a comparison with the model results, thus being one of the very rare opportunities to make a comprehensive comparison of model simulations and observations for a mesocyclone development in the Antarctic.

The satellite image composite at 0200–0400 GMT 31 July (Figure 5.61a) showed a synoptic-scale cyclone (L) with its main centre close to the Antarctic coast at 24° W, 71° S and two mesocyclones: the well-developed mesocyclone M1, which had moved southeastward during the previous 10 h to 20° W, 60° S,

490   5 Numerical simulation

(a)

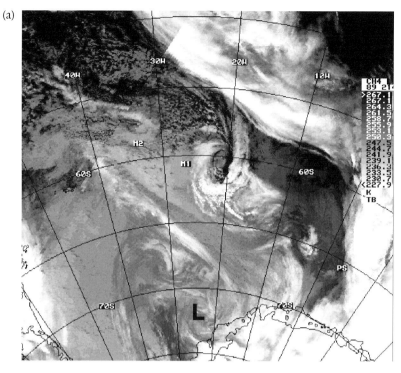

(b)

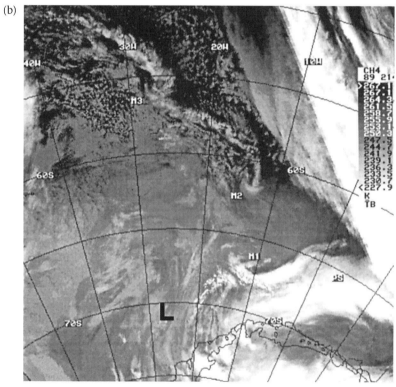

and a mesocyclone in its initial stage at 32° W, 57° S (M2 in the following). Both mesocyclones were close to the northern extent of the sea ice and the distance between their centres about 800 km. During the next few hours, a third mesocyclone developed at the cloud band west of M2 at 40° W, 55° S (M3 in the following, Figure 5.61b), i.e. about 600 km west of M2. The diameter of M1 could be estimated at about 700 km, and the cloud structure revealed similarities to a short baroclinic wave, and cloud top temperatures of the cloud band east of the centre were about $-40\,°C$. M2 seemed to result from a boundary layer front (BLF) development, which formed under conditions of large-scale flow almost parallel to the sea ice edge and was advected over the open water (see Section 5.2.2). During its further development M1 moved over the sea ice and towards the Antarctic coast (Figure 5.61b).

The development of the mesocyclone M1 was well described in the NORLAM simulations. Figure 5.62 shows the MSLP for a subsection of the LAM50 domain (see Figure 5.36a) for 24 and 42 h of simulation valid at 0000 GMT and 1800 GMT 31 July, respectively. After 12 h (not shown) a weak trough was simulated to move eastwards along the sea ice front at 32° W, 57° S. After 24 h (Figure 5.62a), the synoptic-scale cyclone (L) had remained almost stationary at 27° W, 70° S, and the closed isobar of 968 hPa at 20° W, 60° S indicated the centre of the mesocyclone M1 in the simulation. The simulation results corresponded well to the satellite image at the same time (Figure 5.61a) with respect to the position of the centre and the diameter. The formation of the less significant mesocyclone M2 was not captured by the model. The simulations during the following 12 h showed the southeastward movement of M1 over the sea ice and its further intensification during 31 July. The simulated development again agreed well with the satellite imagery. Figure 5.62b displays the MSLP after 42 h simulation time (valid at 1800 GMT 31 July). L had filled (central pressure 946 hPa), but M1 had deepened considerably (pressure at the centre of about 955 hPa) and its centre lay at 10° W, 69° S. A pronounced trough extended from the centre of M1 to the northeast, which was associated with the cold front passing over the position of the R/V *Polarstern* later on in the simulation (see Figure 5.61b).

The mesocyclone simulated after 24 h showed the initial circulation around its centre, associated with strong cold advection in its northwestern part. An

Figure 5.61. AVHRR channel 4 images (mosaics, reduced resolution) from global area coverage (GAC) data (polar stereographic projection, brightness temperature scale in K) for (a) 0200–0400 GMT and (b) 1600–1800 GMT 31 July 1986. The mesocyclone is designated as M1, and L marks the centre of a synoptic-scale cyclone (from Heinemann, 1998).

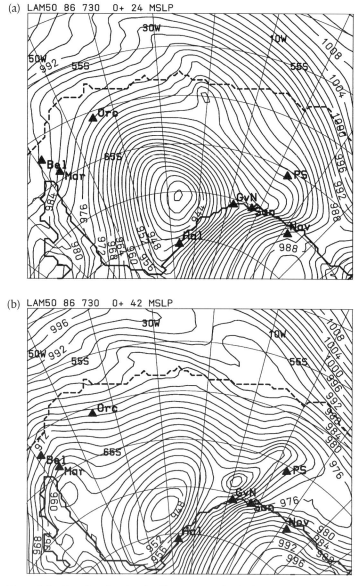

Figure 5.62. Simulated MSLP (isolines every 2 hPa) after (a) 24 h, (b) 42 h simulation time for the subsection of the LAM50. The simulations are valid at (a) 0000 GMT and (b) 1800 GMT 31 July 1986 (from Heinemann, 1998).

area of pronounced low-level convergence was present at a frontal zone in its southeastern part. After 36 h, M1 had moved over the sea ice and the low-level circulation was well developed (Figure 5.63a), and strong advection was present at the frontal zones associated with M1. A mid-level forcing was found, since a pronounced 700 hPa short-wave trough lay upstream of the surface

5.2 The Antarctic 493

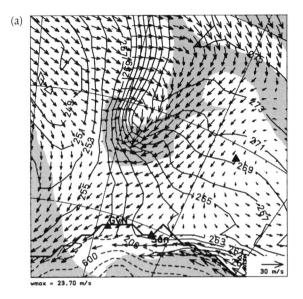

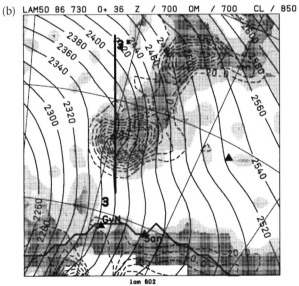

Figure 5.63. (a) Potential temperature (solid lines, isolines every 2 K) and wind vectors (every grid point) at 950 hPa and Antarctic topography for heights exceeding 200 m (broken lines, contour intervals of 400 m) for a sub-area of the LAM50 domain after 36 h (valid at 1200 GMT 31 July 1986). Areas with a relative humidity at 700 hPa exceeding 90% are shaded. A scaling vector is shown in the lower right corner. Small triangles indicate positions of GvN, Sanae (San) and *Polarstern* (PS). (b) As (a), but for the geopotential height (solid lines, isolines every 20 gpm) and upward vertical velocity in pressure coordinates (broken lines, isolines every 20×10^{-2} Pa s^{-1}) at 700 hPa. Shaded areas indicated simulate cloud coverage at 850 hPa (lighter shading, 40–90%; darker shading, >90%) (from Heinemann, 1998).

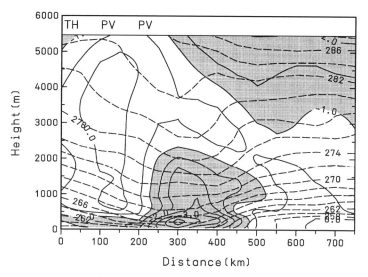

Figure 5.64. Vertical cross-section along line 3–3′ from Figure 5.63b through the centre of M1 after 36 h for the potential temperature (broken lines, contour interval 2 K) and potential vorticity (isolines every 0.5 PVU). Areas with PV lower than −1.0 PVU are shaded (modified from Heinemann, 1998).

centre (Figure 5.63b). The maximum of upward velocity (2.2 Pa s^{-1}) lay southeast of the surface centre, i.e. the conditions for a further baroclinic development and translation were given. The simulated cloud formed as bands in the regions of strong upward motion. At 700 hPa an area with high relative humidity could be seen as a comma-shaped structure similar to the clouds in the satellite image (Figure 5.63a).

The cross-section along the line 3–3′ (indicated in Figure 5.63b, length *c.* 800 km) is displayed in Figure 5.64. The cross-section was chosen to go approximately from south to north through the surface circulation centre of M1. The potential temperature cross-section reflected the presence of a warm sector and the shallow cold air layer behind the cold front. The upper-level forcing discussed above can be seen clearly in the cross-section of the potential vorticity (PV). A low-level cyclonic PV maximum (correlated with the surface cyclone) extended with a −1 PVU isoline up to about 2 km. On the other hand, a mid-level cyclonic PV anomaly was present with a −1 PVU isoline extending down to about 3 km. In accordance with the dynamics associated with PV anomalies (Hoskins *et al.*, 1985), this phase relationship between mid- and low-level PV anomalies is favourable for a further mutual development of the mid-level and surface disturbance.

Several sensitivity experiments were carried out in order to study the physical processes responsible for the development of M1, including a variation of

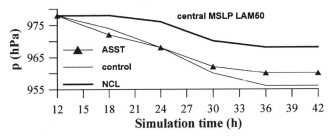

Figure 5.65. Simulated central pressures (MSLP) for M1 for the control run (thin line), runs with an altered SST (ASST) (thin line with triangles), and the 'no cloud physics' run (NCL) (bold line; see text) (from Heinemann, 1998).

the SST, latent heat release and boundary layer processes. In all experiments the mesocyclone M1 developed, but with different intensities, positions and structures. The effect of two experiments on the central pressure of M1 is shown in Figure 5.65. In a first experiment (ASST), the SST was altered to a distribution valid for April conditions (instead of July), but the correct sea ice distribution was kept. As a result, the SST close to the sea ice front had values about 3 K higher compared with the control run. The second experiment (NCL) was performed with the same boundary conditions as for the ASST run, but the cloud physics were switched off (i.e. no heating/cooling due to condensation or evaporation). While the control and ASST runs simulated the rapid deepening of M1 during the first 24 h, the NCL run yielded only a very weak vortex. During the next 24 h, the intensification in the ASST run was slightly weaker than for the control run, and the centre of M1 was simulated more northeastward, but the associated winds reached approximately the same intensity. The differences between the control run and process studies neglecting the surface fluxes of sensible or latent heat are of the same amount as for the ASST run with respect to the central pressure and the structure of M1; but, of course, significant differences can be found for the BLF development close to the sea ice front. In contrast, the NCL run produced only a weak vortex also after longer simulation times.

In the control run, the development phase of M1 (taking place close to the sea ice front) was associated with a precipitation rate of about 1 mm h^{-1}, which reflects the importance of the condensational heating for the intensification of M1 during the first 24 h. The NCL run yielded only a weak mesocyclone even after longer simulation times. In the PV cross-section corresponding to Figure 5.64 for the NCL run (not shown) no low-level or mid-level PV signatures were present, indicating again the importance of the condensational processes for the development of M1 and the coupling between lower and mid-levels.

496   5 Numerical simulation

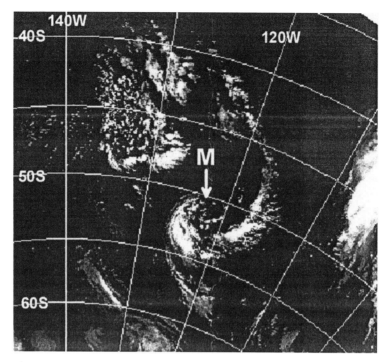

Figure 5.66. An AVHRR infra-red composite image at 1220 GMT 6 July 1994 with the centre of the mesocyclone (M) at 123° W, 51° S indicated (from Heinemann and Claud, 1997).

A case over the Ross and Amundsen Sea

This wintertime case was described in Lieder and Heinemann (1998) by means of satellite data, but additional numerical model simulations using the NORLAM model have been carried out. Figure 5.66 shows an AVHRR infra-red image of the Amundsen Sea for 1220 GMT 6 July 1994. A mesocyclone with a diameter of about 1000 km can be observed at 123° W, 51° S, about 2000 km north of the Antarctic coast. The development of the mesocyclone started in the early morning of 5 July in the area around 150° W, 42° S. After approximately 60 h lifetime the mesocyclone decayed in the Bellingshausen Sea (100° W, 62° S) in the afternoon of 7 July. The mesocyclone could be monitored in the SSM/I-retrieved WS and CLW fields, and in the central part of the mesocyclone high wind speeds of more than 16 m s^{-1} were found. A detailed view of the mesocyclone and the prevailing winds are shown in Figure 5.67, which combines the SSM/I-derived CLW of 0700 GMT 6 July and the ERS-1 scatterometer winds for 0800 GMT. This is a rare case when the narrow 500 km ERS-1 swath fully crossed a mesocyclone. The cyclonic rotation with strong winds around the centre is evident. South of the mesocyclone, moderate winds prevailed and a sharp convergence line could be found near the cloud band extending

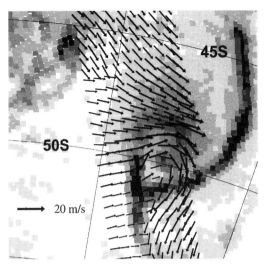

Figure 5.67. ERS-1 scatterometer wind vectors at 0800 GMT 6 July 1994 superimposed on SSM/I-derived CLW (shaded) at 0700 GMT 6 July (from Heinemann and Claud, 1997).

southward from the centre of the mesocyclone with winds turning from westerly to northerly directions.

A signature of the mesocyclone development was also found in the NORLAM simulations. Figure 5.68 shows the MSLP and the 500 hPa geopotential height for a subsection of the LAM50 domain over the Amundsen Sea for 42 h of simulation valid at 0600 GMT 6 July. A surface trough was simulated downstream of the upper-level trough. A belt of upward vertical velocities at 700 hPa lay ahead of the surface centre and was approximately coincident with the simulated cloudiness at 500 hPa (not shown). The observed mesocyclone centre (from ERS-1 winds) lay about 300 km south of the simulated one. The PV cross-section (Figure 5.69) along the line B–B′ (indicated in Figure 5.68, length 750 km) through the surface circulation centre of the mesocyclone showed again the phase shift between a mid-level cyclonic PV anomaly and the low-level cyclonic PV maximum correlated with the surface cyclone. The potential temperature cross-section reflected the mid-level baroclinicity, which corresponded to a reduced static stability between the two cyclonic PV maxima. Although being much weaker than the simulated wintertime mesocyclone over the Weddell Sea, this phase relationship between mid- and low-level PV anomalies looks similar for both cases.

5.2.6 Current status

In recent years, mesoscale numerical models have proved to be capable of successfully simulating Antarctic mesocyclones. Idealized numerical

498 5 Numerical simulation

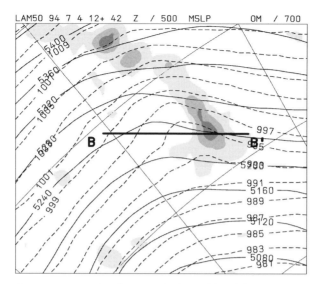

Figure 5.68. NORLAM forecasts (50 km grid) valid at 0600 GMT 6 July 1994. Fields indicated are for the MSLP (broken lines, isolines every 2 hPa), the geopotential height at 500 hPa (solid lines, isolines every 40 gpm) and upward vertical velocity in pressure coordinates (shaded, every 40×10^{-2} Pa s^{-1}) at 700 hPa. The shaded areas are also almost coincident with the simulated cloud coverage at 500 hPa.

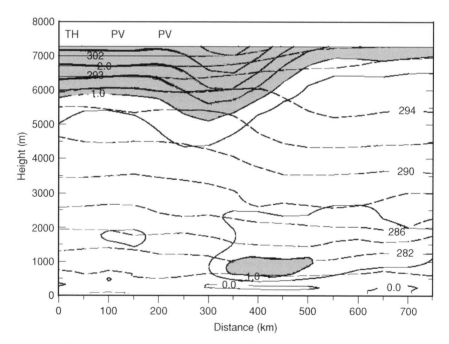

Figure 5.69. Vertical cross-section along line B–B' from Figure 5.68 through the surface centre of the mesocyclone valid at 0600 GMT 6 July 1994 for the potential temperature (broken lines, contour interval 2 K) and potential vorticity (isolines every 0.5 PVU). Areas with a potential vorticity lower than -1.0 PVU are shaded.

simulations using full-physics mesoscale models have contributed significantly to the current understanding of mesocyclogenesis in the Antarctic. For coastal mesocyclone developments the role of the katabatic wind in the formation process was investigated using the NORLAM and MAR high resolution mesoscale models. Simplified initial conditions allowed also the investigation of BLF mesocyclone developments, but the more complex intensification by upper and mid-level forcing, which leads to intense polar lows in reality, was not shown by the present idealized simulations. For several mesocyclone case studies over the Antarctic oceans or near the sea ice front, this kind of upper-level forcing was shown to be present, i.e. these mesocyclones can be regarded as short baroclinic waves. But, the intensification process may also depend on diabatic processes, such as latent heat release, as was shown for one wintertime case. Near-coastal developments show the interaction between the synoptic-scale environment, the katabatic wind over the ice slope and the mesocyclogenesis. Under favourable synoptic flow, a barotropic spin-up of cyclonic vorticity corresponding to an initial mesocyclone at the coast occurs, which may be further intensified by other processes.

Compared with the Arctic, where the forecasting of polar lows is one of the routine tasks of the weather services, the number of investigated cases for the Antarctic is relatively small. In addition, the tools for validating the numerical forecasts (or even the initial analyses) are much more restricted in the Antarctic due to the sparse observational network. As a consequence, retrievals from satellites have to be used for validation purposes in most cases, which has attained increasing importance with the advent of microwave imagers and scatterometers. These sensors allow the determination of the near-surface wind and integrated water parameters for almost all weather conditions with a resolution corresponding to that of mesoscale numerical models.

An important aspect of current and future mesocyclone research in the Antarctic is the combining of satellite data, direct observations and numerical model data. As discussed above, the retrievals of geophysical parameters from satellite data and direct observations can be used for the validation of model simulations. But, the observational data (including satellite retrievals) can also be assimilated into a model in the form of a mesoscale objective analysis, which describes the mesoscale atmospheric state in a dynamically consistent way. Both approaches have their strengths and weaknesses. For example, the intermittent data assimilation may suppress mesoscale structures generated by the model or may generate unrealistic structures, in particular when the data inhomogeneity in the model area is large. Even in the case when microwave satellite data are continuously available (SSM/I data since 1987, ERS data since 1991), the poor areal coverage and restriction to open water areas for the retrievals contribute to these problems. Because no improvement is in sight for the Antarctic, special observational programmes such as FROST (Turner *et al.*, 1996b) are the only way of getting a slightly improved observational network as well as data bases of conventional and satellite data, which are easily accessible for the research community.

J. TURNER, E.A. RASMUSSEN AND B. RØSTING

6

Forecasting of polar lows

6.1 Aspects common to both polar regions

6.1.1 Introduction

Polar lows can have a severe impact on maritime operations and can cause considerable disruption when the more severe systems make landfall. In areas such as the North Sea, there are many gas and oil platforms and it is necessary to have good forecasts of the arrival of severe mesoscale lows to minimize the impact on operations. Within the Antarctic, most of the mesoscale lows are not as vigorous as their counterparts in the north. Nevertheless, they can still cause severe problems during the summer relief operations at the research stations and affect work in the deep field.

Polar low forecasting is an integral part of the general forecasting problem and the results are dependent on the success of the overall forecast. Polar lows often result in a rapid deterioration of the weather at a specific location, and accurate time indications in the forecasts are important. In this chapter we will examine the means by which forecasters attempt to predict the formation and development of mesocyclones and polar lows. The output from numerical weather prediction (NWP) analysis and forecast systems can be used to try and infer where and when mesocyclone developments may take place in a particular region. Satellite imagery and other satellite data are indispensable in identifying and predicting the movement of existing mesoscale vortices a few hours ahead. Such a nowcasting approach is certainly valid for about 6 hours ahead, but can be applicable for 12 hours or more in exceptional cases. In the following sections we discuss these approaches and present selected examples of polar low forecasts. Here we are concerned with the more general aspects of forecasting mesoscale lows that can be applied in both polar regions. Information specific to the Arctic and Antarctic is dealt with in Sections 6.2 and 6.3 respectively.

In their discussion of polar low forecasting, Midtbø and Lystad (in Lystad, 1986) pointed out that forecasting polar lows cannot be separated from the forecasting of other weather phenomena. The accuracy which can be obtained in polar low forecasting will obviously depend on the forecasts of the other weather systems present. The scale of the polar low may change substantially from case to case, and other subsynoptic-scale systems can be present in the cold air outbreaks in addition to polar lows. A weather forecast in a polar low situation may therefore consist of forecasts of a variety of atmospheric scales. For this reason the forecaster will have to combine methods when preparing a weather forecast.

6.1.2 Synoptic and climatological rules for predicting polar low developments

Climatological studies have proved to be a valuable source of information for operational forecasting. Polar low tracks for several seasons have been established (see Chapter 2) showing the forecaster common genesis regions and likely tracks of polar lows.

Polar lows develop during cold air outbreaks and the synoptic-scale flow is often dominated by pronounced cold advection. Thus air may be descending and there will be rising pressure on the synoptic-scale. Polar lows which are embedded in regions of large scale descent, are frequently accompanied by rather modest pressure falls. Occasionally a region of slightly smaller pressure rise than elsewhere, or zero tendency may be an indication of a polar low development. Typically the pressure rise after the passage of a polar low is very large and confined to a fairly small area (Wilhelmsen, 1985). The large pressure rise is accompanied by strong winds, often reaching storm force 9–11 (23–31 m s^{-1}). Cold advection, which in the Northern Hemisphere appears as the combination of low-level northeasterly flow and upper (e.g. 500 hPa) northwesterly flow, is recognized amongst forecasters as the kind of situation where polar lows are likely to develop, particularly when upper positive potential vorticity (PV) anomalies, or alternatively troughs, with 500 hPa temperatures of $-40\,°C$ or below are involved. The last condition ensures that the vertical lapse rate allows deep convection to take place.

6.1.3 Applications of model forecast fields

The analyses and forecasts produced by NWP models have improved greatly over the last few decades and weather systems on the synoptic scale are now generally handled well, even in the polar regions. Also, an increasing number of small-scale systems, such as polar lows, are forecast with high resolution numerical models up to several days ahead. Striking examples of this were discussed by Nielsen (1997) (see Section 5.1.2, Woetmann Nielsen's case of 13–16 October 1993), and Woetmann Nielsen (1998) in Section 3.1.5), in which a polar low was correctly forecast more than $4^{1}/_{2}$ days ahead. However, in general, the representation of mesoscale systems is still rather poor for various reasons, including low resolution of the models, the lack of *in situ* initial data and the difficulties of handling processes such as convection and atmosphere–ocean–sea ice interactions in the polar regions.

Although prediction of specific mesocyclones and polar lows may not be possible (especially around the Antarctic) from the model output alone, it is possible to forecast areas where such vortices may be a threat. At the most basic level, the forecast charts of mean sea level pressure can be used to determine when and where cold air outbreaks over the ice-free ocean will occur, suggesting the possible development of mesoscale vortices. The satellite imagery from these areas can then be monitored for the early indications of cyclogenesis.

The synoptic-scale flow can provide additional information on where mesoscale developments may take place through the use of such quantities as the advection of upper level PV anomalies, which can act as precursors to polar low developments. An alternative approach is to consider upper-level cyclonic vorticity advection and regions of low-level maximum temperature advection. Fortunately these features associated with the synoptic-scale flow are often forecast correctly by the NWP models, which means that the formation of polar lows as noted above occasionally may be forecast several days ahead. An upper synoptic-scale vortex associated with large-scale cyclogenesis may advect upper PV anomalies to a favourable position for triggering a polar low development. It also tends to contribute to an increased Rossby penetration height due to the comparatively low static stability in such regions and the large average relative vorticity as expressed by relation (4.15) in Section 4.4.4. Examples of the use of such quantities in the forecasting of Arctic polar lows will be considered in Section 6.2.2.

6.1.4 A nowcasting approach using satellite data

In the nowcasting period of up to about 12 h ahead it is possible to use a variety of satellite tools to predict the development and movement of mesocyclones over the oceans and land areas. In this section we will examine how each form of data can be applied and then consider integrating the data within the forecasting process. Also, in the following sections a number of polar low developments, as observed by satellites, are discussed in order to assist forecasters in the development of their satellite interpretation skills, especially with regard to polar low forecasting.

Visible and infra-red satellite imagery

Most polar lows and mesocyclones occur over the ice-free ocean so that during the summer season the cloud signatures of the vortices are easily observed in the visible imagery against the low albedo ocean. This can be seen in Figure 6.1, which shows a mesocyclone over Prydz Bay, Antarctica on 4 March 1998. During the winter period, when little or no solar radiation is received, use must be made of infra-red imagery. This has the advantage of consistent imagery being received both day and night, although it can make the detection of some low-level vortices rather difficult if cloud-top temperatures are similar to the ocean or ice surface temperature. However, the texture of the cloud is often different from that of the surface so that it is usually possible to identify the cloud, especially if the contrast of the imagery is enhanced. But deep polar lows are easily observed because of the large temperature contrast between the cloud and the surface.

From the structure of the cloud field seen in the satellite imagery, i.e. the 'cloud signature', it is possible to get an idea of the physical processes involved in the development of the low, the likely strength of the low and its further development (see Section 1.6, 'Cloud signatures').

In the tropics and mid-latitude areas up to around 60° the geostationary satellites provide images at hourly or more frequent intervals allowing fast-moving developments to be monitored closely. Such data cannot be employed in the high latitudes, but the frequent overpasses of the polar orbiting satellites do allow sequences of images to be constructed that can provide guidance on the track of mesocyclones. Since the polar orbiting satellites will observe a particular location from a different viewing angle on each pass it is best to re-map the data onto a common area and map projection, although this is not essential if time is not available to process the data.

Another major use of infra-red imagery in forecasting mesocyclones is in obtaining an estimate of the height of the cloud associated with a low. In the more usual situation where there is a reasonable difference between the cloud top temperatures of the low and the surface, the height of the system can be estimated provided that some knowledge of the lapse rate is available. This can be from a nearby radiosonde ascent, a model temperature profile or the climatological conditions. Such a height estimate is clearly not going to be very accurate but should give a reasonable indication of the general depth of a system.

A further application of satellite imagery is to try and get an estimate of the winds associated with mesocyclones. Sequences of images from geostationary satellites have been used for many years to derive cloud track winds over the tropical and mid-latitude areas and such data are used routinely in the analysis process. However, similar techniques can be used when several images from

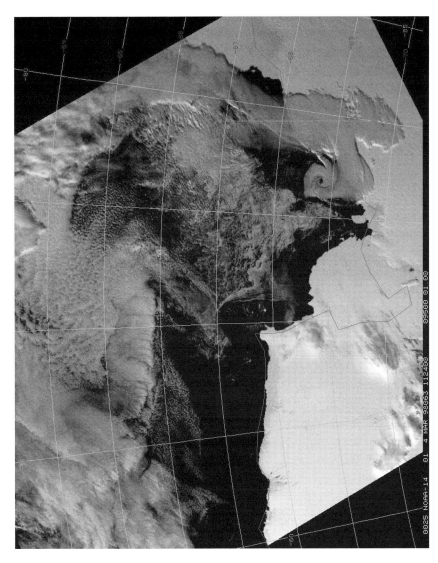

Figure 6.1. A visible wavelength satellite image of a mesocyclone over Prydz Bay, Antarctica on 4 March 1998.

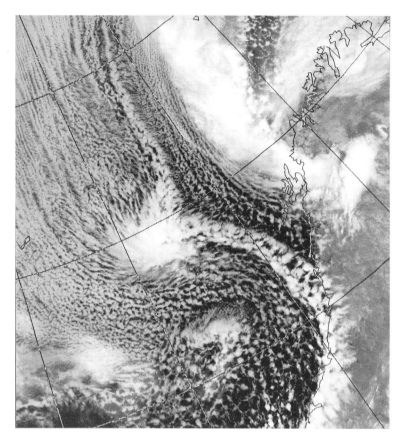

Figure 6.2. An infra-red satellite image at 0240 GMT 27 March 1995 showing a polar low off the Norwegian coast. (Image courtesy of the NERC Satellite Receiving Station, University of Dundee.)

the polar orbiting satellites are available, provided that the time separation of the images is not too great. Of course only winds at the level of the cloud tops can be produced, but such data have proved to be of value in mesocyclone research (Turner *et al.*, 1993b; Turner and Ellrott, 1992, Rasmussen *et al.*, 1996) and potentially have value in operational forecasting.

The surface wind speed may also be estimated from the general shape of the cloud elements in visible or infra-red satellite imagery. Figure 6.2, which shows a polar low in the Norwegian Sea, illustrates elongated cloud features indicative of very strong winds, exceeding 30 kt according to the schematic patterns presented in Figure 6.3. A more general, empirical rule, applicable for active polar lows, states that the wind speed may exceed the general, 'background' flow by 3 to 4 units on the Beaufort scale (Wilhelmsen, 1985). Thus if the synoptic-scale flow is Beaufort force 5, a polar low is likely to produce winds reaching force 8.

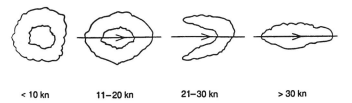

Figure 6.3. Schematic cloud shapes for different wind speeds (in knots). The arrows show the wind direction (from Pearson and Stogaitis, 1988).

Finally the use should be mentioned of night time visual images. During periods with moonlight, excellent satellite images in visible light have been obtained by the USAF DMSP satellites. Examples of such images are shown later in this chapter.

Surface winds from satellite scatterometers

Estimating the near-surface wind speeds associated with a mesocyclone or polar low is extremely important in order to determine whether the system is likely to disrupt air or maritime operations. However, as most mesoscale vortices occur well away from the synoptic reporting stations and automatic weather stations (AWSs) it is necessary to use the satellite data to get some idea of the winds associated with the lows. As discussed earlier, sequences of images can give a general impression of the winds at cloud-level which, provided the clouds are situated in a well mixed boundary layer, may give a good approximation to the surface wind (Rasmussen *et al.*, 1996). Over ice-free ocean areas surface wind vectors can be determined using measurements from the wind scatterometers carried on a number of polar orbiting satellites. These instruments are radars and measure the surface roughness of the ocean from a number of different viewing angles as the satellite passes over an area. With suitable processing, the backscatter measurements can be converted into wind vectors for cells with a resolution of about 50 km across a swath several hundred kilometres wide (Cavanie and Gohin, 1995).

Raw data from the scatterometers are processed into wind vectors by the satellite operators and some other organizations and distributed to end users on the meteorological Global Telecommunications System (GTS). The scatterometer winds are used in operational assimilation schemes and also plotted on some surface analyses. With their resolution of about 50 km the fields of scatterometer winds are able to provide information on the circulation of mesocyclones and the troughs associated with frontal bands. However, the data assimilation schemes are principally designed to produce analyses on the synoptic scale so that for forecasting of mesocyclones it is much better to use the scatterometer wind vectors directly rather than analyses that have had the vectors assimilated. For use in the polar regions, the vectors can be sent directly to the high latitude stations for plotting or transmitted in the form of charts with the vectors drawn.

The scatterometer winds are useful for showing whether a particular vortex has a surface circulation and for determining the maximum surface wind speed associated with a system. Figure 6.4 illustrates the case of a mesocyclone over the Scotia Sea on 5 June 1993 via infra-red imagery and a swath of scatterometer surface wind vectors. The cloud associated with this system consisted mainly of a linear cloud band with only a slight indication of rotation at the northern end. However, the scatterometer winds shown in Figure 6.4b indicate that the low had a tight surface circulation with wind speeds of up to 30 kt, showing that it was an active mesoscale low. This case also shows two problems that are associated with scatterometer data. First, there is considerable loss of data caused by the dealiasing algorithm failing, due mainly to the unavailability of backscatter data from one antenna. Second, the block of winds with a uniform speed and direction at the southern end of the swath are due to a problem known as 'granularity', which can result in non-meteorological fields of winds if the wind speeds are low. Overall, though, scatterometer winds are a valuable supplement to conventional satellite imagery in the forecasting process provided that the problems known to be associated with the data are taken into account. The case discussed above, along with other examples of mesocyclones observed with scatterometer data, is dealt with in more detail by Marshall and Turner (1997a).

Passive microwave data

Data from passive microwave radiometers on the polar orbiting satellites can provide information on a number of geophysical parameters that are of value in forecasting mesocyclones at high latitudes. Instruments such as the Special Sensor Microwave/Imager (SSM/I) on the DMSP satellites observe the Earth at several wavelengths and give imagery with a horizontal resolution of 12.5 or 25 km, depending on the channel. These data can then be processed to give fields of surface wind speed (Goodberlet *et al.*, 1989), integrated water vapour (Claud *et al.*, 1992) and rain rate (Dalu *et al.*, 1993) over the ice-free ocean. The wind speeds can be used in a similar way to the scatterometer observations to find the wind strength associated with particular lows, although no information is available on the circulation (Rasmussen *et al.*, 1996). Being able to determine the amount of precipitation falling from a mesoscale low is particularly valuable, although most of the algorithms used to date have been tuned for mid-latitude conditions. Nevertheless, work is under way to investigate the capabilities of passive microwave data to help in

6.1 Aspects common to both polar regions 509

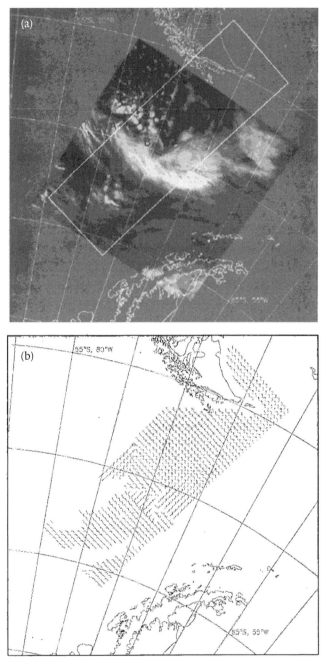

Figure 6.4. Satellite data of a mesocyclone over the Scotia Sea on 5 June 1993. (a) Infra-red imagery for 1037 GMT; (b) a swath of scatterometer surface wind vectors at 1217 GMT (from Marshall and Turner, 1997a).

understanding the precipitation of mesocyclones (Carleton *et al.*, 1993) and further developments of this potentially very valuable tool can be expected in the future.

Synthetic Aperture Radar data

Recently, data from the Synthetic Aperture Radar (SAR) on the RADARSAT-1 satellite have been applied to polar low studies (Chunchuzov *et al.*, 2000; Moore and Vachon, 2002). Spaceborne SAR imagery when combined with conventional meteorological data can provide new insight into the development, structure and dynamics of polar lows and as such assist in their forecasting. SARs are sensitive to changes in the surface roughness induced by the short wavelength surface gravity-capillary waves on the ocean surface. As such they are able to sense fluctuations in the surface wind field. Data from RADARSAT-1 are of particular interest to meteorologists as the instrument has a swath width of up to 500 km at a spatial resolution of 100 m. Due to the high resolution of the imagery, as well as the ability of the SAR to penetrate clouds, the SAR images contain a wealth of information on the structure of polar lows.

6.2 Aspects of forecasting specific to the Arctic

6.2.1 Introduction

Most forecasters working in high northern latitudes have been concerned with polar lows, the more vigorous class of mesoscale vortex, since such systems can have winds of 60 kt (30 m s^{-1}) or stronger, and can pose a great danger to shipping and coastal communities. Traditionally, forecasters have mainly used conventional meteorological data and satellite imagery in forecasting the development and movement of such systems, but numerical models are now starting to have some success in predicting these lows (see Section 5.1).

Remote sensing techniques and NWP model output constitute indispensable tools in forecasting, particularly when they are combined with traditional methods, such as subjective analysis of synoptic weather charts. In this section we will assess how models handle Arctic polar lows through a series of case studies, and consider the use of satellite data at high northern latitudes.

6.2.2 The use of model data in the Arctic

Polar lows accompanied by severe weather present a serious threat to offshore operations in polar and Arctic regions. An important subproject under the Norwegian Polar Lows Project (NPLP) carried out during 1982–85 was dedicated to documenting and evaluating of methods of forecasting polar lows. The results of these studies were published in the technical report, *Polar Low Forecasting*, Part I: *Methods and Evaluation* (Midtbø, 1986), and the main findings can be found in the final report of the project Lystad (1986). A summary of this still relevant work on polar low forecasting is presented below.

Methods for polar low forecasting were developed on the basis of experience gained in the Norwegian weather service in accordance with the goal that the methods should be useful in the operational community. A number of the these methods were tested during a quasi-operational routine in which a group of meteorologists assessed methods for monitoring polar lows. Detailed analyses of situations in which polar lows were observed were a valuable source of information in improving polar low forecasting as well. A number of such *a posteriori* case studies were carried out as part of the project and published in technical reports from the project.

> Identification of weather situations favourable for polar low developments (adjacent to Norway)

As part of the polar low forecasting project a coarse classification scheme based on the location of the polar low developments relative to the large-scale pressure systems was used, rather than a classification scheme based on the differences in the fundamental mechanisms. According to this classification scheme (see Lystad *et al.*, 1986) based on a study of several developments, polar lows could be classified as belonging to one of four groups, A-D as listed below:

> *Type A*[1] *: Polar lows developing in major cold air outbreaks*

This type of system develops in a more or less stationary synoptic situation with a northeasterly surface flow and the thermal wind directed opposite to the surface wind. An example of such a situation, i.e. a 'reverse shear example', was illustrated on the synoptic chart shown on Figure 3.4.

> *Type B: Polar lows developing in troughs behind a synoptic-scale low*

This type of polar low is often connected to a synoptic-scale cyclone on the polar front, or develops in *warm* troughs (Figure 6.5). A number of such cases studied by Grønås *et al.* (1987a) were associated with a 'remarkable tongue

---

[1] The type A and type B polar lows, according to the classification by Lystad *et al.* (1986), should not be confused with the Type A and Type B baroclinic developments discussed in Section 4.2.

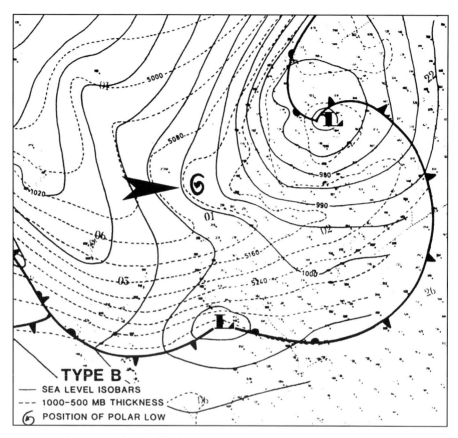

Figure 6.5. A chart indicating a type B polar low development (indicated by an arrow) according to the classification scheme by Lystad (1996). (Figure used courtesy of the Polar Lows Project, Norwegian Meteorological Institute.)

of warm air in the boundary layer southwest of Spitsbergen', probably formed by differential heating connected with a similar tongue of warm sea surface water in the area (see Figure 2.13). Also, the edge of the sea ice has a shape like a tongue or a triangle northwest of Bear Island.

In other cases the warm tongue will be connected to an occlusion associated with a synoptic-scale cyclone. An example of a 'warm trough development' on 29 February–1 March 1984 illustrating this important type of system was discussed by Grønås *et al.* (1987a) and Rabbe (1987). In this case a warm surface trough formed as an initial warm temperature tongue and was steered southwards over the Norwegian Sea, as illustrated on Figure 6.6. On Figure 6.7, showing a satellite image from 1456 GMT 29 February 1984, the warm trough is seen as a cloud band (shown by an arrow), north of which a system of long

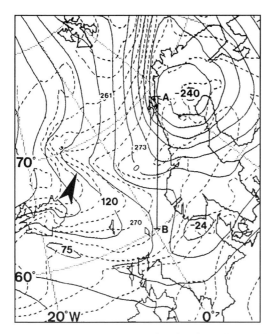

Figure 6.6. Prediction of 1000 hPa height and potential temperature (interval 3 K) showing a warm trough (indicated by an arrow) off the Norwegian coast.

regular cloud streets indicated a strong cold outbreak. During the night of 29 February/1 March the cloud band broke up into several cloud clusters, two of which moved south. As the clusters approached the Norwegian coast they started to rotate. The main development resulting in the formation of the polar low seen on Figure 6.8 took place over 6–10 h as the low approached the coast near Ålesund. The wind record from Ålesund (Figure 6.9) shows that the mean wind increased rapidly from 13 m s^{-1} to almost 26 m s^{-1}, with gusts up to 36 m s^{-1} while the wind direction varied almost 180°.

Type C: The comma cloud type of polar low

This type of polar low is closely connected to an upper-level trough within the cold air mass north of, but generally relatively close to, the polar front (see Figures 1.4 and 1.5). This type of polar low is relatively rare at high latitudes.

Type D: Mesoscale baroclinic waves

This group constitutes polar lows for which baroclinic instability is the major mechanism of development. An example of this type of development will be discussed in Section 6.2.6.

514   6 Forecasting of polar lows

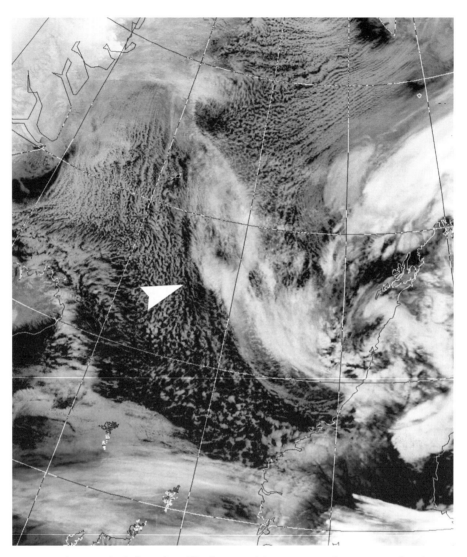

Figure 6.7. An infra-red satellite image at 1456 GMT 29 February 1984 showing a cloud band (indicated by an arrow) associated with the warm trough shown on Figure 6.6 (from Grønas *et al.*, 1987). (Image courtesy of the NERC Satellite Receiving Station, University of Dundee.)

Data from short, medium and even long range numerical forecasts can be applied to evaluate whether any of the four types of weather situations described above is likely to occur within the time span of the forecast.

Monitoring polar lows using data from a numerical model
The term 'monitoring' was defined in the Norwegian study as the forecaster's consideration and interpretation of all relevant data, and the fitting of

6.2 Aspects specific to the Arctic 515

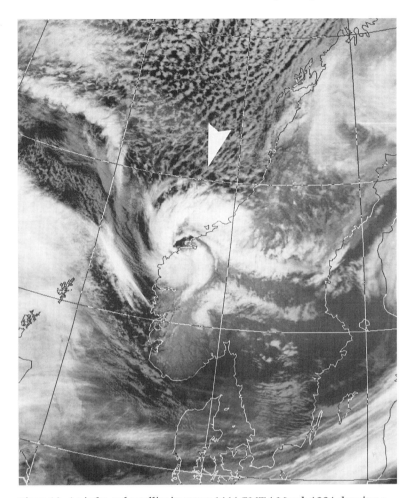

Figure 6.8. An infra-red satellite image at 1444 GMT 1 March 1984 showing a well-developed spiral polar low (indicated by an arrow) at the Norwegian coast. (Image courtesy of the NERC Satellite Receiving Station, University of Dundee.)

those data into the most probable three-dimensional model of the atmosphere, including changes to be expected within the immediate future (*c.* 3 h). This synthesis of all data available is more than a mere analysis, and a prognosis of the expected changes a few hours ahead is essential for the monitoring.

Methods for monitoring polar lows were divided into several categories, briefly described in the following.

The first step is to find the areas where there is a significant probability for observing polar lows. The second step is the initial identification of the individual polar lows.

516 6 Forecasting of polar lows

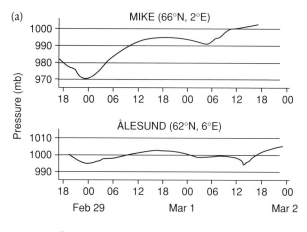

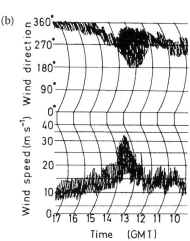

Figure 6.9. (a) Surface pressure record for 29 February–1 March 1984 at weather ship Mike and at Ålesund. (b) wind record at Ålesund airport for 1000–1700 GMT 1 March (after Rabbe, 1987).

Monitoring polar lows using a mesoscale numerical model

The Norwegian Polar Lows Project (NPLP) Report recommended the use of numerical analyses/prognoses from a mesoscale model as a basis for the monitoring of polar lows. In some cases the model will give a correct or nearly correct forecast of the time and location of the polar low development (e.g. Woetmann Nielsen's case: see Section 3.1.5). In other cases the model, while not actually succeeding in forecasting the polar low, on the other hand gives indications as to where a polar low will most likely form, as for example in troughs (see the Section below, 'Forecasting polar lows as troughs or vortices in the numerical model').

Some basic aspects which, according to the NPLP Report, should be considered when numerical output is used for polar low monitoring are listed below.

- The use of a *mesoscale* numerical model to provide the forecaster with the basic meteorological fields, such as height and temperature fields is recommended. These models have demonstrated a potential for forecasting warm surface troughs, the presence of which is an indication of the possibility of polar lows.
- Knowledge of the weather situations associated with polar lows gives the forecaster background for limiting the area where polar lows might be expected.
- A careful investigation of all baroclinic zones should be undertaken, and the possibility for growing baroclinic disturbances should be investigated, including disturbances in reverse shear flow.
- Systematic, *objective* forecasting methods such as described in the following have shown themselves useful for the monitoring process. Such methods seem to identify the majority of cases where polar lows develop, but the areas indicated are often large so that a more specific location is desirable.

To the points mentioned above should be added that the forecaster should pay close attention to the presence of upper-level troughs or vortices which often, as discussed at several places in the preceding chapters, may trigger the development of polar lows.

Polar low track forecasting using trajectories from a numerical model
As mentioned earlier, polar lows usually give severe weather in a short period of time at a fixed location and accurate time indications in the forecasts will therefore be of special importance. In this connection, the forecasts of the tracks of the polar lows are of special importance. The future track of a polar low may be estimated in various ways.

One method of forecasting the track is to extrapolate the observed movement of the centre of the polar low by simply extrapolating the velocity. A refinement to this is to extrapolate an acceleration as well. A second method is to use 850 or 700 hPa observations of the wind, or a numerical analysis of the same, assuming that the polar low moves with the wind velocity at one of these levels.

The two methods may give rather good forecasts some hours ahead (*c.* 6 h) but, in the experience of Norwegian forecasters, the quality of the forecasts decreases rapidly beyond that time.

Eidsvik (1985) used a statistical model to estimate the time prediction for the probabilistic structure of a polar low trajectory and demonstrated that forecasts of polar low tracks based on past motion and climatology were quite inaccurate more than a few hours ahead.

The forecasting method described below is based on the assumption that it is possible to find a steering current in the numerical prognoses, i.e. a level in which the horizontal component of the air motion approximates the motion of the polar low at the surface. Experience has shown that polar lows often follow the observed wind at the 700 or 850 hPa level.

Based on the Norwegian operational limited area model with a mesh length of 150 km (LAM150), trajectories were examined for most polar lows observed during the winter season 1984–85. In many cases large differences between the observed and prognostic tracks were observed. It was concluded that it was difficult to use the trajectories from the LAM150 model in forecasting the tracks of polar lows with great accuracy.

The trajectories of a number of polar lows associated with the strongest winds were also computed using a model (LAM50) with a higher horizontal resolution, i.e. 50 km. However, a comparison with the results from the LAM150 experiment did not reveal any significant differences between the forecasts of polar low tracks based on trajectories from the two models.

No specific comments concerning the movement of polar lows formed in reverse shear conditions were mentioned in the Norwegian studies. For this important group of polar lows, which tend to move very slowly or be almost stationary, the steering level should most likely be sought at a relatively high altitude.

Forecasting polar lows as troughs or vortices in the numerical models

A straightforward way of forecasting polar lows using a numerical model is to track individual lows as either a trough or vortex in the model output fields. In an evaluation based upon a number of forecasts with the Norwegian LAM50 model it was found that the majority of the polar lows were seen as troughs in the prognoses, but that some were seen neither as troughs nor as vortices. Most of the forecast troughs were too weak, and the intense inner cores of the polar lows were never fully described. On the other hand, only a few weak troughs were forecast by the model in situations when they were *not* verified by either troughs or polar lows. Since the middle of the 1980s when the Norwegian study was carried out, the ability of the numerical models to forecast polar lows has increased significantly, as shown by the excellent polar low forecast discussed by Nielsen (1997; see Section 5.1.2).

Use of numerical models in selecting areas where polar lows have a significant probability of occurrence

Based on experience gained during the Norwegian Polar Lows Project, a method was described (Midtbø *et al.*, 1986) for selecting areas where polar lows have a significant probability of developing over the seas around Norway in connection with outbreaks of cold air at the surface. The criteria chosen for selecting these areas were:

1. There must be cold air advection at the surface.
2. The thickness of the 850–500 hPa layer must be below 3960 m. This limit should vary with sea surface temperature: when the sea surface temperature is low, the limit should be lowered.
3. The curvature of the contour lines in the 500 to 700 hPa levels must be cyclonic or neutral.

These criteria select areas in some of the cold air outbreaks, and may be applied to analyses as well as prognoses. The areas selected are in principle independent of the range of the numerical predictions, but may vary between different regions.

The first requirement reflects the point of view taken by most Norwegian meteorologists that a northerly flow is a necessary condition for polar lows to develop in the Norwegian and Barents Seas. Grønås and Kvamstø (1995), for example, state that 'In the Norwegian and Barents Seas, northerly flow and the Arctic inversion seem to be a necessary condition for polar lows to develop'. It should be noted though that exceptions to this rule can be found, e.g. in connection with the strong development on 13 October 1971 illustrated in Figure 3.1 (p. 153) and the corresponding radiosonde ascent shown on Figure 4.26 (p. 338), when an Arctic inversion was *not* present.

The second criterion gives a rough indication of the static stability, as the sea surface tends to determine the near-surface air temperature. The minimum value of the 850–1000 hPa thickness can be replaced with other parameters, such as the 500 hPa temperature, to represent the stability.

The third criterion reflects the experience that polar lows are generally triggered by an upper-level disturbance in the form of a cold, short-wave trough.

As an evaluation of the method of selecting areas where polar lows have a significant probability of developing using the rules mentioned above, all available short-range prognoses from the LAM50 model were investigated for a period of two and a half months during which areas where the criteria were fulfilled were marked on the 500 hPa charts. A rough verification was carried out by comparing the time when a polar low was observed, to the time when a polar

Table 6.1. *The results of an assessment of the ability of LAM 50 forecast model data to predict the occurrence of polar lows in the ocean areas adjacent to Norway (see the text for details).*

		Analyses		
		Yes	No	Sum
Prognoses	Yes	40	54	94
	No	0	43	43
	Sum	40	97	137

low was forecast within one large area composed of the seas adjacent to Norway. The results based on 137 prognoses are given in Table 6.1.

As can be seen from Table 6.1, 40 out of the 94 prognoses were confirmed by observed polar lows. *All* the observed polar lows were forecast, but the method selected more than twice the number of verified cases. To limit the number of cases selected some tighter criteria are therefore desirable.

As mentioned above, accurate time and location indications are of special importance for polar low forecasting. To test the ability of the method discussed above, the sea areas adjacent to Norway were divided into subareas before the verification was carried out. The results of this showed that the method was not successful. The areas within which polar lows were forecast to develop were often wrong, and in many cases too many subareas were selected. However, in all cases when a polar low was observed in an area which was not selected by the forecast, a neighbouring area was always selected.

The results discussed in the preceding reflect the state of the art of forecasting polar lows in the mid-1980s. Since then the performance of numerical models has improved, and model output for determining the formation and development of polar lows can, on many occasions, be used successfully in the forecasting process. For example, the propagation and development of upper synoptic-scale troughs are generally well predicted by the models.

Grønås and Kvamstø (1994) have demonstrated that charts mapping the height of the tropopause, defined as the 2 PVU surface, combined with charts depicting the height of the convective boundary layers are useful in identifying regions susceptible to polar low cyclogenesis (see Section 5.1.2, Grønås and Kvamstø's case).

The regions of deep convective boundary layers are generally easily identified from satellite imagery. The satellite infra-red image presented in Figure 6.10 shows pronounced 'streets' of organized convective clouds. The cold (and high level) cloud tops indicate a deep mixed layer, in contrast to the

Figure 6.10. An infra-red satellite image at 0829 GMT 26 March 1995 showing pronounced 'streets' of organized convective cloud over the Norwegian Sea. (Image courtesy of the NERC Satellite Receiving Station, University of Dundee.)

low-level cloud tops seen in the surrounding areas where a capping inversion prevents a deep convective layer developing. The low static stability associated with the convective boundary layer contributes to a large Rossby depth. If a NWP model forecasts advection of an upper PV anomaly (or upper positive vorticity advection) over such a region, cyclogenesis is likely to take place.

In addition to the several case studies presented in Chapter 5, a case study from March 1995 is discussed below in order to highlight the use of the theory discussed in Section 4.4 and some of the forecasting tools referred to in the preceding sections.

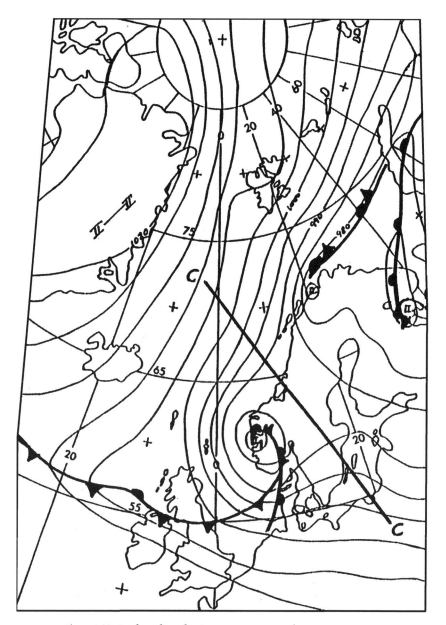

Figure 6.11. Surface chart for 1200 GMT 26 March 1995.

Case study: Norwegian Sea 26–27 March 1995

Figures 6.11 and 6.12 show respectively a synoptic analysis at 1200 GMT 26 March 1995 and an infra-red satellite image from 1825 GMT 26 March. A deep cyclone over Scandinavia established a cold northeasterly and northerly flow over the Norwegian Sea. High cloud tops, easily identified on

6.2 Aspects specific to the Arctic 523

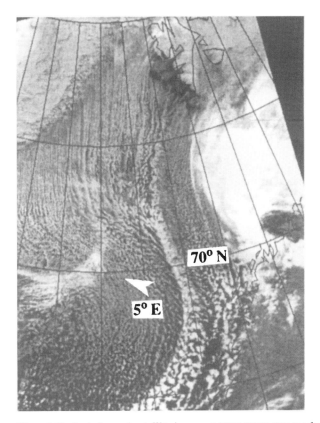

Figure 6.12. An infra-red satellite image at 1825 GMT 26 March 1995 showing a comma-like cloud (indicated by an arrow) west of a line of deep convection associated with a zone of convergence caused by a deflection of the low-level flow around Svalbard.

the satellite image, showed a region of strong convection stretched from the area southwest of Spitsbergen and across the Norwegian Sea. Due to the wedge-shaped area of open water between Spitsbergen and Greenland the Arctic air coming from this region travelled across a longer path of open water than elsewhere. The air also crossed over the region of high SSTs (associated with the northern branch of the North Atlantic current), which enhanced the transfer of sensible heat and moisture to the Arctic air mass. As indicated in the 1200 GMT analysis (Figure 6.11), the northeasterly flow was deflected by the Svalbard islands which created a convergence zone that contributed to the organized convection further south. Figure 6.13 shows the 500 hPa height and 1000–500 hPa thickness analyses at 1200 GMT on 26 March. A short-wave trough over northeast Greenland was moving southeastwards. However, the associated vorticity advection took place in a region of cold advection and this makes a qualitative

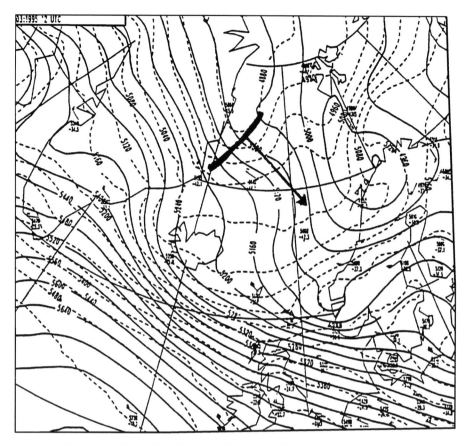

Figure 6.13. The 500 hPa height and 1000–500 hPa thickness analyses for 1200 GMT 26 March 1995. The thick black line indicates the position of a cold-short wave trough over the Greenland Sea.

assessment of the dynamic forcing terms in the quasi-geostrophic omega equation rather difficult. Ascent was forced by the upper vorticity advection while the maximum of cold advection in the same area forced descent. From a PV perspective, the dynamical upper forcing becomes clearer. The PV chart for the same time, shown in Figure 6.14, indicates a PV anomaly on the 290 K isentropic surface (at about 500 hPa in the region of interest) moving southeast and with pronounced PV advection associated with it.

As an indication of upper PV advection, the satellite image from 1825 GMT (Figure 6.12) showed a comma-shaped cloud, indicated by an arrow, west and northwest of the convergence zone. A comparison with the numerical analyses of height and PV fields shown in Figures 6.13 and 6.14 respectively gives a fairly strong indication that the NWP analyses were correct in predicting the presence of the upper trough or PV anomaly. Judging from the satellite image

Figure 6.14. PV analysis for 1200 GMT 26 March 1995 on the 290 K isentropic surface (from Røsting *et al.*, 1996).

from 1825 GMT, the PV anomaly at this time was rapidly approaching the convergence zone and a polar low development was expected to take place within a few hours. The satellite image at 0240 GMT 27 March (Figure 6.2) shows that the two cloud systems had merged by this time. A wave had formed on the convergence zone as the polar low developed. Figure 6.15 shows the satellite image at 1054 GMT 27 March as the polar low approached the Norwegian coast and Figure 6.16 shows the corresponding synoptic analysis. The elongated cloud structures located mainly west and south of the polar low are an indication of strong winds. According to the description given by Pearson and Strogaitis (1988) and reproduced in Figure 6.3, the wind may be estimated to be more than 30 kt.

6.2.3 The use of satellite data in the Arctic

Satellite images are an important tool for detecting polar lows. A regular supply of images is considered to be indispensable and, according to Fett (1989a) is a requirement in detecting polar lows and monitoring their evolution. The forecaster may know in which regions polar lows are possible without having enough conventional observations to verify that belief. In this case a satellite image will often be the only tool to determine whether there are any polar lows or not.

526   6 Forecasting of polar lows

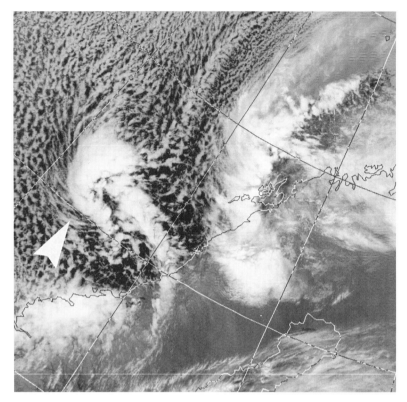

Figure 6.15. An infra-red satellite image at 1054 GMT 27 March 1995 showing a polar low (indicated by an arrow) approaching the Norwegian coast. (Image courtesy of the NERC Satellite Receiving Station, University of Dundee.)

Experience in the interpretation of satellite images is important for polar low forecasting and a considerable number of such images, illustrating different types of polar lows, were shown in Section 1.6 and Chapter 3. Through an examination of these images the forecaster should get an idea as to what polar lows look like at different stages in their development. In this connection, the analogy often referred to between tropical cyclones and polar lows should be mentioned. In a few cases, as illustrated on Figure 5.4, this analogy is striking indeed. However, while mature tropical cyclones have a structure surprisingly similar from case to case, partly due to forcing from a uniform sea surface (Anthes, 1982), polar lows (as seen from satellites) have a rather variable structure reflecting the many forcing mechanisms contributing to their development.

Upper PV anomalies can be detected in water vapour imagery available from the Meteosat geostationary satellite for developments south of 65° N. They appear as comparatively dark features in these images, an indication of

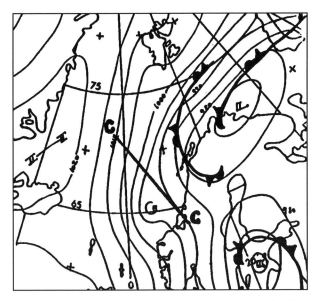

Figure 6.16. Surface analysis for 1200 27 March 1995. The line C–C is discussed in the paper (from Røsting *et al.*, 1996).

the lowering of the tropopause associated with upper PV anomalies or upper troughs (Mansfield, 1994; Røsting *et al.*, 1996). However, the contrast in these images between the dry areas associated with upper positive PV anomalies and the surrounding air, even high clouds, tends to be poor due to the very cold tropospheric air north of the polar front zone.

Although high quality satellite images are essential in many ways they should not be used as the only tool for polar low monitoring and forecasting. It may be difficult on the basis of an image alone to determine whether a polar low is present or not. Also the strength of the low, expressed by the strength of the associated surface wind speed, is difficult to evaluate based on satellite images alone.

6.2.4 The use of weather radar data for monitoring and forecasting polar lows

Weather radars are widely used for monitoring subsynoptic weather phenomena, such as thunderstorms, and in principle have a great potential for monitoring polar lows due to their high resolution of data in time and space.

Monk *et al.* (1986) used precipitation data derived from radar to clarify the speed and direction of movement of two polar lows and to estimate the likely locations of very strong surface winds. A close relationship between cell speeds as observed by radar and peak surface gusts was apparent in the study, and evidence derived directly from radar data led forecasters to correctly predict the places likely to experience gales.

Weather radars were not available in the areas studied during the Norwegian Polar Lows Project, but the possibilities for installing radars at the coast or at the drilling sites at sea were considered. However, compared with satellites, a radar monitors a very limited region, typically an area of radius only 300 km, and for obvious reasons 'conventional radars' are not likely to play an important role in polar low monitoring and forecasting in the near future. On the other hand, synthetic aperture radars (SARs) may, as briefly discussed in Section 6.1.4, play an important role in the future.

6.2.5 Forecasting at specific locations in the Arctic

The Nordic Seas

The forecasting rules for the Nordic Seas discussed in the preceding section were all based on experience gained during the Norwegian Polar Lows Project. These rules should be supplemented with material from a discussion of several polar low developments (illustrated by numerous high quality satellite images) in the region presented by Fett (1989a).

Fett stressed the role of boundary layer fronts (BLFs) for polar low development noting that these fronts were 'apparent in a number of examples noted in the Greenland, Norwegian and Barents Seas' (for a discussion of BLFs see Section 3.1.4).

Among the cases discussed by Fett was a development in northeast Greenland (19–23 December 1983). This case, which illustrates many of the forecasting rules presented in this chapter was also discussed in some details by Hoem (1985), who in his final comments to his study wrote:

> This particular weather situation is an excellent example of the value of using satellite pictures together with weather charts etc. in the analysis. In this case the upper air charts in particular were of invaluable help in selecting the area where a development was possible. Because this region is data sparse, the satellite images have made it possible to reveal the first development of the polar low about 24 hours before it had been possible by conventional methods (charts and radiosonde ascents)…

The first satellite image available for Hoem in this study was from 1627 GMT 20 December 1983 which showed the first stage of a baroclinic wave development at the northeast Greenland coast. Hoem in his analysis exclusively applied satellite images received in either Tromsø or Oslo. Fett on the other hand had access to DMSP satellite images including night-time visual images which enabled him to follow the very early stages of this development. The two studies give, when combined, an unusually well documented description of a significant polar low development. In the following we will present a summary of the two studies which, apart from being a well-documented case study, illustrates the way a forecaster may use his material in order to predict the development of a polar low.

The initial development from a cold surge in the Fram Strait

BLFs in the Fram Strait and other locations move in the direction of the low-level flow or 'surge'. Figure 6.17 shows such a surge through the Fram Strait in a DMSP infra-red image for 2318 GMT 19 December 1983. An enhanced cloud line looking very much like a rope-cloud delineates the frontal position. Note that the open-celled cumulus clouds in the strait do not exist all the way up to the frontal boundary but change form to overcast stratus 40–60 nm (nautical miles; *c.* 70–110 km) behind the front. Winds are believed to turn strongly anticyclonically in this region to a northeasterly direction. This caused a change from open to closed cell or overcast stratus and helped create the cyclonic vorticity maximized at the frontal location. The cold surge apparently commenced as the high pressure ridge seen along the northeast Greenland coast at 0600 GMT 20 December and the pressure gradient over the Fram Strait intensified as illustrated on the surface chart shown on Figure 6.18.

The series of upper air charts in Figure 6.19 starting at 0000 GMT 20 December (Figure 6.19a) reveal the movement of an upper-level cold trough, which moved from a position northeast of Greenland, south and southeastwards over the Greenland Sea on 20 and 21 December 1983. At 0000 GMT 22 December a well-developed upper-level closed low had formed in the region around Jan Mayen at the southern end of the trough (Figure 6.19c).

Concerning the surface analysis from 1200 GMT 20 December, the Fleet Numerical Oceanography Center (FNOC) surface analysis (shown by Fett) ignored a ship observation located near 76° N, 0°, which reported a surface southeast wind of 15 m s^{-1} and a pressure of 985.2 hPa. Nor was the observation plotted on the Norwegian post-analysis from 1200 GMT 20 December shown by Hoem (1985) and reproduced here as Figure 6.20 (the said ship observation on Figure 6.20 being entered by the editors). The Norwegian analysis, together with the satellite images, indicates that the wind reported from the ship might be true. On the other hand, the extremely low surface pressure reported from the ship is obviously wrong, a point of view which was confirmed by the following surface analysis.

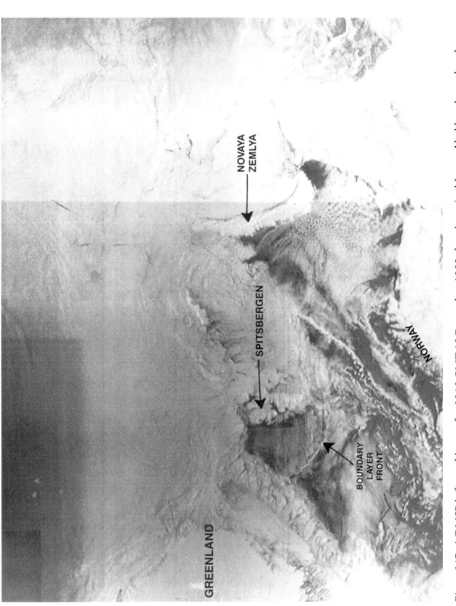

Figure 6.17. A DMSP infra-red image for 2318 GMT 19 December 1983 showing a 'cold surge' led by a boundary layer front (indicated by an arrow), through the Fram Strait (from Fett, 1989a). (Figure provided courtesy of the Naval Research Laboratory, Marine Meteorology Division, Monterey, CA, USA.)

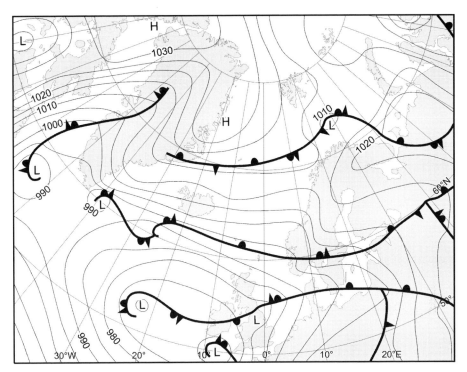

Figure 6.18. Surface analysis for 0700 GMT 20 December 1983 (based on the Berliner Wetterkarte, Freie Universität Berlin).

A satellite image from 1421 GMT 20 December (Figure 6.21), corresponding to the surface chart shown on Figure 6.20a, verified the existence of a developing system with a shallow cloud deck off the sea ice of northeast Greenland. The position of the 500 hPa trough at 1200 GMT (taken from the 500 hPa analysis shown on Figure 6.20b) has been superimposed on the image. It can be seen that the development occurred in advance of this trough in a region of upper-level positive vorticity advection. The low-level vortex was still not well developed through the middle troposphere at this time, as can be seen by the relatively warm temperatures of the overlying cloudiness. The structure of the cloud field (no deep convection was observed), together with the fact that the initial development took place within a pronounced baroclinic zone (see Figure 6.20b) as an upper-level trough was approaching the region, strongly indicates that the development was a baroclinic type B development.

Early on 21 December a satellite image (not shown) revealed a much colder cloud deck associated with the vortex indicating that a further development had taken place.

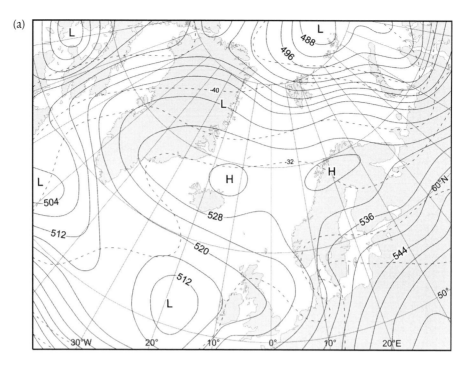

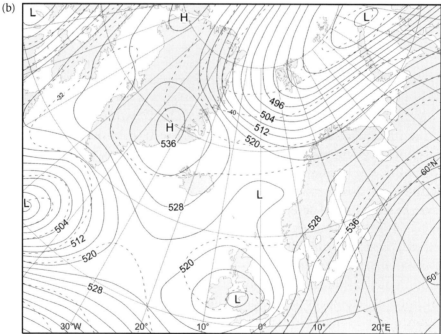

Figure 6.19. 500 hPa analyses for (a) 0000 GMT 20 December 1983, showing an upper-level trough north of the Fram Strait between northeast Greenland and Svalbard; (b) 0000 GMT 21 December, showing the upper-level trough further south over the Greenland Sea; (c) 0000 GMT 22 December, showing an upper-level cold low which formed at the base of the trough shown on the two previous figures (based on the Berliner Wetterkarte, Freie Universität Berlin).

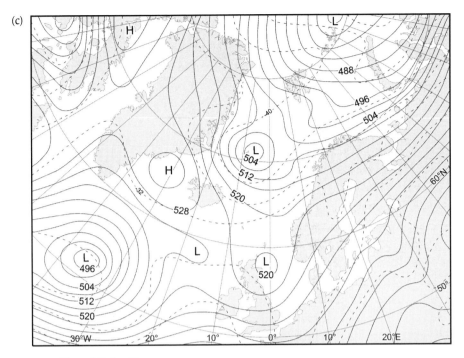

Figure 6.19 (cont.).

A *night-time visual* image (Figure 6.22) from 0401 GMT 21 December shows that two secondary vortices had developed east of the major vortex. It is not clear from the satellite image which, if any, of the vortices was destined to undergo the intensification normally associated with a polar low.

New satellite images during the morning and early afternoon, an example of which is shown on Figure 6.23, revealed that the cloud system apparently intensified and took on the appearance of a small comma cloud, which is often noticed during rapid intensification of baroclinic waves (see Figure 1.21c). Around this time winds of 20–25 m s^{-1} should be anticipated under the cloud shield in the northern quadrant of the storm with lesser winds to the south (Fett, 1989a).

An impressive view of the polar low was gained via another night-time visual DMSP satellite image (Figure 6.24) of the polar low on the evening of 21 December. The polar low had intensified and strong winds were suggested particularly in the western sector of the storm where long and tightly packed cloud lines were indicative of winds in the 20–25 m s^{-1} range. Note the resemblance between the cloud system seen on Figure 6.24 and that associated with the Labrador Sea polar low in Figure 3.53.

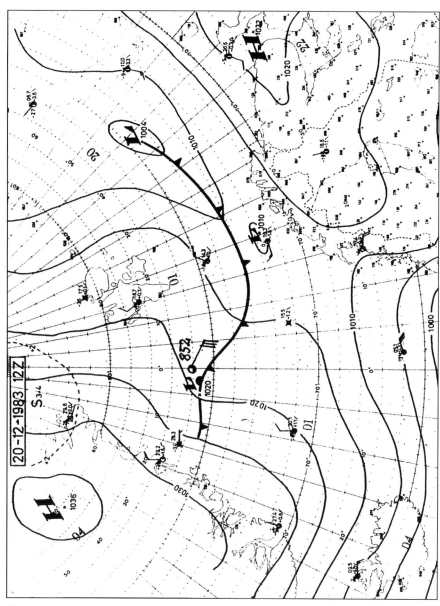

Figure 6.20. Analyses for 1200 GMT 20 December 1983. (a) Surface pressure, (b) 700 hPa height (solid line) and 1000–700 hPa thickness (broken line) (after Hoem, 1985). (Figure used courtesy of the Polar Lows Project, Norwegian Meteorological Institute.)

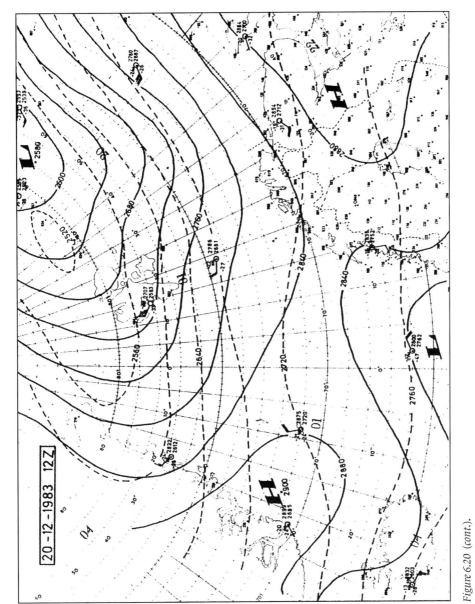

Figure 6.20 (cont.).

536   6 Forecasting of polar lows

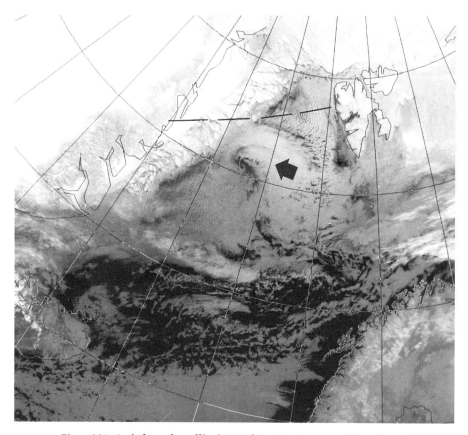

Figure 6.21. An infra-red satellite image for 1421 GMT 20 December 1983, showing a developing system in the form of a low-level 'cloud-vortex' (indicated by an arrow) south of the Fram Strait, off the coast of northeast Greenland. The broken line indicates the position of the 500 hPa trough at 1200 GMT. (Image courtesy of the NERC Satellite Receiving Station, University of Dundee.)

A satellite image from 0403 GMT 22 December (Figure 6.25) shows that the low-level vortex at that time was ringed by spiral cloud bands and was the dominant feature of the Greenland/Norwegian Sea. The post-analysis (Hoem, 1985) from 0000 GMT 22 December shown on Figure 6.26 shows that the diameter of the closed circulation at the surface was rather small so that detection was only feasible when satellite data were utilized to assist the analysis. Typically some of the major analysis centres failed to resolve the storm on their synoptic analyses.

The 500 hPa height analysis from 0000 GMT 22 December was shown on Figure 6.19c and the position of the centre of the upper-level cold vortex was determined precisely by the two nearby radiosonde ascents. The system at

Figure 6.22. A DMSP night-time visible image of the Iceland–Greenland area at 0401 GMT 21 December 1983, showing the polar low in the form of a multiple vortex system (from Fett, 1989a). (Figure provided courtesy of the Naval Research Laboratory, Marine Meteorology Division, Monterey, CA, USA.)

538   6 Forecasting of polar lows

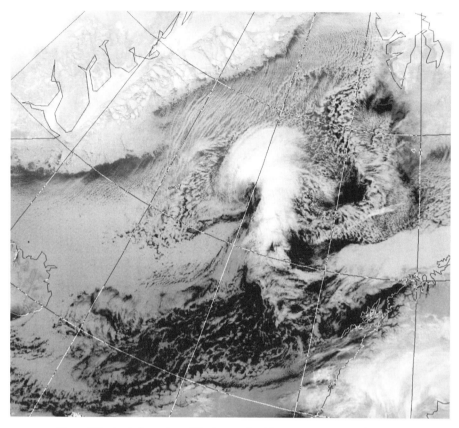

Figure 6.23. An infra-red satellite image for 1409 GMT 21 December 1983, showing a comma cloud associated with the development of the baroclinic polar low. (Image courtesy of the NERC Satellite Receiving Station, University of Dundee.)

that time could be described as a deep, vertically-aligned, cold core vortex, intensifying with height. The strong upper-level vortex seen on Figure 6.19c formed during 21 December as the result of a baroclinic development within the strong baroclinic zone seen on Figure 6.20b.

A final satellite image from this development (1357 GMT 22 December 1983) is shown on Figure 6.27. The cloud field at this time was characteristic of a cold core system, but as explained above, the system in this case did not originally develop out of a pre-existing upper-level cold core system as is often observed, but instead formed as the end product of a baroclinic development.

Later, on 23 December, the polar low merged with a low centre that had been moving northward from Great Britain, but, according to Fett, coastal stations

6.2 Aspects specific to the Arctic   539

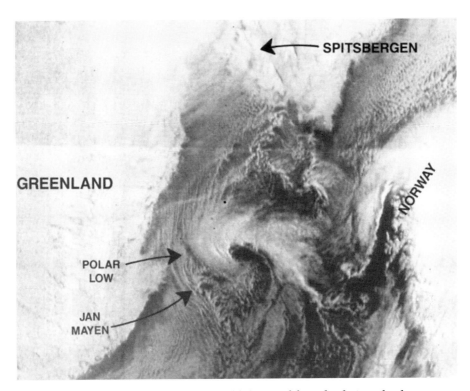

Figure 6.24. A DMSP night-time visible image of the Iceland–Greenland area at 1913 GMT 21 December 1983, showing the polar low around the time that it reached its maximum intensity (from Fett, 1989a). (Figure provided courtesy of the Naval Research Laboratory, Marine Meteorology Division, Monterey, CA, USA.)

in Norway were later surprised by the unexpected intensity of the winds affecting the region as the storm moved onshore.

Based on this case, Fett drew the following important conclusions:

1. Cold surge events and the progress of the resulting BLF should be monitored because of their role in polar low developments.
2. High quality satellite data are required at frequent intervals (every 2–3 h) to succesfully monitor evolutionary patterns of polar low development.
3. Observations from ships and island stations locations in the Arctic should be closely monitored for unusually pronounced 3 h pressure changes. These often signal the approach or passage of a polar low.
4. Before disregarding conventional observations showing unusually strong winds from odd directions or unusually low pressures, a check with satellite data should be made to ensure that there are no nearby unexpected storm developments.

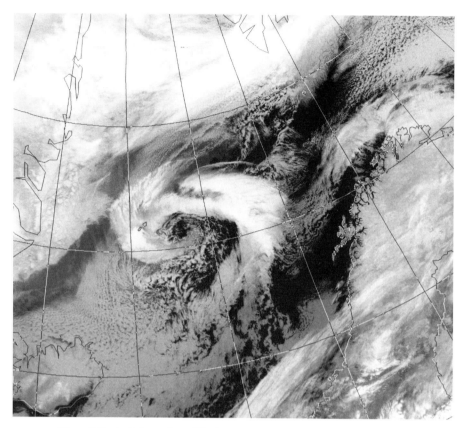

Figure 6.25. An infra-red satellite image at 0403 GMT 22 December 1983, showing the polar low discussed in Fett (1989). (Image courtesy of the NERC Satellite Receiving Station, University of Dundee.)

5 Upper cold lows and upper cold troughs appear to be intimately involved in polar low developments. During spring, autumn and winter, whenever such systems are forecast to move over Greenland, Norwegian or Barents Seas, the possibility of polar low developments should be considered.

In the preceding analysis of the development of 20–22 December 1983 an unusually large number of satellite images were available, permitting a detailed monitoring of the whole of this significant development. Using satellite images together with routine meteorological data, such as upper air charts, the forecaster may, even in cases when the numerical models fail to forecast the polar low, anticipate the formation of the low and evaluate its future development.

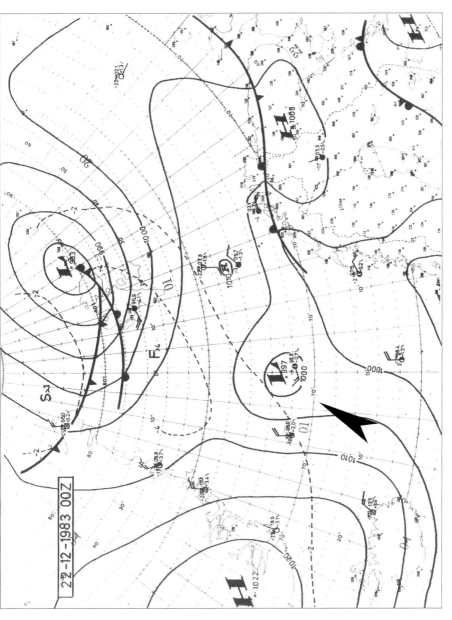

Figure 6.26. Surface analysis at 0000 GMT 22 December 1983, showing the polar low (indicated by an arrow) corresponding to the cloud spiral seen on Figure 6.25 (from Hoem, 1985). (Figure used courtesy of the Polar Lows Project, Norwegian Meteorological Institute.)

Figure 6.27. An infra-red satellite image at 1357 GMT 22 December 1983, showing the polar low in its decaying phase. (Image courtesy of the NERC Satellite Receiving Station, University of Dundee.)

Other polar low developments in the Nordic Seas region

Fett (1989a) discussed a number of other polar low developments, including the 'ACE polar low' of 27 February 1984 and the Bear Island polar low development of 13–14 December 1982. Since these cases have already been discussed in Chapters 3 and 5 only some of the conclusions as drawn by Fett are given in a slightly modified form.

1. Several examples of BLFs were observed in the Greenland, Norwegian and Barents Seas region.
2. Vorticity tends to be maximized along the BLF so that vortex formation is common. Normally such vortices are weak and do not undergo intense development. However, when an upper cold low or trough moves in the vicinity of a BLF, intense and often explosive polar low developments can occur.
3. It is unlikely that this is the only mode of polar low development in the region but it appears to be a frequent and often favoured mode of development.
4. Conventional analyses are generally not adequate in terms of detecting BLFs or polar low formation and movement.

5 Satellite data at frequent intervals (every 2 or 3 h) are required to definitively monitor such developments.

Forecasting polar low developments in the Denmark Strait

As discussed in Chapter 3, a secondary genesis regions for polar lows can be found around Iceland and the Denmark Strait. Fett (1989a), in a discussion of polar lows in the Denmark Strait, described the region in the following way:

> The region off the southeast tip of Greenland is a notable area favourable for cyclogenesis. Under conditions of westerly flow the high plateau of the ice cap, with elevations approaching 10,000 ft, acts as a moutain barrier similar to the Rockies in promoting the development of a lee trough. Additionally, low-level thermal baroclinicity is enhanced because of the juxtaposition of the marginal ice zone in that region and a branch of the warm Gulf Stream, which turns counter clockwise to meet the cold southward-flowing East Greenland Current. This baroclinicity extends through the Denmark Strait where the Irminger Current flows northwards past Iceland.

Fett, as an example of polar low developments in this particular region, discussed briefly the formation of a polar low near Iceland, within the central region of a large, synoptic-scale extratropical cyclone on 13 December 1986 (Figure 6.28). The polar low did not develop near the centre of the low but rather as an 'offset feature' moving in the cyclonic flow around the periphery of the centre. The vortex developed, according to Fett, out of a region of enhanced convection associated with an upper-level, short-wave trough as illustrated on the satellite image shown in Figure 6.29 from 1051 GMT 13 December 1986. Note that this development took place far away from the Greenland coast and did not seem to be associated with any downslope flow from the ice cap. The cloud feature developed further into the hook-shaped cloud with a nearly cloud-free eye seen on Figure 6.30. Such vortices are quite often observed at this location and are a class of polar low of moderate severity. As the main low pressure centre moved westwards from near Iceland towards the coast of Greenland, the polar low was steered northwards from its position in Figure 6.29 towards the west coast of Iceland (Figure 6.30), and then further northwards through the Denmark Strait. Since this particular system was a feature embedded in a larger circulation that had a strong pressure gradient extending out for hundreds of miles, its individual intensity is difficult to assess. Similar vortices, however, have produced sustained winds at Keflavik of over 25 m s^{-1}.

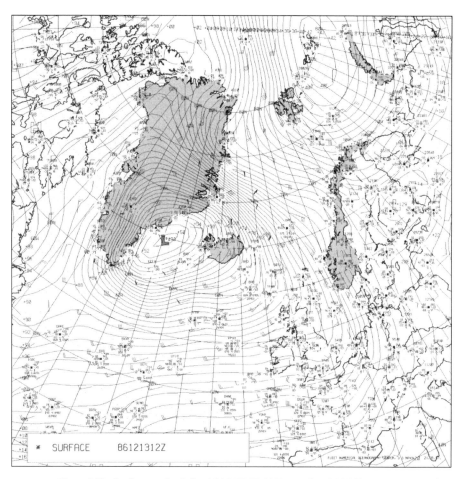

Figure 6.28. Surface analysis for 1200 GMT 13 December 1986 (from Fett, 1989a). (Figure provided courtesy of the Naval Research Laboratory, Marine Meteorology Division, Monterey, CA, USA.)

Based on this and similar cases, Fett drew the following conclusions about polar low developments in the Denmark Strait.

1. The Denmark Strait region is a favoured region for cyclogenesis.
2. Polar lows can develop in this region in 14 h or less.
3. Polar lows are mesoscale developments that may and often do occur within a larger scale circulation as an offset or short-wave development some distance from the main low pressure centre.
4. Polar lows often develop within or adjacent to a field of especially enhanced open-celled cumulonimbus. Such developments signal the favourable upper-level cold trough support noted in many polar low developments.

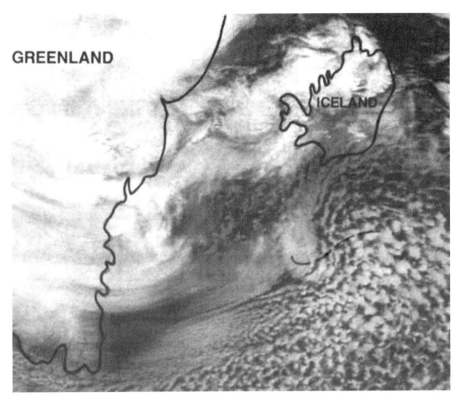

Figure 6.29. A NOAA 10 infra-red image of the Denmark Strait at 1051 GMT 13 December 1986, showing a region of enhanced convection south of Iceland associated with the short wave trough indicated by the broken line (from Fett, 1989a). (Figure provided courtesy of the Naval Research Laboratory, Marine Meteorology Division, Monterey, CA, USA.)

5 By combining the satellite evidence with other known factors favourable for polar low production the forecaster can be alerted to the exact areas where this potential exists.

Forecasting polar lows in the Bering Sea, around the Aleutian Islands and in the Gulf of Alaska

This topic was discussed by Fett *et al.* (1993). Using a distinction similar to that used in tropical meteorology between easterly waves and tropical depressions (the latter being defined as a closed tropical circulation having a wind speed of less than 17 m s^{-1}), a cloud vortex in the cold air behind a frontal system in the Arctic was referred to only as a polar vortex or cloud vortex until the forecaster could ascertain that (1) a closed circulation and pressure minimum exists, and (2) that wind speeds of 35 kt (18 m s^{-1}) or more had been attained. At that point, the polar vortex or cloud vortex could be referred to as a polar low.

546   6 Forecasting of polar lows

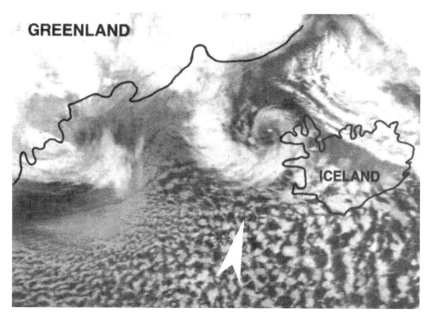

Figure 6.30. A NOAA 10 infra-red image of the Denmark Strait at 2045 GMT 13 December 1986 showing the polar low (indicated by an arrow) west of Iceland (from Fett, 1989a). (Figure provided courtesy of the Naval Research Laboratory, Marine Meteorology Division, Monterey, CA, USA.)

This, they argue, 'seems especially fitting since the term polar low has, in recent years, come to mean an intense Arctic storm, often of hurricane intensity'. Finally, Fett *et al.* (1993) limit their definition to spiraliform cloud systems (see Section 1.6, 'Cloud signatures'), excluding the comma-shaped systems that are found in close proximity to major frontal boundaries.

In the following we have quoted a number of paragraphs from the handbook by Fett *et al.* (1993) 'intended to provide forecasters with the background information necessary to assist them in recognizing and forecasting polar low activity' in the northeastern Pacific region:

> Observations in the region revealed that a favourable location for polar lows is in the western and northwestern sector of dissipating synoptic lows. Because the polar lows develop some distance away from the center of the synoptic low, they are likely to have moderate to strong steering flow above them. As a result they will move at a velocity representative of the steering flow and not with the movement of the dissipating synoptic low. The more intense the polar low and related vertical development, the higher in the atmosphere will be the representative steering level. Steering is actually an integrated result of the flow throughout the layer of the atmosphere interacting with the low, but in general, the flow near

the midpoint of the vertical development of the low will be representative of the integrated or mean steering flow. Because of the suppressed atmosphere of high latitudes, even strong polar lows will likely have a representative steering level closer to 850 hPa rather than the 500 hPa that is generally used as a rule of thumb for synoptic-scale mid-latitude features. A speed of advance of 30 kt is not unusual for well-developed polar lows. Because of the relatively small horizontal extend of the Bering Sea and Gulf of Alaska and the tendency to dissipate rapidly over land, the average duration of polar lows is only a day or two. Only under optimum forcing and migration path can polar lows persist for several days. Because their forcing is localized and they are small in size, polar lows can develop rapidly, frequently reaching maximum development within 12 to 24 h from initial formation, and dissipate rapidly.

Fett *et al.* (1993) continue:

Because of their small scale and localized forcing, polar lows are usually below the resolution of existing operational numerical models [*editors' remark*: in this connection it should be remembered that the handbook was published in 1993, and that the numerical models nowadays have improved so much that present, up-to-date models in many cases are able to also forecast polar lows as demonstrated in Section 5.2]. However, the models can depict the favourable synoptic-scale patterns. Forecasters should monitor closely the numerical analyses and prognoses for existing and developing favourable synoptic conditions. Equally important is the utilization of satellite imagery and conventional observations.

The following features are, according to Fett *et al.*, favourable for polar low *development*:

- A cold air outbreak and formation of low-level cloud street patterns.
- Large negative anomalies in upper-level temperature and heights.
- Development of a low-level baroclinic zone, such as an Arctic BLF, due to convergence of air streams with different overwater flow length or wrap-around warm air slots to the west and south of occluding frontal systems.
- Upper-level short-wave troughs and/or jet maxima moving over low-level features.

After initial development occurs as a result of low-level baroclinic conditions, the polar low clouds that are generally of convective form begin to organize into the classic comma shape. If favorable upper-air conditions exist, further intensification will occur, which can result in winds reaching hurricane force. During the development process of polar lows certain sequences of cloud signatures can be seen in satellite imagery. Initial development typically occurs along Arctic fronts or other convergence or baroclinic zones that exhibit recognizable enhanced cloud development and organization.

Intensifying polar lows tend to have highly symmetric, spiral-shaped cloud signatures with vigorous cumulonimbus surrounding a clear area or eye. Polar lows at this stage of development will have central pressures 4 to 8 hPa below ambient surface pressure (SLP) and winds in the 30–40 kt (15–21 m s^{-1}) range, and their cloud pattern will be 185–370 km in diameter. To reach Arctic hurricane status, increased low-level convergence or upper-level divergence is required, resulting in the cloud pattern becoming asymmetric with an anticyclonic curved cirrus shield reflecting the outflow region. The cirrus shield may cover the cumulonimbus-surrounded eye. At this stage the cloud pattern will have a diameter of 556–741 km, the central pressure will be near 970 hPa and winds will be in the 20–30 m s^{-1} range

(see the discussion of 'Bresch *et al.*'s case of 7 March 1977' in Section 5.1.2).

The following features are, according to Fett *et al.*, favourable for polar low *intensification*:

- Development or movement of an upper-air negative height anomaly of about 200 m over an oceanic region. Over the Bering Sea and Gulf of Alaska this in general means 500 hPa heights below 5000 m.
- Negative 500 hPa temperature anomalies of $-6\,°C$ equating to a 500 hPa temperature of generally $< -36\,°C$ for this area.
- Approach of an upper-level jet streak or area of PVA.

In connection with a discussion of an episodic event of polar vortex and polar low development over a period of 7 days, Fett *et al.*, noted that polar vortices associated with polar low developments appear in a diversity of sizes and intensities and exhibit a fairly wide range of speeds of advance and lifetime. Other characteristics include the following, according to Fett *et al.*

1. Polar vortices generally develop within a synoptic-scale, cold air outbreak characterized by cloud streets and open celled cumulus patterns under the influence of an upper-level cold low or trough.
2. The initial development of polar vortices and lows occurs hundreds of kilometres away from the synoptic-scale frontal cloud band.

3 The more intense systems tend to form on lines of enhanced convective activity located some 360–556 km south of the ice edge.
4 A second regime of small, weaker appearing vortices tend to form near the ice edge in areas of relatively weak off-ice flow.

During the 7-day study by Fett *et al.* (1993), three large polar vortices were observed to develop and track eastwards, the first of them, a comma cloud near 57° N, 176° E, as shown on Figure 6.31. A BLF extending from over the ice edge southward along 170° W to beyond the Aleutian chain is also seen in

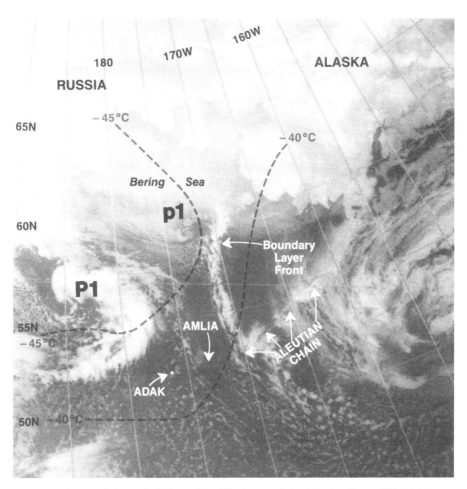

Figure 6.31. A DMSP infra-red satellite image of the Aleutian Sea at 0625 GMT 4 February 1990 showing polar low P1 over the Bering Sea (from Fett *et al.*, 1993). (Figure provided courtesy of the Naval Research Laboratory, Marine Meteorology Division, Monterey, CA, USA.)

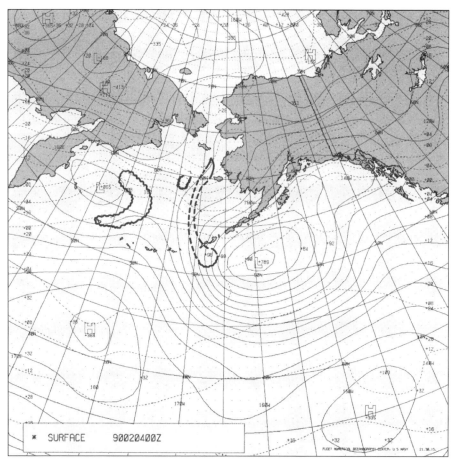

Figure 6.32. Surface analysis for 0000 GMT 4 February 1990. The outline of the clouds associated with the polar low and the BLF seen on Figure 6.31 are indicated (from Fett *et al.*, 1993). (Figure provided courtesy of the Naval Research Laboratory, Marine Meteorology Division, Monterey, CA, USA.)

the figure. The surface and 500 hPa analyses for 0000 GMT 4 February 1990 (Figures 6.32 and 6.33), about 6 h before the time of Figure 6.31, show the synoptic-scale pattern. The comma head of P1 at this time was located over the southern portion of the steepest surface air temperature gradient. The 500 hPa chart (Figure 6.33) shows a region of −45 °C air covering most of the Bering Sea. Over the position of P1 there were anomalies of *c.* 12 °C negative temperature and 210 m negative 500 hPa height. Figure 6.34 shows satellite imagery for 5 February, about 39 h after Figure 6.31 when a tight vortex, P2, had formed near 58° N, 173° E close to the location where P1 formed. Note that P2 formed

6.2 Aspects specific to the Arctic 551

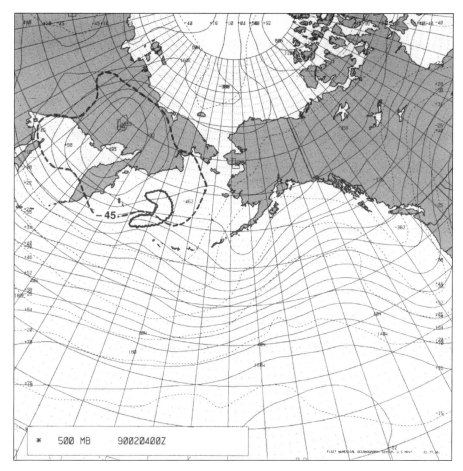

Figure 6.33. The 500 hPa analysis for 0000 GMT 4 February 1990. The outline of the clouds associated with the polar low seen on Figure 6.31 is indicated (from Fett *et al.*, 1993). (Figure provided courtesy of the Naval Research Laboratory, Marine Meteorology Division, Monterey, CA, USA.)

very near the coastline of Kamchatka in an intense cold air outbreak shown by the cloud streets in that region.

The western Bering Sea is a region where both synoptic and local circulation patterns exist that are conducive to formation of BLFs. Figure 6.35 shows a BLF that developed between the low-level cold flow off the land and the ambient maritime air. Vortices can be seen near both ends of the BLF. BLFs are low-level baroclinic zones and the low-level thermal advection pattern over the western Bering Sea around the time of Figure 6.35 is illustrated by Figure 6.36. Both the surface analysis and the 850 hPa analyses (not shown), indicate strong

552   6 Forecasting of polar lows

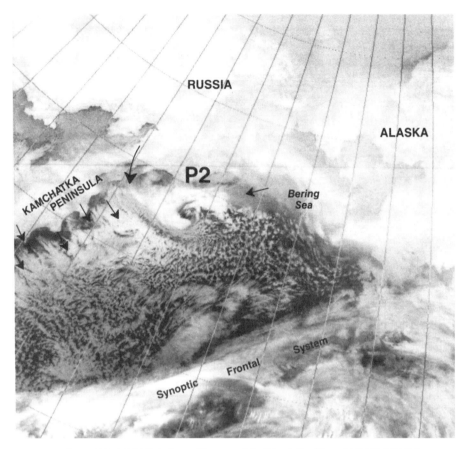

Figure 6.34. A DMSP infra-red image of the North Pacific at 2107 GMT 5 February 1990 showing polar low P2 in the form of a tight vortex close to the coast of Kamchatka (from Fett *et al.*, 1993). (Figure provided courtesy of the Naval Research Laboratory, Marine Meteorology Division, Monterey, CA, USA.)

cold air advection over the western and central Bering Sea where the BLF formed.

Forecasting polar low developments in the East Siberian/Chukchi/Beaufort Seas

Polar lows occasionally develop over the East Siberian/Chukchi/Beaufort Seas region. One of the few documented cases of polar low development in this region was briefly discussed in Section 3.1.10. In this case a US Coast Guard icebreaker operating in the vicinity of Point Barrow was suddenly and unexpectedly struck with winds in excess of 60 kt and associated with strong surface pressure falls. At the same time a forecast was received on

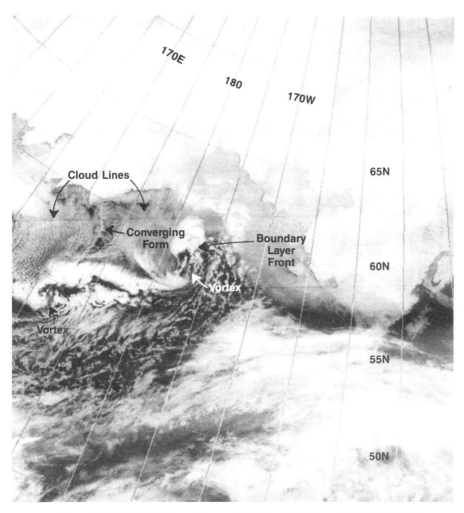

Figure 6.35. A DMSP infra-red image of the North Pacific at 1956 GMT 6 February 1990 showing a BLF over the western Bering Sea, along which vortices have formed (from Fett *et al.*, 1993). (Figure provided courtesy of the Naval Research Laboratory, Marine Meteorology Division, Monterey, CA, USA.)

the ship indicating that they could expect generally undisturbed weather with southerly winds of 15–20 kt for the next 24 h.

Although being an excellent example of faulty analysis and forecast guidance, the forecast was not illogical (Fett, 1992). The reason for the discrepancy between the forecast and the actual weather was the unexpected formation of a strong polar low. In order to elucidate the special features that favoured the intense polar low development, Fett (1992) presented a number of satellite images as well as surface and 500 hPa charts.

554   6 Forecasting of polar lows

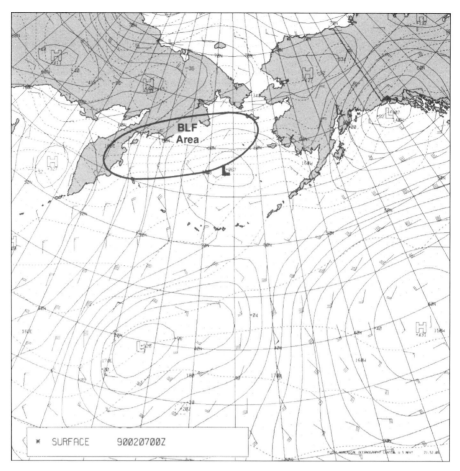

Figure 6.36. Surface analysis for 0000 GMT 7 February 1990 (from Fett *et al.*, 1993). (Figure provided courtesy of the Naval Research Laboratory, Marine Meteorology Division, Monterey, CA, USA.)

On 11 October 1985 a small cloud vortex (not shown) could already be seen over the ice-covered sea northwest of Wrangel Island, the cloud vortex being located below the centre of an upper-level cold core low. The 500 hPa analysis for 0000 GMT 12 October is shown on Figure 6.37 and a corresponding satellite image from 0634 GMT 12 October on Figure 6.38. The satellite image shows two cloud systems, one cloud spiral (marked LOW) within the central region of the 500 hPa low, and another (marked VORTEX), the 'vortex' further east associated with a 500 hPa trough. A similar combination of a spiraliform cloud within the centre of a cold low and a comma-shaped vortex associated with a trough on the periphery of the low was previously discussed in connection with another cold low type development (see Section 3.1.4, 'Example: 14–18 October 1980').

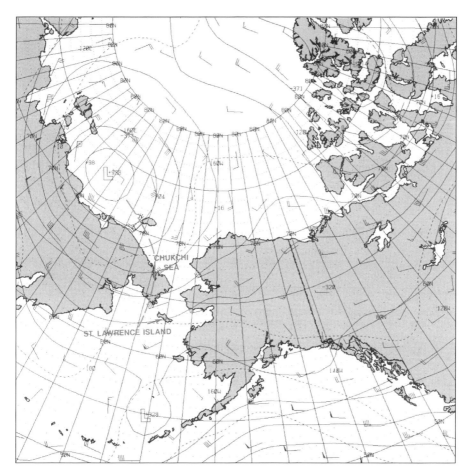

Figure 6.37. The 500 hPa analysis for 0000 GMT 12 October 1985 (from Fett, 1992). (Figure provided courtesy of the Naval Research Laboratory, Marine Meteorology Division, Monterey, CA, USA.)

The 500 hPa chart from 0000 GMT 14 October, i.e. around the time when the US icebreaker was hit by the polar low, is shown on Figure 6.39. The outlines of the cloud field, as observed from the satellite image from 2002 GMT 13 October shown on Figure 6.40, have been added. The two figures show that the developing polar low was situated slightly south of the upper-level circulation in a region of low 500 hPa temperatures between −40 and −45 °C. A cluster of deep convective clouds associated with the upper-level trough can be seen east of the polar low. A final satellite image of the low from 0548 GMT 14 October is shown in Figure 6.41. The surface centre of the polar low at this time was located at the northwest coast of Alaska, southwest of Point Barrow, and the two cloud systems had begun to merge. A surface analysis from 6 hours

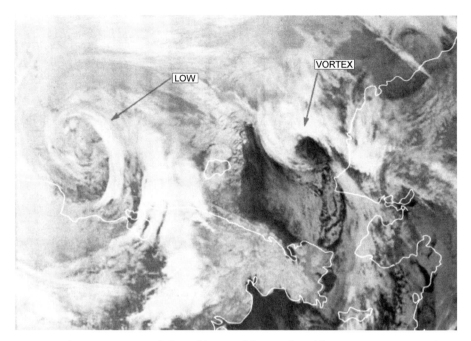

Figure 6.38. A DMSP infra-red image of the North Pacific at 0634 GMT 12 October 1985 (from Fett, 1992). (Figure provided courtesy of the Naval Research Laboratory, Marine Meteorology Division, Monterey, CA, USA.)

later at 1200 GMT 14 October is shown in Figure 6.42. Fett does not include any detailed discussion of which mechanisms led to the formation of the polar low, but, based on the development, he draws some general conclusions, as follows:

- Intense polar lows of the type experienced in the Greenland/Norwegian/Barents Seas also develop in the region of the East Siberian/Chukchi/Beaufort Seas.
- Satellite data at frequent intervals (every 2–3 h) are required to monitor the development of polar lows. Coverage should start in the vicinity of Wrangel Island and extend eastward continously over the Beaufort Sea.
- Surface analyses should incorporate satellite evidence of vortex development at the earliest opportunity. Hurricane-force winds associated with polar lows, however, can occur prior to the formation of a recognizable cloud vortex. Precursor signs can give warning prior to a development.
- A cold 500 hPa low appears to be an essential ingredient in severe polar low development. Cold temperatures aloft act to destabilize the atmosphere and give rise to open-cellular cumulonimbus production over water south or west of the convective cloud band in which

6.2 Aspects specific to the Arctic   557

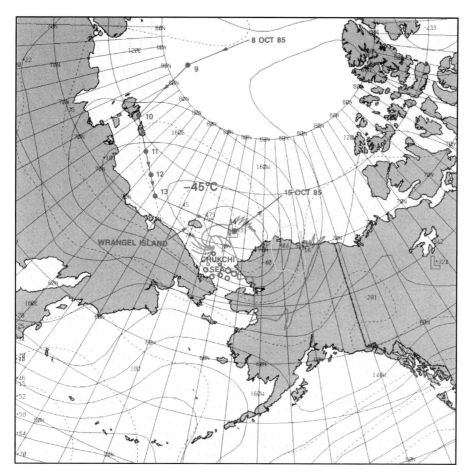

Figure 6.39. The 500 hPa analysis for 0000 GMT 14 October 1985 with superimposed cloud outlines from Figure 6.40 (from Fett, 1992). (Figure provided courtesy of the Naval Research Laboratory, Marine Meteorology Division, Monterey, CA, USA.)

development will occur. Such developments serve as a useful precursor of polar low development.
- Whenever surface lows and upper cold lows track together in eastward movement over the pack ice toward Alaska, they should be continously monitored for signs of potential polar low development.

6.2.6 Weather and ocean waves

Weather

Needless to say, the 'weather' which accompanies polar lows is of primary interest for the forecaster. Over the years many lives have been lost,

Figure 6.40. A DMSP infra-red image of the Arctic Ocean north of Alaska at 2002 GMT 13 October 1985. The polar low (indicated by the thin black arrow) was situated west of a cluster of deep convective clouds (from Fett, 1992). (Figure provided courtesy of the Naval Research Laboratory, Marine Meteorology Division, Monterey, CA, USA.)

especially when small vessels in the Norwegian waters unexpectedly encountered polar low with strong winds of gale or storm force. More recently, severe weather connected with polar lows has presented a major problem to offshore activities at high latitudes in the region north of Alaska and along the Norwegian coast, in the Barents Sea.

Eye-witness accounts of polar lows are rare in the literature and some claims about exceptional weather events, such as almost instantaneous wind increases to over 90 kt in connection with polar lows have not been verified. However, the extreme weather associated with polar lows has been known for many years. Harold, in her PhD thesis, quoted a World War II pilot, Bill van Horne, who had encountered these systems while flying across the Atlantic.

> It may be that the features 'discovered' via satellites were the detailed meteorological structures, but one thing is sure: every pilot who flew across the Atlantic in the 50 or more years of propeller planes knew about them. Any who didn't are probably long since dead!

Figure 6.41. A DMSP infra-red satellite image of the Arctic Ocean north of Alaska at 0548 GMT 14 October 1985 (from Fett, 1992). (Figure provided courtesy of the Naval Research Laboratory, Marine Meteorology Division, Monterey, CA, USA.)

...We carried radar and, so long as the radar operator was alert and giving the pilot directions to 'vector around' storms, we were OK. But a couple of times he missed and we got bounced around a bit. At least once we thought we were going to swim home!

We flew through the storm areas, and the density of thunderstorms in them have to be believed. We saw them on the radar scope; now you can see them from above. I remember one that was between Newfoundland and the tip of Greenland that was 200-miles N–S and 50-miles E–W. It was completely filled with thunderheads so closely packed as nearly to touch. We had no way to measure the vertical heights, but the tops looked like they reached 50 or 60 000 feet; but that may have been an illusion.

The storm area was made up of a group of thunderstorms. They were similar to what we see on land, in 'squall lines' that are made up of many cumulo-nimbus, spaced close together...

(Harold, 1997).

A further account of the effects of the extreme weather in connection with polar lows at sea mentioned earlier was given by Fett (1992), who described the problems of the US Coast Guard Icebreaker *Polar Sea* when this vessel encountered a polar low near Point Barrow, Alaska. The ship had just completed a helicopter recovery operation that required good weather conditions. Less than 5 hours later, the ship measured winds at 75 kt (37 m s^{-1}), seas estimated at

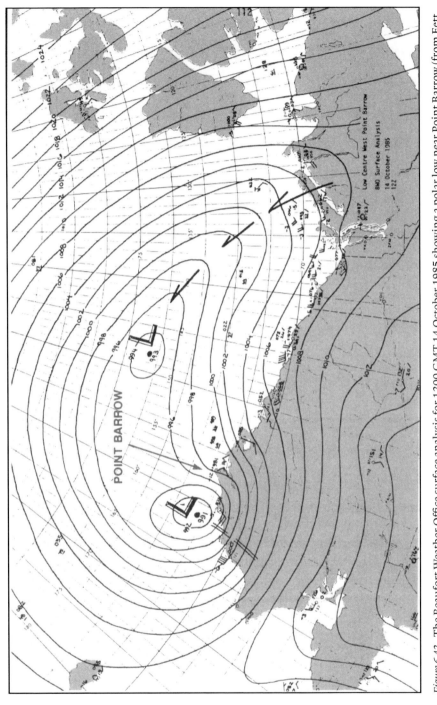

Figure 6.42. The Beaufort Weather Office surface analysis for 1200 GMT 14 October 1985 showing a polar low near Point Barrow (from Fett, 1992). (Figure provided courtesy of the Naval Research Laboratory, Marine Meteorology Division, Monterey, CA, USA.)

15 feet (4–5 m) high, icing of superstructure, low visibility in snow and the ship sustained damage from the unpredicted storm.

Businer and Baik (1991), in an article 'An Arctic Hurricane over the Bering Sea', described, based on a similar real event, the weather associated with what they called 'the closest cousin to the cyclones of the tropics' in the following vivid way:

> On 7 March 1977, a Soviet icebreaker plied the western Bering Sea near the edge of the Arctic winter ice. The day dawned clear and cold; the first rays of a feeble Arctic sun failed to offset the sting of a gusty northeast wind. Suddenly, with the approach of midday, a dark band of cloud obscured the horizon. The winds, at first light and northerly, veered to the southwest and rapidly increased to gale force, gusting to hurricane strength over 64 knots (32 m s^{-1}) and engulfing the ship in a confused onslaught of 30 foot (9 m) waves. Snow and sea swirled thickly; frequent lightning illuminated a deck thickly encrusted with ice. The barometric pressure, already low, plunged further as evening temperatures rose to their warmest in a month.

Businger and Baik's description was of a polar low in the Bering Sea but could just as well apply to one in the Barents Sea or elsewhere that these dangerous storms develop.

Environmental condition under polar lows in the Norwegian sea areas

Houmb *et al.* (1986), utilizing data from satellites, ships, coastal meteorological stations, buoys and the weather ship AMI (71.5° N, 19.5° E) carried out a study of the environmental conditions for the seas around Norway under 153 polar lows from 1971 to 1985. The study included waves, wind, air temperature, icing, visibility and precipitation. We summarize their conclusions in the following.

Wind speed

Based on the wind data obtained within the period, Houmb *et al.* concluded that the data indicated 1, 10 and 100 year extreme values of *c.* 26, 33 and 40 m s^{-1} respectively which means, for example, that on average a wind speed value of 33 m s^{-1} will be exceeded once during a period of 10 years in association with a polar low occurring in the seas around Norway. In the period considered in the study, on one occasion a wind speed of 35 m s^{-1} had been recorded.

Air temperature and icing

Usually, the wind force increases before the temperature rises, but in the Norwegian sea areas, where the temperature is relatively high, the combination of strong winds and very low temperatures is rare during polar lows.

Observations from AMI confirm that the temperature in general was rising when a polar low was passing. Using observations during 26 polar low cases close to the weather ship it was found that the most frequent combination of wind and temperature was a wind speed between 10 and 17.5 m s^{-1} and a temperature of -2 to $-6\,°C$. Lower temperatures and higher wind speeds were of rather short duration, which emphasizes the abrupt nature of polar lows.

Icing presented a minor problem at AMI during passages of polar lows, probably because of the relative high sea surface temperatures in the region and the short duration of severe weather associated with the lows.

Precipitation and visibility

According to the study by Houmb *et al.* the strong ascent in the polar low results in intense heavy showers, most often coming as snow, the precipitation lasting longer than the 'ordinary' showers.

Houmb *et al.* (1986) concluded that

- Polar lows do not seem to contribute to increasing present extreme estimates of wind speed, wave height, air temperature, ice loading or precipitation.
- Extremely poor visibility may occur during polar lows.
- Polar lows are associated with rapid changes in environmental conditions that are difficult to predict.

Environmental conditions under polar lows over the British Isles and in the North Sea region

In the section above, conditions under high latitude polar lows over the Norwegian and the Barents Sea were considered. However, also further south, over the British Isles and in the North Sea region, polar lows occasionally may cause disruption of traffic and other problems. The following description of the weather typically observed during the passage of polar lows over the British Isles was given in the *Handbook of Weather Forecasting*, Vol. II (Meteorological Office 1964):

> The approach of a polar low is heralded by falling tendencies (which may be only very slight), a backing and sometimes a decrease of the surface winds and a pronounced change in type, and increase of amount, of medium and high clouds. These changes in upper clouds are usually very noticeable because the preceding cold air is often virtually free from

upper clouds, apart from anvil cirrus clouds, which are either the residue of or blown off from the tops of shower clouds.

The Handbook continues:

> Although showers, which may be both severe at times and frequent, are often features of a typical polar outbreak, a polar low causes either a period of countinous precipitation or such extensive and frequent showers that for most practical purposes it is little different from continous precipitation. The sky is usually almost completely covered by upper clouds and the dull weather, with precipitation in the polar low, contrasts unpleasantly with the bright, showery, but at times exhilarating, weather of the pure polar outbreak. In winter, and even in spring, polar lows may bring considerable falls of snow, particularly on high ground and at times on low ground. The falls of snow from a polar low, being of longer duration and of greater horizontal extent than those from the showers of a simple polar outbreak, may disrupt transport and cause considerable difficulties both to the general public and to some special sections of the nation, notably farmers maintaining livestock out of doors, particularly in hilly country where there may be deep drifts.
>
> Clearance of the sky after the passage of the polar low is often rapid and complete. Incoming solar radiation during the day may wholly or partially melt any frozen precipitation which is lying. Strong outgoing radiation may produce hard frosts at night. Any subsequent hard freezing of partially melted snow often causes very great difficulties for transport.
>
> Not all polar lows in winter give solid precipitation, but the current temperatures in the lower troposphere should always be carefully considered. In polar lows, cooling by evaporation of precipitation may occasionally be an important factor in determining the nature of precipitation at the ground. This cooling should generally be considered when indicating the level of the $0\,°C$ isotherm for aircraft flying at relatively low-levels near or through a polar low. As the $0\,°C$ isotherm is often within 2,000 or 3,000 feet of mean sea level in winter or early spring, any lowering of the $0\,°C$ isotherm may leave little or no room for manoeuvre between flight at a level with temperature above $0\,°C$ and the ground. As the precipitation and cloud system of a polar low might extend horizontally over a distance of about 100 or 150 miles, and vertically to at least 700 hPa, an aircraft may have to traverse a substantial extent of cloud in a search for a path free from the risk of ice accretion.

Several accounts of polar lows crossing the British Isles and the weather associated with their passages can be found in articles in *Weather* published in the

1960s and 1970s (see Section 1.2, A brief historical review). Stevenson (1968), for example, described the passage of a polar low over southern England when 11 inches (c. 30 cm) of snow fell around Brighton and gave rise to chaotic traffic conditions.

Ocean waves under polar lows

Practical experience tells us that the waves generated by polar lows have had serious consequences in the form of shipwrecks, which have given polar lows a reputation as particularly violent storms (Houmb and Lönseth, 1986).

Polar lows differ from large-scale storms through their spatial and temporal scales. Whereas synoptic-scale cyclones have a horizontal scale of 1000–2000 km and typical durations of several days, the polar low has a space scale of a few hundred km and a duration of one or two days. The polar low's capacity for wave generation is thus limited by a small scale in space (a short fetch) and in time as well, and if they were more or less stationary, one would not expect very high waves even for extreme winds. For example, a wind speed of 35 m s^{-1} would produce a significant wave height of approximately 5.5 m over 100 km (Dysthe and Harbitz, 1987). However, taking the motion of a polar low into account, a different picture emerges.

Since the polar low is not stationary but generally is advected in a northerly large-scale airstream, a wave packet can be developed following the polar low. The wave packet will be dominated by waves in a frequency band centred around waves having a group velocity equal to the travelling speed of the low, since these waves are going to stay in the enhanced wind field for a long time. Also they will be confined to the sector of the low where the wind direction is aligned with the direction of motion of the polar low. The priciples of this process are illustrated by the simplified model shown in Figure 6.43. In this way the polar low can be accompanied by a wave packet with a sharp front and at a station one may experience a very rapid increase of wave height as the front passes. The lock-in between the low and certain waves has been referred to as 'resonant fetch enhancement'.

The waves associated with polar lows are not expected to reach the 100-year extreme significant wave height which is around 15 m for the offshore areas outside Norway. Furthermore, the waves will not be as high in the northern areas, e.g. the Barents Sea, as one may expect further south along the Norwegian coast and in the North Sea. This is due to different lengths of fetch since the polar lows reaching the southern areas will be much older (Houmb and Lönseth, 1986).

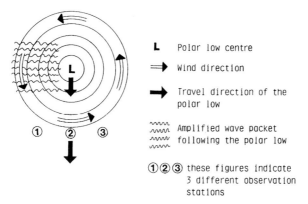

Figure 6.43. A simplified model of wave generation under a polar low (from Houmb and Lönseth, 1986). (Figure used courtesy of the Polar Lows Project, Norwegian Meteorological Institute.)

Extreme observed significant wave heights at weather ship AMI during 18 polar low passages are shown on Figure 6.44. Each polar low was tracked during the time it affected AMI, resulting in a total of 134 observations of wind and waves. Significant wave heights seem to concentrate in the interval between 2 and 5 m (65%), but several cases (23%) of wave heights above 5 m were observed. Wave heights of 11 m were observed in a case with the highest observed wind speeds (26 m s^{-1}).

The evolution of the wave parameters T_p (peak period), H_c (significant wave height), and H_{max} (maximum wave height) of a polar low which passed weather ship AMI on 17–18 February 1978 are shown on Figure 6.45. The wind speed associated was estimated to be in the range 25–28 m s^{-1} and the speed of the low at 10 m s^{-1}. A remarkably steep rise in all the parameters was seen between 1700 and 2000 GMT.

6.3 Aspects of forecasting specific to the Antarctic

6.3.1 Introduction

In the Antarctic, mesoscale lows and the more intense polar lows can have a severe impact on research and logistical operations being carried out by both ships and aircraft through the strong winds, extensive cloud and precipitation that the systems can bring. For example, the mesocyclone described by Turner *et al.* (1993b) lasted for over three days and gave very poor weather that stopped the logistical re-supply operation that was taking place at Halley Station on the eastern side of the Weddell Sea. Halley obtains all its supplies from a support vessel moored about 10 km away from the station, with all the

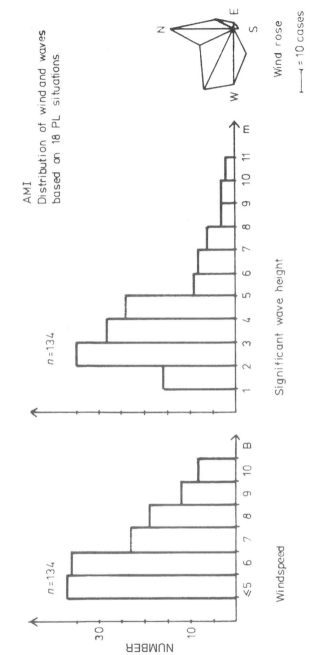

Figure 6.44. The frequency distributions of wind speed, wave height and wind direction at ship AMI based on 3-hourly observations (18 cases) (from Lystad *et al.*, 1986). (Figure used courtesy of the Polar Lows Project, Norwegian Meteorological Institute.)

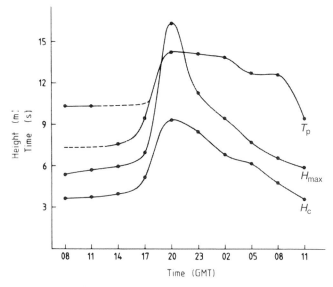

Figure 6.45. Evolution of the wave parameters, T_p (peak period), H_c (significant wave height) and H_{max} (maximum wave height) at AMI, 17–18 February 1978 (from Dysthe and Harbitz, 1987).

cargo being transferred by tracked vehicle. During much of the period when this mesocyclone was close to the station surface winds were close to gale force with prolonged falling and blowing snow, effectively stopping all work. In the short Austral summer it is essential to maximize the use of resources and those concerned with planning operations require advance guidance of poor weather conditions.

Mesocyclones can also have a severe impact on air operations, which are often concerned with deploying and recovering field parties that are working in remote areas in the interior of the continent. When a party is first to be deployed there is usually no-one on the ground in the area of interest who can provide meteorological observations. This means that the forecaster responsible for advising the pilot on conditions at the landing site has to make decisions on the basis of only satellite imagery and model output. As discussed below, a number of mesocyclones occur over the high Antarctic plateau and although they are not usually very vigorous, low cloud associated with the systems can make landing an aircraft near these systems impossible. As all fuel has to be transported into the Antarctic by ship or aircraft, it must be used efficiently and it is essential to try and minimize aborted flights caused by poor weather.

There is also a great deal of air activity in the region of the Ross Ice Shelf and McMurdo Station. As discussed in Section 3.2, many mesocyclones are found on the ice shelf and some of these are vigorous enough to affect air operations, both within the Antarctic and the flights to and from New Zealand. Accurate forecasting of mesoscale weather systems is therefore an important element in Antarctic weather prediction for air operations.

6.3.2 The use of model fields in the Antarctic

In recent years there has been a steady improvement in the quality of the NWP analyses and forecasts for the Antarctic (Turner *et al.*, 1996b, 1999) as new models have been introduced and better use has been made of the available data. Today few synoptic-scale lows are missing from the analyses, although the detailed structure of some systems is still poorly represented. However, most mesoscale lows are missing from the analyses since such systems are only represented by a few grid points and the satellite sounder observations, which are the main form of data used to produce the analyses over the Southern Ocean, still have a coarse horizontal and vertical resolution. The forecasting of individual mesocyclones is therefore very difficult given the poor initial conditions on the mesoscale that are used for the NWP integrations.

One case of a surge of cold, unstable air over the eastern Bellingshausen Sea associated with a vigorous depression just to the north of the Antarctic Peninsula occurred on 17 July 1995. The infra-red satellite imagery for 1837 GMT on this day is shown in Figure 6.46. On this occasion convective cloud was present over much of the ocean area to the west of the Peninsula and a mesocyclone was present just to the west of the tip of the Peninsula. The 36 h pressure at mean sea level (PMSL) forecast from 0000 GMT 16 July which verified at 1200 GMT 17 July is shown in Figure 6.47. It can be seen that the model gave an excellent forecast of the low pressure system close to 60°S, 60°W and correctly predicted the cold air outbreak to the west of the synoptic-scale low. The model even predicted a sharp trough of low pressure at approximately the location where the mesocyclone developed, although experience suggests that it is not wise to put too much emphasis on such small-scale features. Nevertheless, the model on this occasion gave a very good prediction of the broad-scale synoptic flow and provided good guidance on the possibility of mesocyclone development.

6.3.3 The use of imagery in the Antarctic

Figure 6.48 shows a pair of images of a mesocyclone that developed on the high interior plateau of the Antarctic on 9 December 1997. Both these images are AVHRR channel 3 (3.7 μm) data since in this channel clouds composed of water droplets stand out very clearly against the snow surface, which appears black at this wavelength. In Figure 6.48a the first signs of development are apparent as the linear cloud band lying east–west begins to show signs

6.3 Aspects specific to the Antarctic 569

Figure 6.46. An infra-red satellite image for 1837 GMT 17 July 1995.

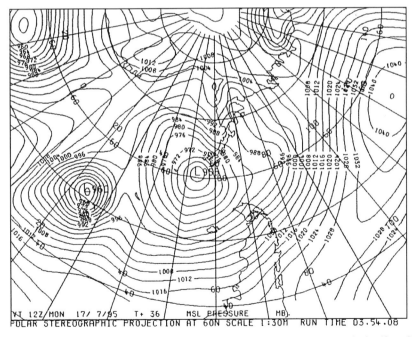

Figure 6.47. A 36-hour PMSL forecast from the UK Meteorological Office global model verifying at 1200 GMT 17 July 1995.

570   6 Forecasting of polar lows

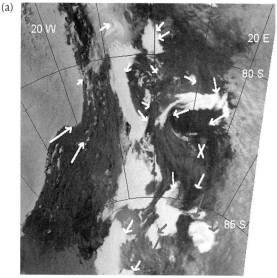

Figure 6.48. AVHRR satellite imagery (3.7 μm) showing a mesocyclone over the Antarctic plateau on 9 December 1997. (a) Imagery for 0542 GMT with cloud motion vectors for the period 0406–0542 GMT indicated as arrows. The position of the mesocyclone at 1524 GMT is indicated by a cross. (b) Data for 1710 GMT with the mesocyclone indicated by an arrowhead.

of rotation. During 9 December many images of this mesoscale low were received at Rothera Station, allowing its track to be followed closely. Between 0406 GMT, when the developing low was first observed, and 1524 GMT the low moved southwards and developed a spiraliform cloud signature. The vectors indicating the movement of various cloud elements associated with the low and other nearby clouds are indicated on Figure 6.48a, along with the location of the low at the latter time, which is indicated by a cross. This figure shows that during this period of a little less than 2 hours there were clear indications of cyclonic rotation in the clouds associated with the developing low. Clouds in the southern half of the image are all moving in a south-southwesterly direction, indicating the general direction of low-level flow in this area.

Between 1524 and 1710 GMT the mesocyclone continued to move generally southwards and the imagery for this later time, shown in Figure 6.48b, indicates the well-developed spiraliform cloud signature of the system in its mature phase. After this time the cloud at the heart of the mesocyclone began to decline and the vortex dissipated early on 10 December. More details of this case can be found in Turner and Ladkin (1998).

6.3.4 The use of automatic weather station (AWS) surface data

In recent years large numbers of AWSs have been deployed in some parts of the Antarctic as part of major research programmes concerned with the meteorology of the continent, including the investigation of mesocyclones. In particular, many AWSs have been installed on the Ross Ice Shelf which has almost a mesoscale network of observing systems that can provide data on surface pressure, the wind field, temperature and humidity (Stearns and Wendler, 1988). Similarly, there are about 12 US and Italian AWSs in the vicinity of Terra Nova Bay, which is now well covered by observing systems. Although primarily for research, the data from these AWSs are of immense value in forecasting. Many of the data are put onto the GTS and used in the preparation of numerical analyses, but the data can be used to produce manual analyses with mesoscale detail that can be used to monitor the development of and predict the future track of mesocyclones. For example, Figure 6.49a (taken from Carrasco and Bromwich, 1993) shows a mesoscale low on the edge of the Ross Ice Shelf as observed by infra-red satellite imagery. The mesoscale regional analysis in Figure 6.49b was prepared using data from the extensive network of AWSs on the ice shelf, along with the observations from McMurdo Station. Such analyses are of great value in monitoring and predicting the movement of these small-scale weather systems.

6.3.5 Forecasting mesocyclones in particular areas of the Antarctic

Mesocyclones are found at one time or another in all parts of the Antarctic, but the prediction of such systems presents different problems depending on the area under consideration. Over the ice-free waters of the Southern Ocean, where most mesocyclones are found, the model fields are now very good and the full range of satellite observing systems can be employed. Satellite

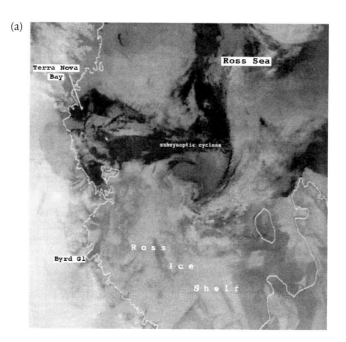

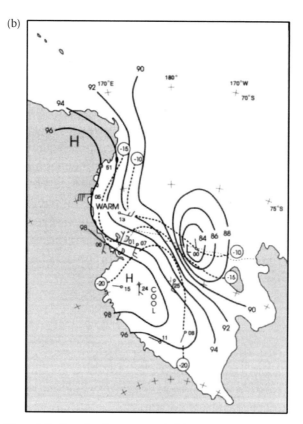

Figure 6.49. Data for the Ross Ice Shelf area at 1438 GMT 20 February 1988. (a) Thermal infra-red imagery, (b) surface analysis (from Carrasco and Bromwich, 1993).

imagery from the polar orbiting satellites is not received very frequently at these more northerly latitudes, but in some areas imagery from the geostationary satellites can be employed. In this zone there is usually a good contrast between the cloud-top temperatures of the mesocyclones and the sea, allowing easy detection of most lows.

In the sea ice zone, scatterometer data and products derived from passive microwave measurements are not available and there is less contrast in temperature and reflectivity between the clouds and the sea ice. Nevertheless, the lows can usually be detected against the characteristic lead pattern of the sea ice and followed in sequences of satellite imagery.

Some areas of persistent open water close to the edge of the continent (coastal polynyas) have many mesocyclone developments and, as discussed in Section 2.2, the eastern side of the Weddell Sea close to Halley research station and the southern part of the Ross Sea near Terra Nova Bay are particularly prone to mesocyclones. Because of the relatively ice-free conditions in these areas they have many logistical operations taking place and mesoscale lows can cause severe disruption. Although the lows can still only be predicted using a nowcasting approach, since these areas are monitored closely by forecasters on the stations the early signs of development, such as indications of cyclonic rotation of cloud bands can be detected at an early stage.

Although there are few mesocyclones over the interior of the Antarctic, the ice shelves, and in particular the Ross and Ronne/Filchner Ice Shelves, do have many such developments throughout the year. As in other interior parts of the Antarctic, the models give poor predictions on the mesoscale and the systems have to be detected and their future development predicted using satellite imagery, and in the case of the Ross Ice Shelf, the AWS observations. As the surfaces of the ice shelves are so featureless the detection of mesocyclones is very easy since the cloud vortices stand out well against the ice surface. Prediction of the movement of the lows can be carried out using sequences of images and any *in situ* measurements employed to monitor the surface pressure anomaly of the system.

6.4 Future work required to improve forecasts

The most important advances in the operational prediction of mesocyclones and polar lows will come from the development and routine use of high resolution, limited area NWP models. Such models, covering the whole of the Antarctic and with a horizontal resolution of about 10 km have already been proposed, to be run with an embedded 1 km resolution model for particular sectors of the continent. It remains to be seen whether resources are available to run these models on a routine basis. However, the cost of aviation activities within the Antarctic and from outside the continent is so great that the costs of running high resolution models can soon be saved if fewer flights are aborted.

Experimental runs of fairly high resolution models (Lieder and Heinemann, 1999) have shown that when the models have a resolution of around 25 km they can predict the development of individual mesoscale lows and resolve much of their structure. However, in parallel with the development of the models there must also be advances in assimilation of observations that can help in the preparation of mesoscale analyses.

Over the interior of the Antarctic the emphasis will still be on nowcasting the movement and development of vortices through the use of satellite data. However, in the coming decades improved radiometers will be able to give imagery with a higher horizontal resolution and at more wavelengths allowing improved identification and tracking of these lows.

In the Arctic there has always been more *in situ* data and generally higher quality analyses because of the closer proximity to major population centres. Limited area models covering Europe and the Norwegian Sea already produce very good forecasts on the synoptic scale and are now of value in predicting some polar lows. With increases in computer power over the coming years and the use of higher resolution models it is anticipated that many more systems will be represented. However, further work is required to improve the representation of processes that are important at high latitude, such as the fluxes of heat and moisture from the ocean, the various types of convection found and the details of the sea ice.

7

Conclusions and future research needs

It will have become clear from the preceding chapters that major advances have been made over the last few decades in our understanding of the distribution, occurrence, formation and development mechanisms of polar lows and other high latitude mesoscale vortices. In this final chapter we summarize our present understanding of this family of weather systems and consider the requirements for future research.

7.1 The spatial distribution of polar lows and other high latitude mesoscale vortices

Polar lows were first investigated over the Nordic Seas and when systematic research started into such systems in the 1960s the case studies were concerned with lows occurring in this region. Subsequently, such investigations were gradually extended to new areas in the Northern Hemisphere, such as the Labrador Sea, the Gulf of Alaska and the Sea of Japan, where similar active mesoscale vortices were identified. Since the 1970s isolated cases have also been reported of polar lows over the marginal seas around the Arctic Ocean, including the Beaufort Sea, the Chukchi Sea and the Kara Sea. However, it is now unlikely that significant numbers of active systems will be found in any new areas in the Northern Hemisphere.

In the Southern Hemisphere, polar lows have only been thoroughly investigated in a limited number of regions. The lows found around New Zealand seem to have many of the characteristics of the systems occurring in the Nordic Seas, but much more work needs to be done on the polar lows in this area. Satellite imagery also shows a number of what appear to be polar lows at relatively northerly latitudes, in areas such as around the tip of South America. However, the lack of *in situ* data across the Southern Hemisphere makes it difficult to determine whether these systems are vigorous enough to be classified as polar lows, so a priority for future research must be to carry out detailed observational and modelling studies of systems in these areas using data such as surface wind measurements from satellite-borne scatterometers, which can provide more insight into the nature of these lows. Vigorous polar lows have also been studied in the Weddell and Ross Seas, but at these southerly latitudes there is no deep convection and the systems are quite different from most of the polar lows of the Nordic Seas. Nevertheless, further north near the edge of the sea ice, studies have found several polar lows that were similar to the Northern Hemisphere systems, including the feature of deep convection.

Satellite imagery has shown that minor cloud vortices are a year-round feature of both polar regions and that these systems usually have a weak surface circulation or no surface circulation at all in many cases. Many of the vortices observed around the Antarctic are of this type, with winds well below gale force.

Over the years many studies have been carried out into the climatological occurrence of polar lows based on satellite imagery. However, most of these have been concerned with fairly limited areas and researchers have used different criteria in determining what constitutes a polar low. Many studies have just considered all small-scale vortices seen in satellite imagery, perhaps splitting the vortices into 'comma clouds' and 'spiraliform lows' depending on the cloud features observed. As multi-sensor satellite data sets become available and operational numerical weather prediction (NWP) systems achieve horizontal resolutions that can resolve polar lows, it will become possible to produce global climatologies of these systems based on more objective data so that greater insight will be obtained into the environment in which polar lows develop.

7.2 Possible climatic effects of polar lows

For some time it has been recognized that the most vigorous polar lows can cause considerable damage and even loss of life if they affect inhabited coastal communities. Thus it is particularly important to be able to forecast these lows and give adequate warning of their impending arrival. However, it is still not clear whether they have a significant *climatological* impact. With their relatively small horizontal scale and short life cycle it is tempting to dismiss these systems as transitory disturbances that have no long-term effects on the climate system. However, recent research has suggested that this may not be the case. Researchers have noted that the most active polar lows, which are associated with strong winds and very cold air masses over relatively warm waters, can give rise to large fluxes of heat into the lowest layers of the atmosphere. This can result in marked cooling of the surface layer of the ocean, which has the potential to destabilize the ocean. If a particular area is affected by one or more polar lows over an extended period of time then oceanic deep convection may be triggered, possibly affecting the deeper ocean layers. However, further research based on modelling experiments is needed to determine how often this occurs and the climatological impact.

7.3 Observational and modelling studies

The many detailed case studies carried out since the 1960s have proved tremendously valuable in helping us understand the structure and evolution of polar lows, and aided the theoreticians in formulating their ideas about the development of the systems. Over the years the case studies have become increasingly sophisticated. The early studies used the available *in situ* data and possibly some satellite imagery, but during the 1980s new forms of satellite data, such as temperature soundings and scatterometer winds were used increasingly. By the 1990s the operational NWP systems had advanced to the point where they had sufficient horizontal resolution to resolve some polar lows and studies using a combination of *in situ*, satellite and model data became increasingly common. We believe that this trend towards combined observational/modelling studies is set to continue as high resolution mesoscale models are run more frequently and new forms of satellite data become available. These tools will allow the conduct of detailed investigations of the structure of polar lows and the environment in which they develop.

While some flights through polar lows have been undertaken by research aircraft in both the Arctic and Antarctic, such campaigns are very expensive and there is always the difficulty of having the aircraft in the right place at the right time in order to be able to investigate a suitable polar low. To date, some of the most detailed information on the broadscale three-dimensional structure of polar lows and temporal evolution has been obtained via aircraft. It is hoped that in the future further campaigns can be mounted to investigate polar lows in both hemispheres, especially as the high resolution data collected can now be assimilated into mesoscale models, thereby aiding numerical investigations.

7.4 Formation and development mechanisms

Originally, polar lows were thought to be driven by convection in a manner similar to tropical cyclones. However, case studies showed that some vortices were baroclinic in structure leading to the debate, in the late 1970s and 1980s, over whether polar lows were baroclinic disturbances or whether convection played a role via CISK. However, it is now generally accepted that there is a spectrum of high latitude, mesoscale weather systems, from purely baroclinic polar lows, that resemble small mid-latitude depressions, through the 'comma clouds' found close to the polar front, to lows dominated by deep convection, where the main energy source is latent heat release. Producing rigid definitions and classification schemes for polar lows has not been entirely successful, as 'hybrid' systems can always be found and individual lows can change significantly during their lifetimes as different forcing mechanisms come into play.

While there is generally acceptance of the wide variety of systems that come under the broad umbrella of the polar low definition, there is still a debate going on over the means by which convection acts to intensify the lows. While CISK has been linked to polar low developments since the 1970s, the theory of ASII/WISHE (see Chapter 4) was put forward in the late 1980s. The relative importance of CISK and ASII/WISHE for the development of polar lows is still unclear, but most likely both mechanisms are important in the polar regions.

Recently, theoretical investigations have moved away from the CISK versus WISHE problem and focused more upon the way in which heating affects the development of polar lows. As discussed in Chapters 4 and 5, this approach, in conjunction with the PV concept has proved to be highly valuable, and has led to important new insights, including a number of new conceptual models for the development of polar lows.

While most of the recent theoretical debate about the development of polar lows has been concerned with convection, the importance of baroclinic processes should not be forgotten. Many polar lows start as minor baroclinic disturbances. Baroclinic systems with strong surface winds are rather rare, but there are some well-documented cases in both polar regions. Baroclinic polar lows can be subdivided into various categories that appear to be strongly related to geographical location. Reverse shear baroclinic polar lows were first observed in the Nordic Seas region, where they constitute an important class of system.

The 'comma clouds' first discussed in detail by Reed are now recognized as basically deep baroclinic systems and quite different from the many convective polar lows found at more northerly latitudes in the Northern Hemisphere. While certainly falling within the definition of a polar low, the name is unfortunate since the term 'comma cloud' is often used to denote the cloud signature of many hooks of cloud seen in satellite imagery. However, it is unlikely that this confusion will be resolved in the short term.

It will be clear from Chapter 4 that much work still has to be done to resolve all the issues regarding our theoretical understanding of the formation and development of polar lows. With the development of new theoretical tools, such as potential vorticity, and the insights obtained from modelling experiments it is anticipated that advances will be made over the coming decades.

7.5 Forecasting of polar lows

Until quite recently, in most cases it was only possible to predict the development and track of polar lows using a nowcasting approach, whereby the movement of existing systems were extrapolated over a period of up to about 12 hours ahead. But as some of the most significant advances over the last couple of decades in the field of polar low research have been in the successful representation of these systems in numerical models, we are now in the position of having some polar lows explicitly represented in the operational NWP analyses and forecasts. These current state-of-the-art NWP systems have a resolution of about 50 km or even less, which is high enough to represent the larger polar lows. New satellite temperature sounding instruments are also providing much better data for the NWP systems than earlier generations of instruments, so the analyses are much more likely to have existing polar lows represented.

Regional mesoscale models with a resolution of 10 km or higher are being run increasingly for various parts of the world. From research integrations it has been shown that such models can handle polar lows well, even in data-sparse areas such as the Antarctic. As these models are applied to areas prone to polar lows our ability to predict the development and track of the lows should increase dramatically, allowing earlier warning to be provided to maritime operators and to coastal communities.

7.6 Final thoughts

Around 1980 there was considerable confusion regarding the meaning of the term 'polar low', and even, occasionally, a dispute over the very existence of such a phenomenon. While occasionally papers still appear questioning whether polar lows are not just small Arctic or Antarctic counterparts of the larger mid-latitude depressions, the body of evidence suggests that there is a broad spectrum of lows in the polar regions that develop because of processes unique to the polar regions. We still do not have a complete theoretical understanding of why all polar lows develop. But major advances have been made over the half century since they were first observed and we have every reason to believe that the next 50 years will bring even greater insight.

References

Aakjær, P.D. (1992). Polar lows affecting Denmark. *Tellus* **44A**, 155–72.
Aarhus, B. and Raustein, E. (1998). Simulation of a polar low development over the Greenland Sea/Norwegian Sea. In *Proceedings The Rossby 100 Symposium, 8–12 June 1998, Stockholm, Sweden*, pp. 17–19.
Adams, N. (1997). Model prediction performance over the Southern Ocean and coastal region around East Antarctica. *Aust. Met. Mag.* **46**, 287–96.
Adolphs, U. and Wendler, G. (1995). A pilot study on the interactions between katabatic winds and polynyas at the Adélie Coast, eastern Antarctica. *Antarct. Sci.* **7**, 307–14.
Albright, M.D., Reed, R.J. and Ovens, D.W. (1995). Origin and structure of a numerically simulated polar low over Hudson Bay. *Tellus* **47A**, 834–48.
Allan, T.D. and Guymer, T.H. (1984). SEASAT measurements of wind and waves on selected passes over JASIN. *Int. J. Rem. Sens.* **5**, 379–408.
Allison, I., Tivendale, C.M., Akerman, G.J., Tann, J.M. and Wills, R.H. (1982). Seasonal variations in the surface energy exchange over Antarctic sea ice and coastal waters. *Ann. Glaciol.* **3**, 12–16.
Allison, I., Wendler, G. and Radok, U. (1993). Climatology of the East Antarctic ice sheet (100° E to 140° E) derived from automatic weather stations. *J. Geophys. Res. – Atmos.* **98**, 8815–23.
Alvarez, J.A. (1958). An anomalous July over the southern parts of South America. *Notos* **7**, 3–5.
Anderson, R.K., Ashman, J.P., Bittner, F., Farr, G.R., Ferguson, E.W., Oliver, V.J. and Smith, A.H. (1969). *Application of meteorological satellite data in analysis and forecasting*. ESSA, Washington, DC.
Anderssen, E.C. (1965). A study of atmospheric long waves in the Southern Hemisphere. *Notos* **14**, 57–65.
Andreas, E.L. (1985). Heat and moisture advection over Antarctic sea ice. *Mon. Wea. Rev.* **113**, 736–46.
Andreas, E.L. and Markshtas, A.P. (1985). Energy exchange over Antarctic sea-ice in the spring. *J. Geophys. Res.* **90**, 7199–212.
Anthes, R.A. (1971). A numerical model of the slowly varying tropical cyclone in isentropic coordinates. *Mon. Wea. Rev.* **99**, 617–35.

Anthes, R.A. (1982). *Tropical Cyclones. Their Evolution, Structure and Effects*, 19th edition. American Meteorological Society, Boston, 208 pp.

Anthes, R.A. and Warner, T.T. (1978). Development of hydrodynamic models suitable for air pollution and other mesometeorological studies. *Mon. Wea. Rev.* **106**, 1045–78.

Anthes, R.A., Hsie, E.-Y. and Warner, T.T. (1987). *Description of the Penn/NCAR Mesoscale Model Version 4 (MM4). Tech. Note No. 282*. NCAR, Boulder, CO.

Arakawa, A. (2000). Future development of general circulation models. In *General Circulation Model Development*, ed. D.A. Randall, pp. 721–80. Academic Press, London.

Asai, T. and Miura, Y. (1981). An analytical study of meso-scale vortex-like disturbances observed around Wakes Bay area. *J. Met. Soc. Jap.* **59**, 832–43.

Audunsdóttir, A. (1999). On the formation of deep water during cold air outbreaks. MSc thesis, Dept of Geophysics, University of Copenhagen.

Auer, A.H. (1986). An observational study of polar air depressions in the Australian region. In *Preprint volume Second International Conference on Southern Hemispheric Meteorology, 1–5 December 1986*, pp. 46–9. American Meteorological Society, Boston.

Austin, J.M. (1951). Mechanism of pressure change. In *Compendium of Meteorology*, ed. T.F. Malone, pp. 630–638. American Meteorological Society, Boston.

Bader, M.J., Forbes, G.S., Grant, J.R., Lilley, R.B. and Waters A.J. (1995). *Images in Weather Forecasting: A Practical Guide for Interpreting Satellite and Radar Imagery*. Cambridge University Press, Cambridge.

Bannon, P.R. (1995). Hydrostatic adjustment: Lamb's problem. *J. Atmos. Sci.* **52**, 1743–52.

Bannon, P.R. (1996). Nonlinear hydrostatic adjustment. *J. Atmos. Sci.* **53**, 3606–17.

Barlow, L.L., White, J.W.C., Barry, R.G., Roger, J.C. and Grootes, P.M. (1993). The North Atlantic Oscillation signature in deuterium and deuterium excess signals in the Greenland Ice Sheet Project 2 ice core, 1840–1970. *Geophys. Res. Lett.* **20**, 2901–4.

Barnston, A.G. and Livezey, R.E. (1987). Classification, seasonality and persistence of low-frequency atmospheric circulation patterns. *Mon. Wea. Rev.* **115**, 1083–126.

Barry, R.G. (1970). A framework for climatological research with particular reference to scale concepts. *Trans. Inst. Brit. Geogr.* **49**, 61–70.

Barry, R.G. (1983). Arctic Ocean ice and climate: perspectives on a century of polar research. *Ann. Assoc. Amer. Geogr.* **73**, 485–501.

Barry, R.G. and Chorley, R.J. (1992). *Atmosphere, Weather and Climate*. Routledge, London.

Bell, G.D. and Bosart, L.F. (1989). A 15-year climatology of Northern Hemisphere 500 mb closed cyclone and anticyclone centers. *Mon. Wea. Rev.* **117**, 2142–63.

Bergeron, T. (1954). Reviews of tropical hurricanes. *Quart. J. Roy. Met. Soc.* **80**, 131–64.

Betts, A.K. (1982). Saturation point analysis of moist convective overturning. *J. Atmos. Sci.* **39**, 1484–505.

Betts, A.K. (1986). A new convective adjustment scheme. I. Observational and theoretical basis. *Quart. J. Roy. Met. Soc.* **112**, 677–91.

Billing, H., Haupt, I. and Tonn, W. (1983). Evolution of a hurricane-like cyclone in the Mediterranean Sea. *Beitr. Phys. Atmos.* **56**, 508–10.

Bjerknes, J. (1966). A possible response of the atmospheric Hadley circulation to equatorial anomalies of ocean temperature. *Tellus* **18**, 820–9.

Bjerknes, J. and Solberg, H. (1922). Life cycle of cyclones and the polar front theory of atmospheric circulations. *Geophys. Publ.* **9**, 30–45.

Black, R.I. (1982). Beaufort Storm of July, 1982 (unpublished manuscript). Satellite and Beaufort Office, Edmonton, Canada, 18 pp.

Bleck, R. (1990). Detection of upper/lower vortex interaction associated with extratropical cyclogenesis. *Mon. Wea. Rev.* **118**, 573–85.

Blier, W. (1996). A numerical modeling investigation of a case of polar airstream cyclogenesis over the Gulf of Alaska. *Mon. Wea. Rev.* **124**, 2703–25.

Bluestein, H.B. (1993). *Synoptic-dynamic meteorology in midlatitudes* (2 vols.). Oxford University Press, Oxford.

Blumen, W. (1979). On short-wave baroclinic instability. *J. Atmos. Sci.* **36**, 1925–33.

Bond, N.A. and Shapiro, M.A. (1991). Polar lows over the Gulf of Alaska in conditions of reverse shear. *Mon. Wea. Rev.* **119**, 551–72.

Bratseth, A.M. (1985). A note on CISK in polar air masses. *Tellus* **37A**, 403–6.

Bresch, J.F., Reed, R.J. and Albright, M.D. (1997). A polar-low development over the Bering Sea: analysis, numerical simulation, and sensitivity experiments. *Mon. Wea. Rev.* **125**, 3109–30.

Bretherton, F.P. (1966). Baroclinic instability and the short wave cut-off in terms of potential vorticity. *Quart. J. Roy. Met. Soc.* **92**, 335–45.

Bromwich, D.H. (1986). Boundary layer studies in Terra Nova Bay, Antarctica. *Antarctic Climate Research* **1**, 9–13.

Bromwich, D.H. (1987). A case study of mesoscale cyclogenesis over the southwestern Ross Sea. *Ant. J. of the US* **22**, 254–56.

Bromwich, D.H. (1988). Snowfall in high southern latitudes. *Rev. Geophys.* **26**, 149–68.

Bromwich, D.H. (1989). Satellite analysis of Antarctic katabatic wind behaviour. *Bull. Amer. Met. Soc.* **70**, 738–49.

Bromwich, D.H. (1991). Mesoscale cyclogenesis over the southwestern Ross Sea linked to strong katabatic winds. *Mon. Wea. Rev.* **119**, 1736–52.

Bromwich, D.H. and Parish, T.R. (1988). Mesoscale cyclone interactions with the surface windfield near Terra Nova Bay, Antarctica. *Ant. J. of the US* **23**, 172–75.

Bromwich, D.H., Carrasco, J.F., Liu, Z. and Tzeng, R.Y. (1993). Hemispheric atmospheric variations and oceanographic impacts associated with katabatic surges across the Ross Ice Shelf, Antarctica. *J. Geophys. Res. – Atmos.* **98**, 13045–62.

Bromwich, D.H., Carrasco, J.F. and Turner, J. (1996). A downward developing mesoscale cyclone over the Ross Ice Shelf during winter. *Global Atmos. Ocean Sys.* **4**, 125–47.

Bromwich, D.H., Robasky, F.M., Cullather, R.I. and Van Woert, M.L. (1995). The atmospheric hydrologic cycle over the Southern Ocean and Antarctica from operational numerical analyses. *Mon. Wea. Rev.* **123**, 3518–38.

Browning, K.A. (1990). Organization of clouds and precipitation in extratropical cyclones. In *Extratropical Cyclones: The Erik Palmén Memorial Volume*, ed. C.W. Newton and E.O. Holopainen. pp. 129–53. American Meteorological Society, Boston.

Browning, K.A. and Dicks, E.M. (2001). Mesoscale structure of a polar low with strong upper-level forcing. *Quart. J. Roy. Met. Soc.* **127**, 359–75.

Businger, S. (1985). The synoptic climatology of polar-low outbreaks. *Tellus* **37A**, 419–32.

Businger, S. (1987). The synoptic climatology of polar-low outbreaks over the Gulf of Alaska and the Bering Sea. *Tellus* **39A**, 307–25.

Businger, S. and Baik, J.J. (1991). An Arctic hurricane over the Bering Sea. *Mon. Wea. Rev.* **119**, 2293–322.

Businger, S. and Reed, R.J. (1989a). Cyclogenesis in cold air masses. *Wea. and Forecasting* **2**, 133–56.

Businger, S. and Reed, R.J. (1989b). Polar lows. In *Polar and Arctic Lows*, ed. P.F. Twitchell, E. Rasmussen and K.L. Davidson. pp. 3–45. A Deepak, Hampton, VA.

Callaghan, J. and Betts, M.S. (1987). Some cases of westward moving disturbances in the Mawson–Davis area, Antarctica. *Aust. Met. Mag.* **35**, 79–85.

Carleton, A.M. (1979). A synoptic climatology of satellite observed extratropical cyclone activity for the Southern Hemisphere winter. *Arch. Met. Geophys. Biokl. B* **27**, 265–79.

Carleton, A.M. (1981a). Climatology of the 'instant occlusion' phenomenon for the Southern Hemisphere winter. *Mon. Wea. Rev.* **109**, 177–81.

Carleton, A.M. (1981b). Ice–ocean–atmosphere interactions at high southern latitudes in winter from satellite observations. *Aust. Met. Mag.* **29**, 183–95.

Carleton, A.M. (1981c). Monthly variability of satellite-derived cyclonic activity for the Southern Hemisphere winter. *J. Clim.* **1**, 21–38.

Carleton, A.M. (1985a). Satellite climatological aspects of the 'polar low' and 'instant occlusion'. *Tellus* **37A**, 433–50.

Carleton, A.M. (1985b). Synoptic cryosphere–atmosphere interactions in the northern hemisphere from DMSP image analysis. *Int. J. Rem. Sens.* **6**, 239–61.

Carleton, A.M. (1986). Synoptic-dynamic character of 'bursts' and 'breaks' in the Southwest U.S. summer precipitation singularity. *J. Clim.* **6**, 605–23.

Carleton, A.M. (1988a). Meridional transport of eddy sensible heat in winters marked by extremes of the North Atlantic Oscillation, 1948/49–1979/80. *J. Clim.* **1**, 212–23.

Carleton, A.M. (1988b). Sea ice–atmosphere signal of the Southern Oscillation in the Weddell Sea, Antarctica. *J. Clim.* **1**, 379–88.

Carleton, A.M. (1989). Antarctic sea-ice relationships with indices of the atmospheric circulation of the Southern Hemisphere. *Clim. Dyn.* **3**, 207–20.

Carleton, A.M. (1992). Synoptic interactions between Antarctica and lower latitudes. *Aust. Met. Mag.* **40**, 129–47.

Carleton, A.M. (1995). On the interpretation and classification of mesoscale cyclones from satellite infrared imagery. *Int. J. Rem. Sens.* **16**, 2457–85.

Carleton, A.M. (1996). Satellite climatological aspects of cold air mesocyclones in the Arctic and Antarctic. *Global Atmos. Ocean Sys.* **5**, 1–42.

Carleton, A.M. and Carpenter, D.A. (1989a). Satellite climatology of 'polar air' vortices for the Southern Hemisphere winter. In *Polar and Arctic Lows*, ed. P.F. Twitchell, E.A. Rasmussen, and K.L. Davidson. pp. 401–13. A Deepak, Hampton, VA.

Carleton, A.M. and Carpenter, D.A. (1989b). Intermediate-scale sea ice–atmosphere interactions over high southern latitudes in winter. *Geoj.* **18**, 87–101.

Carleton, A.M. and Carpenter, D.A. (1990). Satellite climatology of 'polar lows' and broadscale climatic associations for the southern hemisphere. *Int. J. Climatol.* **10**, 219–46.

Carleton, A.M. and Fitch, M. (1993). Synoptic aspects of Antarctic mesocyclones. *J. Geophys. Res. – Atmos.* **98**, 12997–3018.

Carleton, A.M. and Song, Y. (1997). Synoptic climatology, and intrahemispheric associations, of cold air mesocyclones in the Australasian sector. *J. Geophys. Res. – Atmos.* **102**, 13873–87.

Carleton, A.M. and Song, Y. (2000). Satellite passive sensing of the marine atmosphere associated with cold-air mesoscale cyclones. *Professional Geographer (Special Focus Issue on Remote Sensing in Hydroclimatology)* **52**, 289–306.

Carleton, A.M., Carpenter, D.A. and Weser, P.J. (1990). Mechanisms of interannual variability of the Southwest U.S. summer rainfall maximum. *J. Clim.* **3**, 999–1015.

Carleton, A.M., John, G. and Welsch, R. (1998). Interannual variations and regionality of Antarctic sea-ice–temperature associations. *Ann. Glaciol.* **27**, 403–8.

Carleton, A.M., McMurdie, L.A., Katsaros, K.B., Zhao, H., Mognard, N.M. and Claud, C. (1995). Satellite-derived features and associated atmospheric environments of Southern Ocean mesocyclone events. *Global Atmos. Ocean Sys.* **3**, 209–48.

Carleton, A.M., McMurdie, L.A., Zhao, H., Katsaros, K.B., Mognard, N. and Claud, C. (1993). Satellite microwave sensing of Antarctic ocean mesocyclones. In *Proceedings of the Fourth International Conference on Southern Hemisphere Meteorology and Oceanography. March 29–April 2, 1993, Hobart, Australia*, pp. 497–8. AMS, Boston.

Carlson, T.N. (1991). *Mid-latitude Weather Systems*. HarperCollins Academic, London.

Carrasco, J.F. and Bromwich, D.H. (1991). A case study of katabatic wind-forced mesoscale cyclogenesis near Byrd Glacier. *Ant. J. of the US* **26**, 258–61.

Carrasco, J.F. and Bromwich, D.H. (1993). Mesoscale cyclogenesis dynamics over the southwestern Ross Sea, Antarctica. *J. Geophys. Res.* **98 D7**, 12973–95.

Carrasco, J.F. and Bromwich, D.H. (1994). A survey of mesoscale cyclonic activity near McMurdo Station, Antarctica. *Ant. J. of the US* **29**, 298–301.

Carrasco, J.F. and Bromwich, D.H. (1995). A case study of a midtropospheric subsynoptic-scale cyclone that developed over the Ross Sea and Ross Ice Shelf of Antarctica. *Antarct. Sci.* **7**, 199–210.

Carrasco, J.F. and Bromwich, D.H. (1996). Mesoscale cyclone activity near Terra Nova Bay and Byrd Glacier, Antarctica during 1991. *Global Atmos. Ocean Sys.* **5**, 43–72.

Carrasco, J.F., Bromwich, D.H. and Liu, Z. (1997a). Mesoscale cyclone activity over Antarctica during 1991. 1. Marie Byrd Land. *J. Geophys. Res. – Atmos.* **102**, 13923–37.

Carrasco, J.F., Bromwich, D.H. and Liu, Z. (1997b). Mesoscale cyclone activity over Antarctica during 1991. 2. Near the Antarctic peninsula. *J. Geophys. Res. – Atmos.* **102**, 13939–54.

Carroll, J.J. (1982). Long-term means and short-term variability of the surface energy balance components at the South Pole. *J. Geophys. Res.* **87**, 4277–86.

Cavanie, A. and Gohin, F. (1995). The AMI-Wind scatterometer. *Ocean. Apps. Rem. Sens.* **1**, 457–60.

Charney, J.G. (1955). The use of the primitive equations of motions in numerical prediction. *Tellus* **7**, 22–6.

Charney, J. and Eliassen, A. (1964). On the growth of the hurricane depression. *J. Atmos. Sci.* **21**, 68–75.

Charnock, H. (1955). Wind stress over a water surface. *Quart. J. Roy. Met. Soc.* **81**, 630–40.

Chen, B., Smith, S.R. and Bromwich, D.H. (1996). Evolution of the tropospheric split jet over the South Pacific Ocean during the 1986–89 ENSO cycle. *Mon. Wea. Rev.* **124**, 1711–31.

Chunchuzov, I., Vachon, P.W. and Ramsay, B. (2000). Detection and characterization of mesoscale cyclones in RADARSAT Synthetic Aperture Radar images of the Labrador Sea. *Can. J. Rem. Sens.* **26**, 213–30.

Claud, C., Katsaros, K.B., Mognard, N.M. and Scott, N.A. (1996). Comparative satellite study of mesoscale disturbances in polar regions. *Global Atmos. Ocean Sys.* **4**, 233–73.

Claud, C., Katsaros, K.B., Petty, G.W., Chedin, A. and Scott, N.A. (1992). A cold air outbreak over the Norwegian Sea observed with the TIROS-N Operational Vertical Sounder (TOVS) and the Special Sensor Microwave Imager (SSM/I). *Tellus* **44A**, 100–18.

Claud, C., Mognard, N.M., Katsaros, K.B., Chedin, A. and Scott, N.A. (1993). Satellite observations of a polar low over the Norwegian Sea by special sensor microwave imager, Geosat, and TIROS-N operational vertical sounder. *J. Geophys. Res. – Oceans* **98**, 14487–506.

Coughlan, M.J. (1983). A comparative climatology of blocking action in the two hemispheres. *Aust. Met. Mag.* **31**, 3–13.

Craig, G. and Cho, H.R. (1989). Baroclinic instability and CISK as the driving mechanism for polar lows and comma clouds. In *Polar and Arctic Lows*, ed. P.F. Twitchell, E. Rasmussen and K.L. Davidson. pp. 131–40. A Deepak, Hampton, VA.

Craig, G.C. and Gray, S.L. (1996). CISK or WISHE as the mechanism for tropical cyclone intensification. *J. Atmos. Sci.* **53**, 3528–40.

Cullather, R.I., Bromwich, D.H. and van Woert, M.L. (1996). Interannual variations in Antarctic precipitation related to El Niño southern oscillation. *J. Geophys. Res.* **101**, 19109–18.

Dalu, G., Prabhakara, C. and Nuccarione, J. (1993). Optimization of an algorithm for the estimation of rainfall from the SSM/I data. *J. Met. Soc. Jap.* **71**, 419–24.

Danielsen, E.F. (1990). In defense of Ertel's potential vorticity and its general applicability as a meteorological tracer. *J. Atmos. Sci.* **47**, 2013–20.

Dannevig, P. (1954). *Meteorologi for Flygere* [in Norwegian]. Aschehoug, Oslo.

Davidova, N.G. (1967). Types of synoptic process and associated wind fields in oceanic regions of the Southern Hemisphere. In *Polar Meteorology, WMO Tech. Note 87, Proceedings WMO/SCAR/ICPM Symp. on Polar Meteorology, Geneva, 5–9 Sept. 1966 (WMO-No. 211.TP.11)*. Secretariat of the WMO, Geneva.

Davis, C.A. (1992). A potential-vorticity diagnosis of the importance of initial structure and condensational heating in observed extratropical cyclogenesis. *Mon. Wea. Rev.* **120**, 2409–28.

Davis, C.A. and Emanuel, K.A. (1991). Potential vorticity diagnostics of cyclogenesis. *Mon. Wea. Rev.* **119**, 1929–53.

Davis, C.A., Grell, E.D. and Shapiro, M.A. (1996). The balanced dynamical nature of a rapidly intensifying oceanic cyclone. *Mon. Wea. Rev.* **124**, 3–26.

Deser, C. and Blackmon, M.L. (1993). Surface climate variations over the north-atlantic ocean during winter – 1900–1989. *J. Clim.* **6**, 1743–53.

Dickson, R.R. and Namias, J. (1976). North American influences on the circulation and climate of the North Atlantic sector, *Mon. Wea. Rev.* **104**, 1255–65.

Douglas, M.W., Fedor, L.S. and Shapiro, M.A. (1991). Polar low structure over the northern Gulf of Alaska based on research aircraft observations. *Mon. Wea. Rev.* **119**, 32–54.

Douglas, M.W., Shapiro, M.A., Fedor, L.S. and Saukkonen, L. (1995). Research aircraft observations of a polar low at the east Greenland ice edge. *Mon. Wea. Rev.* **123**, 5–15.

Drewry, D.J. (1983). *Antarctica: Glaciological and Geophysical Folio*. Scott Polar Research Institute, Cambridge.

Dudhia, J. (1993). A nonhydrostatic version of the Penn State–NCAR Mesoscale Model: validation tests and simulation of an Atlantic cyclone and cold front. *Mon. Wea. Rev.* **121**, 1493–513.

Duncan, C.N. (1977). A numerical investigation of polar lows. *Quart. J. Roy. Met. Soc.* **103**, 255–68.

Duncan, C.N. (1978). Baroclinic instability in a reversed shear flow. *Met. Mag.* **107**, 17–23.

Durran, D.R. (1999). *Numerical Methods for Wave Equations in Geophysical Fluid Dynamics*. Springer-Verlag, New York.

Dysthe, K.B. and Harbitz, A. (1987). Big waves from polar lows? *Tellus* **39A**, 500–8.

Eady, E.T. (1949). Long waves and cyclone waves. *Tellus* **1**, 33–52.

Edouard, S., Vautard, R. and Brunet, G. (1997). On the maintenance of potential vorticity in isentropic coordinates. *Quart. J. Roy. Met. Soc.* **123**, 2069–94.

Eidsvik, K.J. (1985). *Polar low trajectories stochastic model identification, Technical Report No. 15, The Polar Lows Project*. Norwegian Meteorological Institute, Oslo.

Eliassen, A. (1952). Slow thermally or frictionally controlled meridional circulations in a circular vortex. *Astrophysica Norvegica* **5**, 19–60.

Ellett, D.J. (1993). The north-east Atlantic: a fan-assisted storage heater? *Weather* **48**, 118–26.

Emanuel, K.A. (1985). Frontal circulation in the presence of small symmetric stability. *J. Atmos. Sci.* **42**, 1062–71.

Emanuel, K.A. (1986a). An air–sea interaction theory for tropical cyclones. Part I: steady-state maintenance. *J. Atmos. Sci.* **43**, 585–604.

Emanuel, K.A. (1986b). A two stage air–sea interaction theory for polar lows. In *Preprints, The International Conference on Polar Lows, Oslo, Norway, 20–23 May 1986*. pp. 187–200. Norwegian Meteorological Institute, Oslo.

Emanuel, K.A. (1988). Observational evidence of slantwise convective adjustment. *Mon. Wea. Rev.* **116**, 1805–16.

Emanuel, K.A. (1994). Sea–air heat transfer effects on extratropical cyclones. In *The Life Cycles of Extratropical Cyclones Volume III. Proceedings of an International Symposium 27 June–1 July 1994, Bergen, Norway*, ed. S. Grønås and M.A. Shapiro. pp. 67–72. American Meteorological Society, Boston.

Emanuel, K.A. and Rotunno, R. (1989). Polar lows as arctic hurricanes. *Tellus* **41A**, 1–17.

Emanuel, K.A., Neelin, J.D. and Bretherton, C.S. (1994). On large-scale circulations in convecting atmospheres. *Quart. J. Roy. Met. Soc.* **120**, 1111–43.

Engels, R. and Heinemann, G. (1996). Three-dimensional structures of summertime Antarctic meso-scale cyclones. Part II: Numerical simulations with a limited area model. *Global Atmos. Ocean Sys.* **4**, 181–208.

Enomoto, H. and Ohmura, A. (1990). The influence of atmospheric half-yearly cycle on the sea ice extent in the Antarctic. *J. Geophys. Res.* **95**, 9497–511.

Ernst, J. A. and Matson, M. (1983). A Mediterranean tropical storm? *Weather* **38**, 332–37.

Ese, T., Kanestrom, I. and Pedersen, K. (1988). Climatology of polar lows over the Norwegian and Barents Seas. *Tellus* **40A**, 248–55.

Fantini, M. (1990). The influence of heat and moisture fluxes from the ocean on the development of baroclinic waves. *J. Atmos. Sci.* **47**, 840–55.

Farrara, J.D., Ghil, M., Mechoso, C.R. and Mo, K.C. (1989). Empirical orthogonal functions and multiple flow regimes in the Southern Hemisphere winter. *J. Atmos. Sci.* **46**, 3219–23.

Farrell, B. (1985). Transient growth of damped baroclinic waves. *J. Atmos. Sci.* **42**, 2718–27.

Fett, R.W. (1989a). *Navy Tactical Applications Guide. Volume 8, Part 1. Arctic–Greenland/Norwegian/Barents Seas. Weather Analysis and Forecast Applications*. Science and Technology Corporation, Hampton, VA.

Fett, R.W. (1989b). Polar low development associated with boundary layer fronts in the Greenland, Norwegian and Barents Seas. In *Polar and Arctic Lows*, ed. P.F. Twitchell, E. Rasmussen, and K.L. Davidson. pp. 313–22. A Deepak, Hampton, VA.

Fett, R.W. (1992). *Navy Tactical Applications Guide. Volume 8, Part 2. Arctic–East Siberian/Chukchi/Beaufort Seas. Weather Analysis and Forecast Applications*. Naval Research Laboratory, Monterey, CA.

Fett, R.W., Englebretson, R.E. and Perryman, D.C. (1993). *Forecasters Handbook for the Bering Sea, Aleutian Islands and Gulf of Alaska*. Naval Research Laboratory, Monterey, CA.

Fitch, M. and Carleton, A.M. (1992). Antarctic mesoscale regimes from satellite and conventional data. *Tellus* **44A**, 180–96.

Forbes, G.S. and Lottes, W.D. (1985). Classification of mesoscale vortices in polar air streams and the influence of the large-scale environment on their evolutions. *Tellus* **37A**, 132–55.

Forbes, G.S. and Merrit, J.H. (1984). Mesoscale vortices over the Great Lakes in wintertime. *Mon. Wea. Rev.* **112**, 377–81.

Fujiyoshi, Y., Kodama, Y., Tuboki, K., Nishimura, K. and Ono, N. (1996). Structure of cold air during the development of a broad cloud band and a meso-α-scale vortex: simultaneous two-point radiosonde observations. *J. Met. Soc. Jap.* **74**, 281–97.

Fujiwhara, S. (1923). On the growth and decay of vortical systems. *Quart. J. Roy. Met. Soc.* **49**, 287–93.

Fujiwhara, S. (1931). Short note on the behaviour of two vortices. *Proc. Phys. Math. Soc. Japan Ser. 3*, **13**, 106–10.

Gallée, H. (1995). Simulation of the mesocyclonic activity in the Ross Sea, Antarctica. *Mon. Wea. Rev.* **123**, 2051–69.

Gallée, H. (1996). Mesoscale atmospheric circulations over the southwestern Ross Sea sector, Antarctica. *J. Appl. Met.* **35**, 1129–41.

Gallée, H. and Schayes, G. (1994). Development of a 3-dimensional meso-γ primitive equation model – katabatic winds simulation in the Area of Terra Nova Bay, Antarctica. *Mon. Wea. Rev.* **122**, 671–85.

Gill, A.E. (1982). *Atmosphere and Ocean Dynamics*. Academic Press, London.

Giovinetto, M.B., Yamazaki, K., Wendler, G. and Bromwich, D.H. (1997). Atmospheric net transport of water vapor and latent heat across 60°S. *J. Geophys. Res.* **102**, 11171–79.

Gloersen, P. (1995). Modulation of hemispheric sea-ice cover by ENSO events. *Nature* **373**, 503–6.

Goodberlet, M.A., Swift, C.T. and Wilkerson, J.C. (1989). Remote sensing of ocean surface winds with the Special Sensor Microwave/Imager. *J. Geophys. Res.* **94**, 14547–55.

Gray, W.M., Scheaffer, J.D. and Landsea, C.W. (1997). Climate trends associated with multi-decadal variability of intense Atlantic hurricane activity. In *Hurricanes, Climate Changes and Socioeconomic Impacts: A Current Perspective*, ed. H. Diaz and R.S. Pulwarty. pp. 15–53. Springer-Verlag, New York.

Grell, G.A., Dudhia, J. and Stauffer, D.R. (1994). A description of the fifth-generation Penn State/NCAR mesoscale model (MM5). NCAR Technical Note, NCAR/TN-398 + STR, Boulder, CO. 138 pp.

Grønås, S. (1995). The seclusion intensification of the New Year's Day storm, *Tellus* **47A**, 733–46.

Grønås, S. and Hellevik, O.E. (1982). *A limited area prediction model at the Norwegian Meteorological Institute. Tech. Rep. No. 61.* Norwegian Meteorological Institute, Oslo.

Grønås, S. and Kvamstø, N.G. (1994). Synoptic conditions for arctic front polar lows. In *Proceedings of the International Conference on the Life Cycles of Extratropical Cyclones. Vol III. 27 June to 1 July 1994, Bergen, Norway*, 3rd edition, pp. 89–95.

Grønås, S. and Kvamstø, N.G. (1995). Numerical simulations of the synoptic conditions and development of Arctic outbreak polar lows. *Tellus* **47A**, 797–814.

Grønås, S., Foss, A. and Lystad, M. (1987a). Numerical simulations of polar lows in the Norwegian Sea. *Tellus* **39A**, 334–54.

Grønås, S., Foss, A. and Lystad, M. (1987b). The Norwegian mesoscale NWP system. *Proceedings of the Symposium on Mesoscale Analysis and Forecasting, Vancouver, Canada*. ESA SP-282.

Guest, P.S., Davidson, K.L., Overland, J.E. and Frederickson, P.A. (1995). Atmosphere–ocean interactions in the marginal ice zones of the Nordic seas. *Arctic Oceanography: Marginal Ice Zones and Continental Shelves. Coastal and Estuarine Studies* **49**, 51–95.

Häkkinen, S. (1995). Simulated interannual variability of the Greenland Sea deep water formation and its connection to surface forcing. *J. Geophys. Res.* **100**, 4761–70.

Hanley, D. and Richards, W.G. (1991). *Polar Lows in Canadian Waters 1977–1989. Report: MAES 2-91.* Scientific Services Division, Atlantic Region, Atmospheric Environment Service, Canada.

Harangozo, S.A. (1997). Atmospheric meridional circulation impacts on contrasting winter sea ice extent in two years in the Pacific sector of the Southern Ocean. *Tellus* **49**, 388–400.

Harley, D.G. (1960). Frontal contour analysis of a 'polar' low. *Met. Mag.* **89**, 141–7.

Harold, J.M. (1997). Characteristics of polar mesocyclones over the north-east Atlantic region. PhD thesis, University of East Anglia, 249pp.

Harold, J.M., Bigg, G.R. and Turner, J. (1999a). Mesocyclone activity over the North-East Atlantic. Part 1: Vortex distribution and variability. *Int. J. Climatol.* **19**, 1187–204.

Harold, J.M., Bigg, G.R. and Turner, J. (1999b). Mesocyclone activity over the Northeast Atlantic. Part 2: An investigation of causal mechanisms. *Int. J. Climatol.* **19**, 1283–99.

Harrold, T.W. and Browning, K.A. (1969). The polar low as a baroclinic disturbance. *Quart. J. Roy. Met. Soc.* **95**, 710–23.

Hattle, J.B. (1968). Polar fronts of the Southern Hemisphere. *Notos* **17**, 15–22.

Haugen, J.E. (1986). Numerical simulations with an idealized model. In *Proceedings of the International Conference on Polar Lows*, pp. 151–60. Norwegian Meteorological Institute, Oslo.

Haynes, P.H. and McIntyre, M.E. (1987). On the evolution of vorticity and potential vorticity in the presence of diabatic heating and frictional or other forces. *J. Atmos. Sci.* **44**, 828–41.

Haynes, P.H. and McIntyre, M.E. (1990). On the conservation and impermeability theorems for potential vorticity. *J. Atmos. Sci.* **47**, 2021–31.

Heinemann, G. (1990). Mesoscale vortices in the Weddell Sea region (Antarctica). *Mon. Wea. Rev.* **118**, 779–93.

Heinemann, G. (1996a). On the development of wintertime meso-scale cyclones near the sea ice front in the Arctic and Antarctic. *Global Atmos. Ocean Sys.* **4**, 89–123.

Heinemann, G. (1996b). Three-dimensional structures of summertime Antarctic meso-scale cyclones: Part I: Observational studies with aircraft, satellite and conventional data. *Global Atmos. Ocean Sys.* **4**, 149–80.

Heinemann, G. (1996c). A wintertime polar low over the eastern Weddell Sea (Antarctica): a study with AVHRR, TOVS, SSM/I and conventional data. *Met. Atmos. Phys.* **58**, 83–102.

Heinemann, G. (1997). Idealized simulations of the Antarctic katabatic wind system with a three-dimensional mesoscale model. *J. Geophys. Res. – Atmos.* **102**, 13825–34.

Heinemann, G. (1998). A mesoscale model-based study of the dynamics of a wintertime polar low in the Weddell Sea region of the Antarctic during the Winter Weddell Sea Program field phase 1986. *J. Geophys. Res.* **103**, 5983–6000.

Heinemann, G. and Claud, C. (1997). Report of a workshop on 'Theoretical and observational studies of polar lows' of the European Geophysical Society Polar Lows Working Group. *Bull. Amer. Met. Soc.* **78**, 2643–58.

Heinemann, G. and Rose, L. (1990). Surface energy balance parameterizations of boundary-layer heights and the application of resistance laws near an Antarctic ice shelf front. *Bound.-Layer Meteorol.* **51**, 123–58.

Herman, Y. (1989). *The Arctic Seas. Climatology, Oceanography, Geology and Biology*. Van Nostrand Reinhold, New York.

Hewson, T.D., Craig, G.C. and Claud, C. (2000). Evolution and mesoscale structure of a polar low outbreak. *Quart. J. Roy. Met. Soc.* **126**, 1031–63.

Heygster, G., Burns, B., Hunewinkel, T., Künzi, K., Meyer-Lerbs, L., Schottmüller, H., Thomas, C., Lemke, P., Viehoff, T., Turner, J., Harangozo, S., Lachlan-Cope, T. and Pedersen, L. (1996). *PELICON – Project for Estimation of Long-term variability in Ice CONcentration. Final report to the EC.* University of Bremen, Bremen.

Hoem, V. (1985). *Polar low case studies II, December 1982–December 1985. Technical Report No. 6, Polar Lows Project.* Norwegian Meteorological Institute, Oslo.

Hogan, A.W. (1997). A synthesis of warm air advection to the south polar plateau. *J. Geophys. Res.* **102**, 14009–20.

Hollinger, J., Lo, R., Poe, G., Savage, R. and Pierce, J. (1987). *Special sensor microwave/imager user's guide.* Naval Research Laboratory, Washington, DC.

Holton, J.R. (1979). *An Introduction to Dynamic Meteorology*, 2nd edition. Academic Press, New York.

Holton, J.R. (1992). *An Introduction to Dynamic Meteorology*, 3rd edition. Academic Press, New York.

Horel, J.D. and Wallace, J.M. (1981). Planetary-scale atmospheric phenomena associated with the Southern Oscillation. *Mon. Wea. Rev.* **109**, 813–29.

Hoskins, B.J., McIntyre, M.E. and Robertson, A.W. (1985). On the use and significance of isentropic potential vorticity maps. *Quart. J. Roy. Met. Soc.* **111**, 877–946.

Houmb, O.G. and Lönseth, L. (1986). Ocean waves under polar lows. In *Polar lows in the Norwegian, Greenland and Barents Sea. Final report of the Polar Lows Project*, pp. 173–96. Norwegian Meteorological Institute, Oslo.

Houmb, O.G., Lönseth, L., Schjölberg, P. and Vollan, B. (1986). Environmental conditions under polar lows. In *Proceedings The International Conference on Polar Lows, Oslo, Norway*, pp. 343–57.

Howard, D.A. (1983). An analysis of the variability of cyclones around Antarctica and their relationship to sea ice extent. *Ann. Assoc. Am. Geogr.* **73**, 519–37.

Hurrell, J.W. (1995). Decadal trends in the North-Atlantic oscillation – regional temperatures and precipitation. *Science* **269**, 676–79.

Hurrell, J.W. (1996). Influence of variations in extratropical wintertime teleconnections on Northern Hemisphere temperature. *Geophys. Res. Lett.* **23**, 665–68.

Hurrell, J.W. and van Loon, H. (1994). A modulation of the atmospheric annual cycle in the Southern Hemisphere. *Tellus* **46A**, 325–38.

Hurrell, J.W. and van Loon, H. (1997). Decadal variations in climate associated with the North Atlantic Oscillation. *Clim. Change* **36**, 301–26.

Hurrell, J.W., Kushnir, Y. and Visbeck, M. (2001). The North Atlantic Oscillation. *Science* **291**, 603–5.

Jacobs, S.S. and Comiso, J.C. (1997). Climate variability in the Amundsen and Bellingshausen Seas. *J. Clim.* **10**, 697–709.

James, I.N. (1988). On the forcing of planetary scale Rossby waves by Antarctica. *Quart. J. Roy. Met. Soc.* **114**, 619–37.

James, I.N. (1989). The Antarctic drainage flow: implications for hemispheric flow on the southern hemisphere. *Antarctic Sci.* **1**, 279–90.

Jones, D.A. and Simmonds, I. (1993). A climatology of Southern Hemisphere extratropical cyclones. *Clim. Dyn.* **9**, 131–45.

Katsaros, K.B., Bhatti, I., McMurdie, L.A. and Petty, G.W. (1989). Identification of atmospheric fronts over the ocean with microwave measurements of water vapor and rain. *Wea. and Forecasting* **4**, 449–60.

Kellogg, W.W. and Twitchell, P.F. (1986). Summary of the workshop on Arctic lows, 9–10 May 1985, Boulder, Colorado. *Bull. Amer. Met. Soc.* **67**, 186–93.

Kibe, S. (1988). Small scale low in the vicinity of the west coast of Hokkaido Island [in Japanese]. *Tenki* **35**, 146–51.

Kidson, J.W. (1988). Interannual variations in the Southern Hemisphere circulation. *J. Clim.* **1**, 1177–98.

Kidson, J.W. (1991). Intraseasonal variations in the Southern Hemisphere circulation. *J. Clim.* **4**, 939–53.

King, J.C. (1994). Recent climate variability in the vicinity of the Antarctic Peninsula. *Int. J. Climatol.* **14**, 357–69.

Klein, T. (1996). Idealized simulations of mesocyclonic activity in the eastern Weddell Sea region. Extended abstracts, workshop on *Theoretical and observational studies of polar lows*, St Petersburg, 23–26 September 1996. Meteorologisches Institut der Universität Bonn, pp. 12–16.

Klein, T. and Heinemann, G. (2001). On the forcing mechanisms of mesocyclones in the eastern Weddell Sea region: process studies using a mesoscale numerical model. *Meteorol. Zeitschr.* **10**, 113–22.

Kleinschmidt, E. (1951). Principles of the theory of tropical cyclones. *Arch. Met. Geophys. Biokl. B* **4a**, 53–72.
Kottmeier, C. and Hartig, R. (1990). Winter observations of the atmosphere over Antarctic sea ice. *J. Geophys. Res.* **95**, 16551–60.
Krauss, W. (1986). The North Atlantic current. *J. Geophys. Res.* **91**, 5061–74.
Kuo, H.L. (1965). On the formation and intensification of tropical cyclones through latent heat release by cumulus convection. *J. Atmos. Sci.* **22**, 40–63.
Kuroda, Y. (1992). The convergence cloud band and the shipwreck in the Japan Sea [in Japanese]. *Umi to Sora* **67**, 261–79.
Kushnir, Y., Cardone, V.J., Greenwood, J.G. and Crane, M.A. (1997). The recent increase in North Atlantic wave heights. *J. Clim.* **10**, 2107–13.
Lamb, H. (1908). On the theory of waves propagating vertically in the atmosphere. *Proc. London Math. Soc.* **84**, 551–72.
Lamb, H. (1932). *Hydrodynamics* (6th Edition 1945). Dover, New York.
Lamb, P.J. and Peppler, R.A. (1987). North Atlantic Oscillation: concept and an application. *Bull. Amer. Met. Soc.* **68**, 1218–25.
Large, W.G. and van Loon, H. (1989). Large-scale low frequency variability of the 1979 FGGE surface buoy drifts and winds over the Southern Hemisphere. *J. Phys. Ocean.* **19**, 216–32.
LeMarshall, J.R., Kelly, G.A. and Karoly, D.J. (1985). An atmospheric climatology of the Southern Hemisphere based on the ten years of daily numerical analyses (1972–1982): I. Overview. *Aust. Met. Mag.* **33**, 65–85.
Lieder, M. and Heinemann, G. (1998). Antarctic mesocyclone events over the southern Pacific during FROST SOP1 and SOP3: A meso-scale analysis using AVHRR, SSM/I, ERS-Scatterometer and numerical model data. In *Proceedings of The 1998 Meteorological Satellite Data Users' Conference, Paris*, pp. 317–18.
Lieder, M. and Heinemann, G. (1999). A summertime Antarctic mesocyclone event over the Southern Pacific during FROST SOP3: a mesoscale analysis using AVHRR, SSM/I, ERS, and numerical model data. *Wea. and Forecasting* **14**, 893–908.
Lindzen, S., Lorenz, E. N. and Platzman, G.W. (1990). *The Atmosphere – A Challenge. The Science of Jule Gregory Charney*. American Meteorological Society, Boston.
Lyall, I.T. (1972). The polar low over Britain. *Weather* **27**, 378–90.
Lyons, S.W. (1983). Characteristics of intense Antarctic depressions near the Drake Passage. In *Preprints, First International Conference Southern Hemisphere Meteorology, 31 July–6 August 1983*, pp. 238–40. American Meteorological Society. Boston.
Lystad, M. (1986). *Polar lows in the Norwegian, Greenland and Barents Seas. Final report, Polar Lows Project*. Norwegian Meteorological Institute, Oslo.
Lystad, M., Hoem, V. and Rabbe, A. (1986). Case studies. In *Polar lows in the Norwegian, Greenland and Barents Sea. Final Report, Polar Lows Project*, ed. M. Lystad, pp. 63–109. Norwegian Meteorological Institute, Oslo.
Mailhot, J., Hanley, D., Bilodeau, B. and Hertzman, O. (1996). A numerical case study of a polar low in the Labrador Sea. *Tellus* **48A**, 383–402.
Mak, M. (1982). On moist quasi-geostrophic baroclinic instability. *J. Atmos. Sci.* **39**, 2028–37.
Mansfield, D.A. (1974). Polar lows: the development of baroclinic disturbances in cold air outbreaks. *Quart. J. Roy. Met. Soc.* **100**, 541–54.

Mansfield, D.A. (1994), The use of potential vorticity in forecasting cyclones: operational aspects. In *The Life Cycles of Extratropical Cyclones. Volume 3. Proceedings of an International Symposium, 27 June–1 July 1994, Bergen, Norway*, ed. S. Grønås and M.A. Shapiro, pp. 326–31. American Meteorological Society, Boston.

Mapes, B.E. (1997). Equilibrium vs. activation control of large-scale variations of tropical deep convection. In *The Physics and Parameterization of Moist Convection. NATO-ASI Series C, vol. 505*, ed. R.K. Smith, pp. 321–58. Kluwer Academic Publishers, Amsterdam.

Marshall, G.J. and King, J.C. (1998). Southern Hemisphere circulation anomalies associated with extreme Antarctic Peninsula winter temperatures. *Geophys. Res. Lett.* **25**, 2437–40.

Marshall, G.J. and Turner, J. (1997a). Surface wind fields of Antarctic mesocyclones derived from ERS-1 scatterometer data. *J. Geophys. Res.* **102**, 13907–21.

Marshall, G.J. and Turner, J. (1997b). Katabatic wind propagation over the western Ross Sea observed using ERS-1 scatterometer data. *Antarct. Sci.* **9**, 221–6.

Martin, D.W. (1968). *Satellite studies of cyclonic developments over the Southern Ocean, Tech. Rept. No. 9*. International Antarctic Meteorological Research Centre, Bureau of Meteorology, Melbourne.

Mayes, P.R. (1985). Secular variations in cyclone frequencies near the Drake Passage, southwest Atlantic. *J. Geophys. Res.* **90**, 5829–39.

McMurdie, L.A. and Katsaros, K.B. (1991). Satellite-derived integrated water-vapor distribution in oceanic midlatitude storms: variation with region and season. *Mon. Wea. Rev.* **119**, 589–605.

McMurdie, L.A., Claud, C. and Atakturk, S. (1997). Satellite-derived atmospheric characteristics of spiral and comma-shaped southern hemisphere mesocyclones. *J. Geophys. Res. – Atmos.* **102**, 13889–905.

Meehl, G.A. and van Loon, H. (1979). The seasaw in winter temperatures between Greenland and northern Europe. Part III: Teleconnections with lower latitudes. *Mon. Wea. Rev.* **107**, 1095–106.

Melander, M.V., McWilliams, J.C. and Zabusky, N.J. (1987). Axisymmetrization and vorticity-gradient intensification of an isolated two-dimensional vortex through filamentation. *J. Fluid Mech.* **178**, 137–59.

Meteorological Office (1962). *A Course in Elementary Meteorology*. HMSO, London.

Meteorological Office (1964). *The Handbook of Weather Forecasting*. Meteorological Office, Bracknell, UK.

Meteorological Office (1972). *The Meteorological Glossary*. Meteorological Office, Bracknell, UK.

Midtbø, K.H. (1986). Polar low forecasting. In *Proceedings of the International Conference on Polar Lows, Oslo 1986*, pp. 257–71. Norwegian Meteorological Institute, Oslo.

Midtbø, K.H., Naustvik, M., Hoem, V. and Smits, J.C. (1986). *Polar Low Forecasting. Part 1: Methods and Evaluation. Technical Report 19, Polar Lows Project*. Norwegian Meteorological Institute, Oslo.

Miner, T., Sousounis, P.J., Wallman, J. and Mann, G. (2000). Hurricane Huron. *Bull. Amer. Met. Soc.* **81**, 223–36.

Mitchell, H.L., Charett, C., Chouinard, C. and Brasnett, B. (1990). Revised interpolation statistics for the Canadian data assimilation procedure: their derivation and application. *Mon. Wea. Rev.* **118**, 1591–614.

Miyazawa, S. (1967). On the vortical mesoscale disturbances observed during the period of heavy snow or rain in the Hokuriku districts. *J. Met. Soc. Jap.* **45**, 166–76.

Mo, K.C. and Ghil, M. (1987). Statistics and dynamics of persistent anomalies. *J. Atmos. Sci.* **44**, 877–901.

Mo, K.C. and White, G.H. (1985). Teleconnections in the Southern Hemisphere, *Mon. Wea. Rev.* **113**, 22–37.

Mo, K.C. and van Loon, H. (1984). Some aspects of the interannual variation of mean monthly sea level pressure on the Southern Hemisphere. *J. Geophys. Res.* **89**, 9541–6.

Möller, J.D. and Montgomery, M.T. (2000). Tropical cyclone evolution via potential vorticity anomalies in a three dimensional balance model. *J. Atmos. Sci.* **57**, 3366–87.

Monk, G.A., Browning, K.A. and Jonas, P.R. (1984). Examples of the operational utility of radar observations of cold air vortices. In *Proceedings of the International Conference on Polar Lows*, Oslo, Norway 20–23 May 1986.

Montgomery, M.T. and Farrell, B.F. (1992). Polar low dynamics. *J. Atmos. Sci.* **49**, 2484–505.

Montgomery, M.T. and Enagonio, J. (1998). Tropical cyclogenesis via convectively forced vortex Rossby waves in a three-dimensional quasi-geostrophic model. *J. Atmos. Sci.* **55**, 3176–207.

Moore, G.W. and Peltier, W.R. (1987). Cyclogenesis in frontal zones. *J. Atmos. Sci.* **44**, 384–409.

Moore, G.W.K. and Peltier, W.R. (1989). On the development of polar low wavetrains. In *Polar and Arctic Lows*, ed. P.F. Twitchell, E. Rasmussen and K.L. Davidson, pp. 141–53. A Deepak, Hampton, VA.

Moore, G.W., Reader, M.C., York, J. and Sathiyamoorthy, S. (1996). Polar lows in the Labrador Sea – a case study. *Tellus* **48A**, 17–40.

Moore, G.W.K. and Vachon, P.W. (2002). A polar low over the Labrador Sea: Interactions with topography and an upper level potential vorticity anomaly, and an observation by RADARSAT-1 SAR. *Geophys. Res. Lett.* 10.1029/2001 GL014007.

Moses, T., Kiladis, G.N., Diaz, H. and Barry, R.G. (1987). Characteristics and frequency of reversals in mean sea level pressure in the North Atlantic sector and their relationship to long-term temperature trends. *J. Clim.* **7**, 13–30.

Motoki, T. (1974). A small cyclonic echo pattern formed in the Ishikari Plain [in Japanese]. *Tenki* **21**, 245–50.

Mullen, S.L. (1979). An investigation of small synoptic cyclones in polar air streams. *Mon. Wea. Rev.* **107**, 1636–47.

Mullen, S.L. (1982). Cyclone development in the polar airstreams over the wintertime continent. *Mon. Wea. Rev.* **110**, 1664–76.

Mullen, S.L. (1983). Explosive cyclogenesis associated with cyclones in polar air streams. *Mon. Wea. Rev.* **111**, 1537–53.

Münzenberg-St Denis, A. (1994). Quasilineare instabilitätsanalyse und ihre Anwendung auf die strukturaufklärung von mesozyklonen im östlichen Weddellmeergebiet. PhD thesis, University of Bonn, 131pp.

Münzenberg-St Denis, A. and Schilling, H. D. (1996). Dynamical features of mesocyclones in the East Weddell Sea region, an instability analysis. *Global Atmos. Ocean Sys.* **4**, 209–31.

Nagata, M. (1987). On the structure of a convergent cloud band over the Japan Sea in winter; a prediction experiment. *J. Met. Soc. Jap.* **65**, 871–83.

Nagata, M. (1992). Modeling case study of the Japan Sea convergent cloud band in a varying large-scale environment; evolution and upscale effect, *J. Met. Soc. Jap.* **70**, 649–71.

Nagata, M. (1993). Meso-α-scale vortices developing along the Japan Sea polar airmass convergence zone cloud band: numerical simulation. *J. Met. Soc. Jap.* **71**, 43–57.

Nakamura, N. and Oort, A.H. (1988). Atmospheric heat budgets of the polar regions. *J. Geophys. Res.* **93**, 9510–24.

Nielsen, N.W. (1997). An early-autumn polar low formation over the Norwegian Sea. *J. Geophys. Res. – Atmos.* **102**, 13955–73.

Ninomiya, K. (1976). Wind profile and the kinetic energy budget in the mixed layer of polar air-mass transformed over Kuroshio region. *J. Met. Soc. Jap.* **54**, 361–9.

Ninomiya, K. (1989). Polar/comma-cloud lows over the Japan Sea and the northwestern Pacific in winter. *J. Met. Soc. Jap.* **67**, 83–97.

Ninomiya, K. (1991). Polar low development over the east coast of the Asian Continent on 9–11 December 1986. *J. Met. Soc. Jap.* **69**, 669–85.

Ninomiya, K. (1994). A meso-scale low family formed over the northeastern Japan Sea in the northwestern part of a parent polar low. *J. Met. Soc. Jap.* **72**, 589–603.

Ninomiya, K. and Hoshino, K. (1990). Evolution process and multi-scale structure of a polar low developed over the Japan Sea on 11–12 December 1985. Part II. Meso-β-scale low in meso-α-scale polar low. *J. Met. Soc. Jap.* **68**, 306–18.

Ninomiya, K., Fujimori, J. and Akiyama, T. (1996). Multi-scale features of the cold air outbreak over the Japan Sea and the northwestern Pacific. *J. Met. Soc. Jap.* **74**, 745–61.

Ninomiya, K., Hoshino, K. and Kurihara, K. (1990). Evolution process and multi-scale structure of a polar low developed over the Japan Sea on 11–12 December 1989. Part I: Evolution process and meso-α-scale structure. *J. Met. Soc. Jap.* **68**, 293–306.

Ninomiya, K., Wakahara, K. and Okubo, H. (1993). Meso-a-scale low development over the northeastern Japan Sea under the influence of a parent large-scale low and a cold vortex aloft. *J. Met. Soc. Jap.* **71**, 73–91.

Nordeng, T.E. (1986). *Parameterization of physical processes in a three dimensional numerical weather prediction model. Technical Report No. 65*. The Norwegian Meteorological Institute, Oslo.

Nordeng, T.E. (1987). The effect of vertical and slantwise convection on the simulation of polar lows. *Tellus* **39A**, 354–76.

Nordeng, T.E. (1990). A model-based diagnostic study of the development and maintenance of two polar lows. *Tellus* **42A**, 92–108.

Nordeng, T.E. and Rasmussen, E.A. (1992). A most beautiful polar low – a case study of a polar low development in the Bear Island region. *Tellus* **44A**, 81–99.

Nordeng, T.E., Foss, A., Grønås, S., Lystad, M. and Midtbø, K.H. (1989). On the role of resolution and physical parameterization for numerical simulations of polar lows. In *Polar and Arctic Lows*, ed. P. F. Twitchell, E. Rasmussen and K.L. Davidson. pp. 217–32. A Deepak, Hampton, VA.

O'Connor, W.P., Bromwich, D.H. and. Carrasco, J.F (1994). Cyclonically forced barrier winds along the Transantarctic Mountains near Ross Island. *Mon. Wea. Rev.* **122**, 137–50.

Offiler, D. (1994). The Calibration of ERS-1 Satellite Scatterometer Winds. *J. Atmos. Ocean. Tech.* **11**, 1002–17.

Ogura, Y. (1964). Frictionally controlled, thermally driven circulation in a circular vortex with application to tropical cyclones. *J. Atmos. Sci.* **21**, 610–21.

Økland, H. (1977). *On the intensification of small-scale cyclones formed in very cold air masses heated over the ocean. Institute Report Series No. 26*. Institutt for Geofysikk, Universitet, Oslo.

Økland, H. (1983). Modelling the height, temperature and relative humidity of a well-mixed planetary boundary layer over a water surface. *Bound.-Layer Meteorol.* **25**, 121–41.

Økland, H. (1987). Heating by organized convection as a source of polar low intensification. *Tellus* **39A**, 397–408.

Økland, H. (1989). On the genesis of polar lows. In *Polar and Arctic Lows*, ed. P.F. Twitchell, E. Rasmussen and K. L. Davidson. pp. 179–90. A Deepak, Hampton, VA.

Økland, H. and Schyberg, H. (1987). On the contrasting influence of organized moist convection and surface heat on a barotropic vortex. *Tellus* **39A**, 385–90.

Ooyama, K. (1964). A dynamical model for the study of tropical cyclone development. *Geofis. Intern.* **4**, 187–98.

Ooyama, K.V. (1969). Numerical simulation of the life cycle of tropical cyclones. *J. Atmos. Sci.* **26**, 3–40.

Orlanski, I. (1975). A rational subdivision of scales for atmospheric processes. *Bull. Amer. Met. Soc.* **56**, 527–30.

Orvig, S. (1970). *Climates of the Polar Regions, World Survey of Climatology*, Vol. **14**. Elsevier, Amsterdam.

Palmén, E. and Newton, C.W. (1969). *Atmospheric Circulation Systems*. Academic Press, New York.

Parish, T.R. (1982). Surface airflow over East Antarctica. *Mon. Wea. Rev.* **110**, 84–90.

Parish, T.R. (1992). On the interaction between Antarctic katabatic winds and tropospheric motions in the high southern latitudes. *Aust. Met. Mag.* **40**, 149–67.

Parish, T.R. and Bromwich, D.H. (1986). The inversion wind pattern over West Antarctica. *Mon. Wea. Rev.* **114**, 849–60.

Parish, T.R. and Bromwich, D.H. (1987). The surface windfield over the Antarctic ice sheets. *Nature* **328**, 51–54.

Parish, T.R. and Bromwich, D.H. (1998). A case study of Antarctic katabatic wind interaction with large-scale forcing. *Mon. Wea. Rev.* **126**, 199–209.

Parker, M.N. (1989). Polar lows in the Beaufort Sea. In *Polar and Arctic Lows*, ed. P.F. Twitchell, E. Rasmussen and K.L. Davidson. pp. 323–30. A Deepak, Hampton, VA.

Parker, N. (1991). *Polar Low Handbook for Canadian Meteorologists*. Environment Canada, Edmonton.

Parker, N. (1997). *Cold air vortices and polar low handbook for Canadian meteorologists*. Environment Canada, Edmonton.

Pearson, G.M. and Strogaitis. G. (1988). Satellite imagery interpretation in synoptic and mesoscale meteorology (unpublished). Atmospheric Environment Service, Downsview, Ontario, Canada.

Pedersen, T.S. and Rasmussen, E.A. (1985) On the cut-off problem in linear CISK-models, *Tellus* **37A**, 394–402.

Pettersen, S. (1950). Some aspects of the general circulation or the atmosphere. In *Cent. Proc. Roy. Met. Soc.*, pp. 120–55. Royal Meteorological Society, London.

Petterssen, S. and Smebye, S.J. (1971). On the development of extratropical cyclones. *Quart. J. Roy. Met. Soc.* **97**, 457–82.

Petterssen, S., Bradburg, L. and Pederson, K. (1962). The Norwegian cyclone models in relation to heat and cold sources. *Geophys. Norwegica* **24**, 243–80.

Phillips, N.A. (1951). A simple three-dimensional model for the study of large-scale extratropical flow patterns. *J. Meteorol.* **8**, 381–94.

Pittock, A.B. (1980). Patterns of climatic variations in Argentina and Chile – I. Precipitation, 1931–60. *Mon. Wea. Rev.* **108**, 1347–61.

Pittock, A.B. (1984). On the reality, stability and usefulness of Southern Hemisphere teleconnections. *Aust. Met. Mag.* **32**, 75–82.

Pytharoulis, I., Craig, G.C. and Ballard, S.P. (1999). Study of the hurricane-like Mediterranean cyclone of January 1995. *Phys. Chem. Earth (B)* **24**, 627–32.

Rabbe, A. (1975). Arctic instability lows. *Met. Annaler* **6**, 303–29.

Rabbe, A. (1987). A polar low over the Norwegian Sea, 29 February–1 March 1984. *Tellus* **39A**, 326–33.

Radok, U., Smith, C.A. and Wendler, G. (1998). Southern Ocean synoptics – observations and analyses. *Met. Apps.* **5**, 33–6.

Ralph, F.M., Shapiro, M.A., Neiman, P.J. and Miletta, J. (1994). Observations of multiple mesoscale cyclones (50–700 km) within synoptic-scale cyclones. *Proceedings, International Symposium on the life Cycles of Extratropical Cyclones, 27 June–1 July 1994, Bergen, Norway*. Vol. III, pp. 192–8.

Rasmussen, E. (1977). *The polar low as a CISK phenomena*. University of Copenhagen, Institute for Theoretical Meteorology, Copenhagen.

Rasmussen, E. (1979). The polar low as an extratropical CISK disturbance. *Quart. J. Roy. Met. Soc.* **105**, 531–49.

Rasmussen, E. (1981). An investigation of a polar low with a spiral cloud structure. *J. Atmos. Sci.* **38**, 1785–92.

Rasmussen, E. (1983). A review of meso-scale disturbances in cold air masses. In *Mesoscale Meteorology – Theories, Observations and Models*, ed. D.K. Lilly and T. Gal-Chen, pp. 247–83. D Reidel, Boston.

Rasmussen, E. (1985a). A case study of a polar low development over the Barents Sea. *Tellus* **37A**, 407–18.

Rasmussen, E. (1985b). *A Polar Low Development over the Barents Sea*. Norwegian Meteorological Institute, Oslo.

Rasmussen, E.A. (1987). How satellite imagery describes the evolution and structure of polar lows. In *Satellite and Radar Imagery Interpretation*, pp. 205–22. EUMETSAT, Darmstadt.

Rasmussen, E.A. (1989). A comparative study of tropical cyclones and polar lows. In *Polar and Arctic Lows*, ed. P.F. Twitchell, E. Rasmussen and K.L. Davidson. pp. 47–80. A Deepak, Hampton, VA.

Rasmussen, E.A. (1990). On the application of satellite data for forecasting/nowcasting and research of polar lows. In *Proceedings of the Fifth Conference on Satellite Meteorology and*

Oceanography, September 3–7 1990, London, England, pp. 384–5. American Meteorological Society, Boston.

Rasmussen, E.A. and Aakjær, P.D. (1992). Two polar lows affecting Denmark. *Weather* **47**, 326–38.

Rasmussen, E.A. and Cederskov, A. (1994). Polar lows: a critical appraisal. In *Proceedings, International Symposium on the life Cycles of Extratropical Cyclones*, 27 June–1 July 1994, Bergen, Norway, Vol. III, pp. 199–203.

Rasmussen, E. and Lystad, M. (1987). The Norwegian polar lows project: a summary of the international conference on polar lows. *Bull. Amer. Met. Soc.* **68**, 801–16.

Rasmussen, E.A. and Purdom, J.F. (1992). Investigations of a polar low using geostationary satellite data. In *Preprints, Sixth Conference on Satellite Meteorology and Oceanography, Atlanta, Georgia*, pp. 120–5. American Meteorological Society, Boston.

Rasmussen, E. and Zick, C. (1987). A subsynoptic vortex over the Mediterranean Sea with some resemblance to polar lows. *Tellus* **39A**, 408–25.

Rasmussen, E.A., Claud, C. and Purdom, J.F. (1996). Labrador Sea polar lows. *Global Atmos. Ocean Sys.* **4**, 275–333.

Rasmussen, E.A., Guest, P.S. and Davidson, K.L. (1997). Synoptic and mesoscale atmospheric features over the ice-covered portion of the Fram Strait in spring. *J. Geophys. Res. – Atmos.* **102**, 13975–86.

Rasmussen, E.A., Pedersen, T.S., Pedersen, L.T. and Turner, J. (1992). Polar lows and arctic instability lows in the Bear Island region. *Tellus* **44A**, 133–54.

Rasmussen, E.A., Turner, J. and Twitchell, P.F. (1993). Report of a workshop on the applications of new forms of satellite data in polar low research. *Bull. Amer. Met. Soc.* **74**, 1057–73.

Rasmussen, L. (1989). Greenland winds and satellite imagery. *VEJRET, Special issue in English* 32–7.

Rayner, J.N. and Howarth, D.A. (1979). Antarctic sea ice: 1972–1975. *Geog. Rev.* **69**, 202–23.

Reed, R.J. (1979). Cyclogenesis in polar airstreams. *Mon. Wea. Rev.* **107**, 38–52.

Reed, R.J. (1987). Polar lows. In *Conference proceedings, The Nature and Prediction of Extra Tropical Weather Systems 7–11 September 1987*, ECMWF, Reading. pp. 213–36.

Reed, R.J. (1992). An Arctic hurricane over the Bering Sea – comment. *Mon. Wea. Rev.* **120**, 2713–15.

Reed, R.J. and Blier, W. (1986a). A case study of a comma cloud development in the Eastern Pacific. *Mon. Wea. Rev.* **114**, 1681–95.

Reed, R.J. and Blier, W. (1986b). A further case study of comma cloud development in the Eastern Pacific. *Mon. Wea. Rev.* **114**, 1696–708.

Reed, R.J. and Duncan, C.N. (1987). Baroclinic instability as a mechanism for the serial development of polar lows: a case study. *Tellus* **39A**, 376–85.

Reed, R.J., Kuo, Y.H., Albright, M.D., Gao, K. and Huang, W. (2001). Analysis and modelling of a tropical-like cyclone in the Mediterranean Sea. *Met. Atmos. Phys.* **76**, 183–202.

Renfrew, I.A., Moore, G.W.K. and Clerk, A.A. (1997). Binary interactions between polar lows. *Tellus Ser. A* **49**, 577–94.

Renfrew, I.A., Moore, G.W.K., Holt, T.R., Chang, S.W. and Guest, P. (1999). Mesoscale forecasting during a field program: meteorological support of the Labrador Sea deep convection experiment. *Bull. Amer. Met. Soc.* **80**, 605–20.

Roch, M., Benoit, R. and Parker, N. (1991). Sensitivity experiments for polar low forecasting with the CMC mesoscale finite-element model. *Atmos. - Ocean.* **29**, 381–419.

Rockey, C.C. and Braaten, D.A. (1995). Characterization of polar cyclonic activity and relationship to observed snowfall events at McMurdo Station, Antarctica. In *Proceedings of the Fourth Conference on Polar Meteorology and Oceanography, January 15–20 1995, Dallas, Texas*, pp. 244–5. American Meteorological Society, Boston.

Rogers, J.C. (1981). Spatial variability of seasonal sea level pressure and 500 mb height anomalies. *Mon. Wea. Rev.* **109**, 2093–106.

Rogers, J.C. (1984). The association between the North Atlantic Oscillation and the Southern Oscillation in the Northern Hemisphere. *Mon. Wea. Rev.* **112**, 1999–2015.

Rogers, J.C. (1997). North Atlantic storm track variability and its association to the north Atlantic oscillation and climate variability of northern Europe. *J. Clim.* **10**, 1635–47.

Rogers, J.C. and Mosley-Thompson, E. (1995). Atlantic arctic wave cyclones and the mild Siberian winters of the 1980s. *Geophys. Res. Lett.* **22**, 799–802.

Rogers, J.C. and van Loon, H. (1979). The seasaw in winter temperatures between Greenland and northern Europe. Part II: Some oceanic and atmospheric effects in middle and high latitudes. *Mon. Wea. Rev.* **107**, 509–19.

Rogers, J.C. and van Loon, H. (1982). Spatial variability of sea level pressure and 500 mb height anomalies over the Southern Hemisphere. *Mon. Wea. Rev.* **110**, 1375–92.

Rogers, J.C., Bolzan, J.F. and Pohjola, V.A. (1998). Atmospheric circulation variability associated with shallow-core seasonal isotopic extremes near Summit Greenland. *J. Geophys. Res. – Atmos.* **103**, 11205–19.

Røsting, B., Sunde, J. and Midtbø, K.H. (1996). Monitoring of NWP models by use of satellite data. *Met. Apps.* **3**, 331–40.

Rutherford, G.T. (1969). Occlusion sequences south of Australia. In *Proceedings of Inter-Regional Seminar on the Interpretation of Meteorological Satellite Data*, pp. 49–53. Bureau of Meteorology for WMO, Melbourne.

Sanders, F. and Gyakum, J.R. (1980). Synoptic-dynamic climatology of the 'bomb'. *Mon. Wea. Rev.* **108**, 1589–606.

Sardie, J.M. and Warner, T.T. (1983). On the mechanism for the development of polar lows. *J. Atmos. Sci.* **40**, 869–81.

Sardie, J.M. and Warner, T.T. (1985). A numerical study of the development mechanism of polar lows. *Tellus* **37**, 460–77.

Sater, J.E., Ronhovde, A.G. and Van Allen, L.C. (1971). *Arctic Environment and Resources*. The Arctic Institute of North America, Washington, DC.

Sawyer, J.S. (1947). Notes on the theory of tropical cyclones. *Quart. J. Roy. Met. Soc.* **73**, 101–26.

Scherhag, R. and Klauser, L. (1962). Grundlagen der Wettervorhersage. In *Meteorologisches Taschenbuch*, ed. F. Bauer. Akademische Verlagsgesellschaft, Leipzig.

Schubert, W.H. and Alworth, B. (1987). Evolution of potential vorticity in tropical cyclones. *Quart. J. Roy. Met. Soc.* **113**, 147–62.

Schubert, W.H. and Hack, J.J. (1982). Inertial stability and tropical cyclone development. *J. Atmos. Sci.* **39**, 1687–97.

Schubert, W.H., Hack, J.J., Silva Dias, P.L. and Fulton, S.R. (1980). Geostrophic adjustment in an axisymmetric vortex. *J. Atmos. Sci.* **37**, 1464–84.

Schwerdtfeger, W. (1960). The seasonal variation of the strength of the southern circumpolar vortex. *Mon. Wea. Rev.* **88**, 203–8.
Schwerdtfeger, W. (1975). The effect of the Antarctic Peninsula on the temperature regime of the Weddell Sea. *Mon. Wea. Rev.* **103**, 45–51.
Schwerdtfeger, W. (1979). Meteorological aspects of the drift of ice from the Weddell Sea toward the mid-latitude westerlies. *J. Geophys. Res.* **84**, 6321–8.
Schwerdtfeger, W. (1984). *Weather and Climate of the Antarctic*. Elsevier, Amsterdam.
Scorer, R.S. (1952). Sonic and advective disturbances. *Quart. J. Roy. Met. Soc.* **78**, 76–81.
Scorer, R.S. (1986). *Cloud Investigation by Satellite*. Ellis Horwood, Chichester.
Scorer, R.S. (1988). Sunny Greenland. *Quart. J. Roy. Met. Soc.* **114**, 3–29.
Sechrist, F.S., Fett, R.W. and Perryman, D.C. (1989). *Forecasters Handbook for the Arctic. Technical Report TR 89–12*. Naval Environmental Prediction Research Facility, Monterey, CA.
Serreze, M.C., Kahl, J. and Schnell, R. (1992). Low-level temperature inversions of the Eurasian Arctic and comparisons with Soviet ice island data. *J. Clim.* **5**, 599–613.
Serreze, M.C., Rogers, J.C., Carsey, F. and Barry, R.G. (1997). Icelandic low cyclone activity: climatological features, linkages with the NAO and relationships with recent changes in the Northern Hemisphere circulation. *J. Clim.* **10**, 453–64.
Shapiro, L.J. (2000). Potential vorticity asymmetries and tropical cyclone evolution in moist three-layer model. *J. Atmos. Sci.* **57**, 3645–62.
Shapiro, M.A. and Fedor, L.S. (1989). A case study of an ice-edge boundary layer front and polar low development over the Norwegian and Barents Seas. In *Polar and Arctic Lows*, ed. P.F. Twitchell, E. Rasmussen and K.L. Davidson. pp. 257–77. A Deepak, Hampton, VA.
Shapiro, M.A. and Keyser, D. (1990). Fronts, jet streams and the tropopause. In *Extratropical Cyclones, The Erik Palmén Memorial Volume*, ed. C.W. Newton and E.O. Holopainen. pp. 167–91. American Meteorological Society, Boston.
Shapiro, L.J. and Willoughby, H.E. (1982). The response of balanced hurricanes to local sources of heat and momentum. *J. Atmos. Sci.* **39**, 378–94.
Shapiro, M.A., Fedor, L.S. and Hampel, T. (1987). Research aircraft measurements of a polar low over the Norwegian Sea. *Tellus* **39A**, 272–306.
Shapiro, M.A., Hampel, T. and Fedor, L.S. (1989). Research aircraft observations of an Arctic front over the Barents Sea. In *Polar and Arctic Lows*, ed. P.F. Twitchell, E. Rasmussen and K.L. Davidson. pp. 279–89. A Deepak, Hampton, VA.
Shaw, W.J., Pauley, R.L., Gobel, T.M. and Radke, L.F. (1991). A case study of atmospheric boundary layer mean structure for flow parallel to the ice edge, aircraft observations from CEAREX. *J. Geophys. Res.* **96**, 4691–708.
Simmonds, I. and Jacka, T.H. (1995). Relationships between the interannual variability of Antarctic sea ice and the Southern Oscillation. *J. Clim.* **8**, 637–47.
Simmonds, I. and Jones, D.A. (1998). The mean structure and temporal variability of the semiannual oscillation in the southern extratropics. *Int. J. Climatol.* **18**, 473–504.
Sinclair, M.R. (1994). An objective cyclone climatology for the Southern Hemisphere. *Mon. Wea. Rev.* **122**, 2239–56.
Sinclair, M.R. (1995). A climatology of cyclogenesis for the Southern Hemisphere. *Mon. Wea. Rev.* **123**, 1601–19.
Sinclair, M.R. and Cong, X. (1992). Polar air stream cyclogenesis in the Australasian region: a composite study using ECMWF analyses. *Mon. Wea. Rev.* **120**, 1950–72.

Sinclair, M.R., Renwick, J.A. and Kidson, J.W. (1997). Low-frequency variability of Southern Hemisphere sea level pressure and weather system activity. *Mon. Wea. Rev.* **125**, 2531–43.

Smith, R.K. (1997). On the theory of CISK. *Quart. J. Roy. Met. Soc.* **123**, 407–18.

Smith, R.K. (2000). The role of cumulus convection in hurricanes and its representation in hurricane models. *Rev. Geophys.* **38**, 465–89.

Smith, S.R. and Stearns, C.R. (1993). Antarctic pressure and temperature anomalies surrounding the minimum in the Southern Oscillation index. *J. Geophys. Res.* **98** D7, 13071–83.

Smits, J.C. (1985). *Polar lows observed during the winter 1984/85 and a summary of the three winter seasons 1982–85. Polar Lows Project, Technical Report 13*. DNMI, Oslo.

Song, Y. and Carleton, A.M. (1997). Climatological 'models' of cold air mesocyclones derived from SSM/I data. *Geocarto Int.* **12**, 79–90.

Stearns, C.R. and Wendler, G. (1988). Research results from Antarctic automatic weather stations. *Rev. Geophys.* **26**, 45–61.

Stevens, B., Randall, D.A., Lin, X. and Montgomery, M.T. (1997). Comments on 'On large-scale circulations in convecting atmospheres' by Emanuel et al., (1994). *Quart. J. Roy. Met. Soc.* **123**, 1771–78.

Stevenson, C.M. (1968). The snowfall of early December 1967. *Weather* **23**, 156–61.

Stoelinga, M.T. (1996). A potential vorticity based study of the role of diabatic heating and friction in a numerically simulated baroclinic cyclone. *Mon. Wea. Rev.* **124**, 849–74.

Streten, N.A. (1968). Some features of mean annual windspeed data for coastal East Antarctica. *Polar Rec.* **14**, 315–22.

Streten, N.A. (1975). Satellite derived influences to some characteristics of the South Pacific atmospheric circulation associated with the Niño event of 1972–73. *Mon. Wea. Rev.* **103**, 989–95.

Streten, N.A. (1977). Aspects of the year-to-year variation of seasonal and monthly mean station temperature over the Southern Hemisphere. *Mon. Wea. Rev.* **105**, 195–206.

Streten, N.A. (1980). Some synoptic indices of the Southern Hemisphere mean sea level circulation 1972–77. *Mon. Wea. Rev.* **108**, 18–36.

Streten, N.A. (1983). Antarctic sea ice and related atmospheric circulation during FGGE. *Arch. für Met. Geophys. und Biokl.* **A32**, 231–46.

Streten, N.A. (1990). A review of the climate of Mawson – a representative strong wind site in East Antarctica. *Antarctic Sci.* **2**, 79–89.

Streten, N.A. and Troup, A.J. (1973). A synoptic climatology of satellite observed cloud vortices over the Southern Hemisphere. *Quart. J. Roy. Met. Soc.* **99**, 56–72.

Streten, N.A. and Pike, D.J. (1980). Characteristics of the broadscale Antarctic sea ice extent and the associated atmospheric circulation 1972–1977. *Arch. für Met., Geophys. und Biokl. A* **29**, 279–99.

Sumner, E.J. (1950). The significance of vertical stability in synoptic development. *Quart. J. Roy. Met. Soc.* **76**, 384–92.

Sunde, J., Rosting, B., Breivik, L.A., Midtbø, K.H. and Ulstad, C. (1994). Operational monitoring and forecasting of mesoscale weather phenomena in ocean regions surrounding Norway. *Met. Apps.* **1**, 237–45.

Sundqvist, H. (1970). Numerical simulation of the development of tropical cyclones with a ten-level model, Part 1. *Tellus* **22**, 359–90.

Sundqvist, H., Berge, E. and Kristjansson, J.E. (1989). Condensation and cloud parameterization studies with a mesoscale numerical weather prediction model. *Mon. Wea. Rev.* **117**, 1641–57.

Suttie, T.K. (1970). Portrait of a polar low, *Weather* **25**, 504–7.

Sutton, R.T. and Allen, M.R. (1997). Decadal predictability of North Atlantic sea surface temperature and climate. *Nature* **388**, 563–7.

Taljaard, J.J. (1969). Air masses of the Southern Hemisphere. *Notos* **18**, 79–104.

Taylor, H.W., Gordon, A.L. and Molinelli, E. (1978). Climatic characteristics of the Antarctic Polar Front zone. *J. Geophys. Res.* **83**, 4572–8.

The Lab Sea Group (1998). The Labrador Sea deep convection experiment. *Bull. Amer. Met. Soc.* **79**, 2033–58.

Thompson, D.W.J. and Wallace, J.M. (1998). The Arctic Oscillation signature in the wintertime geopotential height and temperature fields. *Geophys. Res. Lett.* **25**, 1297–300.

Thompson, W.T. and Burk, S.D. (1991). An investigation of an Arctic front with a vertically nested mesoscale model. *Mon. Wea. Rev.* **119**, 233–61.

Thorpe, A.J. (1985). Diagnosis of balanced vortex structure using potential vorticity. *J. Atmos. Sci.* **42**, 397–406.

Thorpe, A.J. (1997). Attribution and its application to mesoscale structure associated with tropopause folds. *Quart. J. Roy. Met. Soc.* **123**, 2377–99.

Thorpe, A.J. and Emanuel, K.A. (1985). Frontogenesis in the presence of small stability to slantwise cojnvection. *J. Atmos. Sci.* **42**, 1809–24.

Tijm, A.B.C. and Van Delden, A. (1999). The role of sound waves in sea-breeze initiation. *Quart. J. Roy. Met. Soc.* **125**, 1997–2018.

Trenberth, K.E. (1980). Planetary waves at 500 mb in the southern hemisphere. *Mon. Wea. Rev.* **108**, 1378–89.

Trenberth, K.E. (1981). Interannual variability of the Southern Hemisphere 500 mb flow: regional characteristics. *Mon. Wea. Rev.* **109**, 127–36.

Trenberth, K.E. (1986). The signature of a blocking episode on the general circulation in the Southern Hemisphere. *J. Atmos. Sci.* **43**, 2061–9.

Trenberth, K.E. (1991). Storm tracks in the Southern Hemisphere. *J. Atmos. Sci.* **48**, 2159–78.

Trenberth, K.E. and Mo, K.C. (1985). Blocking in the Southern Hemisphere. *Mon. Wea. Rev.* **113**, 3–21.

Trenberth, K.E. and Paolino, D.A. (1981). Characteristic patterns of variability of sea level pressure in the Northern Hemisphere. *Mon. Wea. Rev.* **109**, 1169–89.

Trenberth, K.E. and Shea, D.J. (1987). On the evolution of the Southern Oscillation. *Mon. Wea. Rev.* **115**, 3078–96.

Troup, A.J. and Streten, N.A. (1972). Satellite-observed southern hemisphere cloud vortices in relation to conventional observations. *J. Appl. Met.* **11**, 909–17.

Tsuboki, K. and Wakahama, G. (1992). Mesoscale cyclogenesis in winter monsoon air streams: quasi-geostrophic baroclinic instability as a mechanism of the cyclogenesis off the west coast of Hokkaido Island, Japan. *J. Met. Soc. Jap.* **70**, 77–93.

Turner, J. and Ellrott, H. (1992). High latitude moisture structure determined from HIRS water vapour imagery. *Int. J. Rem. Sens.* **13**, 81–95.

Turner, J. and Ladkin, R. (1998). Mesocyclones over the interior of the Antarctic. In *Proceedings of the European Geophysical Society Polar Lows Working Group Workshop on*

'Polar lows – current state and needs of future research', Copenhagen, 17–19 June 1998, EGS Polar Lows Working Group, Bonn. pp. 12–17.

Turner, J. and Row, M. (1989). Mesoscale vortices in the British Antarctic Territory. In *Polar and Arctic lows*, ed. P.F. Twitchell, E. Rasmussen and K.L. Davidson. pp. 347–56. A Deepak, Hampton, VA.

Turner, J. and Thomas, J.P. (1994). Summer-season mesoscale cyclones in the Bellingshausen–Weddell region of the Antarctic and links with the synoptic-scale environment. *Int. J. Climatol.* **14**, 871–94.

Turner, J., Bromwich, D., Colwell, S., Dixon, S., Gibson, T., Hart, T., Heinemann, G., Hutchinson, H., Jacka, K., Leonard, S., Lieder, M., Marsh, L., Pendlebury, S., Phillpot, H., Pook, M. and Simmonds, I. (1996b). The Antarctic First Regional Observing Study of the Troposphere (FROST) project. *Bull. Amer. Met. Soc.* **77**, 2007–32.

Turner, J., Colwell, S.R. and Harangozo, S.A. (1997). Variability of precipitation over the coastal western Antarctic Peninsula from synoptic observations. *J. Geophys. Res.* **102**, 13999–4007.

Turner, J., Corcoran, G., Cummins, S., Lachlan-Cope, T. and Leonard, S. (1996a). Seasonal variability of mesocyclone activity in the Bellingshausen/Weddell region of Antarctica. *Global Atmos. Ocean Sys.* **5**, 73–97.

Turner, J., Lachlan-Cope, T.A. and Thomas, J.P. (1993a). A comparison of Arctic and Antarctic mesoscale vortices. *J. Geophys. Res.* **98 D7**, 13019–34.

Turner, J., Lachlan-Cope, T.A., Thomas, J.P. and Colwell, S. (1995). The synoptic origins of precipitation over the Antarctic Peninsula. *Antarctic Sci.* **7**, 327–37.

Turner, J., Lachlan-Cope, T.A., Warren, D.E. and Duncan, C.N. (1993b). A mesoscale vortex over Halley Station, Antarctica. *Mon. Wea. Rev.* **121**, 1317–36.

Turner, J., Leonard, S., Marshall, G.J., Pook, M., Cowled, L., Jardine, R., Pendlebury, S. and Adams, N. (1999). An assessment of operational Antarctic analyses based on data from the FROST project. *Wea. and Forecasting* **14**, 817–34.

Twitchell, P.F., Rasmussen, E.A. and Davidson, K.L. (ed.) (1989). *Polar and Arctic Lows*. A Deepak, Hampton, VA.

Van Delden, A. (1989). On the deepening and filling of balanced cyclones by diabatic heating. *Met. Atmos. Phys.* **41**, 127–45.

Van Delden, A. (1992). The dynamics of meso-scale atmospheric circulations. *Physics Reports* **211**, 252–376.

Van Delden, A. (2000). Linear dynamics of hydrostatic adjustment to horizontally homogeneous heating. *Tellus* **52A**, 380–90.

van Loon, H. (1956). Blocking action in the Southern Hemisphere, Part 1. *Notos* **5**, 171–5.

van Loon, H. (1962). On the movement of lows in the Ross and Weddell Sea sectors in summer. *Notos* **11**, 47–50.

van Loon, H. (1966). On the annual temperature range over the southern oceans. *Geog. Rev.* **58**, 497–515.

van Loon, H. (1967). The half-yearly oscillations in middle and high Southern latitudes and the coreless winter. *J. Atmos. Sci.* **24**, 472–86.

van Loon, H. (1972). Pressure in the Southern Hemisphere. In *Meteorology of the Southern Hemisphere*, ed. H. van Loon, J.J. Taljaard, T. Sasamori, J. London, D.V. Hoyt, K. Labitzke and C.W. Newton. pp. 59–86. American Meteorological Society, Boston.

van Loon, H. (1984). The Southern Oscillation, Part III: Associations with the trades and with the trough in the westerlies of the South Pacific Ocean. *Mon. Wea. Rev.* **112**, 947–54.

van Loon, H. and Kidson, J.W. (1993). The association between latitudinal temperature gradient and eddy transport, Part III: the southern hemisphere. *Aust. Met. Mag.* **42**, 31–7.

van Loon, H. and Rogers, J.C. (1978). The seesaw in winter temperatures between Greenland and northern Europe. Part I: General description. *Mon. Wea. Rev.* **106**, 296–310.

van Loon, H. and Rogers, J.C. (1984). Interannual variations in the half-yearly cycle of pressure gradients and zonal wind at sea level on the Southern Hemisphere. *Tellus* **36A**, 76–86.

Venzke, S., Allen, M.R., Sutton, R.T. and Rowell, D.P. (1999). The atmospheric response over the North Atlantic to decadal changes in sea surface temperature. *J. Clim.* **12**, 2562–84.

Viebrock, H. (1962). The transfer of energy between the ocean and the atmosphere in the Antarctic region. *J. Geophys. Res.* **67**, 4293–302.

Villalba, R., Cook, E.R., Darrigo, R.D., Jacoby, G.C., Jones, P.D., Salinger, M.J. and Palmer, J. (1997). Sea-level pressure variability around Antarctica since AD 1750 inferred from subantarctic tree-ring records. *Clim. Dyn.* **13**, 375–90.

Wallace, J.M. and Gutzler, D.S. (1981). Teleconnections in the geopotential height field during the Northern Hemisphere winter. *Mon. Wea. Rev.* **109**, 784–812.

Wallace, J.M. and Hobbs, P.V. (1977). *Atmospheric Science. An Introductory Survey*. Academic Press, San Diego.

Walsh, J.E. and Chapman, W.L. (1990). Short-term climatic variability of the Arctic. *J. Clim.* **3**, 237–50.

Walsh, J.E. and Portis, D.H. (1999). Variations of precipitation and evaporation over the North Atlantic Ocean, 1958–1997. *J. Geophys. Res. – Atmos.* **104**, 16613–31.

Weldon, R.B. (1979). *Cloud patterns and the upper air wind field, Part IV*. NOAA, Washington, DC.

Weller, G. (1980). Spatial and temporal variations in the South Polar surface energy balance. *Mon. Wea. Rev.* **108**, 2006–14.

Wendland, W.M. and McDonald, N.S. (1986). Southern Hemisphere airstream climatology. *Mon. Wea. Rev.* **114**, 88–94.

Wendler, G., Adolphs, U., Hauser, A. and Moore, B. (1997a). On the surface energy budget of sea ice. *J. Glaciol.* **43**, 122–30.

Wendler, G., Gilmore, D. and Curtis, J. (1997b). On the formation of coastal polynyas in the area of Commonwealth Bay, eastern Antarctica. *Atmos. Res.* **45**, 55–75.

Wendler, G., Ishikawa, N. and Kodama, Y. (1988). The heat balance of the icy slope of Adélie Land, Eastern Antarctica. *J. Appl. Met.* **27**, 52–65.

Wendler, G., Stearns, C., Weidner, G., Dargaud, G. and Parish, T. (1997c). On the extraordinary katabatic winds of Adélie Land. *J. Geophys. Res. – Atmos.* **102**, 4463–74.

White, W.B. and Peterson, R.G. (1996). An Antarctic circumpolar wave in surface pressure, wind, temperature and sea-ice extent. *Nature* **380**, 699–702.

Whittaker, L.M. and Horn, L.H. (1984). Northern Hemisphere extratropical cyclone activity for four mid-season months. *J. Climatol.* **4**, 297–310.

Wiin-Nielsen, A. (1989). On the precursors of polar lows. In *Polar and Arctic Lows*, ed. P.F. Twitchell, E. Rasmussen and K.L. Davidson. pp. 85–107. A Deepak, Hampton, VA.

Wilhelmsen, K. (1985). Climatological study of gale-producing polar lows near Norway. *Tellus* **37A**, 451–59.

Wilhelmsen, K. (1986a). Climatological study of gale producing polar lows near Norway. In: *Proceedings of the International Conference on Polar Lows, Oslo 1986*, pp. 31–9. The Norwegian Meteorological Institute, Oslo.

Wilhelmsen, K. (1986b). *Climatological study of polar lows near Norway*, Part I. Norwegian Meteorological Institute, Oslo.

Wilson, H.P. (1971). An interesting Arctic storm. (Unpublished manuscript).

Wirth, V. (1995). Diabatic heating in an axisymmetric cut-off cyclone and related stratosphere–troposphere exchange. *Quart. J. Roy. Met. Soc.* **121**, 127–47.

Wirth, V. (2001). Cyclone–anticyclone asymmetry concerning the height of the thermal and the dynamical tropopause. *J. Atmos. Sci.* **58**, 26–37.

Woetmann Nielsen, N. (1998). Om forudsigelighed af polare lavtryk [In Danish]. *Vejret* **20**, 37–48.

Yamagishi, Y., Doi, M., Kitabatake, N. and Kamiguchi, H. (1992). A polar low which accompanied strong gust [In Japanese]. *Tenki* **39**, 27–36.

Yarnal, B. and Henderson, K.G. (1989a). A climatology of polar low cyclogenetic regions over the North Pacific Ocean. *J. Clim.* **2**, 1476–91.

Yarnal, B. and Henderson, K.G. (1989b). A satellite-derived climatology of polar low evolution in the North-Pacific. *Int. J. Climatol.* **9**, 551–66.

Zick, C. (1983). Method and results of an analysis of comma cloud developments by means of vorticity fields from upper tropospheric satellite wind data. *Meteor. Rdsch.*, **36**, 69–84.

Zick, C. (1994). Polar lows in the SW Pacific region and their transition from or into synoptic-scale cyclones. In *The Life Cycles of Extratropical Cyclones, Volume III. Proceedings of an International Symposium 27 June–1 July 1994, Bergen, Norway*, ed. S. Grønås and M.A. Shapiro, pp. 248–55. American Meteorological Society, Boston.

Zillman, J.W. and Johnson, D.R. (1985). Thermally-forced mean mass circulations in the Southern Hemisphere. *Tellus* **37A**, 56–76.

Zillman, J.W. and Price, P.G. (1972). On the thermal structure of mature Southern Ocean cyclones. *Aust. Met. Mag.* **20**, 34–48.

Zillman, J.W. and Dingle, W.R.J. (1969). Shorter contribution: Southern Ocean sea–air energy exchange. *Aust. Met. Mag.* **17**, 166–72.

Index

Adélie Land, Antarctica 31, 112, 134, 149
adjustment
 convective 345, 386
 geostrophic 343, 353–9
 hydrostatic 343, 346–53
 to heating 342–46, 367–80
air masses
 Antarctic 136, 256, 269
 Arctic 20, 62, 64–5, 68, 74–5, 167, 178, 183, 195, 222, 397, 400
 Arctic source region 62
 Asian 232
 modification and transformation 74–5
aircraft experiments 8, 10, 203–6, 229, 266, 279–80, 405, 430, 477, 480
air–sea interaction 75, 67–69, 88–89, 148–9, 191, 327, 337, 409–10, 431–2
Air–Sea Interaction Instability (ASII), *see* wind induced surface heat exchange (WISHE)
air–sea temperature difference 83, 101–3, 107–8
Alaska 100, 229, 555, 559
Ålesund, Norway 513, 516
Aleutian Islands 100, 105, 106, 232, 432, 545, 549
Aleutian low 106
Amundsen/Bellingshausen Sea 48, 115, 116, 122, 126, 127, 134, 136–7, 139, 142, 146, 148, 247, 253, 268, 273, 460, 485–9, 496–7
Antarctic circumpolar current (ACC) (oceanic) 110, 118, 127

Antarctic circumpolar trough (ACT), *see* circumpolar trough
Antarctic circumpolar wave (ACW) 149
Antarctic Peninsula 16, 121, 123, 137, 486, 568
anticyclonic upper-level outflow 45, 216, 261, 548
Arctic
 CISK 183
 fronts 15, 167, 169, 170, 172, 198, 204, 311, 436, 444, 459
 hurricanes 10, 13, 199, 229, 230, 443, 548
 instability low 4–5, 10
Arctic bomb 13
Arctic Cyclone Expedition (ACE) 167, 203–6, 407, 430–2
Arctic Ocean 52–3, 66, 97, 558
Arctic outbreak polar lows 15, 406, 417
Arctic/Antarctic front type of polar low 15, 158, 252, 254, 406, 417
Australian region 9, 134, 137, 142, 260
automatic weather stations (AWSs) 115, 120, 126, 135, 146–7, 250, 255, 275, 571
available potential energy (APE) 315

Baffin Island and Baffin Bay 95–97, 225–7, 445, 455
balanced systems 343–96
Baltic Sea 183
Barents Sea 2, 17, 19, 30, 65, 68–9, 74, 76, 81, 159, 163, 171, 173, 174, 177, 179, 185, 191, 341, 406, 417, 421
baroclinic deepening, *see* baroclinic instability

605

baroclinic instability
  general 8, 83, 86, 101, 155, 287–317, 396, 401, 404, 417, 436, 444, 455
  type A 290, 294–9
  type B 290–1, 309, 404, 409, 415, 531
baroclinic leaf cloud 40, 43
baroclinic polar lows/waves 12, 39–45, 51, 151, 158, 185, 189–90, 259, 287, 485, 513, 538
baroclinic zones 60, 132, 136, 251, 255, 294, 442, 468, 483, 487, 538
baroclinicity
  deep 88, 100, 291, 314, 317, 406
  shallow 137, 145, 148, 241–2, 250–2, 255–6, 259, 261, 409, 416, 435–6, 454, 464, 468, 475, 543
barotropic instability 168, 181, 222, 230, 240–2, 246, 252, 259, 261, 299, 317–21, 406
Bear Island 26–7, 68, 73–5, 80, 82, 168, 179, 182, 191–206, 238, 331, 334–5
Beaufort Sea 52, 97, 228, 552–7
Bellingshausen Sea, see Amundsen/Bellingshausen Sea
Bering Sea 97–8, 100, 199, 228–31, 432, 436, 545, 550, 551, 561
binary interactions of polar lows 201–2
blocking 118–19, 133
Blumen model 306–7
boundary layer fronts (BLFs) 20, 22, 73, 93, 130, 132, 148, 167–72, 193, 197, 200, 249–50, 252, 253, 269, 299, 318, 454, 468–70, 491, 495, 528, 530, 539, 542, 549, 551, 554–5
burst, of polar lows 105
Byrd Glacier 134, 256, 276, 278

CAPE, see convective available potential energy
Cape Farewell 65, 70–2, 188, 190, 214
case studies
  ACE polar low, Aarhus and Raustein's case (27 February 1984) 203–6, 430–2
  Albright's et al.'s case (8–9 December 1988) 444–55
  Bear Island (12–14 December 1982) 191–9, 340, 399, 403
  Bear Island (21–22 November 1983) 200–2

  Blier's case (10–12 January 1987) 432–6
  Bresch et al.'s case (7 March 1977) 436–44
  Greenland Sea development (19–23 December 1983) 528–42
  Grønås and Kvamstø's cases (February 1987, April 1990, April 1991, March 1992) 417–21
  Halley polar low (January 1986) 251–2, 255, 262, 269–71, 273, 283, 292–4, 483–5
  Harrold and Browning's case (9 December 1967) 7–8, 151, 189, 207, 292, 335
  Mailhot et al.'s case (10–12 January 1989) 430, 455–9
  Nordeng and Rasmussen's case, The 'most beautiful polar low' (26 February 1987) 20, 183, 410–17
  Nordeng's cases (25–26 January 1987 and 26 January 1988) 407–10
  Norwegian Sea (26–27 March 1995) 522–5
  The 'James' polar low in the Tasman Sea (21–27 August 1990) 260–2
  The Fett–Hoem case in the Nordic Seas (19–23 December 1983) 528–42
  The Ross and Amundsen Sea (11–12 January 1995) 485–9
  The Ross and Amundsen sea (July 1994) 496–7
  Weddell Sea (30 July–1 August 1986) 489–97
  Woetman Nielsen's case (13–16 October 1993) 421–30
Chukchi Sea 52, 97, 228, 552–7
circumpolar trough (Antarctic, ACT) 113, 116–18, 132, 141
circumpolar vortex 15, 25, 28, 35, 53–60, 73–4, 89, 95, 101, 103, 105, 173, 177, 191, 214
CISK, see conditional instability of the second kind
classification schemes
  Businger–Reed 15, 158, 172, 294, 469
  extended Businger-Reed 15, 158–91
  general 14–16, 86, 511–14
  morphological 14, 99
  Norwegian Polar Lows Project 511–15
climatology (synoptic/dynamic)
  general 76–9, 116–23, 134–7
  use in forecasting 502

cloud
  clusters 196–8, 513, 555
  cumulonimbus/thunderstorms 544, 548, 556, 559
  head 210
  heights 48–51, 86, 101
  streets 17, 167, 178, 224, 250, 547
  types 101, 175
  vortices 11, 83, 536, 545, 554
cloud liquid water 16, 269, 486, 496
cloud signatures
  commas, see comma cloud signatures
  evolution 46–51
  general 16–51, 177, 229
  leaf cloud, see baroclinic leaf cloud
  spiraliform, see spiral cloud signatures
  swan-like 161–2
cold air depressions 6
cold air vortex 225
cold conveyor belt 271–2
cold low type of polar low/mesocyclone 15, 22, 26–8, 32–5, 158, 160, 172–89, 200, 214, 224, 225, 227, 538
cold, upper-level low/vortex (cold core low/vortex) 26–7, 29, 32–5, 45, 56–60, 105, 112, 119, 132, 172–83, 191–2, 200–2, 214, 218, 224–5, 227, 228, 232, 240, 242, 244, 362, 375–80, 393, 538, 540, 554
column stretching, see vortex stretching
comma cloud development: conceptual model 156
comma cloud signatures 14, 17, 19–20, 86, 99–101, 104, 123, 127–32, 141–4, 247, 533, 538
comma clouds (Reed type) 8, 11–15, 37, 41, 88, 151, 154–7, 175, 189, 234, 249, 260, 315, 513, 533, 549, 554
conceptual models of polar lows 270–3
conditional instability of the second kind (CISK) 8, 83, 86, 88, 108, 152–3, 161, 183, 196–7, 261, 293, 313, 335, 337, 343–5, 380–5, 393–404, 410, 429, 432, 435, 459
conditionally unstable lapse rate/instability 74–5, 155, 227, 303, 339–40, 344–5, 387, 395
convection 2, 3, 73, 124, 152–4, 164, 174, 335–41, 345–6, 393–400, 464
convective adjustment, see adjustment, convective

convective available potential energy (CAPE) 152–3, 181, 195, 208, 216, 227, 337–41, 345, 352, 384, 387, 394–6, 398, 399, 404, 429
convective heat release, see latent heat release
convective polar lows 12, 148, 152–3, 157, 197, 215, 219, 230, 260, 321
conveyor belts, see warm and cold conveyor belts
Craig and Cho model 402–3
cut off low 25, 215, 217–18, 227, 260

Davis Strait 32–3, 65, 91, 93, 95–7, 214, 225–6, 455
decay of polar lows 153–4, 181, 216, 400, 410, 422, 430, 547
deep sea (bottom) water 88–9, 92
definition of polar lows 3, 10–12, 79, 545–6
Denmark 7, 44, 83–4, 306
Denmark Strait 70–2, 211, 214, 317, 543–5
diameter of polar lows, see horizontal scale of polar low
down scale development 242
downstream development 167, 186, 203, 432
Drake Passage 137, 140, 254
dropsondes 266, 275

Eady model 292
East Canadian polar lows 95–7, 223–6
East China Sea 99–108
East Siberian Sea 228, 552–7
e-folding time 241–2, 293, 396
El Niño–Southern Oscillation (ENSO) 90–2, 116, 118–22, 145, 146
equivalent barotropic air mass 77, 157, 191–2, 403
European Centre for Medium-range Weather Forecasts (ECMWF) 205, 207, 489
European Polar Lows Working Group 10, 12
extra-tropical hurricanes 13, 177–83
eye-like features 20, 25, 51, 163, 194, 196, 218, 221, 227, 229, 410, 437, 543

Fantini model 401
Filchner–Ronne Ice Shelf 114, 251, 262
filling of polar lows, see decay of polar lows

Fleet Numerical Oceanography Center (FNOC) 529
forecasting of polar lows
  features favourable for polar low development 547–9, 556–7
  forecasting the tracks of polar lows 517–18
  forecasting using synoptic and climatological rules 502
  forecasting wind speed 504–7
  general 3, 205, 420–30, 501–74
  the role of boundary layer fronts 539–42, 547
  using numerical model output 502–3, 518–21
  using satellite imagery 504–7, 525–7
Fram Strait 85, 159, 294, 529
frequency of polar lows, see temporal variability of polar lows
frontal instability 166–7, 186
frontogenesis 21, 23, 194, 209, 235–8, 240, 266–73, 294, 309, 400, 412
fronts and frontal structure 12, 21, 35–9, 266–73, 277, 400
FROST project 247, 500

gale force criterium 81
gales 9, 74, 79, 158, 169, 195, 225, 276, 277, 422, 567
genesis regions of polar lows and mesocyclones 2, 20, 83, 88, 95, 103, 113–14, 117, 141–5, 211, 219, 228
Georg von Neumeyer station, Antarctica 256, 284, 476–85
Germany 18
gravity current 169
Great Lakes, The 226–7
Greenland
  deep sea (bottom) water formation, see deep sea (bottom) water
  general 226, 543, 559
  lee low 42, 70–2, 214
  polar lows near 65, 190, 211, 309–16, 528
Greenland Sea 37, 39, 42, 46–7, 68, 77, 85, 88, 166, 168, 303, 407, 430
Gulf of Alaska 97–8, 185, 228–9, 407, 432, 545

Halley Station, Antarctica 136, 251, 256, 262, 269–71, 277, 283, 465, 476–85, 565

HIRLAM model 223, 325, 421–8
history of polar lows research 3–10
hook clouds 169, 177, 179, 247, 543
horizontal scale of polar lows 3, 11–12, 17, 85–7, 93, 100–1, 104, 123–6, 154, 197–8, 225, 226–7, 229, 238, 241–2, 244, 260, 293, 296–8, 300–2, 308, 318, 411, 419, 422, 427, 464, 469, 473, 476, 477, 483, 485, 491, 496, 548
Hudson Bay 32–5, 55, 58, 95–7, 223–5, 308, 407, 444–55, 459, 472
Hudson Strait 225
hurricane force winds 82, 219, 228, 451, 548, 556, 561
hurricane-like polar lows 4–5, 153, 195–6, 218, 221, 227, 228, 245, 399, 410
hurricanes, see tropical cyclones
hydrostatic adjustment, see adjustment, hydrostatic

ice edge, see marginal ice zone
ice edge polar lows and mesocyclones 14, 31, 85, 100, 113–15, 132, 136, 249, 294–6, 468–72, 489–95
Iceland 18, 32, 36, 39, 41, 65, 70, 88, 91, 152, 190, 208, 211, 214, 543, 546
Icelandic low 72, 214
icing 561–2
initial value method 291, 307, 401, 407
instability lows 4, 11, 12
instant occlusions 35–9, 41–2, 86, 128, 130, 132, 134, 138–9, 155, 175, 249
integrated water vapour 16, 248, 267–9, 486, 508
intensification of cyclones (see also barotropic, baroclinic and convective/thermal instability) 342–6, 375–80
inter-annual variability 101–3, 117–18, 145
inversion, see temperature inversion
invertibility principle 324–7, 353–9, 363–7
isentropic potential vorticity, see potential vorticity
isentropic surface analysis 235, 271

Jan Mayen 82, 203, 430
Japan 100, 167
Japan Sea 2, 55, 99–108, 232–46, 296–9
Japan Sea Polar air mass Convergence Zone (JPCZ) 104, 244, 246, 318

Index   609

jet streaks 15, 148, 164, 205, 206, 258, 548
jet streams 101, 116, 154, 252, 260, 484

Kamchatka 55, 100, 199, 229–31, 436, 551
Kara Sea 176
katabatic winds 69, 111–13, 134, 136, 140, 250–2, 255–9, 271, 463, 466–9, 472–7, 480, 499
Kuroshio current 101, 103, 108

Labrador Sea polar lows 2, 88, 91, 92, 95–7, 219–23, 226, 407, 455–9
lake vortices 226–7
landfall 1, 83, 153–4, 181, 216, 400, 405, 409–10, 501
latent heat release 263, 303, 308, 313, 394, 397, 403, 406, 415–16, 419, 435, 439, 443–4, 454–9, 499
lifetime of polar lows 16, 83, 85–7, 124–6, 176, 226, 244, 259, 297, 477, 485, 496, 547
limited area models 285, 405, 437, 445, 455, 460–3, 573
linear models 290, 292–4, 296–300, 304–7, 312–15, 402
locations of polar low/mesocyclone occurrence, see Amundsen/Bellingshausen Sea; Australian region; Baffin Island; Barents Sea; Bear Island; Beaufort Sea; Bering Sea; Byrd Glacier; Chukchi Sea; Davis Strait; Denmark; Denmark Strait; East China Sea; East Siberian Sea; Filchner–Ronne Ice Shelf; Fram Strait; Great Lakes; Greenland Sea; Gulf of Alaska; Hudson Bay; Hudson Strait; Iceland; Japan Sea; Kamchatka; Kara Sea; Labrador Sea; Mediterranean systems; New Zealand; Nordic Seas; North Sea; Northwestern Pacific; Norwegian Sea; Novaya Zemlya; Ross Ice Shelf; Ross Sea; Svalbard; Tasman Sea; Terra Nova Bay; The Netherlands; United Kingdom; Weddell Sea; Wilkes land; Yellow Sea

marginal ice zone (MIZ)/ice edge 40, 67–8, 73, 89, 92, 110, 127, 143, 149, 294, 452–4, 457
Marie Byrd Land 136, 258
McMurdo Station 110, 134–5, 256, 278, 283, 476

Mediterranean systems 13, 22, 24–5, 45, 69, 93, 214–19, 345
merry-go-round cloud signatures 33, 35, 37, 38, 89, 119, 128, 130, 132, 137, 174, 189, 259, 260
mesocyclone formation areas, see genesis regions of polar lows and mesocyclones
mesocyclone tracks, see storm tracks; tracks of polar lows
mesocyclones 1, 2, 9, 10, 14, 50–1, 84–7, 114, 148, 247
meso-α-scale vortices 12, 104, 244
meso-β-scale vortices 104, 167, 239–46, 318
migration effect 157
moisture structure of polar lows and mesocyclones 266–73
monitoring polar lows 514–17, 539
Montgomery–Farrell model 400–1, 407
'most beautiful polar low', see case study, Nordeng and Rasmussen's case
multiple polar lows and mesocyclones 137, 165–7, 185–9, 203–4, 206, 238–42, 537
multi-scale structure 245

NAO index (NAOI) 90
Netherlands, The 190–1
New Zealand 2, 119, 122, 134, 137, 140–2, 145, 146, 260, 575
nomenclature 12–14
non-modal mechanism, see initial value method
Nordic seas 53, 58, 66–94, 150, 158, 184, 528–43
NORLAM model 411, 417, 460–3, 491, 496
normal mode techniques 290, 299, 307
North Atlantic oscillation (NAO) 89–94
North Cape 22, 49, 169, 177, 180, 181, 202–3
North Pacific 154
North Pole 174
North Sea 7, 8, 18, 41, 65, 73, 83, 190–1, 306, 429, 501, 562–4
Northwestern Pacific 99–108, 232–46
Norway 3–7
Norwegian Polar Lows Project 8, 77, 170, 191–206, 510, 516
Norwegian Sea 2, 20, 37, 39, 42, 46–7, 76, 81, 85, 88, 91, 93, 152, 159, 163–4, 171, 173, 186, 190, 203, 208, 331, 334, 341, 407–17, 421–30, 506, 561–2

Novaya Zemlya 55, 58, 65, 68, 73, 167, 176–7
nowcasting 503–10
numerical modelling of polar lows 207, 229–31, 222–3, 395–500

occlusions, polar lows along 159, 184–8, 190, 229–30
ocean currents 53, 55, 66–9
ocean waves 557–61, 564–7
Okhotsk Sea 104, 232
omega equation 327–9
Ooyama's balanced model 380–5
orographic influences on polar lows, *see* topographic forcing
orographic polar lows 190–1
outbreaks of polar lows and mesocyclones 135–40, 208, 228, 248, 259

pack ice, *see* sea ice
passive microwave data 16, 221, 267, 276–7, 279, 283–4, 248, 269, 496, 506–10
polar air cyclones 155
polar air depressions 3, 9–10, 151
polar airstream cyclones 10, 155
polar depression index 9
polar front 11, 12, 24, 36, 39, 41, 100, 104, 116, 154, 167, 190, 309, 331, 403, 422, 431
polar low spectrum 10, 157–91, 403
polar low/mesoscale low family 98, 198, 203, 238–9, 242, 246, 485
polar mesocyclones 12
polar mesoscale cyclone 12
polartief 4
polynyas 110, 111, 261
positive vorticity advection (PVA) 15, 17, 29, 31, 154–5, 158, 165, 174, 240, 258, 531
potential vorticity (PV)
  general 15, 136, 209–10, 222, 265, 290, 307, 308, 321–35, 359–63, 375, 384, 400–2, 412–16, 419–26, 425, 431, 443–4, 459, 494, 497, 503, 521, 524–5, 527
  structure of a circular symmetric cyclone 361–4
precipitation 151, 210–11, 244, 283–4, 451, 562–4

pressure perturbations of polar lows 14, 179–81, 200, 244, 273, 275, 318, 411, 419, 476, 548
primary polar lows 15
Pritz Bay 505
PV anomalies, *see* potential vorticity

radar altimeters 277–9
radar studies 104, 211, 235, 243, 244, 527–8
radiosonde ascents 62, 110–11, 179, 182, 213, 252–3, 273, 295, 337, 461
rain 227, 283, 508
real polar lows, *see* true polar lows
reverse shear/reverse shear polar lows 42, 46–7, 89, 151–2, 159–63, 170, 223, 226, 229, 291, 299–305, 317, 327, 401, 404, 419, 455, 459, 511, 517, 518
Ronne Ice Shelf, *see* Filchner–Ronne Ice Shelf
rope cloud 169, 179, 529
Ross Ice Shelf 48–9, 135, 140, 146, 250, 255–8, 271, 275, 276, 283, 472, 485–9, 496–7, 571
Ross Sea 16, 110, 112, 114, 115, 119, 120, 127, 135, 136, 139, 145, 146, 255–8, 271–3, 274, 283, 460, 472–3, 475, 485, 489, 496
Rossby height (*see also* Rossby penetration depth) 367, 368, 521
Rossby Number 322
Rossby penetration depth (Rossby Depth) (*see also* Rossby height) 174, 222, 308, 327, 330, 503
Rossby radius of deformation 293

satellite imagery 6, 8, 14, 16, 117, 151, 247
scatterometer, *see* wind scatterometer
Scotia Sea 508
sea ice 53–4, 60, 66, 91, 93, 95, 97, 107, 108, 225, 249, 309, 475
Sea of Japan 9
sea state 282–3
sea surface temperatures 68, 83, 91, 92, 93, 95, 107–8, 148, 152, 215, 249, 330, 337
seclusions 178, 185, 229, 235, 263, 419, 431, 459, 466, 547
secondary polar lows 15, 16

Index   611

self development process 288, 308, 400
semi-annual oscillation 117–20, 134, 137, 143
sensible heat flux, *see* surface heat and moisture fluxes
shallow low-level fronts, *see* fronts and frontal structure
shape of clouds, *see* cloud signatures
shear lines 73, 177, 229–30, 236, 238, 240, 277, 280, 318, 321, 398, 473
shear zones, *see* shear lines
Shetland Islands 422
short-wave, upper-level troughs 74, 136, 148, 157, 159, 164, 167–8, 200, 232, 252, 258, 289–90, 307, 309, 397, 400, 432, 435, 447, 454–5, 457, 472, 483–4, 487, 489, 497, 523, 529, 532, 543, 547
short-wave/jet streak type of polar low 15, 158
showers 83, 210, 226
Siple Coast, Antarctica 112, 136
size of polar lows, *see* horizontal scale of polar lows
slantwise ascent 151, 210–11, 269, 271, 292
slantwise convection 413, 416, 435, 444
snowfall 3, 10, 79, 116, 120, 135, 208, 210, 226, 235, 243, 283, 563
Southern Ocean 9, 140, 247–8, 259, 267, 276, 284
Southern Oscillation, *see* El Niño–Southern Oscillation
Southern Oscillation Index (SOI) 146
spatial distribution of polar lows and mesocyclones 84, 85, 126–34, 220
Special Sensor Microwave/Imager (SSM/I), *see* passive microwave data
speed of movement of polar lows and mesocyclones 141
spin-up effect 112, 149, 165, 215–16, 400, 415, 426, 459, 464, 469, 499
spiraliform cloud signatures 14, 17–32, 36, 86, 97, 98, 99–101, 123, 124, 127–30, 145, 149, 173, 175–7, 181, 194, 227, 229, 244, 265, 312, 318, 422, 515, 554
Spitsbergen, *see* Svalbard
St Paul Island 230, 238, 436–42
static stability 330
steering flow 517–18, 546–7

storm force winds 209, 225, 227, 502
storm tides 228
storm tracks (*see also* tracks of polar lows and mesocyclones) 64–6, 163
stratospheric intrusions 211
streamers, *see* cloud streets
strength of polar lows 220, 238, 469, 480–1, 487, 527
summertime polar lows and mesocyclones, *see* warm season polar lows and mesocyclones
surface heat and moisture fluxes 89, 148, 157–8, 218–19, 227, 229–30, 249–50, 253, 261, 317, 337, 341, 395–6, 398–9, 403, 409, 412, 416–17, 428, 432, 439, 443, 445, 451–3, 459, 464, 466, 469, 470, 481, 483
surface pressure associated with mesocyclones (*see also* pressure perturbations of polar lows) 273–6
surges (low level) 169, 294, 529
Svalbard 22, 68, 73, 167–9, 171, 174–5, 177, 512, 523
Svalbard boundary layer front 20, 222
symmetric instability 435
synoptic scale weather systems 134–7, 139, 140, 183, 235, 248
Synthetic Aperture Radar (SAR) 510

Tasman Sea 9, 122, 137, 145, 260
T-bone structure of polar lows 210, 311
teleconnections 146
temperature inversion 53, 60–4, 110, 249–50, 406, 417
temporal variability of polar lows and mesocyclones 79, 81–2, 86–8, 97–8, 120, 124, 127, 133, 141–8, 220
Terra Nova Bay 134, 255–6, 259, 271, 276, 278, 461, 472–6, 571
thermal instability 335–92, 404
thermal structure of polar lows and mesocyclones 262–6
thermal wind 160, 223, 291, 299
thunderstorms, *see* cloud, cumulonimbus/thunderstorms
Tierra Del Fuego 137
TIROS Operational Vertical Sounder (TOVS) 253–4, 259, 262–5, 485
topographic forcing 190–1, 211, 214, 465, 472–3, 477, 481, 485–6, 523

tracks of polar lows and mesocyclones 80, 83–4, 94–5, 129–30, 132–3, 135, 138–9, 144, 502
Transantarctic Mountains 256, 278
Trans-Polar Index 121
tropical cyclones (hurricanes)
　general 25, 183, 335, 342, 380–5
　resemblance to polar lows 4, 8, 152–3, 218, 227, 412
tropopause folding 334
troughs
　general 163–7, 236, 251 260, 481, 491, 502, 511–13, 517, 519
　short wave, upper-level, *see* short wave, upper-level troughs
　Trailing 163–5
　warm 513–24
true polar lows 11, 13–15, 173, 177, 401
Tushima Current 107
two-stage development 307, 396, 397, 400, 454, 457

United Kingdom 1, 6–7, 18, 21, 151, 206–11, 562–4
upper air winds associated with mesocyclones 168, 273
upper-level cold lows, *see* cold, upper-level low/vortex
upper-level disturbances (*see also* short-wave upper-level troughs) 74, 167, 229, 232, 236, 258, 307, 309, 387–91, 409, 431–2, 544
upper-level troughs (*see also* short-wave upper-level troughs) 232, 235, 240

vacuum cleaner effect 389–90, 419
vertical stability of the atmosphere 76, 101, 107, 124, 224–5, 240, 299, 343, 397–8, 413
Victoria Land, Antarctica 258
vortex stretching 70, 148, 214, 256, 257–8, 464, 466–8, 475, 479–81

vorticity 136, 163, 167, 229, 236, 245, 278, 282, 464, 480, 486, 542
vorticity equation 5, 327, 467

warm (temperature) tongue 511–12
warm conveyor belt 151, 271–2
warm core polar lows 13, 20, 45, 49, 153, 174, 214, 216, 218, 227, 238, 244, 261, 263, 264, 318, 362–4, 375–80, 393, 401, 419, 459, 466
warm season polar lows and mesocyclones 86, 97, 174, 190, 476–89
water vapour imagery 269, 526
wave height (ocean) (*see also* ocean waves) 228, 282–3, 561, 564–5
wave packet 564–5
wave trains of polar lows 42, 45–7, 160, 162, 166, 186, 189, 209, 296, 305
wavelength of polar lows, *see* horizontal scale of polar lows
weather associated with polar lows/mesocyclones 4, 7, 151, 208, 210, 226, 243, 557–64
Weddell Sea 20, 50–1, 112, 121, 131, 134, 136, 249–51, 255–7, 259, 260, 267, 271, 274–5, 284, 460, 461, 463–72, 489–95, 497
Wilhelmsen file 159–91, 339
Wilkes Land, Antarctica 134, 149
Wind Induced Surface Heat Exchange (WISHE) 261, 337, 340, 345, 393–404, 409, 410, 429, 435, 452, 453
wind scatterometer 8, 10, 126, 275, 278, 280–2, 284, 486, 496, 507–9
wind speeds associated with polar lows 13, 80–2, 153, 197, 200, 208, 220–1, 225, 226–9, 236, 238, 243–5, 260, 273, 276–82, 284, 309, 311, 419, 422, 451, 469, 496, 506, 513, 533, 548, 559, 561
winter Weddell Sea project 489

Yellow Sea 99–108